Text-Atlas of
CAT ANATOMY

Text-Atlas of
CAT ANATOMY

JAMES E. CROUCH, Ph.D.

Chairman, Division of Life Sciences

Professor of Zoology

San Diego State College

San Diego, California

Illustrated by

MARTHA B. LACKEY, Staff Artist

Museum of Zoology

University of Michigan

Ann Arbor, Michigan

*with 115 Plates—39 in color
and 18 text figures*

LEA & FEBIGER · Philadelphia · 1969

Frontispiece: Head of the Egyptian goddess Bastet.
Courtesy of the British Museum, London, England.

In ancient Egypt the cat was worshipped as a god.
Bastet, the cat-headed Goddess of Joy, represented
femininity and maternity, and the life-preserving
power of the sun. It is said that every Egyptian
woman desired to be like her—a disturbing creature
with slanty eyes, a mysterious gaze, a supple body,
and regal posture.

Library of Congress Catalog Card Number 68:25206
PRINTED IN THE UNITED STATES OF AMERICA

MRINAL K. DAS

This book is respectfully dedicated
to my mother and father
for the greatest of all gifts

PREFACE

Gross anatomy is one of the important basic sciences in the training of any zoologist, indeed of many biologists. Like other basic sciences it is being crowded by the "explosion" of knowledge and the resulting efforts on the part of educators to streamline the curriculum, to increase the effectiveness of the teaching process and thereby to hasten learning on the part of students. The same problems are faced in training for those professions in which anatomy is a necessary basic course like human and veterinary medicine, nursing, physical therapy, physical education and laboratory technology.

Teaching machines, educational TV, programmed instruction and many other methods and devices are being developed and should be encouraged. It is the purpose of this book to present mammalian anatomy, using the cat as a type, in such a way as to speed learning and understanding by placing emphasis on quality illustrations and reducing the text material to descriptions of the illustrations which highlight the most important features of the anatomy. With the exception of the muscular system, this highlighting of important features is further developed by placing in **boldface type,** both in the text and in the illustrations, the names of those structures which I consider of primary importance. In the muscular system boldface type is used to indicate an illustration or illustrations where a muscle shows to greatest advantage. A student would not be expected to know all of the "boldfaced" muscles. For the more interested or the more advanced students the illustrations are a storehouse of anatomy and of anatomical relationships. One needs only to study them carefully.

The illustrations, except those of the skeletal and muscular systems, represent a regional approach to the study of anatomy. Blood vessels and nerves are not put on just a blank page but always in relationship to neighboring structures. To make those structures stand out from others, when they are the center of focus in an illustration, color is often used.

General descriptions of the basic structures, relationships and functions of each one of the body systems are placed at appropriate places through the book. These were written primarily for the beginning student in the hope that they would help him to learn and understand more from the illustrations. In these sections are included some illustrations but for the most part references are made to the principal plates which constitute the most important part of the book.

The illustrations have been drawn, with few exceptions, from embalmed and injected specimens. Since such specimens are used in most courses, it seemed the best approach and the most helpful for the students. Many decisions had to be made because of the variations which one sees in the anatomy of the cat. Some of the more common variations are discussed in the text. The student should not be too concerned if the anatomy of his cat does not fit exactly the illustrations in the book.

There are a few illustrations which deal with the anatomy of the dog and of man and in the text frequent references are made to comparative anatomy and embryology. The student should have some knowledge of the evolutionary history (*phylogeny*) and the life history (*ontogeny*) of mammals.

The terminology is based largely on the **Nomina Anatomica 1961,** second edition. Some modifications in nomenclature are made in keeping with recommendations suggested by **Nomenclatorial Commissions of the World Association of Veterinary Anatomists** and the **American Association of Veterinary Anatomists.** Some names from previous systems of nomenclature are included, often in parentheses, so that this material can be related to previous works which are still in common use. Since anatomy is always to some degree a language course, a glossary is available in the back of the book. It gives derivation, pronunciation and meaning of many of the terms used.

The index of the book is very complete. Since the illustrations reveal more anatomy than is found in the descriptive materials, the index includes references to many of the labels on the illustrations as well as to terms in the text. The pages for text references are in standard type, those to structures shown on the illustrations are in italics. If a structure is described in any detail in the text, it is indicated in the index in boldface type.

Mrs. Martha Lackey, as illustrator, more than any other person, has made the book possible. She has assisted in the dissection as well as in establishing the arrangement of materials in the book. She, a student herself, has presented many ideas which will clarify the material for all who use the book.

There are many who have helped in the preparation of the manuscript to whom I extend my appreciation: for typing, to the Mrs. Ann Smith and Jean Flores and to Misses Susan Gemeroy and Susan Krudwig; for preparation of the muscle table and glossary to Miss Susan

Gemeroy. Dr. Harry Plymale helped in the writing of some of the descriptions of the skeleton. Again, I am grateful to Dr. Charles Moritz who has advised and encouraged me. The publishers, Lea & Febiger, through their financial support, confidence and cooperation have made this book possible. I extend to them my thanks.

If there are errors or omissions in the book, I alone assume all responsibility.

San Diego, California

JAMES E. CROUCH

CONTENTS

LIST OF PLATES

LIST OF TEXT FIGURES

NATURAL HISTORY

The common cat, the principal subject of this book, is a household pet in many homes of this country and of other civilized nations around the world. Yet it has also retained many of its feral instincts and is one of the most destructive of our predators seeking birds, mammals and even fish as its prey. They see well, even in dim light, and have a keen sense of hearing and a less well-developed sense of smell. They stalk their prey or wait patiently for it to appear; they take it by surprise and swiftly, killing it with sharp teeth and claws. They are animals which average about $2\frac{1}{2}$ feet in length including the tail which makes up about one-third of the total. We are excluding here, of course, the Manx cats of the Isle of Man in the Irish Sea which are high-rumped and tailless except for a tuft of hair. There are tailless cats in Russia also which are believed to have originated in the Orient. The cats are descendants of the saber-toothed tigers. The common cat of our country represents a cross between the European Wildcat and the Egyptian race of the African Wildcat. Blue-eyed Siamese cats with their black or chocolate colored heads and feet and cream-colored bodies are domesticated descendants of the African jungle cats.

The domestic cat usually produces 3 to 6 blind, helpless, and fur-covered young in a litter. Their eyelids separate at about the ninth day after birth. They are cared for by the maternal parent since the male is lacking in parental instincts. The gestation period of the cat is about 63 days; the life span on the average about 10 years.

CLASSIFICATION

The classification of the cat among other animals may be summarized as follows:

Phylum Chordata—notochord
                dorsal hollow nerve cord
                pharyngeal pouches
  Group—Craniata (*Vertebrata*)—brain case
                vertebral column
    Subphylum—Gnathostomata—upper and lower jaws
                endoskeleton
                paired appendages
      Superclass—Tetrapoda—have paired limbs rather than fins
        Class—Mammalia—mammary glands
                hair
          Subclass—Theria—mammary glands provided with teats.
            Infraclass—Eutheria—placenta

              Order—Carnivora—well-developed canines
                Suborder—Fissipedia—limbs typically
                    pentadactyl
                  Family—Felidae—digitigrade; toes
                      5-4; claws retractile;
                      head short and round
                    Genus—Felis—body slender;
                      dentition $\dfrac{3\text{-}1\text{-}3\text{-}1}{3\text{-}1\text{-}2\text{-}1}$
                  Species—domestica—cross
                      between European
                      and African wildcats.

The **scientific name** of the common cat is **Felis domestica,** composed by combining the genus and species names.

One should recall that our classification of organisms is based upon the concept that all species, present and past, were derived from primordial forms by modification or change over long periods of time. This we call **evolution.** It means that all organisms are related, even though remotely in many cases. It means further that in any scheme of classification based upon this concept, that all animals placed in a given category are more closely related to each other than to any organism in another category, *i.e.* members of the Phylum Chordata are more like each other than any member of the Phylum Arthropoda (*insects, crustacea, etc.*). It means finally that through evolution and the influence of the environment upon the evolved animal great diversity of animal forms has appeared ranging from the relatively simple to the complex, the generalized to the specialized and many, failing to compete successfully, have become extinct.

ONTOGENY AND PHYLOGENY

The proper classification of animals requires considerable knowledge of an animal's morphology—its anatomical characteristics. There are three primary sources of such information. One is the individual life history of an animal, including its embryological development—its **ontogeny;** another, a study of adult specimens of living animals of all groups, and a third, the study of the fossil record. These last two approaches constitute a study of the "history of the race"—or **phylogeny.**

In the study of a sequence of steps in the embryological development of a species, structures appear which in many cases aid materially in classifying an animal. Most adult Chordates, for example, show no notochord but one is very evident during development. The same is true of pharyngeal pouches in many chordates. Studies in comparative ontogeny lend more information. Also one soon realizes that as one watches an animal progress embryologically from the single-celled fertilized egg to the multicellular, complex adult that he is in a broad sense viewing its phylogenetic history. This **concept of recapitulation** (*biogenetic law*), usually accredited to Haeckel, "ontogeny recapitulates phylogeny," has long been a subject of controversy. Perhaps the most that can be said for recapitulation is that it stimulates the imagination and may encourage the formulation of hypotheses whose validity can be tested.

In comparative studies of organisms, one must be certain that structures which are being used to establish relationships and lines of descent are truly the same and not merely superficial resemblances. This is something that can be checked through embryological studies. Structures which have a common origin developmentally in different species are considered the same though they may later take on different forms and functions. They are called **homologous structures.** The forelimb of the horse, the dog, the frog and bird (*wing*) are all homologous. The wings of birds and insects are not at all alike in terms of origin but are said to be **analogous,** or alike in function. They do not indicate a close relationship of birds and insects. The wings of birds and bats, on the other hand, are similar developmentally and are therefore both homologous and analagous; similar in basic structure and in general function.

EVOLUTION OF ANIMAL ORGANISMS

It would be logical at this point to survey in outline the evolution of animal life and of the Chordates in particular which led to the Class Mammalia. It is not practical to do so, however, considering the expressed function of this book—a presentation of mammalian anatomy with emphasis upon the common cat. The interested reader should refer to such accounts in general biology or zoology books, some of which are listed under *References* at the end of this volume.

The two schematic, "tree-like" diagrams below give a fair summary of the phylogeny of vertebrates (Figs. 1 and 2). The first one, Figure 1, leads to the earliest amphibians, the Stegocephalians; the second, Figure 2, indicating how, from the early amphibians, existing classes of primarily land-dwelling vertebrates have likely evolved. Note especially the Synapsid branch or line of reptiles from which arose the "mammalian-like reptiles" from which, in turn, modern mammals were derived.

THE CLASS MAMMALIA

Mammals appeared in an age when reptiles were the dominate vertebrates, most likely in the early Triassic times, about 200 million years ago. They were mostly small creatures, roughly the size of the common cat. One wonders how these small mammals were able to compete successfully with the reptiles which were present in such

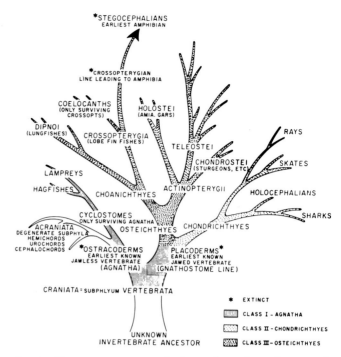

Fig. 1. *Illustrating origins and relationships of vertebrates with reference to the ancestry of the amphibians. Note that the bony fish and cartilaginous types originated as independent lines from placoderm stock and that the particular form of bony fish regarded as being ancestral to amphibians is extinct. Although they are not ancestral to any higher class, the sharks, among existing gnathostomes, appear to have deviated least from placoderm ancestry. The Acraniata are represented here as being degenerate forms derived from an early branch of agnathous vertebrates. (From Leach: Functional Anatomy. Ed 3, McGraw-Hill Book Co., 1961.)*

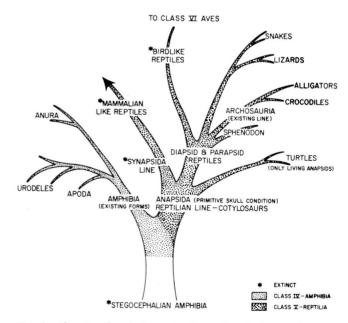

Fig. 2. *Showing the phylogenetic lines of evolution as they appear to have evolved from the earliest tetrapod vertebrate. The cotylosaurs as stem reptiles should also have been indicated as being extinct. (From Leach: Functional Anatomy. Ed 3, McGraw-Hill Book Co., 1961.)*

numbers and which were so diverse in size, form and range of habitats and so highly specialized. Probably the answer lies in fact that mammals were **homoiothermous** (*"warm-blooded"*), *i.e.* capable of maintaining a constant and relatively high body temperature. Reptiles were and are **poikilothermous** (*"cold-blooded"*), *i.e.* their body temperature fluctuates with that of the environment. The homoiothermal condition of mammals enabled them to be consistently more agile in muscular activity and more aware and clever in a wide variety of environmental circumstances, while both the sensory and motor activities of the reptiles were slowed or stopped at critically low or high temperatures. Another characteristic of mammals undoubtedly had survival value and enabled them to become the mentally superior and dominate group which they are today. This was the relatively long association and dependence of the young upon the adults. They were dependent upon the mother for milk and upon both adults in many cases for shelter and protection. This resulted in a forced training period for the young and theoretically better prepared them for competing with other animals and the environment generally. They developed a more intelligent kind of behavior.

As mammals evolved from these small initial forms, they became increasingly **diverse** until today they range in **size** from shrews to elephants; in **general body form** from giraffes to the horse and the whale; in **locomotion** from flying bats, to walking and running antelope, to swimming whales and dolphins and to climbing monkeys. Their **habitats** are equally diverse. Some are **aboreal**; some **terrestrial**; some **aquatic**; and some are **fossorial** (*burrowers*). Individual species show **sexual dimorphism** other than in the genital organs themselves. The males tend to be larger than the females. Many of the sexual differences are considered secondary such as the distribution of hair, the presence of horns, the relative size of mammary glands, and color variations. Mammals also differ in the manner in which they apply their feet to the ground. Some, like the cat, walk on their toes and are said to be **digitigrade**. Man, the bear and others apply the whole sole of the foot to the ground and are **plantigrade**, while the horses, cattle, pigs and others having hoofs which are modifications of the nails and claws are **unguligrade**. Of these types of **"gait"** the plantigrade is the most generalized; the digitigrade is intermediate; the unguligrade is most specialized.

The positions of the paired appendages of mammals require some comment. Their primary position as seen in fish and in some amphibians and reptiles was lateral. This position does not make for efficient support and locomotion on the ground where it is necessary to lift the trunk above the substratum. Rotation and torsion of parts of the limbs are necessary to achieve this. While some progress was made in this direction by amphibians and reptiles, it is carried much farther in mammals where the trunk is carried well above ground. To achieve this both forelimbs and hindlimbs have undergone tortion and rotation of 90 degrees. The humerus has twisted to bring the original dorsal surface of the upper arm to face caudally. This allows the elbow to be directed caudad. To bring the digits (*toes*) forward the forearm is rotated medially resulting in the crossing of the radius over the ulna. This is a fixed position of the forearm in some mammals or it may be rotated at will as in man. In the hindlimb there has been a forward rotation from the primitive condition so that the toes point forward and the original dorsal surface of the limb is directed craniad. These important limb positions and changes should be remembered as you study the cat (Plate 2).

Man, of course, has an erect posture, a condition which we believe was derived from the quadruped position. Man's ancestors presumably went through an arboreal stage when the hands and feet became grasping structures. Descending again to the ground, the hands, having become so highly specialized as graspers, remained so. The feet, less specialized, were used again for walking. The assumption of the erect posture provided numerous advantages and some disadvantages. Perhaps the chief advantage was the freeing of the forelimb for other uses and it is obvious that man has taken full advantage of this "built in" tool. The hand, directed by a well-developed brain and controlled and coordinated by an efficient neuromuscular mechanism, has placed man on a very high plane—the highest of any organism. Only men can write books—a brain to hand undertaking.

While the above remarks have emphasized diversity of mammals, the fact remains that they are more alike in basic structures than they are different. To realize this refer back to page 1 of this Introduction and read the characteristics given under the categories to classify the cat. These characteristics down through the Class Mammalia apply to all mammals. Next move ahead through the text and illustrations of this text-atlas and many more of the common characteristics of mammals will become evident as will some of the differences among the members of this Class of vertebrate animals.

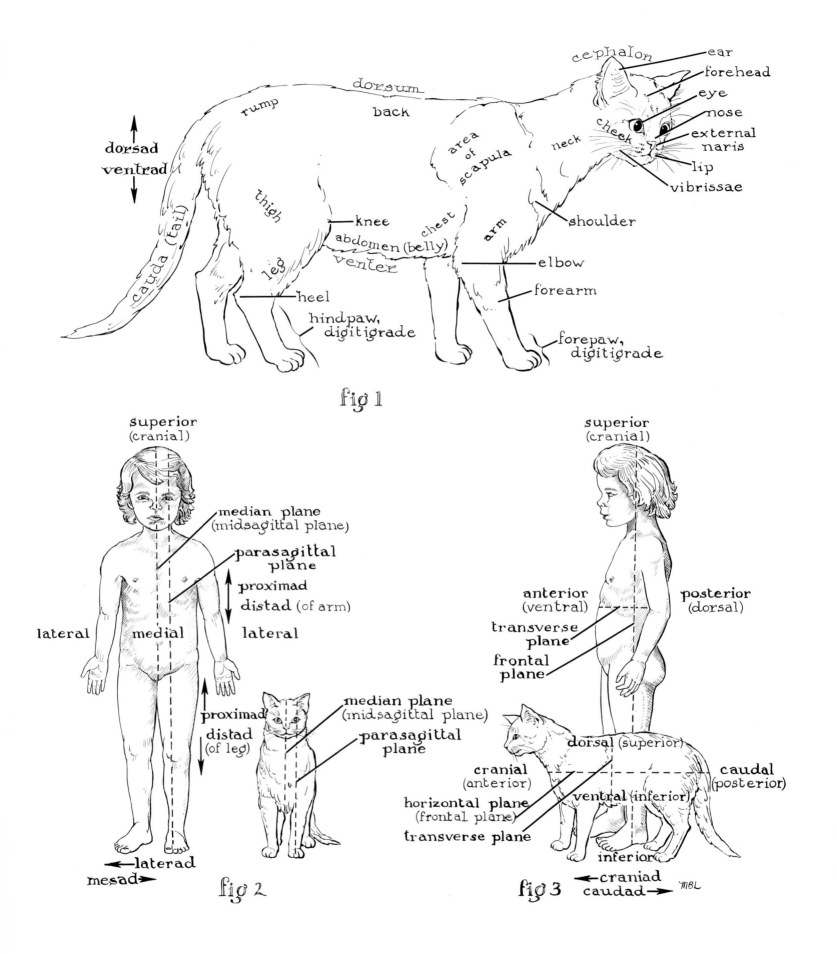

cephalon — ear
forehead
eye
nose
external naris
lip
vibrissae

dorsum

rump   back   area of scapula   neck   cheek

dorsad
ventrad

cauda (tail)

thigh

knee
chest
abdomen (belly)
arm
venter

leg

shoulder

elbow

forearm

heel

hindpaw, digitigrade

forepaw, digitigrade

fig 1

superior (cranial)

median plane (midsagittal plane)

parasagittal plane

proximad

distad (of arm)

lateral   medial   lateral

proximad

distad (of leg)

median plane (midsagittal plane)

parasagittal plane

laterad

mesad

fig 2

superior (cranial)

anterior (ventral)   posterior (dorsal)

transverse plane

frontal plane

dorsal (superior)

cranial (anterior)   caudal (posterior)

horizontal plane (frontal plane)

ventral (inferior)

transverse plane

inferior

craniad
caudad   MBL

fig 3

TERMS DENOTING POSITION, SURFACES
AND DIRECTION

It is essential to have a clear understanding of basic anatomical terms if one is to make progress in learning anatomy. Since the bodies of quadruped animals such as the cat are carried in the horizontal position all surfaces and parts are designated in relationship to that position.

The following terms which are in common usage in comparative vertebrate anatomy are applicable to the cat.

dorsum—the back of the animal or the upper side of an appendage or part

dorsal—refers to the back or upper side

dorsad—means toward the dorsal side

venter—the underside of the animal or of an appendage or part

ventrad—toward the ventral side or venter

cauda—tail

caudal—referring to the tail or posterior end of the body

caudad—toward the tail or posterior part of the body

cephalon—head

cephalic or **cranial**—referring to the head or anterior part of the body

cephalad or **craniad**—toward the head or anterior part of the body

meson—the middle plane of the body dividing it into right and left halves

medial or **mesal**—referring to the meson or to parts situated near it

mediad or **mesad**—toward the meson

lateral—referring to the side of the body or to parts situated near it

laterad—toward the side of the body or part

anterior—the head end of the animal

posterior—the tail end of the animal

ectal or **peripheral**—referring to the outer surface of the body

ental or **central**—referring to the inner or middle part of the body

proximal—referring to the part of a limb or other structure closest to the main mass of the body

proximad—toward the proximal part

distal—away from the main mass of the body; the opposite of proximal

distad—toward the distal end or away from the main mass of the body

The above adjectives and adverbs are sometimes compounded in anatomical descriptions. Caudo-ventrad for example means in the direction of the tail and venter of the animal.

PLANES, SECTIONS AND AXES

The structures of the cat are arranged symmetrically with reference to certain planes and axes of the body or part. Anatomical relationships can be studied by sections cut through the various planes and axes.

Planes and Sections. A plane is a surface, imaginary or real, in which all points, if joined by straight lines, would remain within that one surface. The **median plane** or **section** is vertical passing from cephalic to caudal parts of the body and through its center from dorsum to venter. It divides the body into nearly equal right and left halves. The **sagittal plane** or **section**, like the median plane, is vertical and longitudinal. It includes the median plane and any plane parallel to it. Planes parallel to the median plane are often called **parasagittal**. The medial plane may be called **midsagittal**. The **frontal** (*horizontal*) **plane** or **section** is horizontal and longitudinal. It is at right angles to the median plane and parallel to the dorsum and venter. The **transverse** (*cross*) **plane** or **section** is vertical and at right angles to both the frontal and sagittal planes. It passes through the dorsal, ventral and lateral aspects of the body.

Axes. An axis is a straight line, imaginary or real, passing through a body or part, around which the body or part could be rotated like a wheel on an axle. The **longitudinal** (*anteroposterior; craniocaudal*) **axis** lies in the median plane extending from head to tail. The **dorsoventral** (*sagittal*) **axis** is any line in the median plane extending from dorsum to venter. The **transverse** (*bilateral; mediolateral*) **axis** is any line in the transverse plane running between the lateral surfaces of the body or part.

Many of the above terms must be redefined for man since he carries his body in the erect posture. Descriptions of human structure are made from a formal standard known as the **anatomical position.** In this position the body is erect, the arms hang at the sides and the palms of the hands face forward. Careful study of Plate 1 will enable the student to apply properly the above terminology to man.

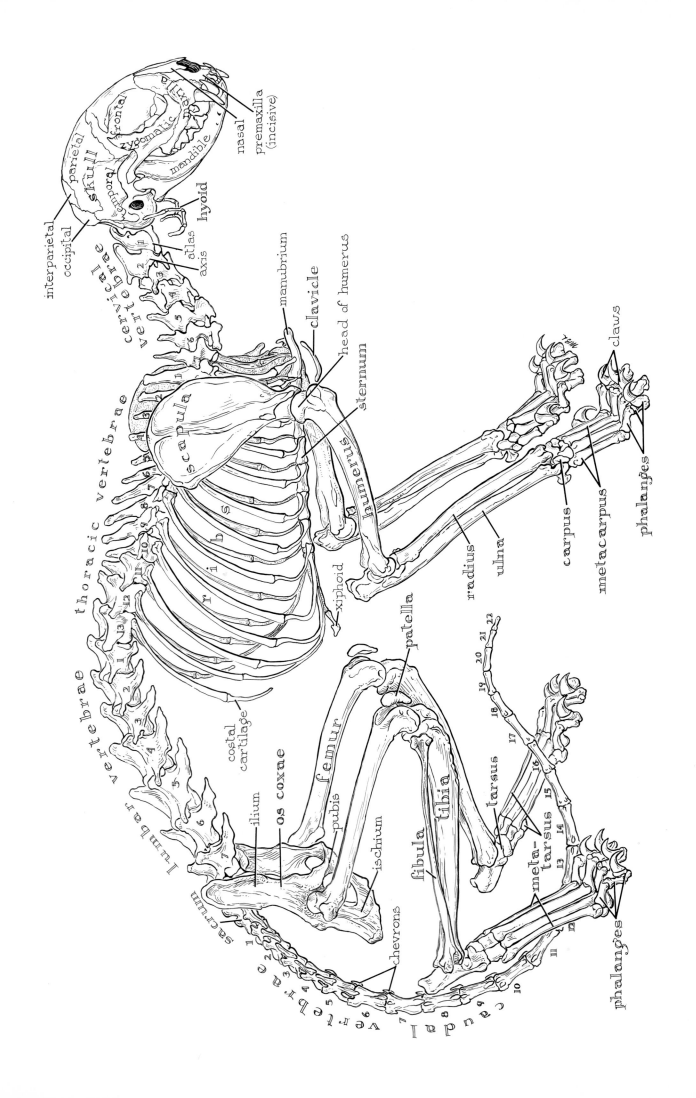

The skeleton may be broadly defined as all of the connective and supportive tissues of the body. More commonly, as in this book, it is considered to be only the bony and cartilaginous components. The skeleton of the cat, as of other vertebrates, is an **endoskeleton;** an internal skeleton as opposed to the **exoskeleton** or outside skeleton as in insects and other Arthropods.

The majority of the bones of the skeleton are preceded in the embryo by cartilage which is then gradually replaced, in part or entirely, by bone through a complicated process called **ossification.** Such bones are called **endochondral.** Among the bones of the skull those forming the roof and side walls of the cranial cavity and those of the face are not preceded by cartilage but ossify directly in membrane. These are called **intramembranous bones.** Some of the bones of the skull, such as the temporal, incorporate bone of both types. Once ossification is complete the endochondral and intramembranous bones show the same histological structure.

The number of bones in the cat's skeleton varies with age. In a kitten there are many separate bones which in the adult have lost their identity by fusing with neighboring bones. The single occipital bone at the back of the skull of an adult cat, for example, is represented by four bones in the kitten—the basioccipital, two exoccipitals, and a supraoccipital. In very old cats there may be still further fusion of bones as seen in the obliteration of the sutures between skull bones. About **287** separate bones may be demonstrated in a young adult cat. Some authors, however, exclude the ear ossicles, sesamoid bones, and chevron bones which would reduce the number to **233.** The 6 **ear ossicles** are housed within the temporal bones. The **sesamoid bones,** of which there are approximately 40, are those formed within tendons and the 8 **chevron bones** are small structures attached to the ventral sides of the caudal vertebrae.

The bones of the skeletal system may be organized and tabulated as follows for purposes of study. Plate 2 should be referred to as you study the table of bones.

TABLE 1

AXIAL SKELETON

SKULL			VERTEBRAL COLUMN		THORAX	
cranial	frontal	2	cervical vertebrae	7	ribs	26
	parietal	2	thoracic vertebrae	13	sternum	1
	interparietal	1	lumbar vertebrae	7		
	occipital	1	sacral vertebrae	1		
	ethmoid	1	caudal vertebrae	21		
	sphenoid	1				
	temporal	2				
facial	premaxilla	2				
	maxilla	2				
	palatine	2				
	vomer	1				
	maxilloturbinal	2				
	nasal	2				
	lacrymal	2				
	malar (jugal, zygomatic)	2				
	mandible	1				
hyoid	tympanohyal (oid)	2				
	stylohyal (oid)	2				
	epihyal (oid)	2				
	ceratohyal (oid)	2				
	basihyal (oid)	1				
	thyrohyal (oid)	2				

APPENDICULAR SKELETON

THORACIC APPENDAGE

shoulder girdle	scapula	2
	clavicle	2
arm	humerus	2
forearm	radius	2
	ulna	2

hand	carpus	scapholunar	2
		triquetral (cuneiform)	2
		pisiform	2
		trapezium	2
		trapezoid	2
		capitate (magnum)	2
		hamate (unciform)	2
	metacarpus		10
	phalanges	proximal	10
		middle	10
		distal	8

PELVIC APPENDAGE

pelvic girdle	innominate	2
leg	femur	2
	patella	2
	tibia	2
	fibula	2

foot	tarsus	astragalus	2
		calcaneus	2
		scaphoid (navicular)	2
		internal cuneiform	2
		middle cuneiform	2
		external cuneiform	2
		cuboid	2
	metatarsus		10
	phalanges	proximal	8
		middle	8
		distal	8

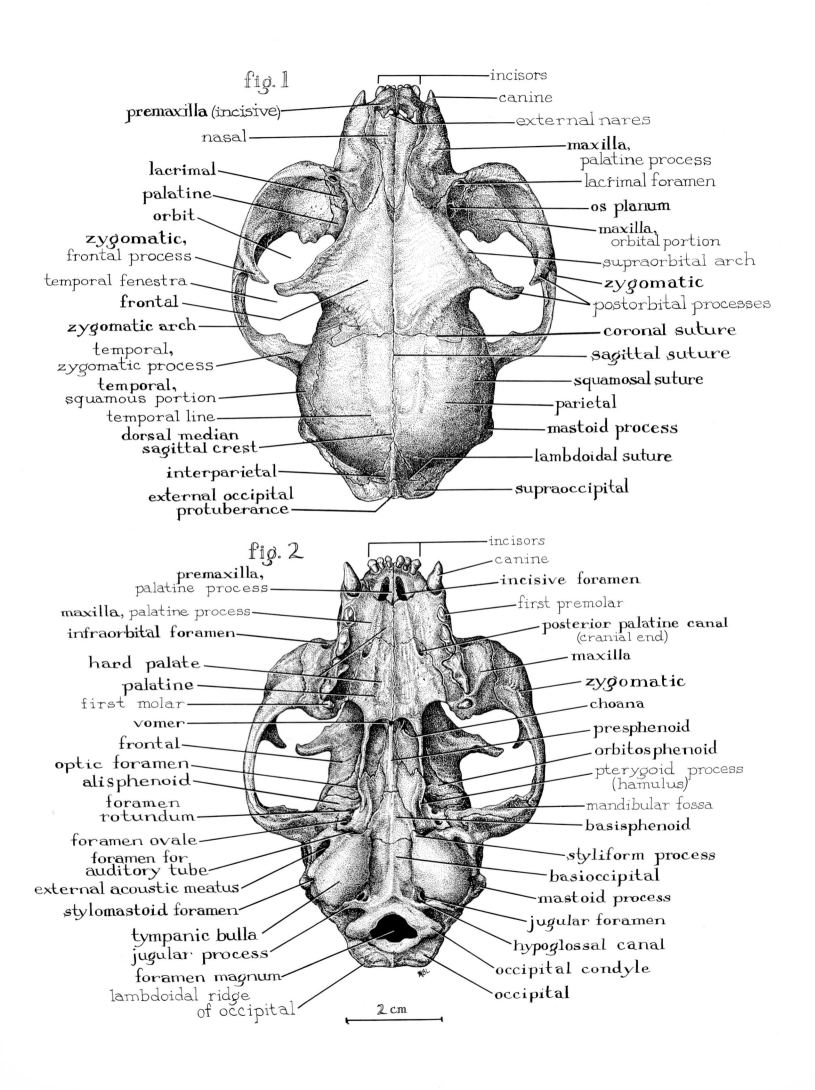

fig. 1

incisors
canine
external nares
premaxilla (incisive)
maxilla,
palatine process
nasal
lacrimal foramen
lacrimal
os planum
palatine
orbit
maxilla,
orbital portion
zygomatic,
frontal process
supraorbital arch
temporal fenestra
zygomatic
frontal
postorbital processes
zygomatic arch
coronal suture
temporal,
zygomatic process
sagittal suture
temporal,
squamous portion
squamosal suture
temporal line
parietal
dorsal median
sagittal crest
mastoid process
interparietal
lambdoidal suture
external occipital
protuberance
supraoccipital

fig. 2

incisors
canine
incisive foramen
premaxilla,
palatine process
first premolar
maxilla, palatine process
posterior palatine canal
(cranial end)
infraorbital foramen
maxilla
hard palate
zygomatic
palatine
choana
first molar
presphenoid
vomer
orbitosphenoid
frontal
pterygoid process
(hamulus)
optic foramen
alisphenoid
mandibular fossa
foramen
rotundum
basisphenoid
foramen ovale
foramen for
auditory tube
styliform process
external acoustic meatus
basioccipital
stylomastoid foramen
mastoid process
tympanic bulla
jugular foramen
jugular process
hypoglossal canal
foramen magnum
occipital condyle
lambdoidal ridge
of occipital
occipital

2 cm

The skull is the anterior expanded portion of the axial skeleton of the cat. It is composed of two parts, the **neurocranium** or **cranium** which houses and protects the brain and organs of special sense and the **splanchnocranium** (*visceral skeleton*) consisting of the jaws, hyoid, and cartilages of the larynx, which are located around the anterior ends of the digestive and respiratory passageways (Plates 2 and 3).

The **neurocranium** consists of a **cranial portion** which encloses the brain and a **facial portion** supporting the face and the olfactory and optic capsules (Table 1).

Dorsal aspect of neurocranium (Plate 3, fig. 1). As seen from the dorsal aspect, the neurocranium is pear-shaped in form with the **zygomatic arches** flaring out laterally. The midline is marked by a prominent **sagittal suture** which is crossed a little over half way back by the **coronal suture**. At the anterior extremity the **premaxillaries** (incisives) are shown bearing the incisor teeth. Above the premaxillaries are the **external nares** leading into the nasal passageways. The nasal area is closed dorsally by the paired nasal bones and laterally by the maxillaries, lacrimals, the os planum of the ethmoids and the palatines. The lacrimal foramina can be seen between the lacrimals and maxillaries.

The next "segment" of the neurocranium dorsally is made up of the prominent **frontal** bones centrally and the **zygomatic** (*malar*) bones laterally. The frontals articulate anteriorly with the frontal processes of the maxillae and with the nasal bones. Posteriorly they articulate with the parietals at the coronal suture. Laterally the frontals have prominent **postorbital processes** which approximate, but do not join, the frontal processes of the zygomatic bones. These bones plus the zygomatic processes of the maxillae form the rims of the prominent **orbits**. Caudad and ventrad of the postorbital processes of the frontals are the **temporal fossae**. The zygomatic processes of the temporal bones extend forward to articulate with the zygomatic bones to complete the **zygomatic arches**. Posterior to the coronal suture the large **parietals** make up the roof of this "segment" of the skull. Laterally they join at the **squamous suture** with the squamous portion of the **temporal bones** which contribute a large part of the lateral wall of the cranial cavity. Posteriorly the parietals articulate with the small **interparietal** medially and with the **occipital bone** laterally. The final and most posterior "segment" of the skull is the **occipital bone,** the superior portion of which articulates along the **lambdoidal suture** with the parietals and interparietal. The posterior boundary of the dorsal surface of the skull is elevated to form the **lambdoidal crest** which centrally is broadened to form an **external occipital protuberance**. Extending cephalad from this protuberance on to the interparietal bone is a sagittal crest.

Ventral aspect of the skull (Plate 3, fig. 2). The ventral aspect of the skull, like the dorsal, is oval in outline but its general surface configuration is flat rather than vaulted. The anterior part of this surface is triangular and consists of the **hard palate** medially; the **maxillary bones** laterally, bearing teeth on their **alveolar margins;** and anteriorly the **premaxillae** bearing the **incisors**. Between the maxillae and premaxillae are the **anterior palatine** (*incisive*) **foramina** through which, in the live subject, pass the **nasopalatine branches of the fifth cranial nerve** and the **nasal arteries**.

The **hard palate** forms the floor of the nasal passageways and the roof of the mouth. It is formed by the **palatine processes** of the **maxillae** and the **horizontal plates** of the **palatines**. Between these two bones are the **posterior palatine canals** through which pass the **palatine nerves** and **arteries**. The zygomatic processes of the maxillae extend dorsad and laterad from the alveolar margins and in their anterior edges contain the **infraorbital foramina** for the passage of the **infraorbital nerves** and **vessels**. They join the zygomatic bones to form the zygomatic arches which extend caudad to be completed by the zygomatic processes of the temporals. This constitutes the widest part of the skull and encloses the **orbits** and **temporal fossae.**

The posterior medial margins of the palatine bones form the ventral border of the **internal nares** or **choanae** which lead into a narrow trough-like part of the skull, the **median fossa**. The median fossa is limited laterally by the **pterygoid processes** and **hamuli** and dorsally by the **palatines, presphenoid, orbitosphenoids, basisphenoid** and **alisphenoids**. The posterior tip of the **vomer** can be seen dorsally as it disappears into the choanae to form a part of the floor of the nasal passageways. The median fossa is covered in the living subject by the soft palate and forms the nasopharynx (Plate 49).

Lateral to the narrow presphenoid are the **orbitosphenoids** which contain the **optic foramina** for the **optic nerves** and the **meningeal arteries**. Posterior to the presphenoid is the **basisphenoid** which articulates laterally with the **alisphenoids**. The pterygoid processes of the alisphenoids have been described above. The alisphenoids articulate anteriorly with the **orbitosphenoids** and extend dorsad into the orbit. They form with the adjacent orbitosphenoids the **orbital** or **sphenoidal fissures** (*foramina*) which carry the **third, fourth,** and **sixth cranial nerves** and the first division of the **fifth nerve**. The alisphenoids contain also the **foramina rotunda** and **ovale** which carry the **second** and **third** divisions of the **fifth cranial nerves** respectively. These complexes of sphenoid bones are represented in adult man by the single sphenoid.

The remainder of the midventral portion of the skull is occupied by the **basioccipital portion** of the **occipital bone.** Lateral to this are the prominent hollow **tympanic bullae** which house the **middle ear** or **tympanic cavity.** They open **laterocephalad** through the **external acoustic meatuses.** At the cephalic border of the bullae there are openings for the **auditory tubes.** The dorsal walls of these openings are formed by the alisphenoids. The tympanic bullae cover the petrous portion of the temporal bones which house the **internal ears.** Lateral to each tympanic bulla is a prominent **mastoid process,** a part of the mastoid portion of the temporal bone, while caudad of each bulla is a less prominent process of the occipital bone the **jugular process.**

Two foramina can be seen at the caudomedial margins of each auditory bulla. The larger one lying between the occipital and temporal bones is the **jugular foramen** for the passage of the **inferior cerebral vein** and the **ninth, tenth,** and **eleventh cranial nerves.** The smaller one, the **hypoglossal canal** is in the exoccipital bone just medial to the former and transmits the **twelfth cranial nerve.**

The posterior end of the skull is composed of the occipital bone in which is seen the conspicuous **foramen magnum** through which the spinal cord passes to join the brain. The foramen magnum is bordered laterally by the prominent **occipital condyles** by which the skull articulates with the first cervical vertebra which is called the atlas.

Lateral and Frontal Views of the Skull
PLATE 4

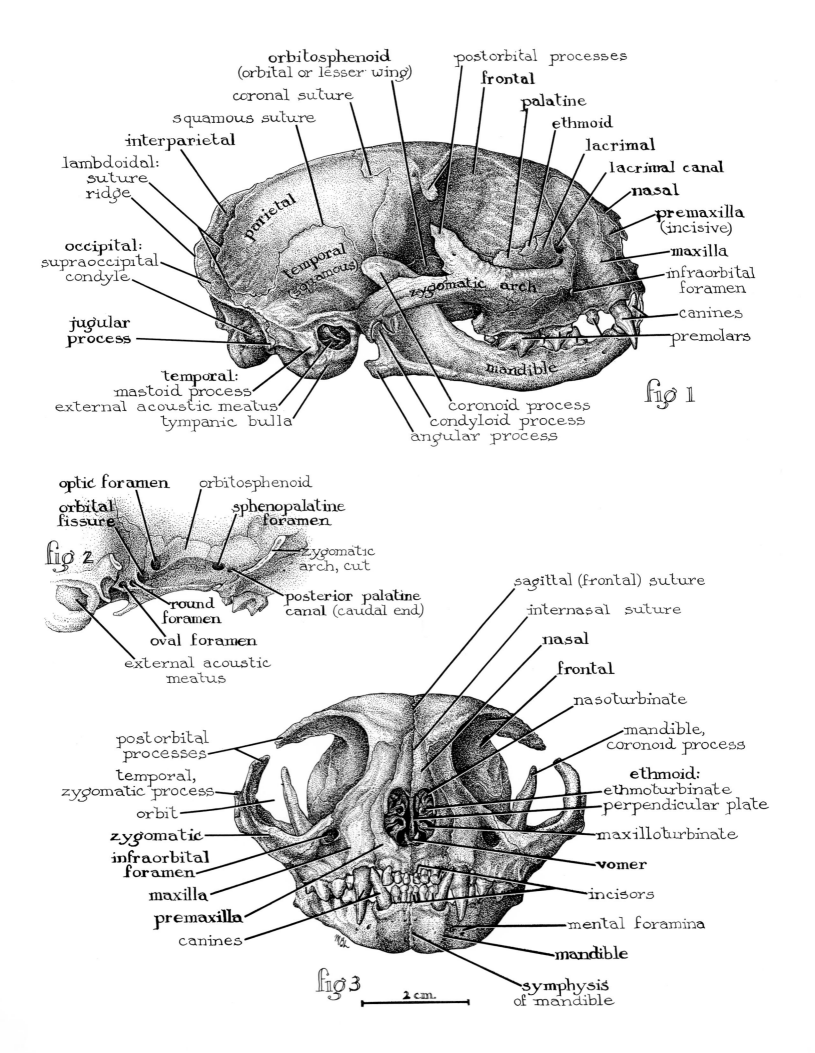

fig 1

orbitosphenoid (orbital or lesser wing)
coronal suture
squamous suture
interparietal
lambdoidal suture ridge
occipital: supraoccipital condyle
jugular process
temporal: mastoid process
external acoustic meatus
tympanic bulla
parietal
temporal (squamous)
postorbital processes
frontal
palatine
ethmoid
lacrimal
lacrimal canal
nasal
premaxilla (incisive)
maxilla
infraorbital foramen
canines
premolars
zygomatic arch
mandible
coronoid process
condyloid process
angular process

fig 2

optic foramen
orbital fissure
orbitosphenoid
sphenopalatine foramen
zygomatic arch, cut
round foramen
oval foramen
external acoustic meatus
posterior palatine canal (caudal end)

fig 3

sagittal (frontal) suture
internasal suture
nasal
frontal
nasoturbinate
mandible, coronoid process
ethmoid: ethmoturbinate perpendicular plate
maxilloturbinate
vomer
incisors
mental foramina
mandible
symphysis of mandible
postorbital processes
temporal, zygomatic process
orbit
zygomatic
infraorbital foramen
maxilla
premaxilla
canines

2 cm.

Lateral aspect of skull (fig. 1). This view, with the mandible in place, is oval in outline. The **orbital** and **temporal fossae**, the **tympanic bulla, occipital condyle** and **zygomatic arch** are the most prominent features.

The **premaxillary** (*incisive*) bones rim the nasal fossa laterally and ventrally at the cranial end of the skull and bear the upper incisor teeth. The **nasal bones** complete the rim of the nasal fossa dorsally fig. 1).

The **maxilla, mandible, zygomatics, lacrimals** and **os planum** of the **ethmoid** complete the facial region of the skull. The maxillae bear the canine, premolar and molar teeth, while the mandible bears these and, in addition, the lower incisors. The space between the canine and premolar on the mandible is the **diastema.** In the maxilla at the cranial root of the zygomatic arch is the large **infraorbital foramen** which in life carries the infraorbital nerve and artery. Above the infraorbital foramen in the cranial part of the orbit is the lacrimal canal for the passage of the tear duct.

The external rim of the orbit is formed by the zygomatic arch, the supraorbital arch and the postorbital processes of the frontal and zygomatic. Since these latter processes do not meet, the orbital rim is open at this point and is continuous with the temporal fossa. In man the orbital and temporal fossae are separated. Ventrally and caudally the limits of the orbit are marked by the optic foramen which is not shown in figure 1.

The remainder of the lateral view lying caudal to the **coronal suture** consists of the **parietal, interparietal, temporal** and **occipital** bones. The **squamosal suture** separates the temporal and parietal bones, while the **lambdoidal suture** lies between the occipital, parietal, interparietal, and temporal. The prominent **zygomatic process** of the temporal bone forms the caudal root of the zygomatic arch and at its base ventrally is the deep **mandibular fossa** for the articulation of the mandible. A prominent **retroglenoid process** marks the caudal boundary of this fossa (Plate 8, fig. 1).

Caudad and ventrad of the postmandibular process is the **tympanic bulla** with the large **external acoustic meatus** opening on its lateral surface. Caudal to the external acoustic meatus and on the lateral side of the bulla is the

mastoid process of the temporal bone. Under the cranial margin of the mastoid process is a small **stylomastoid foramen** for the passage of the seventh cranial nerve. On the caudal surface of the tympanic bulla is the **jugular process** of the occipital bone.

A **lambdoidal ridge** extends from the caudal base of the zygomatic arch, above the tympanic bulla and then turns mediad and caudad to the **external occipital crest.** The **occipital condyles,** for articulation of the skull with the first cervical vertebra or atlas, are at the ventrocaudal part of the skull. They lie to either side of the foramen magnum.

Figure 2 shows a part of the ventrolateral portion of the skull with the mandible removed and a part of the zygomatic arch cut away. This reveals important foramina of the skull. The most cranial one in the series is the caudal opening of the **posterior palatine canal,** the next the large **sphenopalatine foramen.** These are both in the palatine bone. The **optic foramen** lies in the orbitosphenoid bones, the **orbital fissure** (*foramen lacerus anterius*) lies between the orbitosphenoid and the alisphenoid bones. The **round** and **oval** foramina both open through the alisphenoid. Reference to Table 2 will indicate the nerves and blood vessels which traverse these foramina.

Frontal aspect of the skull (fig. 3). From this view the outline of the skull is almost circular. The **orbits,** the **nasal aperture** and **oral aperture** are the most prominent features. Through the heart-shaped nasal aperture the vomer can be seen ventrally and extending dorsad from it is the **perpendicular plate of the ethmoid** (*mesethmoid*) which divides the nasal cavity into right and left portions. Scroll-like **ethmoturbinates** extend into the nasal passageways and ventral to these are the **maxilloturbinates.** These structures are comparable to the superior, middle and inferior nasal conchae (*turbinates*) of man. They increase greatly the surface area of the nasal passageways which in living subjects are covered with a highly vascular and glandular mucous membrane. They serve as an air-conditioning system to warm, moisten and cleanse the incoming air. The olfactory nerve endings are distributed over part of this mucous membrane constituting the receptor for the sense of smell.

TABLE 2—FORAMINA OF THE SKULL

Name	Location (Bones involved)	Traversing Structures
1. Anterior palatine (incisive)	Between maxillary and premaxillary.	V. Trigeminal n., nasopalatine br.; nasal artery.
2. Infraorbital	Maxillary, below orbit, craniad.	V. Trigeminal n., infraorbital br. of maxillary division; infraorbital artery.
3. Lacrimal	Dorsal to infraorbital foramen; lacrimal bone.	Lacrimal duct.
4. Posterior palatine	Palatine	V. Trigeminal n., greater palatine br.; descending palatine artery.
5. Sphenopalatine	Palatine, vertical plate.	V. Trigeminal n., sphenopalatine br.; sphenopalatine artery.
6. Optic	Orbitosphenoid.	Optic n.; ophthalmic artery.
7. Orbital fissure (foramen lacerum anterior; sphenoidal fissure)	Between orbitosphenoid and alisphenoid	III. Oculomotor n.; IV. trochlear n.; VI. abducens n.; V. trigeminal n., ophthalmic division; internal maxillary artery, br.
8. Rotundum	Alisphenoid	V. Trigeminal n., maxillary division.
9. Ovale	Alisphenoid	V. Trigeminal n., mandibular division; internal maxillary artery, br.
10. Olfactory	Ethmoid, cribriform plate.	I. Olfactory n.
11. Mandibular	Mandible, medial surface, caudad.	V. Trigeminal n., inferior alveolar br.; inferior alveolar artery.
12. Mental (may be 2 or 3)	Mandible, lateral surface, craniad.	V. Trigeminal n., dental brs.; dental (mental) vessels.
13. Facial canal	Temporal bone, petrous portion.	VII. Facial n.
14. Stylomastoid	Temporal bone, mastoid portion.	VII. Facial n.
15. Internal auditory meatus	Temporal bone, petrous portion.	VIII. Vestibulo-cochlear n. (auditory or statoacoustic).
16. Jugular	Between temporal and basioccipital.	IX. Glossopharyngeal n.; X. Vagus n.; XI. Accessory n.; inferior cerebral vein.
17. Hypoglossal	Occipital, ventral side.	XII. Hypoglossal n.
18. Condyloid canal	Occipital.	Vein.
19. Eustachian	Tympanic bulla and basisphenoid.	Eustachian tube.
20. Magnum	Occipital.	Spinal cord.

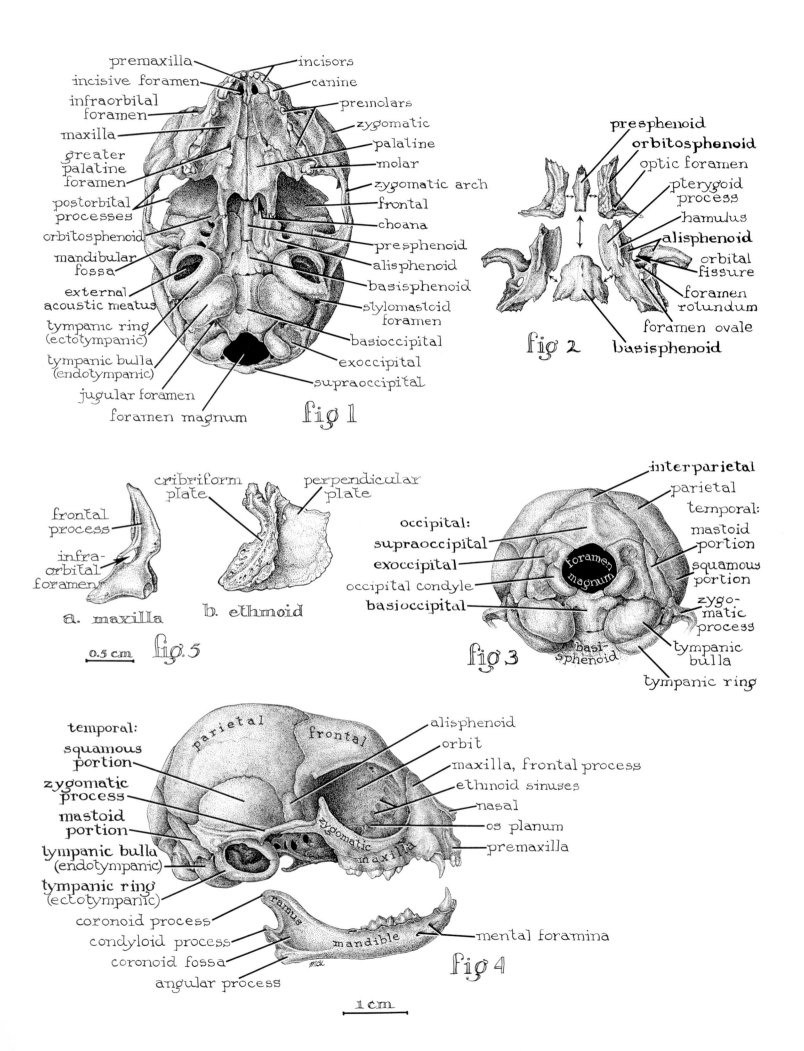

fig 1

premaxilla — incisors
incisive foramen — canine
infraorbital foramen — premolars
maxilla — zygomatic
greater palatine foramen — palatine
postorbital processes — molar
orbitosphenoid — zygomatic arch
mandibular fossa — frontal
external acoustic meatus — choana
tympanic ring (ectotympanic) — presphenoid
tympanic bulla (endotympanic) — alisphenoid
jugular foramen — basisphenoid
foramen magnum — stylomastoid foramen
— basioccipital
— exoccipital
— supraoccipital

fig 2

presphenoid
orbitosphenoid
optic foramen
pterygoid process
hamulus
alisphenoid
orbital fissure
foramen rotundum
foramen ovale
basisphenoid

fig 5

cribriform plate
perpendicular plate
frontal process
infra-orbital foramen

a. maxilla   b. ethmoid

0.5 cm

fig 3

interparietal
parietal
temporal:
mastoid portion
squamous portion
zygomatic process
tympanic bulla
tympanic ring
occipital:
supraoccipital
exoccipital
occipital condyle
basioccipital
foramen magnum
basisphenoid

fig 4

temporal:
squamous portion
zygomatic process
mastoid portion
tympanic bulla (endotympanic)
tympanic ring (ectotympanic)
coronoid process
condyloid process
coronoid fossa
angular process
parietal
frontal
alisphenoid
orbit
maxilla, frontal process
ethmoid sinuses
nasal
os planum
premaxilla
zygomatic
maxilla
ramus
mandible
mental foramina

1 cm

The purpose of this plate is to show more dramatically the composite character of some of the bones whose parts in an adult skull are completely fused as individual bones or are so closely joined as to make separation difficult.

Figure 1 is a ventral view of the skull showing the loose relationship among the components of the occipital, temporal, and sphenoid bones. In figure 2 the sphenoid complex of bones is separated out and since they have been previously described in reference to Plates 9 and 10, no description is repeated here.

The components of the **occipital bone** are clearly shown in both figures 1 and 3. Figure 3 is a view of the caudal surface of the skull. The occipital bone of the kitten is here clearly composed of **basioccipital** ventral to the **foramen magnum,** paired **exoccipital** bones lateral to the foramen magnum and bearing the occipital condyles and a **supraoccipital** completing the rim of the foramen magnum dorsally. These completely fuse into a single bone in the adult—the occipital.

The **temporal bone** is another "bone complex" of considerable importance. It houses the external, middle and internal ear. Its parts are best shown in figures 1 and 4, figure 4 being a lateral view of the skull with the mandible separated from the rest of the skull. The parts of the temporal are the **squamous portion,** a bone of intramembranous origin, *i.e.* ossified in membrane rather than being formed by replacement of cartilage as in endochondral bones. Projecting forward from the squamous portion is a **zygomatic process** of similar origin which joins with the zygomatic bone to form the zygomatic arch. Located around the external auditory canal is a **tympanic ring** or ectotympanic and ventrally a **tympanic bulla** or endotympanic both of intramembranous origin. A **mastoid portion,** shown well on figures 3 and 4, is a bone of endochondral origin and contains the mastoid air cells. Another very vital component of the temporal bone is the **petrous portion** which pushes mediad and craniad into the floor of the skull and houses the inner ear structures such as the cochlea and semicircular canals. It too is endochondral and is not shown in this plate. When fused together these are the temporal bone—a paired bone of the skull and as you may surmise one of the most complicated.

Inside of the middle ear cavity are the three small endochondral bones known as ear ossicles, the malleus, incus and stapes. Only mammals have three ear ossicles. They are derived from the jaw structures of lower vertebrates.

In figure 5 we have introduced two bones of the fetal skull of the cat just for comparison with those of the kitten and the adult (Plates 9 and 11). One of these is the **maxillary,** the other the **ethmoid.** The latter shows only the horizontal or **cribriform plate** with foramina for the olfactory nerve fibers, and the perpendicular lamina or plate which forms a part of the vertical partition in the nose. None of the lateral masses of the ethmoid complex are shown.

It is of interest to note that some of these bones which in mammals become permanently joined to other bones remain separate for the total life of the animal in some of the lower vertebrates. Fish have a separate squamosal bone, for example.

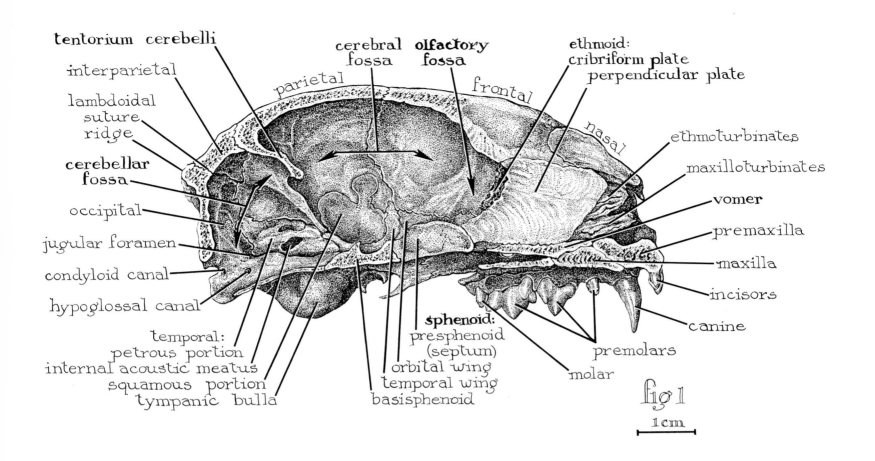

tentorium cerebelli

interparietal

lambdoidal
suture
ridge

**cerebellar
fossa**

occipital

jugular foramen

condyloid canal

hypoglossal canal

temporal:
petrous portion
internal acoustic meatus
squamous portion
tympanic bulla

cerebral
fossa

parietal

**olfactory
fossa**

frontal

nasal

ethmoid:
cribriform plate
perpendicular plate

ethmoturbinates

maxilloturbinates

vomer

premaxilla

maxilla

incisors

canine

premolars

molar

sphenoid:
presphenoid
(septum)

orbital wing
temporal wing
basisphenoid

fig 1

1cm

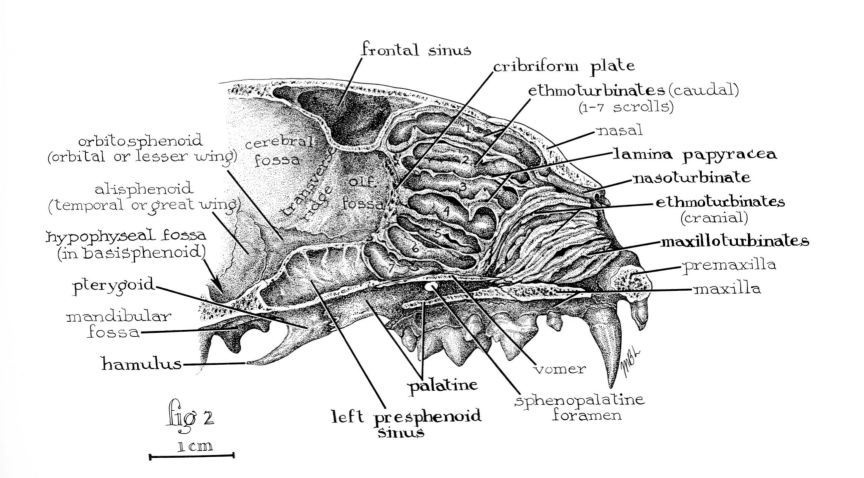

frontal sinus

cribriform plate

ethmoturbinates (caudal)
(1-7 scrolls)

orbitosphenoid
(orbital or lesser wing)

alisphenoid
(temporal or great wing)

hypophyseal fossa
(in basisphenoid)

pterygoid

mandibular
fossa

hamulus

cerebral
fossa

transverse
ridge

olf.
fossa

nasal

lamina papyracea

nasoturbinate

ethmoturbinates
(cranial)

maxilloturbinates

premaxilla

maxilla

vomer

palatine

left presphenoid
sinus

sphenopalatine
foramen

fig 2

1cm

Midsagittal aspect of the skull (fig. 1). The conspicuous features of this view are the **perpendicular plate** (*mesethmoid*) of the ethmoid and the nasoturbinates, **ethmoturbinates** and **maxilloturbinates** cranially followed in a caudad direction by the **olfactory, cerebral** and **cerebellar fossae.** The ventral line of the skull is marked by the premaxillary and maxillary **teeth,** the sharp **hamulus** of the pterygoid process and the **tympanic bulla.** The **foramen magnum** is at the caudal end of the skull.

The **mesethmoid,** which is continued forward in life by cartilage, forms the separation between the right and left nasal passageways. It rests on the **vomer** and **presphenoid** ventrally and dorsally articulates with the median descending plates of the **frontals** and **nasals.** Caudally it is continuous with the **cribriform plate** which separates the nasal passageways from the **olfactory fossa** of the cranial cavity.

In the ventral cranial region one can also see the sequence of bones which form the floor of the nasal passageways, the **premaxillaries, maxillaries, palatines** and **vomer.** The vomer separates from the floor caudally and spreads laterally thus dividing the nasal passageways into dorsal and ventral portions. The ventral portion in life opens caudad into the nasopharynx through the choanae. The bifurcated caudal border of the vomer articulates with the presphenoid.

The small **olfactory fossa** which houses the olfactory bulbs lies at the cranial end of the cranial cavity. Its cranial boundary is formed by the **cribriform plate** of the ethmoid which is perforated by many foramina for the passage of the **olfactory nerve fibers** into the nasal passageways. The roof of the olfactory fossa is formed by the frontal bones which, where they join, form a prominent median crest.

The **cerebral fossa** lies caudal to the olfactory fossa and is separated from it by a transverse ridge. It is bounded caudad by a prominent bony crest the **tentorium cerebelli** which separates it from the cerebellar fossa. At about the level of the coronal suture between the parietal and frontal bones is a ridge which divides the cerebral fossa into anterior and middle fossae. The parietals, frontals, squamous temporal, sphenoid and presphenoid bones form the boundaries of the cerebral fossa. Numerous ridges and furrows mark the inner surfaces of these bones for the reception of the cerebral convolutions of the brain. A deep depression in the floor of the cerebral fossa in the sphenoid bone is the hypophyseal fossa which in the live subject houses the **hypophysis.** Behind it is a prominent elevation, the **dorsum sellae.**

Caudal to the **tentorium cerebelli** is the **cerebellar fossa.** Its roof is formed by the parietals and interparietals; it is limited caudally by the occipital bone in which is located the **foramen magnum.** The floor of the fossa is formed by the petrous portion of the temporals and the basal portion of the occipital, while the lateral walls consist of the mastoid portion of the temporal, and parts of the occipital and parietals. Depressions for the lobes of the cerebellum of the brain mark the lateral, dorsal and caudal walls of the cerebellar fossa.

A few openings or foramina are seen in the floor of the cerebellar fossa. Just lateral to the foramen magnum is the small caudal opening of the **condyloid canal** which carries a small vein forward through the substance of the occipital bone. It opens just caudad of the petrous portion of the temporal. The small opening of the **hypoglossal canal** which carries the **twelfth cranial nerve** can be seen just craniad and ventrad to the condyloid canal. At the caudomedial boundary of the petrous temporal is the large **jugular foramen** for the passage of the **inferior cerebral vein** and the **ninth, tenth,** and **eleventh cranial nerves.** In the petrous itself is the large **internal acoustic meatus** for the passage of the **eighth cranial nerve.** Critical examination of the meatus reveals that it is actually divided into two parts, a ventral portion for the eighth cranial nerve and a dorsal part, the **facial canal,** for the **seventh cranial** or **facial nerve.** The facial canal winds through the petrous laterad to open at the stylomastoid foramen on the ventral side of the skull (Plate 3, fig. 2).

Parasagittal section of anterior two thirds of skull (fig. 2). This figure reveals the positions of the **frontal** and **sphenoidal sinuses** and shows in detail the complexity and extent of the **ethmoturbinates** and **maxilloturbinates** of the nasal fossa. The ethmoturbinates represent the lateral ethmoids and are divided into a **caudal portion** with seven scrolls and a **cranial portion.** The first and seventh scrolls push caudad of the frontal and sphenoidal sinuses respectively. In man the lateral ethmoid is represented by the superior and middle nasal conchae.

The **maxilloturbinates** are paired bones attached to the maxillary and extending into the anterior portion of the nasal cavity. Each consists of a longitudinal plate which bifurcates medially into a dorsal curved lamina and a ventral scroll. The maxilloturbinates correspond to the inferior nasal conchae of man and the space between them and the cranial portion of the lateral ethmoid corresponds to the middle nasal meatus.

The **hamulus** of the pterygoid process of the sphenoid and the **mandibular fossa** are conspicuous structures on the ventral side of figure 2. The **sphenopalatine foramen** in the palatine bone is traversed in the living subject by the sphenopalatine nerve and artery.

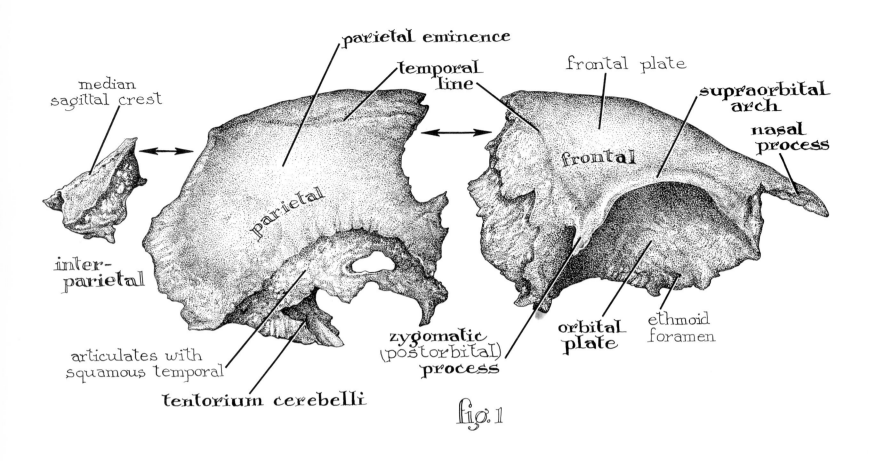

median sagittal crest

interparietal

parietal eminence

temporal line

parietal

articulates with squamous temporal

tentorium cerebelli

frontal plate

supraorbital arch

nasal process

frontal

zygomatic (postorbital) process

orbital plate

ethmoid foramen

fig. 1

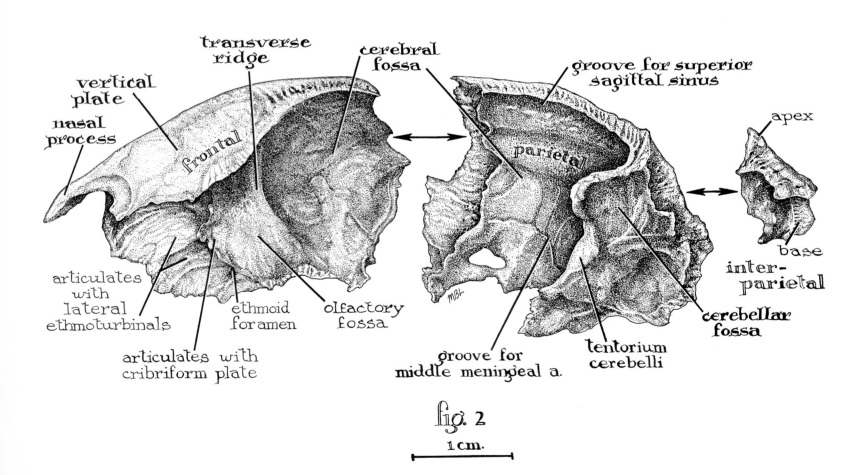

vertical plate

nasal process

transverse ridge

cerebral fossa

frontal

groove for superior sagittal sinus

parietal

apex

articulates with lateral ethmoturbinals

articulates with cribriform plate

ethmoid foramen

olfactory fossa

groove for middle meningeal a.

tentorium cerebelli

cerebellar fossa

base

interparietal

fig. 2

1 cm.

In this plate the bones are separated to show clearly their boundaries but their proper relationships are retained. These are intramembranous bones as indicated by studies in embryology and comparative anatomy.

Lateral view (fig. 1). There are two **frontal bones** in the cat which join at the **sagittal** (*frontal*) **suture** on the median line of the skull. Each bone is divided into frontal and orbital plates by the **supraorbital arch** which continues caudolaterad as the **postorbital** (*zygomatic*) **process.** The **frontal plate** forms the roof of the cranial portion of the cerebral fossa and the olfactory fossa of the cranial cavity and the caudal part of the roof of the nasal passageways. At its mediocranial angle the frontal plate is extended as the **nasal process** which fits between the nasal and maxillary bones. The **orbital plate** forms a large part of the medial, dorsal and caudal wall of the orbital fossa and continues caudad into the **temporal fossa.** A small opening near the ventral border of the orbital plate is the ethmoid foramen which carries an artery of the same name.

The articulations of the borders of this bone with adjacent bones of the skull are clearly indicated on figure 1. Note the bevelled caudal border of the frontal where it articulates with the parietal forming the **coronal** (*fronto-parietal*) **suture.**

The parietal bone is paired and forms a large part of the lateral and dorsal walls of the cranial cavity. These bones join on the median line where they form the caudal portion of the **sagittal suture.** The outer surface of each bone is smooth and convex. The point of greatest convexity is known as the **parietal eminence** and is the point where ossification was initiated. A faint, curved ridge starts near the caudomedial angle and runs craniolaterad on to the frontal bone. This is the **temporal line** and is the line of origin of the temporal muscle. The ventral portion of the lateral surface is roughened and is bevelled to articulate with the squamous portion of the temporal bone. This thin surface in sometimes perforated near the irregularly concave ventrolateral border. The parietal bone articulates with the interparietal and occipital caudally; with the temporal laterally, the alisphenoid at the cranioventral angle, and with the frontal cranially.

The small, triangular **interparietal** bone fits between the parietal bones at their caudomedial angles. Caudally it articulates with the occipital bone at the **lambdoidal suture.**

Medial view (fig. 2). This view of the **frontal bone** presents for examination the anterior portion of the **cerebral fossa,** the **olfactory fossa,** separated from the former by the **transverse ridge,** the **vertical plate** covering the frontal sinus, and a roughened area anteriorly which serves for attachment of the **ethmoid complex** of bone.

The **cerebral fossa** presents a smooth, concave surface marked by low ridges and furrows to accommodate the convolutions of the cerebrum.

Cranial to the transverse ridge and olfactory fossa which houses the olfactory bulbs a sharp irregular ridge marks the point of attachment of the cribriform plate of the ethmoid and anterior to it longitudinal ridges articulate with the lateral ethmoid.

The medial border of the frontal by which it articulates with its opposite number is narrow caudally but widens cranially and is composed of a thin, vertical plate of bone behind which is the frontal sinus and the oval opening by which the sinus communicates with the nasal cavities. One portion of the ethmoturbinal is also located behind the vertical plate where it pushes toward the frontal sinus. These relationships are best illustrated on Plate 6 figure 2.

The inner surface of the **parietal** forms a large portion of the lateral and dorsal wall of the caudal part of the cerebral fossa. This surface is smooth but has ridges and depressions for the convolutions of the cerebrum. A depression adjacent to the medial border forms with the other parietal a groove for the **superior sagittal sinus.** At the cranial border the bone is bevelled on the inner surface for articulation with the frontal. Caudally the cerebral fossa is limited by the **tentorium,** a prominent plate of bone deeply notched at its ventral margin. It fits between the cerebrum and cerebellum in the living subject and separates the cerebral from the cerebellar fossa. It ossifies in the dura mater. In man it remains membranous and hence is not considered a part of the parietal bone. The walls of the cerebellar fossa are marked by ridges and gooves to receive the convolutions of the cerebellum.

The caudal border of the parietal is bevelled internally and in articulating with the interparietal, occipital and temporal forms the **lambdoidal suture.**

The borders of the interparietal are quite thick, porous and bevelled to articulate with the parietals and supraoccipital.

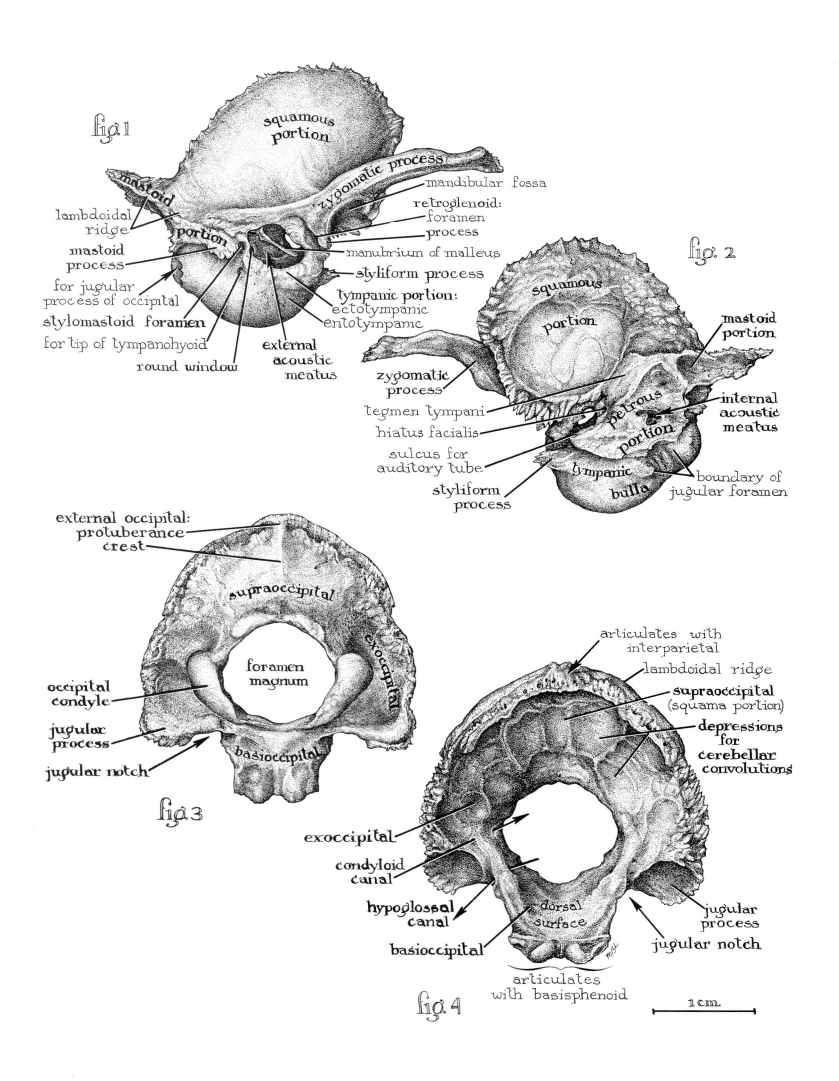

fig. 1

squamous portion

zygomatic process

mandibular fossa

retroglenoid:
foramen
process

mastoid

lambdoidal ridge

portion

mastoid process

for jugular process of occipital

stylomastoid foramen

for tip of tympanohyoid

round window

manubrium of malleus

styliform process

tympanic portion:
ectotympanic
entotympanic

external acoustic meatus

fig. 2

squamous portion

mastoid portion

zygomatic process

tegmen tympani

hiatus facialis

sulcus for auditory tube

styliform process

petrous portion

internal acoustic meatus

tympanic bulla

boundary of jugular foramen

external occipital:
protuberance
crest

supraoccipital

occipital condyle

foramen magnum

exoccipital

jugular process

jugular notch

basioccipital

fig. 3

articulates with interparietal

lambdoidal ridge

supraoccipital
(squama portion)

depressions for cerebellar convolutions

exoccipital

condyloid canal

hypoglossal canal

basioccipital

dorsal surface

articulates with basisphenoid

jugular process

jugular notch

fig. 4

1 cm.

The temporal bone (figs. 1 and 2). The temporal bones form most of the lateral walls of the cranium posterior to the orbit and extend ventrally and medially to articulate with the occipital and sphenoid. Dorsolaterally they join the parietal at the **squamosal suture** and cranially provide articulation for the mandible.

Each temporal consists of a complex of four parts (Plate 5, figs. 1 and 4), the **squamous** and **tympanic** which are composed of **intramembranous bone** and the **mastoid** and **petrous** of **endochondral** origin. The tympanic portion houses the ear ossicles, the **malleus, incus** and **stapes,** in the **middle ear cavity** while the **inner ear** is within the petrous portion (Plate 112). The ear ossicles are also of endochondral origin and are homologous to jaw and hyoid structures of the lower vertebrates.

The **outer surface of the temporal (fig. 1).** The outer surface of the **squamous portion** is smooth and convex with a curved **zygomatic process** arising from its ventral border by caudal and cranial roots. The caudal root is continuous with the lambdoidal ridge which passes caudad between the squama and external auditory meatus; the ventral arises directly out of the ventrocranial angle of the bone. At its base the zygomatic process is horizontal dorsally but twists laterally so that at its cranial end the dorsal surface has become medial. It articulates cranially with the zygomatic bone to form the **zygomatic arch.** On the ventral surface of the base of the zygomatic arch is a transverse, elongated fossa for the articulation of the mandible, **the mandibular fossa.** A sharp transverse ridge caudal to the fossa is the retroglenoid process, while anterior and lateral to the fossa is a small articular tubercle.

The **tympanic portion** of the temporal is in the form of a hollow bulb on the base of the skull, the **tympanic bulla.** It is made up of two parts an outer horse-shoe shaped structure, clearly visible in the skull of the kitten, the ectotympanic (Plate 5, figs. 1 and 4), which surrounds the **external acoustic meatus;** and an inner entotympanic which forms the larger part of the smooth, globular bulla.

Caudad of the external acoustic meatus and in front of the mastoid process is a groove which extends dorsoventrally in the top of which is the **stylomastoid foramen** for the passage of the facial nerve. Just ventrad of this groove is a small pit into which articulates the **tympanohyoid bone** of the **hyoid apparatus.**

On the cranial surface of the bulla a sharp **styliform process** is present which fits into the ventral surface of the basisphenoid. Lateral to this process is a groove for the passage of the auditory (*Eustachian*) tube (Plate 10, fig. 5).

Caudally the bulla is overlapped by the **mastoid process of the temporal.** A roughened area on the bulla caudo-ventrad of the mastoid process is, in the articulated skull, overlaid by the **jugular process** of the occipital.

The **mastoid portion,** includes the **mastoid process** mentioned above and continues caudomediad as a spine-like bone. On its outer surface is a continuation of the lambdoidal ridge.

The **petrous portion** of the temporal can be seen laterally only by looking into the middle ear or tympanic cavity. There the fenestra cochlea (*round window*), and fenestra vestibuli (*oval window*) can be seen caudally the former larger and ventral to the latter. The base of the stapes fits into the fenestra vestibuli in the living subject.

The inner or medial surface of the temporal (fig. 2). The inner surface of the **squamous portion** is smooth and convex except for an extensive border area which is bevelled and roughened for articulation with the parietal dorsally and the alisphenoid cranially. Ventrally it joins with the mastoid, the tympanic, and the petrous portions.

The medial surface of the **tympanum** is largely obscured by the petrous portion. The styliform process can be seen, however, as described above. Caudad of this process are two or three parallel vertical grooves marking the edge of the **jugular foramen** which is completed by the occipital bone in the articulated skull.

The inner surface of the **petrous portion** is marked by the large **internal acoustic meatus** which is divided internally into dorsal and ventral parts. The dorsal part is the facial canal for the passage of seventh cranial or facial nerve. It passes through the petrous bone to emerge at the **stylomastoid foramen.** The ventral part of the meatus is for the eighth cranial or acoustic nerve. Above and caudad of the internal acoustic meatus is a deep appendicular fossa for the appendicular lobe of the cerebellum. Craniad and dorsad of the internal acoustic meatus is a triangular dorsal surface, the tegmen tympani. At the apex of this triangle is a small foramen, the hiatus facialis which transmits the vidian nerve and which leads to the facial canal.

The inner surface of the **mastoid portion** forms a part of the wall of the cranial cavity.

The occipital bone (figs. 3 and 4). The occipital bone consists of four parts which are separate in the kitten but which gradually fuse to form the single bone of the adult (Plate 5, fig. 3). These are the ventral **basioccipital,** the two lateral **exoccipitals** and the dorsal **supraoccipital.** The occipital is of endochondral origin. It forms the caudal wall and part of the base of the cranial cavity and connects the skull to the vertebral column. Its four components form the margins of the large **foramen magnum** through which the spinal cord passes to join the brain.

Caudoventral view of occipital (fig. 3). From this aspect the supraoccipital presents dorsally a prominent **lambdoidal ridge** which extends laterally from a median **external occipital protuberance.** Extending ventrad from this protuberance toward the foramen magnum is the **external occipital crest.** The ventral border of the supraoccipital forms a part of the dorsal rim of the foramen magnum medially and extends laterally into the exoccipitals.

The **exoccipitals** or lateral portions form a part of the rim of the foramen magnum and bear the **occipital condyles** for attachment with the atlas. Lateral to the condyles the bone is elevated into a prominent triangular projection, the **jugular process,** which articulates with the caudal part of the tympanic bulla of the temporal bone. Between the occipital condyle and the jugular process is a deep fossa.

The ventral surface of the **basioccipital** or basilar portion articulates laterally with the **tympanic bulla** and its sharp lateral border touches upon the petrous portion of the temporal. Cranially it articulates with the caudal end of the body of the sphenoid. Caudally it joins the exoc-cipitals laterally while medially it completes the rim of the foramen magnum.

Craniodorsal view of the occipital (fig. 4). The dorsal cranial border of the **supraoccipital** is rough and porous and articulates with the interparietal and parietals. Its inner surface has ridges and depressions for the convolutions of the cerebellum.

The **exoccipitals** are continuous dorsally with the supraoccipital. Their **lateral margins** are rough for articulation with the **mastoid portion of the temporal bone,** while the cranial faces of the **jugular processes** are concave for the reception of the tympanic bullae. Medial to the jugular process is the **jugular notch** which when articulated with the tympanic bulla forms the jugular foramen. The **hypoglossal foramen** is medial to the jugular notch while further dorsad is the opening of the condyloid canal. Arrows have been inserted in these in figure 4.

The dorsal side of the **basioccipital** is concave from side to side for the reception of the pons and medulla of the brain.

Ethmoid, Vomer, Palatine and Sphenoid—Dorsal View
PLATE 9

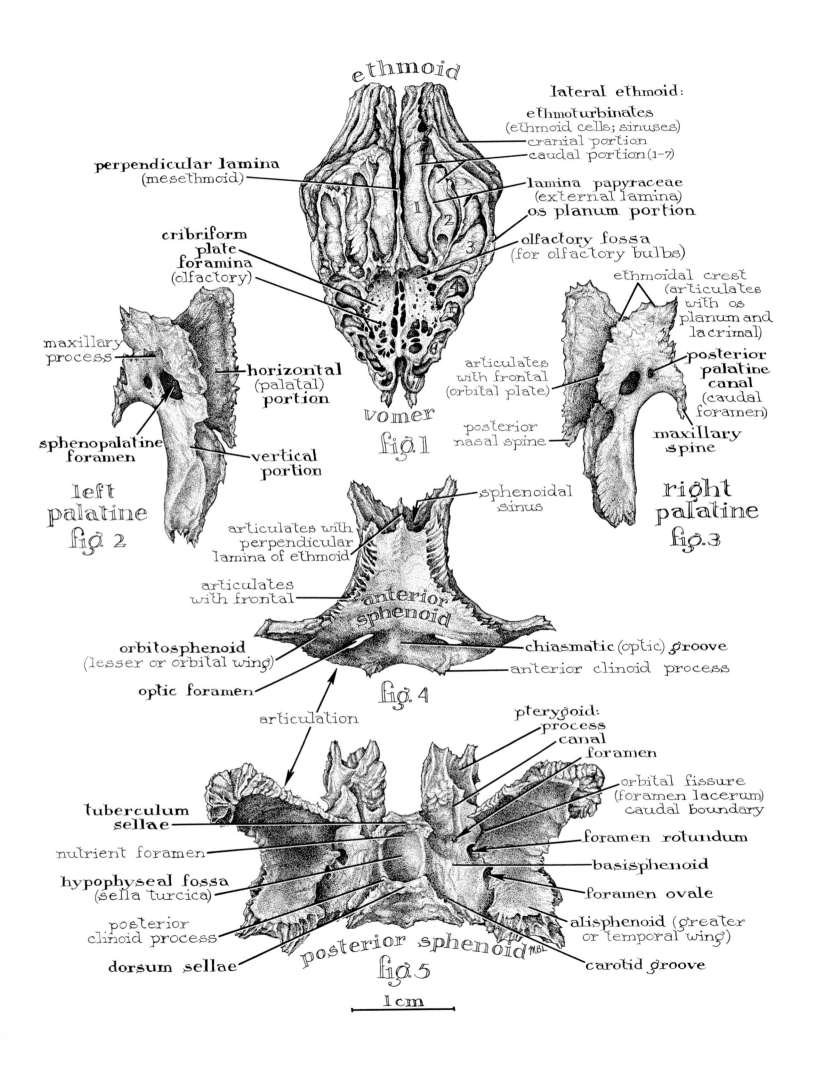

ethmoid

lateral ethmoid:
ethmoturbinates
(ethmoid cells; sinuses)
cranial portion
caudal portion (1-7)

perpendicular lamina
(mesethmoid)

lamina papyraceae
(external lamina)
os planum portion

cribriform
plate
foramina
(olfactory)

olfactory fossa
(for olfactory bulbs)

ethmoidal crest
(articulates
with os
planum and
lacrimal)

maxillary
process

articulates
with frontal
(orbital plate)

posterior
palatine
canal
(caudal
foramen)

horizontal
(palatal)
portion

posterior
nasal spine

maxillary
spine

sphenopalatine
foramen

vertical
portion

vomer
fig. 1

left
palatine
fig. 2

right
palatine
fig. 3

sphenoidal
sinus

articulates with
perpendicular
lamina of ethmoid

articulates
with frontal

anterior
sphenoid

orbitosphenoid
(lesser or orbital wing)

chiasmatic (optic) groove

anterior clinoid process

optic foramen

fig. 4

articulation

pterygoid:
process
canal
foramen

orbital fissure
(foramen lacerum)
caudal boundary

tuberculum
sellae

foramen rotundum

nutrient foramen

basisphenoid

hypophyseal fossa
(sella turcica)

foramen ovale

posterior
clinoid process

alisphenoid (greater
or temporal wing)

dorsum sellae

posterior sphenoid
fig. 5

carotid groove

1 cm

Since this plate and the following one, Plate 10, show the same bones from the dorsal and ventral sides respectively, they should be studied together. Reference should be made also to Plates 3 and 5.

The **ethmoid** is a single bone consisting of four parts; the **perpendicular lamina** (*mesethmoid*) which forms a large part of the median partition of the nasal cavity; the two **ethmoturbinates** or lateral ethmoids, thin scroll-like plates of bone which project into the nasal cavities; and the **cribriform plate** which separates the cranial from the nasal cavity (fig. 1).

The **cribriform plate** forms the caudal portion of the bone and is pierced by many **cribriform foramina** for the olfactory nerve fibers passing from the olfactory epithelium of the nose to the olfactory bulbs of the brain in the cranial cavity. From the mediocranial part of the cribriform plate the mesethmoid projects forward, while from its craniolateral border the ethmoturbinates have their origin. Laterally the cribriform plate joins the frontal bones.

The **ethmoturbinates,** by their scroll-like plates, enclose the **ethmoid cells** and form an extensive surface for the overlying mucous membrane which contains the receptors of the olfactory organ. A part of the lateral surface of the ethmoturbinates appears externally in the medial wall of the orbit between the lacrimals and frontal as the **os planum** (*lamina papyracea of man*). The medial surface is separated from the mesethmoid by a space which is broadest at the junction of the cranial and caudal parts of the ethmoturbinates, an area comparable to the superior meatus of man.

The **mesethmoid** is continuous caudally with the cribriform plate; fits ventrally into a groove on the dorsal side of the vomer; articulates dorsally with the nasal crest and the median vertical lamina of the frontal bone, while its cranial border is continuous with the cartilaginous septum of the nose. Its lateral surfaces are free.

The **vomer** appears in figure 1 as two projections behind the ethmoid, most of the bone being obscured in this view but visible on the following plate.

The **palatine** bones (figs. 2 and 3) are paired each consisting of **horizontal** and **vertical** portions. The horizontal portions join on the midline to complete the hard palate caudal and medial to the maxillae. They, therefore, contribute to the floor of the nasal cavities, the roof of the mouth and form the ventral rim of the **choanae** (*internal nares*). Laterally a prominent **maxillary spine** extends caudad, while the remainder of the lateral border is continuous with the vertical plate. The **vertical plate** has an outer concave surface which forms a part of the orbit, while its convex inner surface faces into the nasal cavity.

A large **sphenopalatine foramen** is found just cranial to the middle of the plate and ventral to it a small foramen, the caudal opening of the **posterior palatine canal.** Caudally the vertical plate articulates inconspicuously with the pterygoid portion of the sphenoid so that this bone may appear, especially in older cats, as the caudal part of the palatine.

The **sphenoid** bone lies in the central part of the floor of the skull. It consists of an **anterior** and a **posterior sphenoid** (figs. 4 and 5). The anterior sphenoid in turn consists of a narrow median **presphenoid** housing the sphenoidal sinuses into which project the seventh (*ventral*) scrolls of the ethmoturbinates and the **orbitosphenoids** which form part of the medial walls of the orbit and contain the **optic foramina** for the passage of the optic nerves. Between these foramina is the **chiasmatic** (*optic*) **groove** in which the optic nerves meet to form a partial crossing over.

The **posterior sphenoid** or sphenoid (*proper*) articulates craniad with the anterior sphenoid and caudad with the basioccipital and temporals (fig. 5). It consists of a median **basisphenoid,** two **alisphenoids** extending dorsolaterad and two **pterygoid processes** with **hamular processes** projecting ventrad.

The **basisphenoid** bears on its dorsal surface the **hypophyseal fossa** which receives the hypophysis (*pituitary*) of the brain. Posterior to this fossa is the **dorsum sellae** (*clinoid plate*) with its lateral projections the **posterior clinoid processes.** These structures and the **tuberculum sellae** constitute the sella turcica. Lateral to the basisphenoid are the **alisphenoids** in the base of which are three foramina. The **orbital** (*sphenoidal*) **fissure** is formed in conjunction with the anterior sphenoid and carries the third, fourth, and sixth cranial nerves and the first division of the fifth cranial nerve. Behind the orbital fissure is the **foramen rotundum** carrying the second division of the fifth cranial nerve and behind it the **foramen ovale** carrying the third division of the fifth cranial nerve. A tiny **pterygoid foramen** penetrates the sphenoid between the body and the alisphenoid. It is extended craniad on the dorsal surface of the pterygoid process as a groove, the two constituting the **pterygoid canal** through which passes the Vidian nerve. Between the basisphenoid and alisphenoid at their caudal margins, a notch marks the position of the **external carotid foramen.** From this notch the **carotid groove** extends forward to the sella turcica. In the living subject it houses the internal carotid artery.

The components of the anterior and posterior sphenoids are present in lower vertebrates as separate bones, while in man all are fused into a single bone.

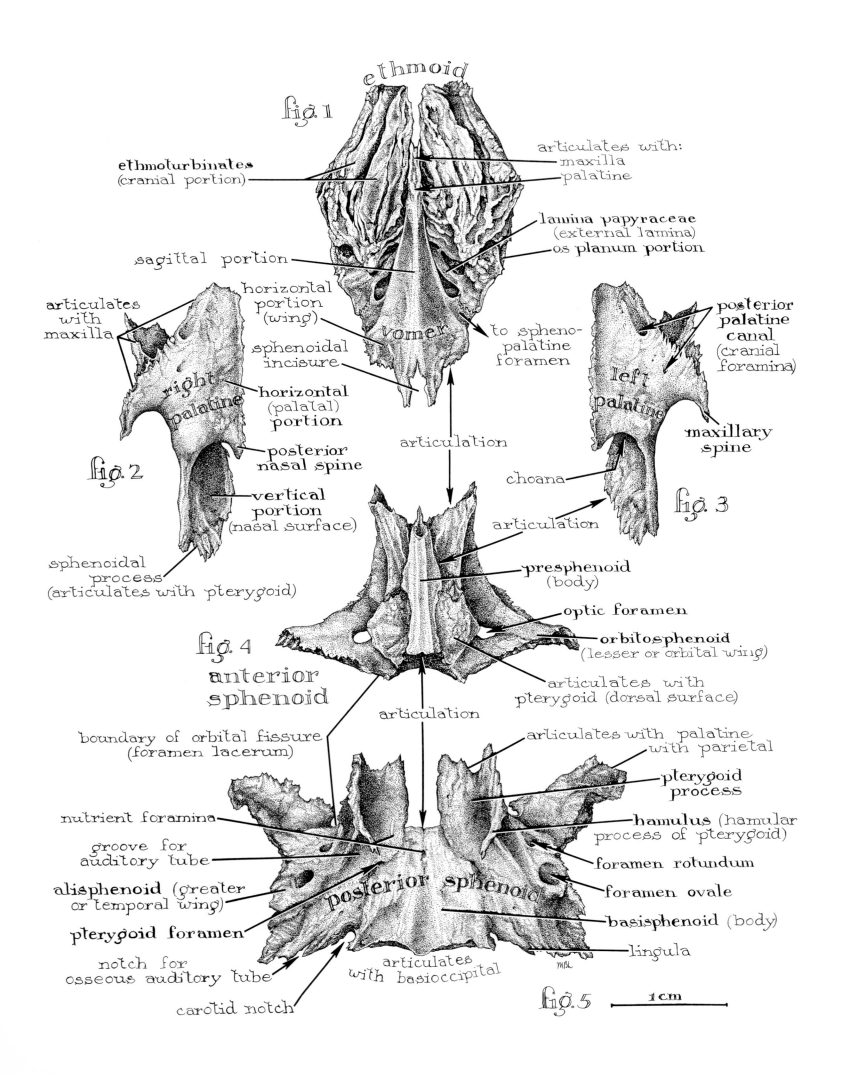

fig. 1

ethmoid

ethmoturbinates
(cranial portion)

articulates with:
— maxilla
— palatine

sagittal portion

lamina papyraceae
(external lamina)
os planum portion

vomer

to spheno-
palatine
foramen

fig. 2

articulates
with
maxilla

right palatine

horizontal
portion
(wing)

sphenoidal
incisure

horizontal
(palatal)
portion

posterior
nasal spine

vertical
portion
(nasal surface)

sphenoidal
process
(articulates with pterygoid)

fig. 3

left palatine

posterior
palatine
canal
(cranial
foramina)

maxillary
spine

choana

articulation

articulation

fig. 4
anterior
sphenoid

presphenoid
(body)

optic foramen

orbitosphenoid
(lesser or orbital wing)

articulates with
pterygoid (dorsal surface)

articulation

boundary of orbital fissure
(foramen lacerum)

articulates with palatine
with parietal

pterygoid
process

nutrient foramina

groove for
auditory tube

alisphenoid (greater
or temporal wing)

pterygoid foramen

notch for
osseous auditory tube

carotid notch

posterior sphenoid

hamulus (hamular
process of pterygoid)

foramen rotundum

foramen ovale

basisphenoid (body)

lingula

articulates
with basioccipital

fig. 5   1 cm

MSL

Ethmoid, Vomer, Palatine and Sphenoid—Ventral View
PLATE 10

Only the **ethmoturbinates** show to advantage in the ventral view of the ethmoid, the other parts being covered by the vomer (fig. 1).

The **vomer** appears from the ventral view as a triangular bone with a median ridge and a bifurcated caudal border for articulation with the presphenoid. Its thin lateral plates cover the ventral ethmoturbinals and enclose some of their air cells and also help to separate the olfactory and respiratory passageways. Its vertical portion contributes dorsally to the nasal septum and its cranial end articulates with the horizontal plates of the maxillae and the palatines.

The ventral view of the **palatines** shows the **horizontal** or **palatal** portion to advantage (figs. 2 and 3). Each horizontal portion fits by its lateral and cranial borders into the maxillary bone for articulation. The lateral border, at about its midpoint, has a prominent spine which projects caudolaterad, the **maxillary spine.** The medial borders unite one with the other and at midpoint and caudally form the **posterior nasal spine.** Near the craniomedial borders are two or more foramina, the cranial openings of the **posterior palatine canals.**

The **vertical portion** of the palatines form the lateral walls of the posterior nares, the cranial part of the wall of the **median pterygoid fossa** and a portion of the medial wall of the orbital cavities.

The **presphenoid** portion of the anterior sphenoid is seen in this plate as an hourglass-shaped median bone with the orbitosphenoids extending laterad (fig. 4). In the articulated skull it is overlapped craniolaterad by the bifurcated caudal border of the vomer followed caudad by the palatines and pterygoids (Plate 3, fig. 2). Its caudal end joins the posterior sphenoid. Internally is found the **sphenoidal**

sinus, divided into two by a median partition. The ventral ethmoturbinals extend into the sinuses and the craniodorsal border of the presphenoid articulates with the ventral margin of the cribiform plate. A slight ridge on the lateral surface separates the **external pterygoid fossa** from the orbital fossa. A part of the external pterygoid muscle originates here.

The triangular **wings** (*lesser*) or **orbitosphenoids** join the presphenoid and project laterally to form a part of the cranial wall dorsally and of the orbitotemporal fossa ventrally. The cranial edge of the wing articulates with the frontal bone, the caudal edge with the alisphenoid (*greater wing*) of the postsphenoid. An **optic foramen** penetrates the base of each orbitosphenoid near the caudal margin. It carries the optic nerve and the ophthalmic artery.

The **postsphenoid** as seen from the ventral aspect shows clearly the triangular, central **body** or **basisphenoid** with the **greater wings** or **alisphenoids** extending craniad and dorsolaterad to the parietal bone and the **pterygoids** (*processes*) with **hamular processes** craniad and ventrad of the basisphenoid.

The **alisphenoid** attaches along most of the lateral border of the basisphenoid. The **foramina rotundum** and **ovale** are clearly visible in its basal portion as is also the partial wall of the orbital fissure which is completed in the articulated skull by the anterior sphenoid.

The **pterygoid processes** face medially into the caudal portion of the **median pterygoid fossa** and laterally into the external and internal pterygoid fossae. From the laterocaudal margin of the pterygoid process a triangular curved spine projects caudad, the **hamulus.**

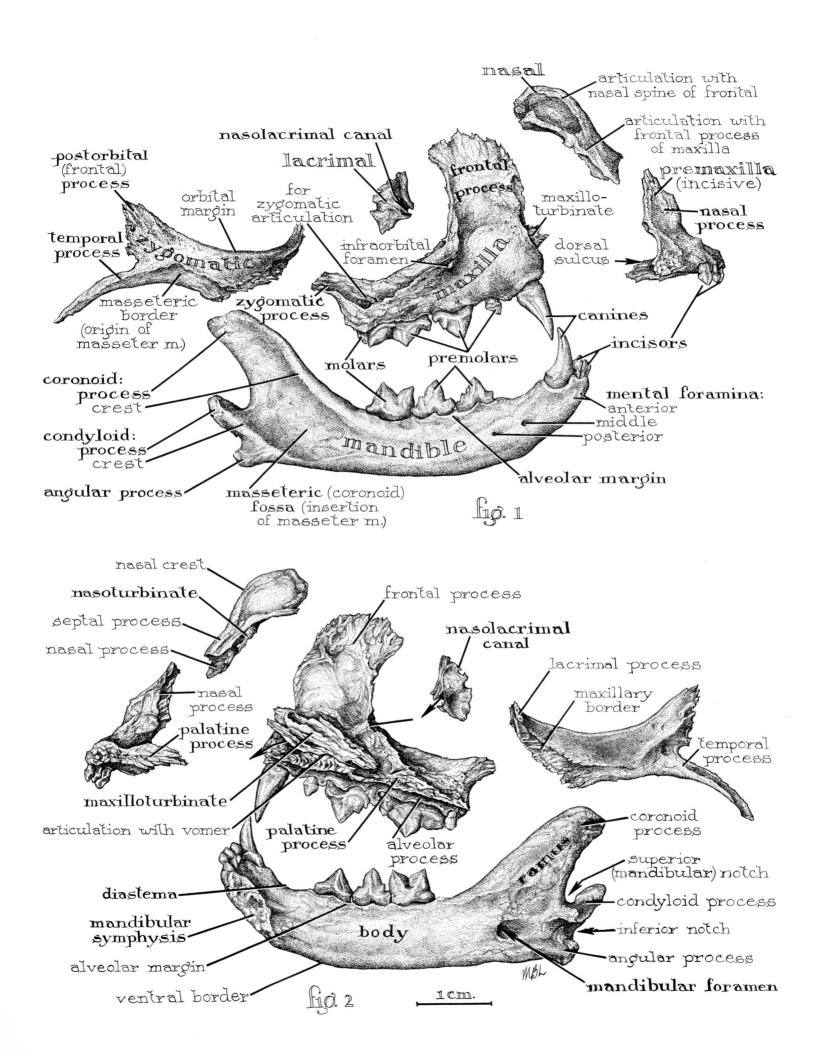

fig. 1

- nasal
- articulation with nasal spine of frontal
- articulation with frontal process of maxilla
- **premaxilla** (incisive)
- nasal process
- postorbital (frontal) process
- orbital margin
- nasolacrimal canal
- *lacrimal*
- frontal process
- maxillo-turbinate
- temporal process
- *zygomatic*
- for zygomatic articulation
- infraorbital foramen
- *maxilla*
- dorsal sulcus
- masseteric border (origin of masseter m.)
- zygomatic process
- canines
- incisors
- coronoid: process crest
- molars
- premolars
- mental foramina: anterior, middle, posterior
- condyloid: process crest
- *mandible*
- angular process
- masseteric (coronoid) fossa (insertion of masseter m.)
- alveolar margin

fig. 2

- nasal crest
- **nasoturbinate**
- septal process
- nasal process
- frontal process
- **nasolacrimal canal**
- lacrimal process
- maxillary border
- nasal process
- palatine process
- temporal process
- maxilloturbinate
- articulation with vomer
- palatine process
- alveolar process
- coronoid process
- diastema
- mandibular symphysis
- *body*
- *ramus*
- superior (mandibular) notch
- condyloid process
- inferior notch
- angular process
- **mandibular foramen**
- alveolar margin
- ventral border

1 cm.

MbL

In this plate the bones are separated to show clearly their boundaries but their relative positions are retained except that the lacrimal has been turned slightly laterad. These are intramembranous bones. The mandible forms around, but not in, Meckel's cartilage of the embryo. With the exception of the mandible these bones are firmly articulated one to another allowing no movement. They form suture or fibrous joints. The mandible articulates with the temporal bone by a synovial joint which is freely movable—the only such joint in the skull. It is shown in Plate 4, figure 1.

The **premaxillae** (*incisive*) together form the ventral and lateral rims of the nares and bear the six incisor teeth. They form also the cranial portion of the hard palate. Each bone is divided therefore into nasal and palatine processes. The **palatine processes** unite medially and form dorsad a trough, the dorsal sulcus, for the reception of the nasal septum including caudad a part of the vomer. The caudal borders of the palatal portions are deeply notched forming the anterior and medial and a part of the lateral walls of the **anterior palatine** or **incisive foramina**. The walls of these foramina are completed laterally and caudally by the maxillae (Plate 3, fig. 2). The **nasal processes** of the premaxillae articulate with the nasals medially and with the maxillae laterally. In adult man the premaxilla and maxilla, though coming from separate ossification centers, are fused into one maxillary bone.

The **ramus** of the **mandible** extends dorsad and has on its lateral surface a **masseteric fossa** for the insertion of the masseter muscle and on its caudal side a transverse **condyloid process** for articulation with the mandibular fossa of the temporal bone (Plate 4, fig. 1). The ramus continues dorsocaudad as the prominent **coronoid process** into which the temporal muscle inserts. On the caudal border between the coronoid and condyloid processes is the **superior notch,** while between the condyloid and angular processes is the **inferior notch.**

The **nasal bones** join by their medial lamellae forming the **nasal crest** which contributes to the nasal septum. Their horizontal lamellae form a part of the roof of the nasal cavities. They articulate externally with the premaxillary, maxillary and frontal bones (Plate 3, fig. 1); internally with the perpendicular plate and **ethmoturbinates** (*superior conchae*) of the ethmoid bone.

The **lacrimal bone** is a thin scale-like bone which forms a part of the cranial or nasal wall of the orbit. It is bordered by the os planum of the ethmoid, the frontal, malar, palatine and maxilla bones. With the maxilla it forms the **nasolacrimal groove** and **canal** which carries the tear duct by which the tears pass from the eye to the nasal passageway.

The **zygomatic** (*malar*) **bone** is the bone of the cheek and forms part of the lateral and cranial wall of the orbit. With the zygomatic process of the temporal it forms the **zygoma** or **zygomatic arch.** Craniad it articulates with the maxilla, while its caudal border has two processes, the **zygomatic process** articulating with the temporal and a **postorbital process** which approximates the postorbital process of the frontal to partially close the orbit caudally. These two postorbital processes are connected in the living animal by an orbital ligament. The smooth lateral surface is marked by a **longitudinal ridge** along which the masseter muscle has its origin.

The **maxillae** meet one another in the median line, contribute to the hard palate and with the premaxillae form the anterior palatine foramina. They bear the remainder of the upper teeth on the **alveolar margin;** the canines, premolars and molars. Each bone consists of a body; a **palatine plate** in the roof of the mouth; a **frontal process** extending dorsocaudad to articulate with the nasal and frontal bones; a **zygomatic process** to articulate with the zygomatic; and an **orbital plate** forming part of the floor of the orbit. Anterior to the orbital plate is the large **infraorbital foramen** which carries the **infraorbital nerve** and **artery.**

The medial side of the frontal process faces into the nasal cavity and has transverse ridges for the articulation with the ethmoid. Ventrally where the frontal process joins the body is a thin, scroll-like ridge for the **ventral** (*inferior*) **nasal concha** or **maxilloturbinate.**

The **mandible** or **inferior maxilla** is formed from two halves which unite anteriorly at the **symphysis** to form an immovable joint. Each half consists of a **horizontal portion** or **body** and a vertical portion, the **ramus.**

The horizontal portion bears the three incisors, one canine, two premolars and one molar on its **alveolar margin.** A space between the canine and the first premolar is the **diastema.** On its smooth lateral surface at the cranial end is a single or sometimes two foramina, the **mental foramina.** The medial surface of the horizontal portion is also smooth except cranially where there is a roughened area for articulation with the other side of the mandible. Caudally, on the medial surface, is a large **mandibular foramen** leading into a mandibular canal which opens anteriorly and laterally through the mental foramen. The **inferior alveolar nerve** and **artery** traverse the mandibular canal from the mandibular to the mental foramina. The ventral border of the horizontal portion is smooth and rounded and terminates caudad at the **angle** or **angular process.**

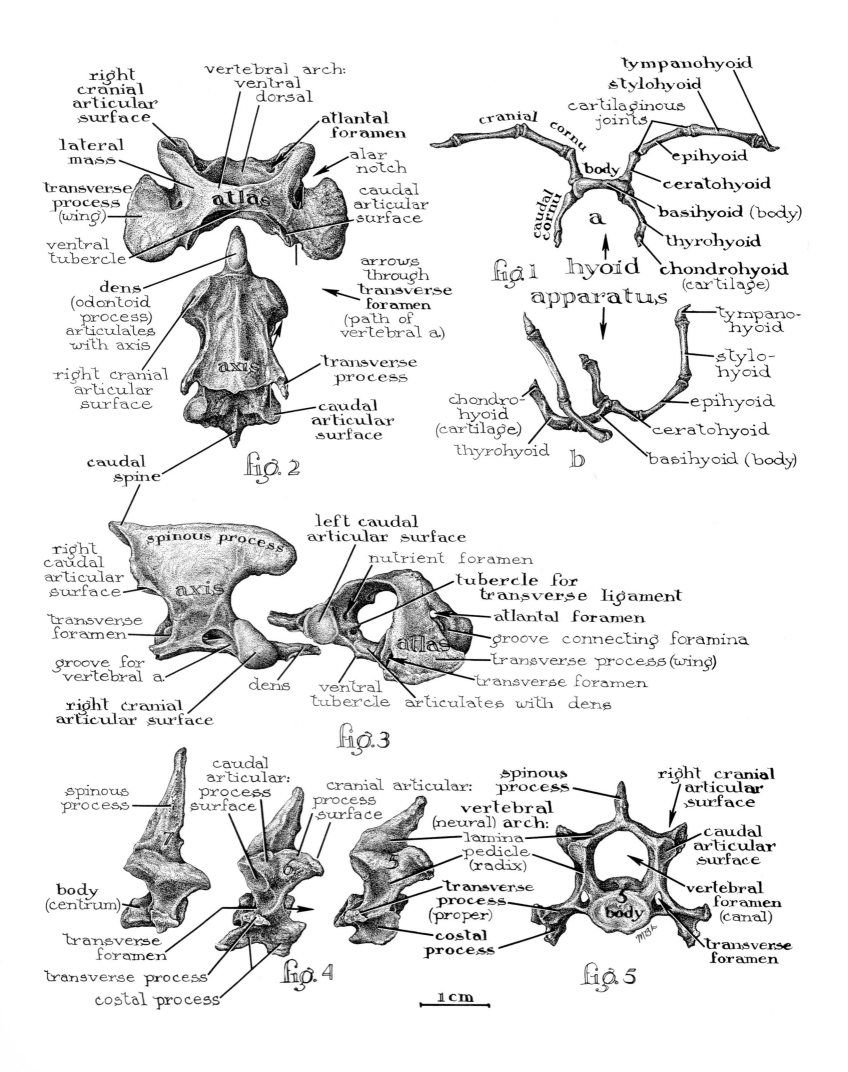

fig. 1 hyoid apparatus

Labels in fig. 1 (a): tympanohyoid, stylohyoid, cartilaginous joints, cranial cornu, body, caudal cornu, epihyoid, ceratohyoid, basihyoid (body), thyrohyoid, chondrohyoid (cartilage)

Labels in fig. 1 (b): tympano-hyoid, stylo-hyoid, epihyoid, ceratohyoid, chondro-hyoid (cartilage), thyrohyoid, basihyoid (body)

fig. 2

Labels: right cranial articular surface, vertebral arch: ventral dorsal, atlantal foramen, lateral mass, alar notch, transverse process (wing), caudal articular surface, ventral tubercle, atlas, dens (odontoid process) articulates with axis, arrows through transverse foramen (path of vertebral a.), right cranial articular surface, axis, transverse process, caudal articular surface, caudal spine

fig. 3

Labels: spinous process, left caudal articular surface, nutrient foramen, tubercle for transverse ligament, right caudal articular surface, axis, atlantal foramen, groove connecting foramina, transverse foramen, transverse process (wing), groove for vertebral a., atlas, transverse foramen, dens, ventral tubercle, articulates with dens, right cranial articular surface

fig. 4

Labels: spinous process, caudal articular: process surface, cranial articular: process surface, body (centrum), transverse foramen, transverse process, costal process

fig. 5

Labels: spinous process, right cranial articular surface, vertebral (neural) arch: lamina, pedicle (radix), caudal articular surface, body, transverse process (proper), costal process, vertebral foramen (canal), transverse foramen

1 cm

The **hyoid apparatus** or **bone** consists of a paired chain of small bones and cartilages (Plate 1, figs. a and b), extending from the tympanic bullae of the temporal bones to either side of the thyroid cartilage of the larynx. A cross-piece, the **basihyal** (*basihyoid*) connects the two sides forming a distorted "H." The upright parts are called **horns** or **cornua.** The **caudal cornua** are made up of the **thyrohyal bones** each tipped by the cartilaginous **chondrohyals** which attach to the thyroid cartilage; the **cranial cornua** are composed of the **ceratohyal, epihyal, stylohyal,** and **tympanohyal bones,** the last of which is embedded in a groove in the tympanic bulla just ventrad of the stylomastoid foramen. Cartilage is seen to connect these bones. This apparatus is formed from the second and third arches of the visceral skeleton and serves to support the tongue and laryngeal cartilages. It gives origin to muscles passing to the tongue and larynx.

The **vertebral column** or spinal column consists of a varying number of serially homologous bony segments, the **vertebrae.** These almost completely replace the notochord which is seen as a precursor of the vertebral column in early embryonic life. The column can be divided into five regions: cervical, thoracic, lumbar, sacral, and caudal. These contain seven, thirteen, seven, three, and four to twenty-six vertebrae respectively in the cat. This plate shows the unique atlas and axis (*epistropheus*) and the more typical fifth, sixth, and seventh cervical vertebrae. The **atlas** is in the form of a ring, devoid of a spinous process and a centrum or body. This latter portion has united with the axis to become the **odontoid process** or **dens.** The atlas has large horizontally expanded wing-like masses, the **transverse processes,** which extend from the **lateral masses.** At the **cranial** end, on its medial surface are two concave **cranial articular surfaces** which articulate with the occipital condyles. Caudally, there are two concave **caudal**

articular surfaces which articulate with the axis. A **transverse foramen** (*foramen transversarium*) is seen in the lateral mass, the cranial opening of which communicates by way of a groove with the **atlantal foramen.** Caudally the transverse foramen exits laterad of the caudal articular surface. These foramina serve to carry the vertebral artery. Within the vertebral canal, a pair of lateral **tubercles** serve as attachment for the **transverse ligament** which holds the dens of the axis in place.

The axis, or second vertebra (figs. 2 and 3), serves as a pivotal center for the atlas and is characterized by the cranial extension of the centrum forming the tooth-like odontoid process or dens, and an elongated, laterally flattened spinous process. Small **transverse processes** project from the centrum, each containing a transverse foramen allowing the passage of the vertebral artery and vein. Convex cranial articular surfaces occur on either side of the base of the dens, and, facing ventrally, is a pair of caudal articular surfaces caudad of the **vertebral arch.**

The remaining cervical vertebrae are rather similar to one another (figs. 4 and 5). The seventh has no transverse foramen, but the others have this distinguishing feature. Another characteristic of the majority is that the spinous processes grow successively longer from the third to the seventh, and the third and fourth are directed dorsad, while the remaining are dorsocraniad. The transverse processes arise from two roots, one from the body and one from the neural process. The former is the **costal element,** the latter, the **transverse element** or **transverse process proper.** These enclose the transverse foramen near their convergence. The articular surfaces are situated at the junction of the **pedicles** and **laminae.** The cranial articular surfaces are very prominent, convex, and are directed dorsomediad, while the caudal articular surfaces are smaller, concave and face ventrocaudolaterad.

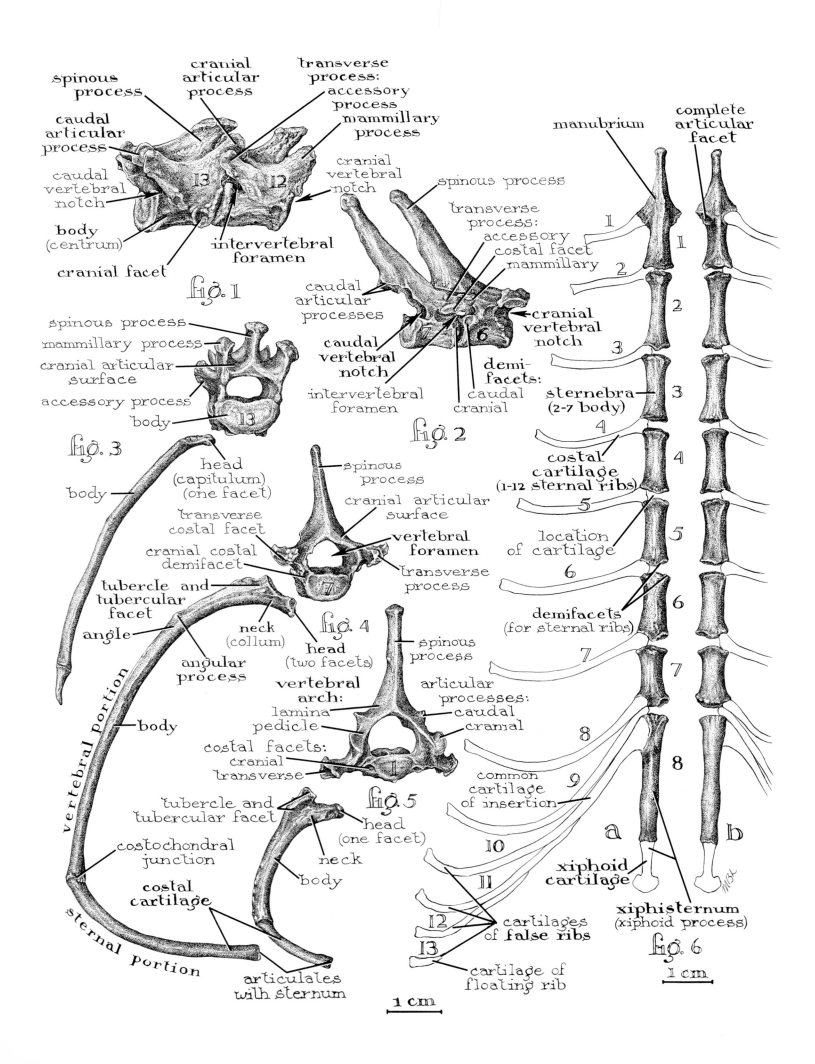

fig. 1

spinous process
cranial articular process
transverse process:
  accessory process
  mammillary process
caudal articular process
caudal vertebral notch
cranial vertebral notch
body (centrum)
cranial facet
intervertebral foramen

13 12

fig. 2

spinous process
transverse process:
  accessory
  costal facet
  mammillary
caudal articular processes
caudal vertebral notch
intervertebral foramen
cranial vertebral notch
demi-facets:
  caudal
  cranial

7 6

fig. 3

spinous process
mammillary process
cranial articular surface
accessory process
body

13

fig. 4

spinous process
cranial articular surface
vertebral foramen
transverse process
transverse costal facet
cranial costal demifacet

7

head (capitulum) (one facet)
body
tubercle and tubercular facet
angle
angular process
neck (collum)
head (two facets)
body
vertebral portion

fig. 5

spinous process
articular processes:
  caudal
  cranial
vertebral arch:
  lamina
  pedicle
costal facets:
  cranial
  transverse

1

common cartilage of insertion
tubercle and tubercular facet
costochondral junction
head (one facet)
neck
body
costal cartilage
sternal portion
articulates with sternum

1 cm

fig. 6

manubrium
complete articular facet
sternebra (2-7 body)
costal cartilage (1-12 sternal ribs)
location of cartilage
demifacets (for sternal ribs)

1
2
3
4
5
6
7
8
9
10
11
12
13

cartilages of false ribs
cartilage of floating rib

a b

xiphoid cartilage
xiphisternum (xiphoid process)

1 cm

This plate shows representative thoracic vertebrae articulated in pairs; the first, seventh, and thirteenth thoracic vertebrae with their accompanying ribs and the sternum with the costal cartilages attached.

The **thoracic vertebrae** serve as points of articulation for the **vertebral portion** of the ribs (figs. 3 to 5). There are two **articular surfaces** or **facets** on each thoracic vertebra except the first, eleventh, twelfth, and thirteenth. These surfaces are the **cranial** and **caudal costal demifacets** of the **body** (**centrum**) into which the head or **capitulum** of the rib fits (fig. 2). Most ribs join two vertebrae rather than one, thus being intervertebral, the rib capitulum articulating with the caudal costal demifacet of one vertebra and the cranial costal demifacet of the vertebra posterior. The first thoracic vertebra bears an entire facet at its cranial end, while a demifacet is seen at its caudal end. The last three thoracic vertebrae bear a complete facet on each side of their centra (fig. 1). Each transverse process has a smooth **tubercular facet** on its ventral side laterally for articulation with the **tuberculum** of the rib. The tuberculum articulates with the **transverse process** of the more caudal of the two vertebrae to which the capitulum unites. The **spinous processes** of the cranial thoracic vertebrae are long, decreasing in length to the twelfth and sloping caudad to the tenth. The eleventh vertebra has a vertical spinous process; the twelfth and thirteenth have spinous processes which slope craniad. Notice also the increasing size of the centra and vertebral foramina and the changes in the appearance of the transverse processes (figs. 1 to 5). The seventh thoracic vertebra and the succeeding ones have transverse processes which are divided into three tubercles (fig. 1); one directed craniad, the **mammillary process**; one caudad—the **accessory process**; and the third looks ventrad and bears a **transverse costal facet.** The eleventh, twelfth, and thirteenth thoracic vertebrae have lost the transverse process bearing the transverse costal facet, therefore the last three ribs are united to their respective centra by their heads alone. On the dorsal surface of each lamina, a smooth oval **cranial articular surface** occurs upon the **cranial articular process** at its cranial border. On the ventral surface of each lamina, a similar area, the **caudal articular surface** occurs upon the **caudal articular process** at its caudal border. When in the natural position, the cranial and caudal articular surfaces

of adjacent vertebrae fit one against the other, thus strengthening the joint between the contiguous vertebrae while permitting slight rotary motion. The last three thoracic vertebrae afford very limited rotary motion because of the interlocking of the articular processes (fig. 1). Notice the **intervertebral foramina** which serve for passage of the spinal nerves.

The ribs (*costae*) number thirteen pairs in the cat. The first nine pairs are **true ribs** (*costae verae*), being attached separately to the **sternum** by their respective **costal cartilages,** the last four pairs are **false ribs** (*costae spuriae*), not being attached separately to the sternum. Three of the false ribs have costal cartilages united to one another at their sternal ends, which join to the sternum by a common cartilage. The last pair are **free** or **floating ribs** since their costal cartilages do not join the sternum (figs. 3 and 6). Ribs can be described as curved flattened rods of bone, attached to the **vertebral column** at their dorsal ends, and attached to a cartilage at their ventral ends. The ribs increase in length to the ninth or tenth, then decrease to the last. Each rib is composed of two parts, the bony **vertebral portion,** and the cartilaginous **sternal portion.** The vertebral portion originates dorsally with a rounded **capitulum** or **head** which articulates with the costal facets or demifacets of the thoracic vertebrae. A constricted area, the **neck** or **collum** is distal to the capitulum, followed by an elevated area, the **tubercle,** which articulates with the transverse process by way of the tubercular facet. The **angle** of the rib is marked by the **angular process** on the lateral border distad of the tubercle. The **shaft** or **body** is that part of the rib seen between the tubercle and the articulation with the sternal portion. The cartilaginous portion is sometimes called the **sternal rib.**

The **sternum** (fig. 6) in the cat retains its segmental character, being composed of eight pieces or **sternebrae** joined by cartilage. It lies on the mid-ventral line of the thorax and serves as the attachment for the costal cartilages of the true ribs. The most anterior segment of the sternum is the **manubrium.** The succeeding six sternebrae make up the **body, gladiolus,** or **corpus sterni.** The last segment is the **xiphisternum** which terminates in a flat cartilaginous tip, the **xiphoid cartilage.** Notice the points of articulation of the sternal portion of the ribs with the sternebrae.

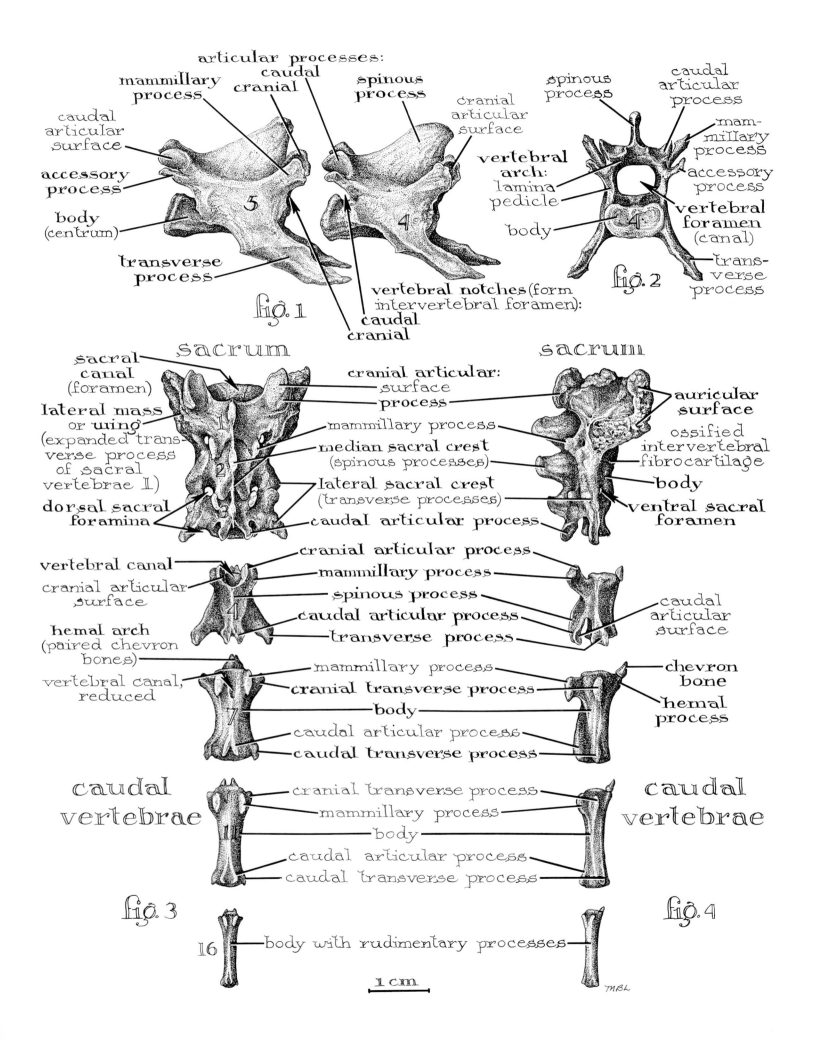

fig. 1

articular processes:
- caudal
- cranial

mammillary process

caudal articular surface

accessory process

body (centrum)

transverse process

spinous process

cranial articular surface

vertebral notches (form intervertebral foramen):
- caudal
- cranial

fig. 2

spinous process

caudal articular process

mammillary process

accessory process

vertebral arch:
- lamina
- pedicle

body

vertebral foramen (canal)

transverse process

sacrum

fig. 3

sacral canal (foramen)

lateral mass or wing (expanded transverse process of sacral vertebrae II)

dorsal sacral foramina

cranial articular:
- surface
- process

mammillary process

median sacral crest (spinous processes)

lateral sacral crest (transverse processes)

caudal articular process

sacrum

fig. 4

auricular surface

ossified intervertebral fibrocartilage

body

ventral sacral foramen

vertebral canal

cranial articular surface

hemal arch (paired chevron bones)

vertebral canal, reduced

cranial articular process

mammillary process

spinous process

caudal articular process

transverse process

caudal articular surface

mammillary process

cranial transverse process

body

caudal articular process

caudal transverse process

chevron bone

hemal process

caudal vertebrae

cranial transverse process

mammillary process

body

caudal articular process

caudal transverse process

caudal vertebrae

body with rudimentary processes

1 cm

The fourth and fifth lumbar vertebrae, the sacrum, and the fourth, seventh, eleventh, and sixteenth caudal vertebrae are shown in this plate.

The **lumbar vertebrae** (figs. 1 and 2) number seven in the cat and are larger, longer, and heavier than the preceding thoracic vertebrae. Their prominent **spinous processes** are flat from side to side and project dorsocraniad; the first five are knobbed at the tips. The flattened **transverse processes** arise from the lateral surface of the centra, and project ventrocraniolaterad. The **accessory processes** diminish in size from the first to the fifth or sixth, and are absent on the seventh and occasionally the sixth. The **cranial articular processes** bear the **cranial articular facets** on their medial surfaces, and prominent **mammillary processes** on their dorsolateral surfaces. The **caudal articular processes** bear the **caudal articular facets** on their lateral surfaces which articulate with the cranial articular facets. Notice that when the lumbar vertebrae are articulated, the caudal articular processes and accessory processes of one interlocks with the cranial articular processes and spinous process of the following in such a way as to prevent rotary motion. An **intervertebral foramen** is seen between articulating vertebrae, each foramen being formed by the **intervertebral notches** of adjacent vertebrae.

In the living cat, **intervertebral fibrocartilages** unite the bodies of the vertebrae, an exception being the atlas and axis. These disk-shaped fibrocartilages consist of two parts, a central pulp-like **nucleus pulposus** (*remnant of the notochord*) which allows flexibility of the column and an outer fibrous portion, the **annulus fibrosus** whose fibrous covering unites with the periosteum of the vertebrae.

The **sacrum** is considered as one bone being made up of three vertebrae which are fused in the adult, but separate in the kitten. The limit of each vertebra is marked by a pair of **dorsal sacral foramina** and a pair of **ventral sacral foramina.** These foramina serve for passage of the dorsal and ventral rami of sacral spinal nerves. The shape of the sacrum is pyramidal, the base of which is directed cranially. It houses a longitudinal canal, the sacral canal, which is a continuation of the vertebral canal. On the dorsal surface, three median spinous processes are seen, and lateral to the first two of these is a pair of tubercles, the result of fusion of the adjacent articular processes. The first sacral vertebra has a major role in supporting the hind limbs. It has a large **lateral mass** on either side bearing **auricular surfaces** for articulation with the **ilium.** This expansion may be a modified transverse process. The transverse processes of all sacral vertebrae fuse and the last vertebra sends forth a pair of transverse processes which project freely in a caudal direction. A pair of cranial articular processes of the first sacral vertebra articulates with the caudal articular processes of the last lumbar vertebra and a pair of caudal articular processes of the last sacral vertebra articulates with the cranial articular processes of the first caudal vertebra. The ventral surface of the sacrum is smooth and convex cranially and concave caudally. It shows the ossified remains of the intervertebral fibrocartilages as two transverse ridges and two pairs of ventral sacral foramina.

The **caudal vertebrae** number from four to twenty-six, the six or seven being typical, the others are gradually reduced to little more than centra. The typical parts of a vertebra, such as neural canal, transverse and articular processes occur only to the eighth or ninth caudal vertebra, the spinous process disappearing at the fourth. On the ventral side, a pair of rounded **hemal processes** begins at the cranial end of the third caudal vertebra and articulates with small pyramidal **chevron bones** forming a **hemal arch** which encloses the **hemal canal.** These structures diminish in size caudally and disappear near the tip of the tail.

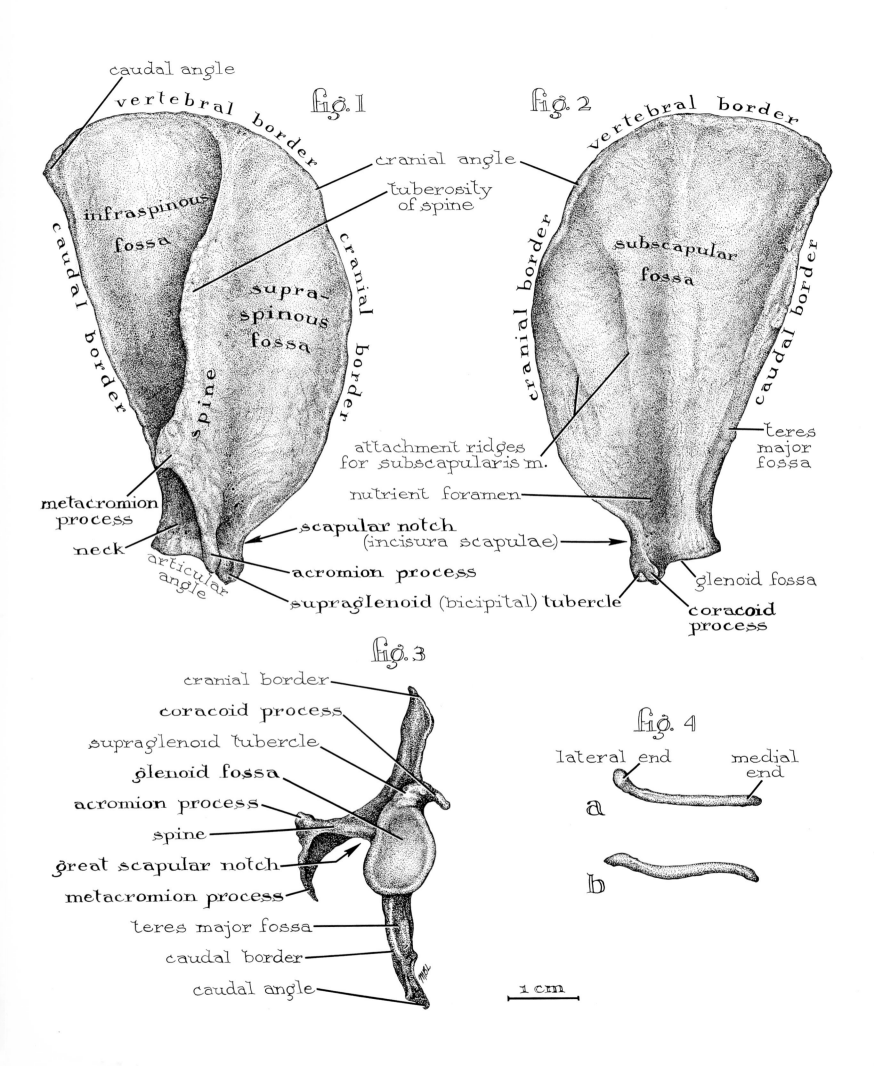

fig. 1

caudal angle

vertebral border

infraspinous fossa

caudal border

spine

supra-spinous fossa

cranial border

metacromion process

neck

articular angle

cranial angle

tuberosity of spine

scapular notch

acromion process

supraglenoid (bicipital) tubercle

fig. 2

vertebral border

cranial border

subscapular fossa

caudal border

attachment ridges for subscapularis m.

nutrient foramen

scapular notch (incisura scapulae)

teres major fossa

glenoid fossa

coracoid process

fig. 3

cranial border

coracoid process

supraglenoid tubercle

glenoid fossa

acromion process

spine

great scapular notch

metacromion process

teres major fossa

caudal border

caudal angle

fig. 4

lateral end

medial end

a

b

1 cm

This portion of the skeleton consists of the forelimbs and hindlimbs and their supporting girdles (Plate 2). The shoulder or pectoral girdle is free from any direct articulation with the axial skeleton but is supported instead by muscles. This provides for the great flexibility and versatility of this girdle and of the forelimbs which it supports. In contrast, the pelvic or hip girdle is firmly articulated to the sacrum of the vertebral column.

Shoulder Girdle—Scapula and Clavicle
PLATE 15

The pectoral or shoulder girdle is formed by the scapula and clavicle. In this plate (Plate 15) three views of the right scapula are shown.

The **scapula** or shoulder blade can be described as a flat, triangular bone with a very prominent ridge, the **spine,** occurring on its lateral surface. This bone does not articulate with the bones of the trunk, but is held in normal position by several important muscles, *i.e.,* the serratus ventralis, levator scapulae, rhomboids, etc. (Plates 30 and 31). It does, however, articulate with the head of the humerus at its ventral end. The scapula presents three obvious borders: the **cranial border;** the **vertebral border** —nearest the vertebral column; the **caudal border** —nearest the axilla or armpit. The spine of the scapula ends distally as the **acromion process** after sending caudad a flat projection, the **metacromion process.** The acromion process articulates with the clavicle in man but not in the cat. The spine divides the lateral surface into two fossae, the **supraspinous fossa** and the **infraspinous fossa;** the former being craniad of the spine, the latter caudad. The entire medial surface of the scapula is occupied by a slightly concave **subscapular fossa.** The ventral end of the scapula terminates in a concave **glenoid fossa** which articulates with the humerus. The extreme cranial aspect of the glenoid fossa bears a tubercle, the **supraglenoidal** or bicipital tubercle for attachment of the tendon of origin of the biceps brachii muscle. Medial to this is a beak-like projection, the **coracoid process,** a vestige of the coracoid bone of lower animals. A slight constriction occurring dorsal to the glenoid fossa is known as the **neck.** The **scapular notch,** or incisura scapulae, is a rounded notch dorsal to the coracoid process and glenoid angle; the **great scapular notch** occurs between the glenoid angle and the inner margin of the acromion.

The **clavicle** is a curved, slender bone imbedded between the cleidotrapezius and cleidobrachialis (*cleidodeltoideus*) muscles (Plate 25) cranial to the shoulder joint. It is vestigial in the cat and has no bony connections, while in man it articulates with the manubrium of the sternum and the acromion of the scapula. It is absent in some mammals such as the horse.

41

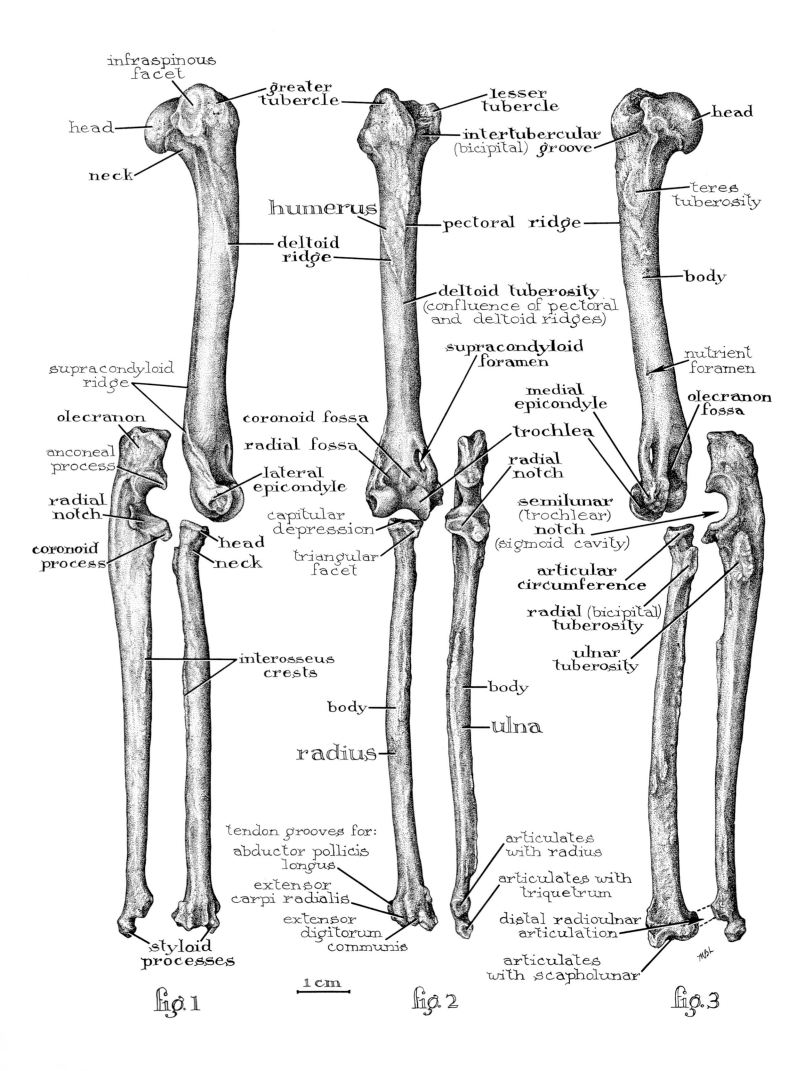

infraspinous facet

greater tubercle

lesser tubercle

head

head

neck

intertubercular (bicipital) groove

teres tuberosity

humerus

pectoral ridge

deltoid ridge

body

deltoid tuberosity (confluence of pectoral and deltoid ridges)

supracondyloid ridge

nutrient foramen

supracondyloid foramen

olecranon

medial epicondyle

olecranon fossa

anconeal process

coronoid fossa

trochlea

radial fossa

radial notch

radial notch

lateral epicondyle

semilunar (trochlear) notch (sigmoid cavity)

coronoid process

head

neck

capitular depression

articular circumference

triangular facet

radial (bicipital) tuberosity

ulnar tuberosity

interosseus crests

body

body

ulna

radius

tendon grooves for:
abductor pollicis longus

articulates with radius

extensor carpi radialis

articulates with triquetrum

extensor digitorum communis

distal radioulnar articulation

styloid processes

articulates with scapholunar

1 cm

fig. 1

fig. 2

fig. 3

The thoracic, pectoral, or forelimb is made up of the humerus, radius, ulna, seven carpal bones, five metacarpal bones, and fourteen phalanges. This plate shows three views of the humerus, radius, and ulna (figs. 1, 2, and 3).

The arm, or **brachium** contains the **humerus.** It articulates proximally with the glenoid fossa of the scapula and distally with the radius and ulna. The rounded proximal **head** is dorsomediad and has two processes adjacent to it, these being the greater tubercle and the lesser tubercle. The **greater tubercle** is a prominent ridge extending cranially from the lateral margin of the head; the **lesser tubercle** is a smaller ridge projecting craniomediad of the head. Between these tuberosities is an **intertubercular** or bicipital groove through which a tendon of the biceps brachii muscle travels. Two ridges extend distally from the greater tubercle, the **deltoid ridge** being lateral, the **pectoral** or **greater tubercular ridge** being ventral. These ridges meet near the middle of the ventral surface of the bone and serve as the insertions for the deltoid and pectoral muscles respectively. The distal end of the bone has two articular portions, the lateral **capitulum** for articulation with the head of the radius, the medial **trochlea** for articulation with the **semilunar notch** of the ulna. Above these surfaces are seen masses, the **lateral** and **medial epicondyles** (*epitrochlea*), from which muscle groups originate. Above the medial epicondyle, an oval **supracondyloid foramen** occurs, for passage of the brachial artery and the median nerve of the limb. From the lateral epicondyle, the **supracondyloid ridge** curves proximad to end near the middle of the dorsal surface of the bone. Proximad of the trochlea and on the dorsal surface is the **olecranon fossa,** a deep cavity which recieves the **olecranon process** of the ulna when the forearm is extended. On the ventral surface are two shallow fossae, the **coronoid fossa** receiving the **coronoid process** of the ulna, and the **radial fossa** receiving a facet on the head of the radius when the forearm is flexed.

The forearm, or **antebrachium** consists of the radius and ulna. The **radius** is the smaller of the two bones and is on the "thumb" side of the forearm. It articulates proximally with the humerus and ulna, distally with the ulna and the **scapholunar** bone of the carpus. When the cat is in the anatomical position, the radius is crossed over the ulna dorsally from the lateral to medial side. It bears two processes, the **radial,** or bicipital **tuberosity** and the **styloid process.** The tuberosity occurs distal to the **neck** of the radius and on the ulnar side and serves as the insertion for the biceps brachii muscle. The proximal **head** of the radius has three surfaces, a depressed oval facet which articulates with the capitulum, a triangular facet which fits into the radial fossa of the humerus, and an **articular circumference** which articulates with the **radial notch** of the ulna. The **styloid process** projects from the distal end of the radius. Longitudinal grooves for tendons are seen also on the distal end, and a small concave facet for articulation with the ulna.

The **ulna** is larger than the radius and occurs caudad of it. Its **semilunar notch** or **great sigmoid cavity** articulates with the trochlea of the humerus proximally. The **olecranon process** forms the proximal end of the ulna and serves as the insertion for the triceps brachii muscle. The process fits into the olecranon fossa of the humerus when the forearm is extended. The distal boundary of the semilunar notch is the coronoid process, which bears on its lateral surface the **radial notch** or **lesser sigmoid cavity,** a concave facet for articulation with the head of the radius. The distal head of the ulna articulates with the radius by way of the styloid process. Notice also that on the middle third of both the radius and ulna there is an area for the attachment of the interosseous membrane.

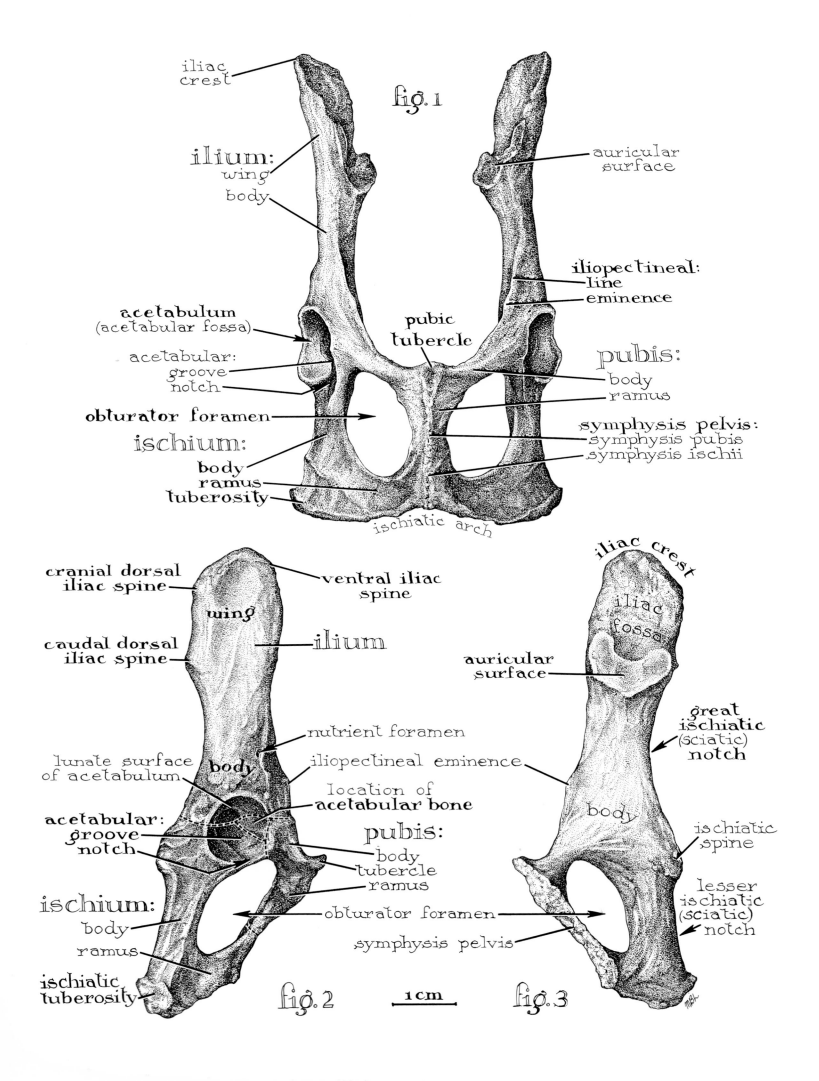

fig. 1

iliac crest

ilium:
wing
body

auricular surface

iliopectineal:
line
eminence

acetabulum (acetabular fossa)

pubic tubercle

acetabular:
groove
notch

pubis:
body
ramus

obturator foramen

ischium:
body
ramus
tuberosity

symphysis pelvis:
symphysis pubis
symphysis ischii

ischiatic arch

cranial dorsal iliac spine

ventral iliac spine

wing

caudal dorsal iliac spine

ilium

iliac crest

iliac fossa

auricular surface

nutrient foramen

lunate surface of acetabulum

body

iliopectineal eminence

great ischiatic (sciatic) notch

location of acetabular bone

acetabular:
groove
notch

pubis:
body
tubercle
ramus

body

ischium:
body
ramus

obturator foramen

symphysis pelvis

ischiatic spine

lesser ischiatic (sciatic) notch

ischiatic tuberosity

fig. 2

1 cm

fig. 3

The **ossa coxae** (*innominate*) are the principal components of the pelvic girdle. They unite ventrally forming a fibrocartilaginous **pelvic** (*ischiopubic*) **symphysis** and articulate dorsally with the sacrum forming the **sacroiliac joint** (fig. 1). The sacrum, a specialized part of the vertebral column, gives a firm base for the attachment of the pelvic girdle which places a limitation upon its movements. This is in marked contrast to the pectoral (*shoulder*) girdle which has no articulation with the axial skeleton but is supported by skeletal muscles. A bowl-like depression, the **acetabulum,** is seen in the middle of the lateral surface of each os coxae into which articulates the head of the femur to form a ball and socket or "universal" joint. A large opening in the ventral part of each bone is the **obturator foramen.**

Each os coxae is composed of three principal parts, the **ilium, ischium,** and **pubis** and a small **acetabular bone** (figs. 2 and 3). These parts are distinct bones in the kitten and are separated by cartilage which becomes ossified in the adult to form a continuum of bone. The ilium, ischium, and acetabular bones all contribute to the walls of the acetabulum. The pubis is excluded from the acetabulum by the acetabular bone.

The **ilium,** the largest component of the pelvic girdle, starts at the acetabulum where it forms a small part of its wall. It extends dorsally and cranially to articulate with the sacrum. Its curved **crest** extends above the sacroiliac joint, delimited by the inconspicuous **ventral iliac spine** and the **cranial dorsal iliac** spine. The dorsal portion of the lateral surface of the ilium is concave and smooth; the ventral acetabular portion is convex and roughened. The medial acetabular surface is smooth but the dorsal or sacral portion has an ear-shaped, roughened area, the **auricular surface,** by which it articulates with a similar surface on the sacrum.

On the dorsal border of the ilium there is a **caudal dorsal iliac spine** and caudad a concavity the **great ischiatic** (*sciatic*) **notch** which extends to the **ischiatic spine.** These structures are more prominent in the skeleton of man. On the ventral border of the acetabular half the **iliopectineal eminence** is seen. This eminence is on the **iliopectineal line** which runs from the ilium onto the pubis to terminate at the symphysis.

The **ischium** lies ventrad and caudad to the acetabulum and ilium. Its **body** forms about two thirds of the wall of the acetabulum. Its caudal extremity is thickened and prominent forming the **ischiatic tuberosity.** From the ischiatic tuberosity a crescent-shaped process the **ramus of the ischium** extends mediad and ventrad joining the one from the opposite side to form the caudal one third of the **pelvic** (*ischiopubic*) **symphysis.** Near the acetabulum on the dorsal border of the bone is the spine of the ischium. Between this spine and the tuberosity of the ischium the dorsal border is concave forming the lesser ischiatic notch.

The pubis is the ventral and anterior part of the os coxae bone. While it does not contribute directly to the wall of the acetabulum, its closely associated acetabular component does form about one sixth of the acetabular circumference. The pubis consists of a **body** and a **ramus,** the latter joining with the one from the opposite side to form about two thirds of the pelvic symphysis. A **pubic tubercle** projects craniad from the point where the pubic bones join ventrally. The concave borders of the pubic and ischial bones together form the **obturator foramen.**

The **acetabulum,** although bowl-shaped, is notched on the ventral one sixth of its border forming the **acetabular notch** (*insicura acetabuli*). This deficiency in the border is closed in the living subject by a transverse ligament. Extending from the acetabular notch to the bottom of the bowl is the **acetabular groove.** From this groove arises the **ligamentum teres** which attaches to the fovea capitis, a depression on the head of the femur. The ligament teres is not as important in supporting the hip joint as it is in carrying nutritive vessels to the joint.

45

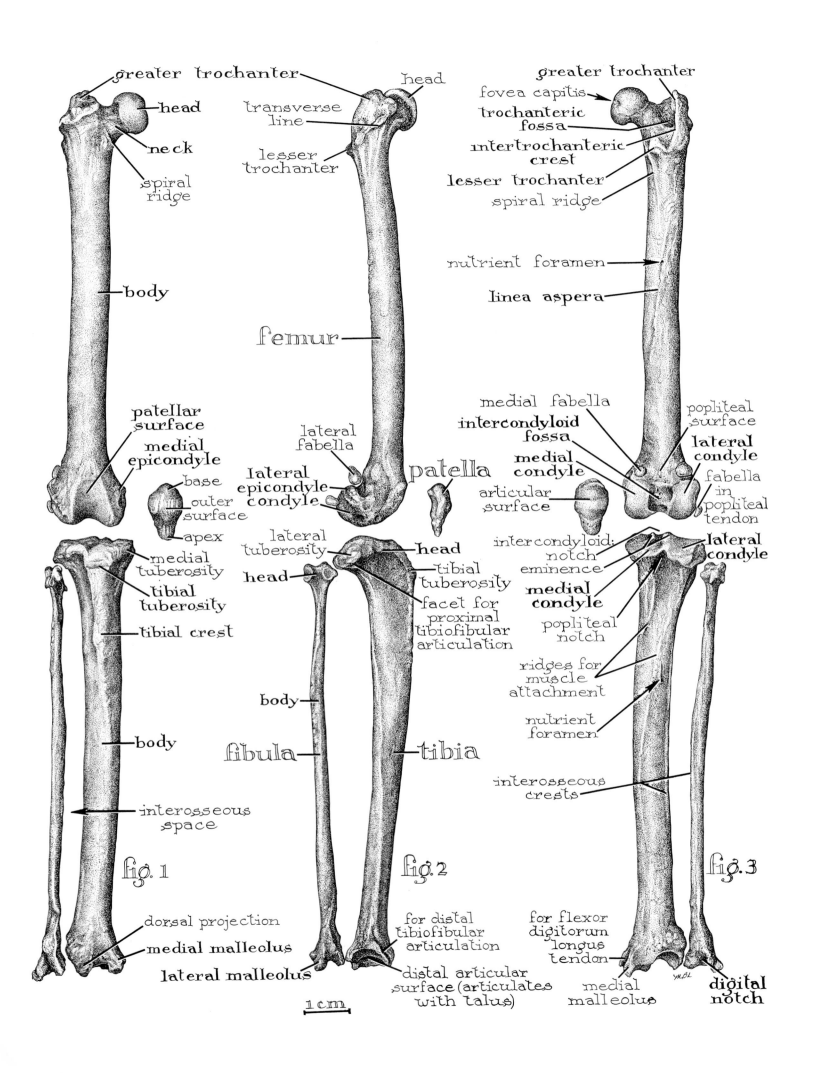

greater trochanter

head

neck

spiral ridge

body

patellar surface

medial epicondyle

base

outer surface

apex

medial tuberosity

tibial tuberosity

tibial crest

body

interosseous space

fig. 1

dorsal projection

medial malleolus

lateral malleolus

head

transverse line

lesser trochanter

femur

lateral fabella

lateral epicondyle condyle

patella

lateral tuberosity

head

tibial tuberosity

facet for proximal tibiofibular articulation

body

fibula

tibia

for distal tibiofibular articulation

distal articular surface (articulates with talus)

1 cm

greater trochanter

fovea capitis

trochanteric fossa

intertrochanteric crest

lesser trochanter

spiral ridge

nutrient foramen

linea aspera

medial fabella

intercondyloid fossa

medial condyle

articular surface

popliteal surface

lateral condyle

fabella in popliteal tendon

intercondyloid notch eminence

medial condyle

lateral condyle

popliteal notch

ridges for muscle attachment

nutrient foramen

interosseous crests

fig. 3

for flexor digitorum longus tendon

medial malleolus

digital notch

The pelvic limbs are made up of the femur, patella, tibia, fibula, seven tarsal bones, five metatarsal bones, and twelve phalanges. This plate shows three views of the femur, patella, tibia and fibula.

The **femur** or thigh-bone is the proximal bone of the pelvic limb. Its proximal **head** is rounded and contains a pit, the fovea capitis, into which the **ligamentum teres** attaches. This ligament attaches at its other end into the acetabulum. Lateral to the head is a constricted portion, the **neck.** A large projection, the **greater trochanter,** is seen laterad of the neck. The **intertrochanteric crest** runs distad and mediad on the posterior surface of the bone to connect the greater trochanter to a pyramidal projection, the **lesser trochanter.** Lying between the intertrochanteric crest and the neck is a depression, the **trochanteric** or digital **fossa.** A second ridge extends from the neck to the lesser trochanter. A spiral ridge or line courses around the neck and on the caudal surface of the shaft, joins the **linea aspera.**

At the distal end of the femur are seen **lateral** (*external*) and **medial** (*internal*) **condyles** which articulate with the tibia. Between these on the ventral surface is a deep notch, the **intercondyloid fossa.** Directly dorsad is the **patellar surface** which articulates with the **patella,** or kneecap. The condyles bear elevations or roughened areas proximally and cranially known as the **lateral** and **medial epicondyles.**

The **patella,** or kneecap, is a **sesamoid** bone. It is a flat, pear-shaped bone and occurs within the tendon of the quadriceps femoris muscle and articulates with the femur. Three other sesamoid bones occur in this region, a pair proximad and caudad of the condyles of the femur in the tendons of the gastrocnemius muscle are the fabellae, and a single one laterad of the lateral condyle of the femur in the tendon of the popliteus.

The **tibia** is the larger bone of the crus, shank, or lower leg and is the longest bone of the body. Its proximal end articulates with the femur by way of oval **lateral** and **medial condyles** which are concave from side to side and convex dorsoventrally. A divided projection, the **intercondyloid eminence** of the tibia with its **intercondyloid notch,** separates these two condyles. Lateral and medial to the condyles are the lateral and medial tuberosities. Below the lateral tuberosity is the facet for articulation with the proximal end of the fibula. On the anterior aspect of the head is the **tibial tuberosity** for the attachment of the **ligamentum patellae.** Below this is a prominent ridge, the **tibial crest.** On the posterior aspect, a deep groove, the **popliteal notch** separates the two condyles. The shaft below is triangular in shape having several rough lines for muscle attachment. The distal end is expanded and has a prolonged **medial malleolus** which bears two gooves for passage of tendons of the flexor longus digitorum and the tibialis posterior. The distal head of the tibia is deeply grooved for articulation with the **talus** (*astragalus*) of the tarsus. Lateral to this groove is a V-shaped projection which, as seen dorsally, improves the articulation of the tibia with the talus and fibula. On the lateral surface of the distal end of the tibia is a facet for the articulation with the distal end of the fibula.

The **fibula** is a slender bone situated laterad of the tibia. It articulates proximally with the tibia and distally with the tibia and talus. On the proximal head is a medial facet for articulation with the tibia, and a longitudinal goove is seen on the lateral surface. The distal end is enlarged to form the **lateral malleolus** and bears a medial facet for articulation with the tibia and a more distal facet for articulation with the talus. A deep **digital notch** is seen on the caudal aspect of the lateral malleolus. This notch allows passage of the tendons of the peroneus brevis and the extensor digitorum lateralis.

Observe the sharp adjacent borders of the tibia and fibula, the interosseous crests, which mark the attachment of the interosseous membrane.

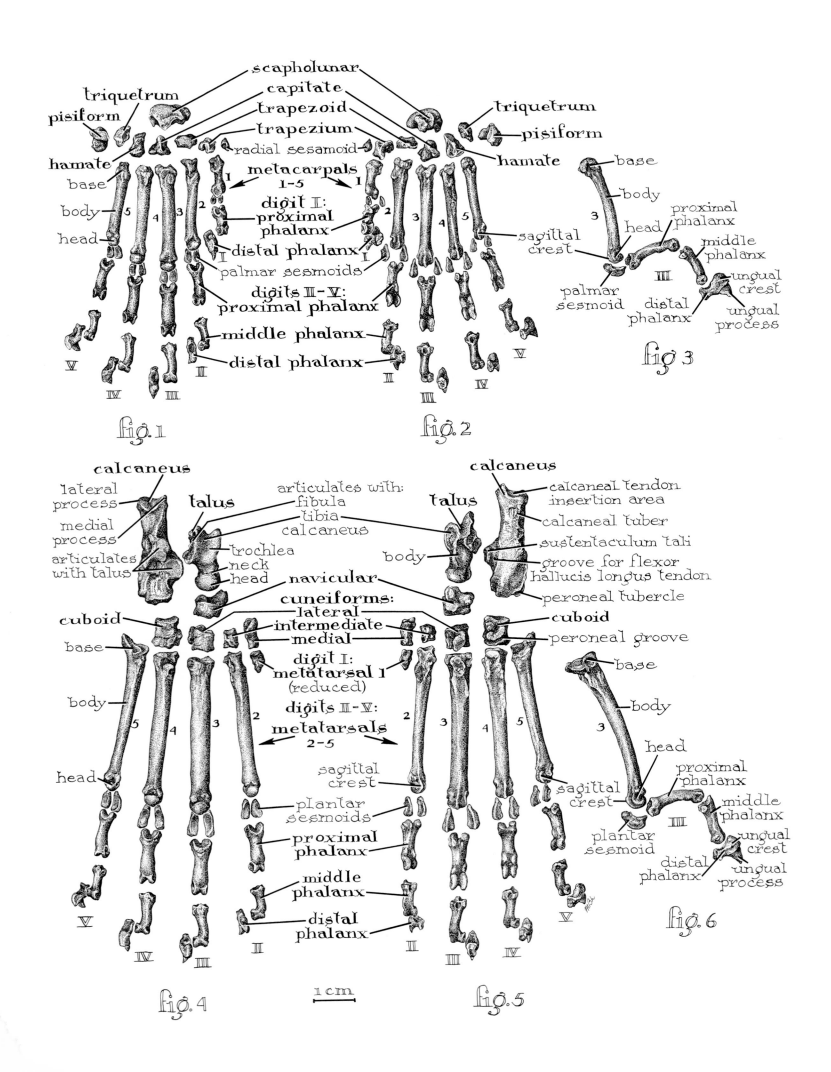

triquetrum
pisiform
scapholunar
capitate
trapezoid
trapezium
radial sesamoid
triquetrum
pisiform
hamate
base
metacarpals 1-5
hamate
body
head
digit I:
proximal phalanx
distal phalanx
palmar sesmoids
digits II-V:
proximal phalanx
sagittal crest
middle phalanx
distal phalanx

base
body
proximal phalanx
head
middle phalanx
palmar sesmoid
distal phalanx
ungual crest
ungual process

fig. 1

fig. 2

fig. 3

calcaneus
lateral process
medial process
articulates with talus
talus
articulates with:
fibula
tibia
calcaneus
trochlea
neck
head
navicular
cuneiforms:
lateral
intermediate
medial
digit I:
metatarsal 1 (reduced)
digits II-V:
metatarsals 2-5
sagittal crest
plantar sesmoids
proximal phalanx
middle phalanx
distal phalanx

calcaneus
talus
body
cuboid
base
body
head

calcaneal tendon insertion area
calcaneal tuber
sustentaculum tali
groove for flexor hallucis longus tendon
peroneal tubercle
cuboid
peroneal groove
base
body
head
proximal phalanx
middle phalanx
sagittal crest
plantar sesmoid
distal phalanx
ungual crest
ungual process

fig. 6

fig. 4

1 cm

fig. 5

The **carpus** or wrist is composed of seven bones and a sesamoid arranged in two rows. The proximal row contains three bones, starting on the radial (*medial*) side, the **scapholunar, triquetrum** (*cuneiform*), and **pisiform.** In the distal row the **trapezium** (*greater multangular*) is medial followed by the **trapezoid** (*lesser multangular*), the **capitate** (*os magnum*) and **hamate** (*unciform*).

The **scapholunar** is a large quadrilateral bone which is the equivalent of the human scaphoid (*navicular*) and lunar. In the kitten, before ossification is complete it consists of scaphoid, lunare and centrale. The **scapholunar** articulates with the radius above and fits into a depression formed by the four distal carpals below. Laterally it joins the triquetrum and radially, the radial sesamoid. The pyramid-shaped **triquetrum,** the median bone of the proximal row, articulates with the hamate, the pisiform and the styloid process of the ulna. The **pisiform,** a long carpal with enlarged ends, articulates with the triquetrum and the styloid process of the ulna.

The **trapezium,** the medial bone of the distal row of carpals, is triangular in shape and articulates proximad with the scapholunar, distad with the first and second metacarpals and laterad with the trapezoid. The **trapezoid** (*lesser multangular*) articulates proximad with the scapholunar, mediad with the trapezium (*greater multangular*), laterad with the capitate and distad with only the second metacarpal. The **capitate** (*os magnum*) articulates proximad with the scapholunar, distad with three metacarpals, the second, third, and fourth, mediad with the trapezoid and laterad with the hamate. The **hamate** (*unciform*) is a small wedge-shaped bone. Proximad it fits closely between the scapholunar and the triquetrum, mediad it joins broadly the capitate, distad it articulates with the fourth and fifth metacarpals.

There are five metacarpals in the **manus** or hand, numbered from one to five starting on the medial or pollex (*thumb*) side. They are all quite similar in shape and size except for the medial one which is very short. Each has a **base** at the proximal end, a **shaft** and at the distal end, a **head.** Their proximal ends or bases are irregular in shape and articulate with the carpus as indicated above. The first metacarpal also articulates with the radial sesamoid. Distad the heads of the bones are more regular in form being convex and articulating with the proximal phalanges of the digits.

The digits, though usually designated just by the numbers one, two, three, four and five, may be named from the radial or medial side the pollex (1), index (2), medius (3), annulus (4), and minimus (5). The bones of the digits are called **phalanges** (*singular-phalanx*) of which there are two in the pollex and three in each of the remaining four digits,

making a total of fourteen. They are named proximal, middle and distal phalanges. The middle one is lacking in the first digit or pollex. The proximal phalanges are the longest and articulate with the metacarpal bones and on the palmar (*volar*) surface of each of these articulations is a pair of **sesamoid bones** (figs. 1–3). The middle phalanges are shorter than the proximal ones and longer than the distal ones. The distal phalanges are the shortest and are quadrangular in shape. Each is provided with a terminal, ensheathed, retractile claw.

The **tarsus** or ankle, like the carpus, is made up of seven bones. The most proximal bone is the large, long calcaneus (*os calcis*) or heel bone. Upon its superior surface toward the medial side is the astragalus (talus) and distal to the astragalus the navicular. These three tarsals make up the irregular distal row of tarsals; the more even proximal row consists of the cuboid laterally followed by the lateral, intermediate and medial cuneiforms.

The **calcaneus** is the bone of the heel and at its proximal end is a groove for the attachment of the tendon of Achilles. Its lateral surface bears a grooved **peroneal tubercle,** while its medial surface has a projection, the sustentaculum tali. The distal end articulates with the cuboid. The dorsal surface of the bone is smooth proximad, but distad it is broadened and has two facets separated by a groove for the articulation with the astragalus.

The **astragalus** is divided into a head, neck and body. Its most conspicuous feature is the **trochlea** on its proximal surface for articulation with the tibia. On its medial and lateral surfaces are facets for articulation with the malleoli of the tibia and fibula. Two facets on the lower surface of the astragalus are for articulation with the calcaneus, while its rounded, smooth, distal head articulates with the navicular.

The **navicular** as the name suggests is boat-shaped and articulates broadly with the head of the astragalus and to a lesser degree with all of the remaining bones of the tarsus.

The **cuboid** is the lateral bone of the distal row of tarsals. Its plantar surface has a deep **peroneal groove** for the tendon of the peroneus longus muscle. The cuboid articulates proximad with the calcaneus, distad with the fourth and fifth metatarsals, and mediad with the navicular and lateral cuneiform.

The **lateral** (*external or third*) **cuneiform** is wedge-shaped and has a hooked process on its plantar surface. It articulates with the navicular proximad; with the cuboid laterad; with the intermediate cuneiform and second metatarsal mediad; and with the third metatarsal distad.

The **intermediate** (*middle or second*) **cuneiform** is a small wedge-shaped bone which fits between the other two cuneiforms. It articulates proximad with the navicular and distad with the second metatarsal.

The **medial** (*internal or first*) cuneiform is a small triangular bone which articulates proximad with the navicular and distad with the rudimentary first metatarsal. Its lateral surface articulates proximad with the intermediate cuneiform and distad with the medial surface of the second metatarsal.

There are five metatarsal bones. The first metatarsal is rudimentary; the others resemble the corresponding metacarpals except that they are longer and heavier. They are, like the metacarpals, divided into **base, body** and **head.**

The bones of the four digits, the **phalanges,** are twelve in number, there being no **hallux** or big toe. They are similar in every detail with the phalanges of the manus. At the articulations between digits and metatarsals are found **sesamoid** bones like those of the manus or hand.

Arthrology is the study of articulations or joints. Articulations are formed where two or more bones come together and are united by fibrous, elastic or cartilaginous tissue. Articulations fall into three main groups on the basis of their most characteristic structural feature. They are fibrous, cartilaginous and synovial.

Fibrous joints are those in which the bones are held together by fibrous or elastic tissue. The union may be close and the fibers short to permit little or no movement as in the **sutures** of the skull. These joints were formerly called synarthroses. In sutures the bones may be fitted together in various ways. **Serrate sutures** are those in which there is an interdigitating of reciprocally alternating processes and depressions as seen in a and b of Figure 3. **Squamous sutures** are those where the reciprocally beveled edges of bones overlap as in the squamosal suture between temporal and parietal bones. **Plane sutures** are those where bones meet at right angled edges or surfaces as in many of the facial bone articulations. **Foliate sutures** are those in which the edge of one bone fits into a fissure of another as between the zygomatic and maxillary bones. Recall that the flat bones of the skull are formed in membrane and that the bones as they grow gradually use up the membrane. Until they do and the sutures are fully formed there remain fontanels or membranous areas as seen to advantage in the skull of a new born human infant.

A fibrous joint where the amount of fibrous tissue between the bones is greater than in suture joints is often called a **syndesmosis.** The distal tibiofibular joint is an example as seen in Figure 3 c. Slight movement is allowed in this kind of articulation. Freer movement is found between the radius and ulna where the interosseous membrane or ligament runs between the adjacent surfaces of the two bones.

A specialized type of fibrous joint is found where the teeth are held in the sockets on the alveolar processes of mandible, incisive and maxillae bones by the peridontal ligament. This is known as a **gomphosis.**

Cartilaginous joints are those in which cartilage occurs between the articulating bones. They are often called **synchondroses** and may involve hyaline cartilage, fibrocartilage or both. Those with **hyaline cartilage** are sometimes called primary joints and are usually temporary (Fig. 4 a). Where temporary, they represent persistent parts of the fetal skeleton as the **epiphyseal** or **growth cartilages** be-

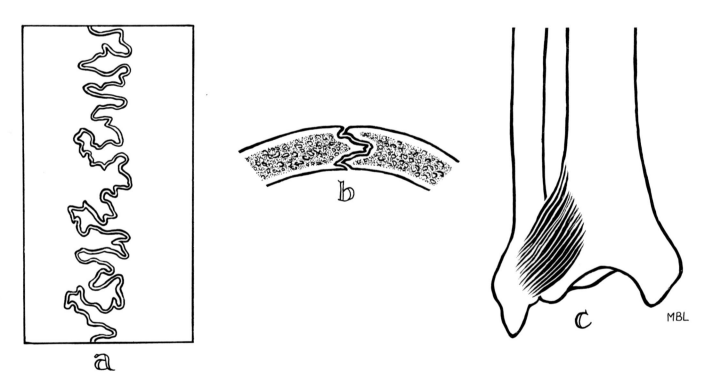

Fig. 3. *Types of fibrous articulations.* a *and* b, *serrate suture joints.* c, *syndesmosis.* (Crouch, Functional Human Anatomy, *Lea & Febiger.)*

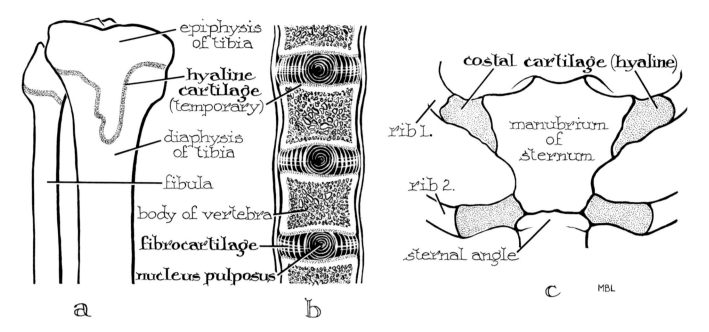

Fig. 4. *Types of cartilaginous articulations.* a, *temporary synchondrosis.* b, *fibrocartilaginous joints (symphysis).* c, *permanent synchondrosis.* (*Crouch,* Functional Human Anatomy, *Lea & Febiger.*)

tween diaphysis and epiphyses of long bones or between the tubercles of the humerus and the shaft, or between the femoral trochanters and the shaft. The sphenooccipital joint is another example. As these cartilages are replaced by bone and there is established an osseous union between the bones or parts of bones it is often referred to as **synostosis.** The hyaline cartilages between the ribs and the sternum, the costochondral articulations, remain throughout life (Fig. 4 c).

Fibrocartilaginous joints, sometimes referred to as secondary joints, are best illustrated by the pelvic symphysis, the joints between the bodies of vertebrae and between the sternebrae. As seen in Figure 4 b hyaline cartilage may also be present between the fibrocartilage and the bone. These joints were once referred to as amphiarthroses.

Synovial joints (Fig. 5) are those which are freely moveable and have a joint or synovial cavity, synovial fluid, a fibrous capsule and articular cartilage. A few of the synovial joints have special structures such as menisci, intraarticular ligaments and fat pads (Fig. 5 a). The knee joint has such special features and is one of the most complex of the body. Synovial joints were once called diarthroses.

The blood and lymph supply are important to these moveable articulations and is supplied by networks of vessels derived from major trunks in the vicinity of the articulations. They supply blood to the joint capsule, the synovial membrane and the epiphyses of the articulating bones. The synovial membrane is also well supplied with lymphatic vessels and through them some substances, as protein, are rapidly removed from the joint cavity. The

nerve supply to the joint is provided by muscular branches, proprioceptive fibers, pain receptor fibers and sympathetics which relate largely to the vasomotor and vasosensory functions.

In synovial joints the ends of the articulating bones are covered with cartilage which is, with few exceptions, hyaline cartilage. Its free surface lacks a perichondrium and the superficial chondrocytes are flattened and arranged in rows. The deeper part of the cartilage is calcified. The thickness of the cartilage varies from one joint to another and in different parts of the cartilage in the same joint. It is thickest in young individuals and in joints which support a lot of weight. Its elasticity and compressability give resiliency to the articulation and thereby protects against fracture of bone by absorbing shock. Lacking blood vessels the hyaline cartilage receives its nutrition from the synovial fluid. It lacks nerves, also (Fig. 5 b).

The articulation is surrounded by a **capsule** composed of an outer fibrous membrane and an inner synovial membrane. The outer **fibrous membrane** is largely white fibrous connective tissue with some elastic fibers. Its fibers are continuous with those of the periosteum of the bones. It serves to protect and to strengthen the joint. It is sometimes called the capsular ligament because the joint ligaments may be thickenings of the capsule as in the case of some collateral ligaments of hinge joints. Other ligaments as the patellar ligament of the stifle or knee joint is outside of and separate from the capsule.

The **synovial membrane** of the capsule is variable in character but is usually described as a vascular connective tissue which lines the fibrous capsule and produces **synovial fluid** (*synovia*). It does not cover the hyaline articular cartilage nor the articular surfaces of the fibrocartilage. It blends with the perichondrium of the cartilage and the periosteum of the bones. It also forms sleeves around ligaments which pass through an articulation and also tendons of muscles such as the biceps in passing through the shoulder joint. The synovial membrane is not always closely attached to the fibrous membrane. It is, in some places, thrown into **synovial folds** which contain fat or it may form numerous processes, the **synovial villi,** which are soft and velvety and often highly vascular. The synovial membrane may extend beyond the joint itself to form fluid filled pads under tendons or ligaments, called **bursae** or may even be in communication with **tendon sheaths.**

The **synovial fluid,** besides furnishing nutrients and removing wastes from the joint cartilages, serves as a lubricant for the contact surfaces of the synovial joints. Leukocytes also circulate in the synovial fluid and phagocytize the products resulting from the wear and tear of the articular cartilages. In composition it is very similar to the tissue fluid containing mucin, albumen, fat droplets, salts and cellular debris. It varies in quantity from joint to joint depending somewhat upon the health of the animal. Usually there is just enough to lubricate the synovial and joint surfaces but in cases of inflammation of the joint it may be produced in painful quantities.

A number of factors contribute to the freedom and direction of movement in a given joint. Among these are the shapes of the articulating surfaces of the bones, the arrangements of ligaments, the looseness of the capsule and, of course, the points of attachment of muscles which are the active agents in producing movements.

Synovial joints may be classified on the basis of the shape of the articulating surfaces of the bones as follows:

1. **Plane joint**—one in which the articulating surfaces are flat and permit a gliding movement. These are found between the articular processes of some vertebrae and in costotransverse joints.
2. **Ball and socket joint** (*spheroidal*)—where a convex head fits into a shallow or deep cavity. The shoulder and hip joints are examples; the shoulder shallow, the hip deep. They permit movement in many axes around one common center. They are sometimes called "universal" joints.
3. **Ellipsoidal joint**—characterized by an elongation of one surface at a right angle to the other thus forming an ellipse. The radiocarpal joint is an example.
4. **Hinge** (*ginglymus*) **joint**—allows movement in only one axis at right angles to the bone as in flexion and extension of the elbow joint. The concave articulating surface of one bone fits a convex surface of the other.
5. **Condylar joint**—a compound joint where convex surfaces of one bone fit into concave surfaces of another bone forming two articular surfaces within one joint capsule. The knee and temporamandibular joints are examples.
6. **Trochoid** (*pivot*) **joint**—allows movement around a single axis. The ring-like atlas which fits over the dens of the axis which acts as a pivot around which the atlas

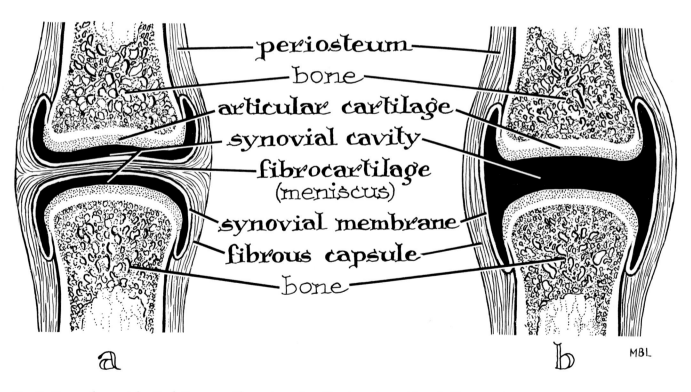

Fig. 5. *Types of synovial articulations.* a, *with meniscus*; b, *without meniscus*. (*Crouch,* Functional Human Anatomy, *Lea & Febiger.*)

turns is an example. The proximal articulation between the radius and ulna is another example.

7. **Saddle joint**—a biaxial articulation the surfaces being reciprocally concavoconvex. The interphalangeal joints are of this type.

The kinds of movement which take place at the various types of joints are considered in reference to the muscular system on page 63.

While no detailed study of joints is contained in this book, the student should be aware of them as he studies the skeletal and muscular systems. A large amount of valuable general information can be gained in this way.

Having studied the skeleton which is an internal supporting and protecting system of the body, it is necessary to look next at a system which is external and, although in a different way than the skeleton, is supporting and protective. Also, it is a system which encloses the whole body and, therefore, one which we must remove before we can proceed to the study of the muscles, the celom and viscera. It is the integumentary system, composed of the skin and its accessory organs like the nails, foot pads, hair and glands.

Functions of the skin. The skin is in the unique position of lying between the organism and its external environment, at the same time that it is a vital part of that organism. Its external cells are in large part dead cells which constantly sluff off but are as constantly replaced by the action of viable dividing cells deeper in the skin. The hair can be cut or burned or otherwise damaged and as long as the living roots remain it can be replaced. The dead parts of the claws can be cut or worn but they continue to grow. Indeed if they are not worn off or trimmed, they may grow in their curved fashion until they penetrate the volar surface of the foot and cause a painful lesion. However, if the skin is extensively and deeply burned, the life of the animal is endangered.

The cat's skin is a protection against the temperature changes in the environment. Its hair traps a layer of air which insulates the body, thereby reducing heat exchange with the environment. By changing the position of the hair, the convection from the surface of the body can be controlled. The thickness of the hair coat is greater in winter than in summer when much hair is shed. In house or domestic cats this cycle is changed and hair loss is a continuous process. Hair is not evenly distributed over the whole surface of the body. Some areas as the nipples, the end of the nose and the pads on the soles of the feet are without hair.

The skin separates the "fluid" body from its dry or wet environment. It keeps the body from drying out and it is waterproof to the extent that it protects it from the entrance of water when the animal is immersed.

The skin is a **secretory** and an **excretory organ** through the agency of certain glands. The cat feeds its young by the secretions of the mammary glands. It produces, through the sebaceous (*holocrine*) and apocrine sweat glands which empty into the hair follicles, secretions which condition the skin and the hair. In some areas there are glands whose secretions and their odors are influential in the reproductive behavior. Eccrine (*merocrine*) sweat glands in the cat are found only in the pads on the **volar** (*plantar*) surfaces of the feet. The evaporation of secretions from the body play a role in temperature regulation but not to the extent that they do in man with his extensive system of sweat glands and his relatively sparsely haired skin. Through the various glands excesses of water and some waste products of metabolism are lost—an excretory process.

The claws of the cat are used for digging, for protection and for aggressive behavior usually associated with reproduction.

The skin contains receptors for stimuli of various kinds like touch, pressure, pain, heat and cold. It is the organ which, through these receptors, plays a large role in enabling the animal to cope with an indifferent external environment. Special long, sensory hairs, the vibrissae, occur on the upper lip, the cheek and above the eyelid. To these, of course, are added the highly specialized "organs of special sense," the eye, ear and nose—not a part of the skin nor as removed from it as one might think. Their sensory surfaces embryologically come from the body surface.

The hair and pigment of the skin protect the animal from actinic exposure and at the same time the skin is an area for vitamin D synthesis. It is also a storage center for some fat and a physical barrier to pathogenic organisms, though not a perfect one.

The skin is continuous at the natural body openings with the mucous membrane of the mouth, anus, nose and urogenital orifices. It is continuous also with the conjunctiva of the eye.

General structure of the skin. The skin is composed of two layers, an outer **epidermis** consisting of stratified squamous epithelium and an inner **dermis** of dense connective tissue. The skin is loosely attached to a **hypodermis** or **superficial fascia** of loose (*areolar*) connective tissue and it in turn connects to the underlying muscles. Study of **Plate 20** with accompanying discussion will present the skin and accessories in greater detail.

The skin accessories, the glands, hair and nails are all products of the epidermis which differentiate by complex processes during its development. Although derived from the epidermis, the completed structures become seated deep in the dermis and in some cases push into the superficial fascia. The exception is the claw which is largely a modification of the outer corneal layer of the epidermis.

Blood is supplied to the skin by many arteries or by branches of arteries whose chief function is to supply the muscles. Many of these, of both types, will be seen in the course of your dissection. These arteries anastomose extensively with one another and form three microscopic plexuses each parallel with the skin surface. One is the **deep or subcutaneous plexus** which lies in the superficial fascia and is made up of the terminal branches of cutane-

55

ous arteries. From this plexus vessels travel outward to a **middle or cutaneous plexus** which is associated with the hair follicles and glands of the dermis. It in turn sends vessels to the **superficial plexus** which lies close beneath the epidermis. Note that the epidermis has no blood vessels within it, but blood is brought to it by the superficial plexus and from this oxygen, nutrients and other material reach the epidermis and waste materials are returned to the blood by way of the tissue fluid.

The veins follow much the same course as the arteries. Arteriovenous anastomoses have been noted. Lymphatic vessels arise from capillary nets in the superficial part of the dermis and around the glands and hair follicles. These vessels drain into a subcutaneous lymphatic plexus.

The nerve supply to the skin comes from a variety of sources. As you proceed with your studies you will note the cutaneous branches of many spinal and cranial nerves. Every hair has a nerve ending. There are nerves to the glands and the arrector pili muscles from the autonomic nervous system. In general the small nerves of the skin enter the subcutaneous tissue and from there are continued by nerve plexuses within the dermis and branches to the epidermis. The sensory nerve endings in the skin range from simple, uncomplicated, terminal fibers to very complex, Pacinian and Meissner's corpuscles. (See Fig. 18, page 319.)

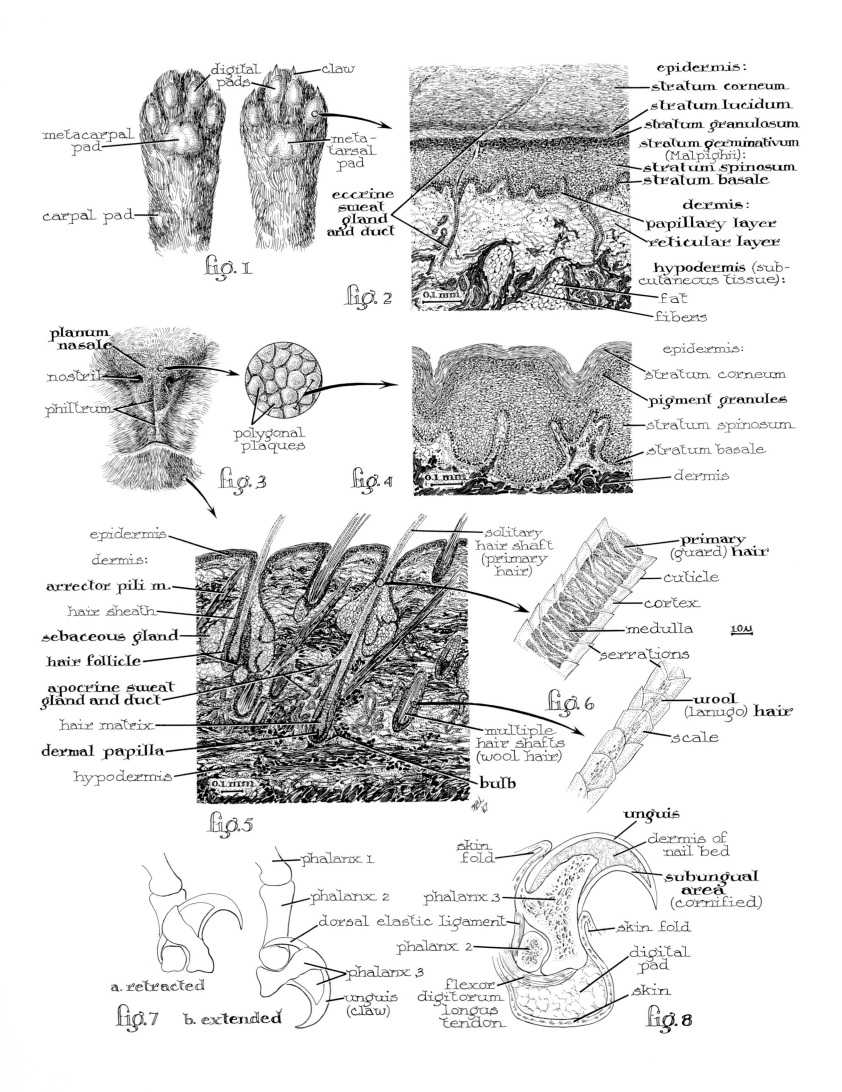

fig. 1

digital pads
claw
metacarpal pad
meta-tarsal pad
carpal pad
eccrine sweat gland and duct

fig. 2

epidermis:
stratum corneum
stratum lucidum
stratum granulosum
stratum germinativum (Malpighii):
stratum spinosum
stratum basale
dermis:
papillary layer
reticular layer
hypodermis (sub-cutaneous tissue):
fat
fibers

0.1 mm

fig. 3

planum nasale
nostril
philtrum
polygonal plaques

fig. 4

epidermis:
stratum corneum
pigment granules
stratum spinosum
stratum basale
dermis

0.1 mm

fig. 5

epidermis
dermis:
arrector pili m.
hair sheath
sebaceous gland
hair follicle
apocrine sweat gland and duct
hair matrix
dermal papilla
hypodermis

solitary hair shaft (primary hair)

multiple hair shafts (wool hair)

bulb

0.1 mm

fig. 6

primary (guard) hair
cuticle
cortex
medulla
serrations
wool (lanugo) hair
scale

10µ

fig. 7

phalanx 1
phalanx 2
phalanx 3
unguis (claw)
a. retracted
b. extended

fig. 8

unguis
dermis of nail bed
subungual area (cornified)
skin fold
phalanx 3
dorsal elastic ligament
phalanx 2
flexor digitorum longus tendon
skin fold
digital pad
skin

This plate serves to extend the brief comments on the structure of the skin. Figure 1 shows the hairless footpads as found on the front and hind feet of the cat. Each digit is provided with one, the **digital pad.** There are four on the hindfoot and five on the forefoot. The forefoot has an additional pad under the pisiform bone called the **carpal pad.** Each forefoot and hindfoot is also provided with a large lobed pad over the metacarpals or metatarsals, called therefore the **metacarpal pad** and **metatarsal pad.**

Figure 2 is a section of the skin through a foot pad. It is the toughest skin of the body and is very thick and usually highly pigmented. Its surface is smoother than that of a dog which has numerous conical papillae. In the cat there are ridges present which often form a branching connected pattern. There are five strata or layers in the epidermis. Starting from the surface inward they are the stratum corneum, stratum lucidum, stratum granulosum, stratum spinosum and stratum basale. The last two strata are called collectively the stratum germinativum (*Malpighii*).

The stratum corneum is thickest in the planum nasale and the digital pad. It is a keratinized anuclear area composed largely of dead cells which constantly sluff off and is capable of withstanding a lot of hard wear. Beneath it is the thin **stratum lucidum** made up of anuclear, homogenous, hyaline material containing refractile droplets called eleidin. The next layer is the **stratum granulosum** easily recognized by the granular appearance of its cells and their spindle-shapes. It is three to five cells thick. The next layer, the **stratum spinosum,** may be as much as sixteen cells thick. Its cells are polygonal in outline and have round to oval nuclei. The cells toward the surface are noticeably flatter than the deeper cells. Intercellular bridges are present. The deepest layer of the epidermis, the **stratum basale** (*cylindricum*) is one cell thick and rests on the dermis. Its cells are cuboidal or columnar and have little cytoplasm and dark-staining nuclei. Mitotic figures can often be seen in this germinal layer.

The **dermis** is composed of a complex of elastic, collagenous (*white*) and reticular fibers enclosing within their framework a variety of mesodermal cells, blood vessels, lymphatics, and nerve endings. It is divided into two portions, an outer papillary and an inner reticular layer. These layers are well differentiated in the hairless areas but not distinctly separated in hairy skin. The muscle is differentiated into smooth as in the arrector pili and striated or skeletal in the case of the integumentary muscles. Mast cells, fibroblasts, histiocytes and melanophores are the principal mesodermal cells. The **papillary layer** is well supplied with closely woven and fine collagenous fibers and bundles. The elastic fibers form a fine network in the papillary layer and increase in number in the reticular layer where the collagenous fibers become coarser.

The **hypodermis,** though not a part of the skin, is closely associated with it structurally and functionally. It is called also the subcutaneous tissue, superficial fascia, or histologically, loose (*areolar*) connective tissue. It contains much fat in adipose individuals.

Notice the **eccrine sweat gland** which has its coiled, tubular secreting portion in the hypodermis and dermis and its duct penetrating through the outer layers to open on the skin surface.

The **planum nasale** is shown in figure 3, another one of the bare areas of skin. It is not quite as thick as the foot pads but has the same layers in its skin including pigment granules. It has no glands. Its surface is marked by shallow grooves which divide it into polygonal plaques which give it a characteristic appearance when examined critically. The plithrum, shown in figure 3, marks the point of joining of the two halves of the upper lip. A section through the skin of the planum nasale is shown in figure 4. Compare it to figure 2 in terms of its thickness and its component layers.

Figure 5 is a section through a piece of hairy skin. Note the thin epidermal layer of the skin and the sleeve of epithelium enclosing each primary hair. The epithelium and the whole hair complex is a product of the epidermis. The **hair follicle** walls are divided into two layers, the outer and inner root sheaths. The hair follicles are thickest at their basal portions where they become expanded to form a bulb. The **dermal papilla** invaginates the bulb. **Arrector pili muscles,** the muscles of the hair, **sebaceous** and **apocrine sweat glands** attach to or empty into the hair follicles. The insert shows a hair follicle with multiple hair shafts, the wool or lanugo hair as contrasted with the primary, solitary or cover hairs just described. A third kind of hair seen in Plates 22 and 23 is the tactile hair (*vibrissae*).

Figure 6 shows the shaft of a **primary hair** with its three layers named from within outward the medulla, cortex and cuticle. Note the serrations caused by the overlapping of the scales of the cuticle. A small part of a **wool hair** (*lanugo*) with its scaly cuticle is also shown in figure 6.

Figures 7 and 8 deal with the claws of the cat, figure 7 showing them in their relationship to the phalanges of the digits and in the retracted and extended condition. Figure 8, a section of a claw, emphasizes some of the details of the structure-a highly specialized development of the epidermis. The **unguis,** which is the horny claw itself, is derived from the corneal layer of the epidermis. It is U-shaped in cross section and is thickest at the top of the U or along the median ridge, the sides are thin.

Distally the claw tapers to a sharp point and bends to form a deeply curved beak-like structure. The horny structure continues ventrally to become the cornified **subungual area.** The claw is supported by the nail or claw bed which is composed of the dermis which here is continuous with the periosteum of the distal phalanx of which the claw is a continuation. Proximally the claw grows out from skin folds which overlap the claw at its base and in which the germinal layers of the epidermis remain active in maintaining the growth of the claw.

Skeletal material is usually provided "ready made" for the student. Therefore, it is when one is ready to start the study of muscles that dissection is required. While this is not a dissection manual, it does seem feasible to approach this complicated system as the student must or should—by doing his own dissection. Even the study of excellent illustrations is no substitute for getting the "feel" of an animal by dissection. However, with good illustrations which do look like the animal you are dissecting, the dissection process can be greatly simplified and directions can be kept to a minimum.

Removing the skin is the first step. There are many ways of skinning a cat. A dorsal approach will be used here. First loosen the skin by picking it up between your fingers and moving it about, especially on the dorsal side where the first incision will be made. Now lift up a piece of skin along the mid-dorsal line and cut a hole in it with your scalpel or scissors large enough so that you can insert your fingers. By keeping your fingers in the incision and lifting the skin away from the underlying muscles, you can make a median incision from the base of the tail to the back of the head with little danger of damaging the underlying structure. Now by lifting the cut edges of skin along the incision and pushing your fingers in between the skin and muscles you can become familiar with the **subcutaneous tissue,** (*also called superficial fascia and loose connective tissue*) a loose fibrous material which holds the skin loosely to the under-lying muscles. It enables the muscles to contract freely within the skin. In some places, as in parts of the head region, the subcutaneous tissue is minimal in quantity and it is more difficult to remove the skin. It is more abundant ventrally than dorsally where you have made your incision and the ventral body wall is more vulnerable, too.

Examine Plate 27 and notice how the skin has been cut around the back of the head, around the forelimb below the elbow, around the thigh and the tail. This is the way your animal should look when properly skinned. Notice, too, that the **integumentary** (*skin*) **muscles,** the cutaneous maximus and the platysma are left on the body. To attain this goal gradually work the skin away from the **subcutaneous tissue** along the dorsal incision and out around the sides of the body. This is important if the integumentary muscles are to be left in place **on the cat.** As you

work the skin loose you will notice some segmentally arranged white cord-like structures which are cutaneous (*skin*) nerves and usually you will see some blood vessels accompanying them. You will have to break them. By making the cuts suggested and patiently loosening the skin you will be able to free it completely and literally pull the cat out of its own skin. The reason for this particular approach is that when you finish dissecting you can put the skin back on the cat as a protective cover to keep the cat from drying out—or when dissecting you can leave it on the part of the body on which you are not working at the time. Drying can be very damaging and make good dissection impossible.

Your next step is to remove carefully the subcutaneous tissue watching out for the integumentary muscles. Your fingers, forceps and the handle of your scalpel, or if used carefully, the blade of the scalpel are your best instruments. Unless you are experienced you should use your scalpel very conservatively. You can do a lot of damage with it quickly. A blunt probe is an excellent instrument for some of your dissection work. It is impossible to put too much emphasis on removing fascia. It reveals and makes clear structures which you might miss entirely or which you could not fully understand otherwise.

Now examine your animal and refer to Plate 27 and the description of it and see how your cat compares. No doubt you have removed some of the integumentary muscles in your dissection. Most of us do, since some of the fibers are so closely tied to the skin, as one would expect.

Since this book contains a complete alphabetical list of the skeletal muscles, their origins, insertions, actions and innervation, no effort will be made to describe each muscle completely in the following illustrations. Rather, only certain special features will be mentioned as well as some directions for dissection given.

Finally, since these are skeletal muscles, constant reference to the bones of the skeleton must be made. If a skeleton is available, work with it in front of you. Even if you feel you know the skeleton, as indeed you should at this point in your study, it helps to have one around. Reference also to the earlier plates of this atlas will help you with skeletal terminology which may have slipped your mind.

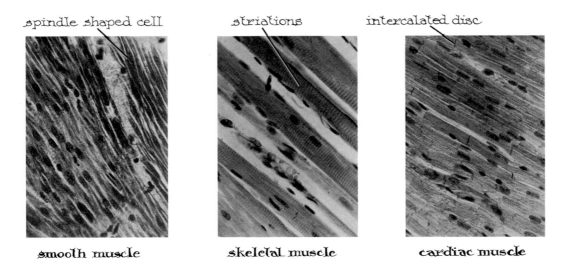

spindle shaped cell striations intercalated disc

smooth muscle skeletal muscle cardiac muscle

fig. 1

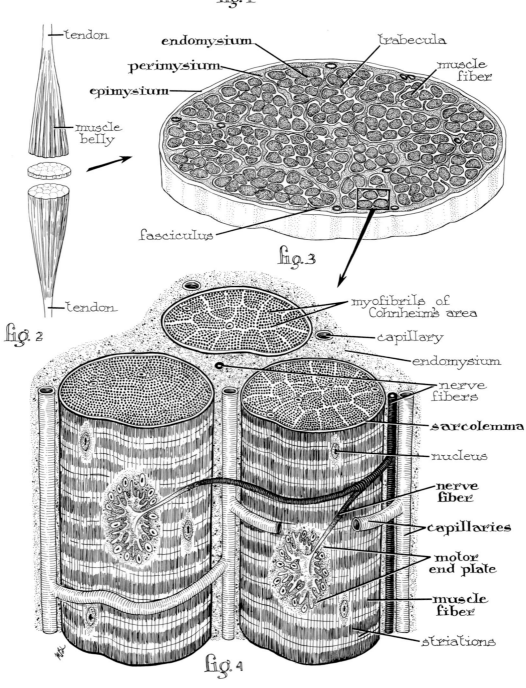

tendon

endomysium trabecula

perimysium muscle fiber

epimysium

muscle belly

fasciculus

tendon

fig. 2

fig. 3

myofibrils of Cohnheim's area

capillary

endomysium

nerve fibers

sarcolemma

nucleus

nerve fiber

capillaries

motor end plate

muscle fiber

striations

fig. 4

Definition. The skeletal system consisting as it does of bones and cartilages and of articulations between these parts is, in terms of movement, a passive structure. The **muscular system** brings action to this system by attaching to the various bones and cartilages, using them as levers and crossing the joints which become the fulcra around which movement is accomplished. The study of the muscular system is called **myology.** There are over five hundred muscles in the cat.

Muscle tissues (Plate 21, fig. 1). Three kinds of muscle tissue are found in the cat; smooth, cardiac and skeletal. **Smooth** or **visceral muscle,** with its spindle-shaped fibers or cells, is laid down in the walls of the various hollow organs of the body, and in the skin. **Cardiac muscle** is found in the walls of the heart and can be recognized microscopically by its faint striations, its branching fibers, **intercalated discs** and large oval nuclei. **Skeletal muscle** makes up about 40 per cent of the body weight and it alone forms the muscle organs of the body which constitute the muscular system. It is a voluntary muscle tissue in contrast to smooth and cardiac muscle which are involuntary. Skeletal muscle tissue is composed of cylindrical, striated fibers ranging from 1 to 15 centimeters in length. They contain numerous, small nuclei many of them located close to the periphery of the fibers. Each muscle fiber is enclosed in a transparent sheath, the sarcolemma (Plate 21, fig. 4).

Muscle organs (Plate 21, figs. 2 to 4). Skeletal muscle fibers are enclosed in a loose connective tissue constituting the **endomysium.** Groups of these fibers are held in bundles, the **fasciculi,** by a continuation of this same loose connective tissue which is now called the **perimysium.** Finally, a number of fasciculi make up the fleshy part or belly of the muscle which is covered by **epimysium,** a deflection or continuation of the perimysium. These continuous connecting and supporting tissues of the muscle belly give strength, protection and support not only to the muscle fibers, but to the blood vessels and nerves with which a skeletal muscle is so richly supplied. At the ends of the belly of the muscle the connective tissues undergo a gradual change to become the white fibrous (*collagenous*) connective tissues of the tendons. The tendons continue into the periosteum thus giving attachment to the muscles and strength to the skeletomuscular complex. When tendons are broad, thin and flat as in the abdominal muscles, they are called **aponeuroses.**

Tendons are referred to as tendons of origin or tendons of insertion. The **tendon of origin** is attached to the more fixed structures where there is less movement during muscle contraction. It is most often the more proximal end of the muscle. The **tendon of insertion** is attached to the more movable structure. Usually it is distal. In the biceps femoris muscle of the hindlimb, for example, the origin is on the ischium, the insertion on the fibula. It operates over the knee joint to flex the leg on the thigh (Plate 40).

Classification. On the basis of function muscles may be classified as flexors, extensors, abductors, adductors, elevators, depressors, rotators, and sphincters. **Flexor** muscles bend a part or reduce the angle between parts as in flexing a limb. An **extensor** muscle has the opposite action of straightening a part or increasing the angle between parts as in extension of the forearm at the elbow by the triceps brachii. **Abductor** muscles are those which move a part away from the median plane of the body or the central axis of a part as when one spreads the fingers of the hand or lifts the arm away from the side of the body. An **adductor** is the antagonist of the abductor having just the opposite action. **Elevators** are muscles which raise parts as the temporalis muscle raises the mandible. Its antagonist would be a **depressor** muscle such as the digastricus which opens the jaw. A **rotator** muscle turns one part around another as the radius around the ulna by supinator or pronator teres; or the rotation of the atlas on the axis of the vertebral column by the obliquus capitis inferior. Finally the **sphincter** muscles are those which occur around orifices such as the orbicularis oris at the mouth or the external anal sphincter at the anus.

Naming of muscles. Muscles are named according to their form, action, position, size or the parts on which they originate and insert. The trapezius and rhomboideus are named because of their form; the latissimus dorsi on the basis of its broad, flat form and its position; the flexor digitorum profundus on the basis of its action as a flexor, its insertions on the digits and its deep position; the sternomastoideus because of its origin on the sternum and its insertion on the mastoid process of the temporal bone; the pectoralis major and minor because of their position and relative sizes. Keeping these ideas in mind enables one to learn and understand the muscles more readily.

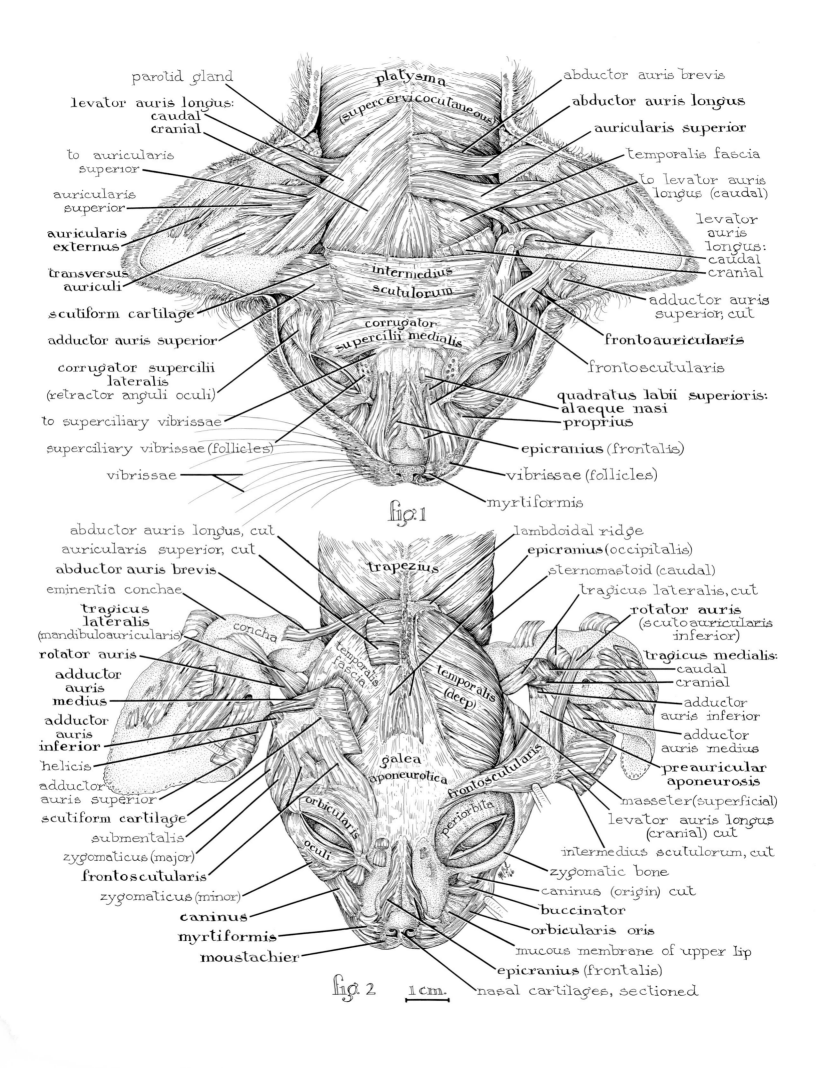

parotid gland

levator auris longus:
caudal
cranial

to auricularis
superior

auricularis
superior

auricularis
externus

transversus
auriculi

scutiform cartilage

adductor auris superior

corrugator supercilii
lateralis
(retractor anguli oculi)

to superciliary vibrissae

superciliary vibrissae (follicles)

vibrissae

platysma
(supercervicocutaneous)

abductor auris brevis

abductor auris longus

auricularis superior

temporalis fascia

to levator auris
longus (caudal)

levator
auris
longus:
caudal
cranial

adductor auris
superior, cut

frontoauricularis

frontoscutularis

quadratus labii superioris:
alaeque nasi
proprius

epicranius (frontalis)

intermedius
scutulorum

corrugator
supercilii medialis

vibrissae (follicles)

myrtiformis

fig. 1

abductor auris longus, cut

auricularis superior, cut

abductor auris brevis

eminentia conchae

tragicus
lateralis
(mandibuloauricularis)

rotator auris

adductor
auris
medius

adductor
auris
inferior

helicis

adductor
auris superior

scutiform cartilage

submentalis

zygomaticus (major)

frontoscutularis

zygomaticus (minor)

caninus

myrtiformis

moustachier

lambdoidal ridge

epicranius (occipitalis)

sternomastoid (caudal)

tragicus lateralis, cut

rotator auris
(scutoauricularis
inferior)

tragicus medialis:
caudal
cranial

adductor
auris inferior

adductor
auris medius

preauricular
aponeurosis

masseter (superficial)

levator auris longus
(cranial) cut

intermedius scutulorum, cut

zygomatic bone

caninus (origin) cut

buccinator

orbicularis oris

mucous membrane of upper lip

epicranius (frontalis)

nasal cartilages, sectioned

trapezius

concha

temporalis
fascia

temporalis
(deep)

galea
aponeurotica

frontoscutularis

orbicularis
oculi

periorbita

fig. 2   1 cm.

If these muscles are to be dissected, a fresh specimen is much easier to use than a preserved one. In most courses these muscles are not dissected by the student since it is a difficult and time-consuming process. A demonstration specimen is most desirable.

If a dissection is done, remove the skin carefully from the top and sides of the head and the face. Be careful not to damage the ears because many of the muscles are associated with them. Save the skin for it may be that some of the muscle tissue has adhered to it. Carefully pick away the fascia from the muscles studying the accompaning illustration as you do so. It took considerable skill and a number of dissections to produce these drawings. The borders of the muscles as seen here are more clearly outlined than they would be on the actual specimen.

In figures 1 and 2 and those on the following plate note that most of the muscles relate to the ear, a few to the eye and the rest to the mouth. Only a few can be mentioned and function will be emphasized. Origins and insertions can be determined from the illustration or from the muscle table in the appendix of this book. The position of the ear and its structure should be noted.

Figure 1 on the right side shows intact the muscles of the dorsal side of the skull. The important **scutiform cartilage** can be seen on the right side lying in the temporal fossa near the cranial part of the ear. It is a narrow cartilage with its long axis craniocaudad from which a thin cartilaginous sheet extends laterad. It serves for the attachment of a number of muscles.

The **platysma** is shown caudally, its cranial border being overlapped by the caudal portion of the levator auris longus.

If you have ever wondered about the versatility of movement of a cat's ears, you can now understand the basis for it by examining this plate. Caudally, on the midline of the neck and sagittal crest is the origin of a two-parted muscle, the **levator auris longus, cranial** and **caudal.** The caudal portion inserts on the auricle and the cranial portion on the **scutiform cartilage.** This muscle pulls the external ear dorsocaudad. On the left side these muscles can be cut and pulled back to reveal the **abductor auris longus** and craniad of it the **auricularis superior.** These muscles draw the ear dorsad and the former also caudad. The **abductor auris brevis** is seen to better advantage on figure 2. It has its origin on the sagittal crest and inserts into the medial surface of the proximal part of the concha. It pulls the ear caudad.

A prominent muscle, the **intermedius scutulorum,** lies between the scutiform cartilages. Its fibers cross the midline and its action is to draw the ears dorsad. Note to the left side a number of muscle bundles called the **frontoauricularis.** This muscle is closely related to the corrugator supercilii medialis and lateralis in origin and inserts in close relationship to the adductor auris superior.

The remaining muscles can be studied by reference to figure 2. If you are doing this dissection, make the cuts indicated leaving only origins and insertions of muscles previously cut. In this way the ears are freed and can be pulled to the sides and the underlying muscles uncovered for study. On the right side we see now the temporalis fascia which is removed on the left to show the deep portion of the **temporal muscle.** The temporalis fascia contains muscle fibers and is considered the superficial temporal. This muscle will be considered with the deep muscles of the head and the muscles of mastication. It inserts on the coronoid process of the mandible.

The **rotator auris** (*scutoauricularis inferior*) is a narrow band of muscle fibers which lies between the auricle and the temporalis muscle. It originates on the scutiform cartilage from which it passes caudad, curves around the base and medial surface of the auricle and inserts on the eminentia concha. It rotates the external ear mediad and caudad.

The **tragicus lateralis** shows well on the right side of the dissection in figure 2. It is sometimes called the mandibuloauricularis. This seems more meaningful a term since it does originate on the mandible between the condyloid and angular process and it inserts on the caudal margin of the tragus of the ear and on the concha just in back of the tragus. It rotates the ear outward and ventrad. On the left side of figure 2 the relationship of the tragicus medialis and lateralis is well illustrated.

The **adductor auris superior, medius** and **inferior** are shown here from a different point of view than on the next plate where their functions are mentioned. Study both views carefully.

The **frontoscutularis** is shown here on the right side in a quite natural position and on the left pulled laterad with the ear. As its name suggests it originates on the frontal bone at the supraorbital margin and inserts on the scutiform cartilage. It draws the ear craniad. It has fibers which mingle with many of the neighboring muscles.

Finally, the **epicranius** (*occipitofrontalis*) **muscle** is one which has a mass of muscle on the occipital bone (*occipitalis muscle*) which joins into a broad flat tendinous sheet, the **galea aponeurotica,** which covers the dorsal surface of the skull between the ears and the eyes and joins another muscle mass on the frontal and nasal bones (*frontalis muscle*). Its function is to move the integument on the dorsal side of the head.

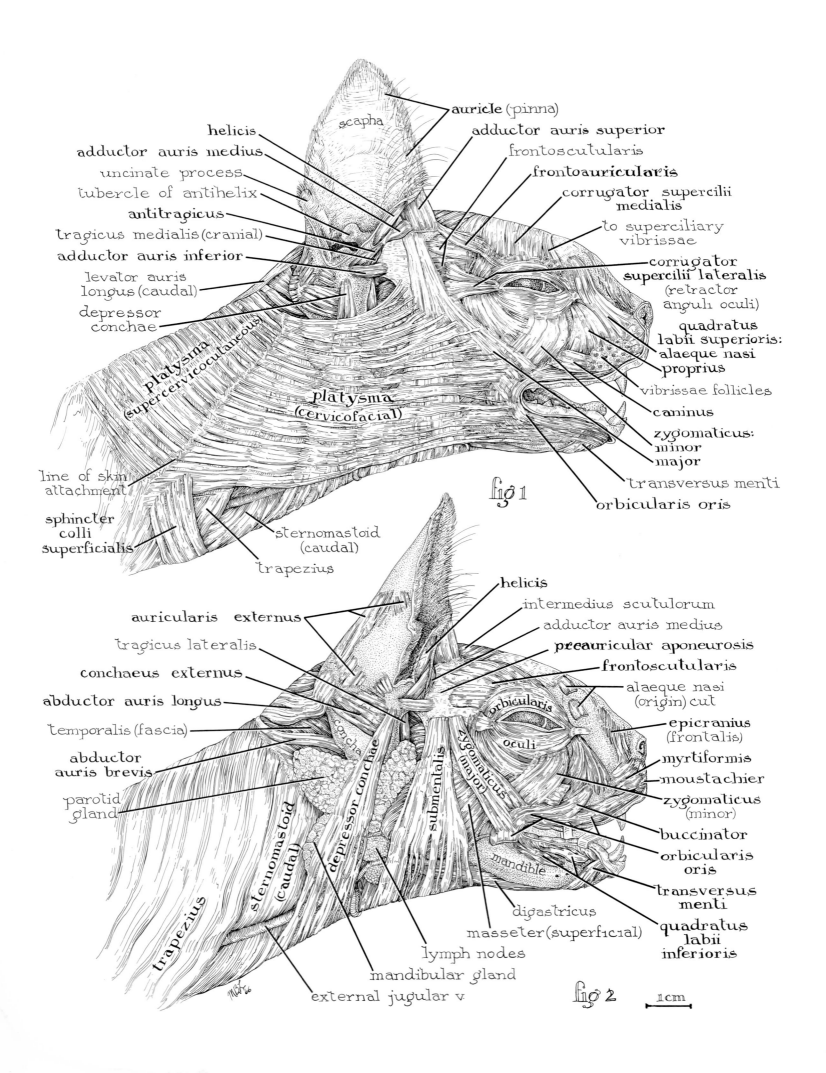

fig 1

auricle (pinna)
scapha
helicis
adductor auris medius
uncinate process
tubercle of antihelix
antitragicus
tragicus medialis (cranial)
adductor auris inferior
levator auris longus (caudal)
depressor conchae
adductor auris superior
frontoscutularis
frontoauricularis
corrugator supercilii medialis
to superciliary vibrissae
corrugator supercilii lateralis (retractor anguli oculi)
quadratus labii superioris: alaeque nasi proprius
vibrissae follicles
caninus
zygomaticus: minor major
platysma (supercervicocutaneous)
platysma (cervicofacial)
line of skin attachment
sphincter colli superficialis
sternomastoid (caudal)
trapezius
transversus menti
orbicularis oris

fig 2

helicis
auricularis externus
tragicus lateralis
conchaeus externus
abductor auris longus
temporalis (fascia)
abductor auris brevis
parotid gland
concha
intermedius scutulorum
adductor auris medius
precauricular aponeurosis
frontoscutularis
alaeque nasi (origin) cut
orbicularis oculi
epicranius (frontalis)
myrtiformis
moustachier
zygomaticus (minor)
buccinator
orbicularis oris
transversus menti
quadratus labii inferioris
zygomaticus (major)
submentalis
mandible
depressor conchae
sternomastoid (caudal)
trapezius
parotid gland
external jugular v.
mandibular gland
lymph nodes
masseter (superficial)
digastricus

1cm

The **platysma** muscle is shown here in its entirety. It is discussed in reference to Plate 27 and that description should be read as this plate is studied.

Three adductors of the ear are shown here, the **adductors auris, superior, medius** and **inferior.** The superior draws the auricle craniad, the medius pulls the concha dorsocraniad and the inferior pulls the ear craniodorsad.

The **helicus, antitragicus** and **tragicus medialis** all lie close to the external ear opening. The helicus draws the cranial margin of the auricle proximad; the antitragicus constricts the opening of the external auditory canal; the tragicus medialis flexes the concha.

Three eye muscles show well in figure 1, the **corrugator supercilii lateralis** which draws the angle of the eye caudad; the **corrugator supercilii medialis** which raises the upper eyelid; the **orbicularis oculi** (labeled in fig. 2) which closes the eye.

Four muscles of the mouth should be emphasized in figure 1. The **quadratus labii superioris** consists of two heads (1) the **alaeque nasi** from the side of the nose and (2) the **angular** (*levator labii superioris proprius*) from the cranial border of the eye. Both heads insert into the upper lip and whisker pad and hence elevate the lip and erect the whiskers or vibrissae. The **caninus** muscle retracts the vibrissae and raises the upper lip. The **orbicularis oris** closes the lips and is a pressor to the labial glands. The **zygomaticus major and minor;** the major is a prominent muscle to which many fibers of the platysma attach. It runs between the ear and the corner of the mouth so in action draws the corner of the mouth dorsocaudad and the external ear ventrocraniad. The zygomaticus minor is of irregular occurrence. Its action is to raise the angle of the mouth.

In figure 2 the platysma has been removed revealing the trapezius and sternomastoid (*caudal*) of the neck region and the depressor conchae, submentalis and zygomaticus major, superficial muscles of the head. The lymph nodes and salivary glands of the head-neck region show through between the depressor conchae and the submentalis. The zygomaticus major and the submentalis have been cut dorsally to reveal the fibrous preauricular aponeurosis craniad of the ear.

The **submentalis** and **depressor conchae** are both ear muscles which draw the external ear ventrad. The **conchaeus externus** constricts the concha. The **auricularis externus** flexes the auricular cartilage.

In the face region the quadratus labii superioris muscle and the whisker pad have been removed as indicated in figure 2. This uncovers the **myrtiformis** muscle which is a small dilator of the nares and elevator of the upper lip, and the **moustachier** which is said to "carry the lip craniad."

Removal of the ventral part of the orbicularis oris uncovers the **quadratus labii inferioris** which depresses the lower lip, and the **transversus menti** which stiffens the lower lip. By cutting away the caninus the **buccinator** muscle is shown which lies against the mucous membrane of the upper lip. It raises the lip and helps to return food from the vestibule to the chewing surfaces of the teeth.

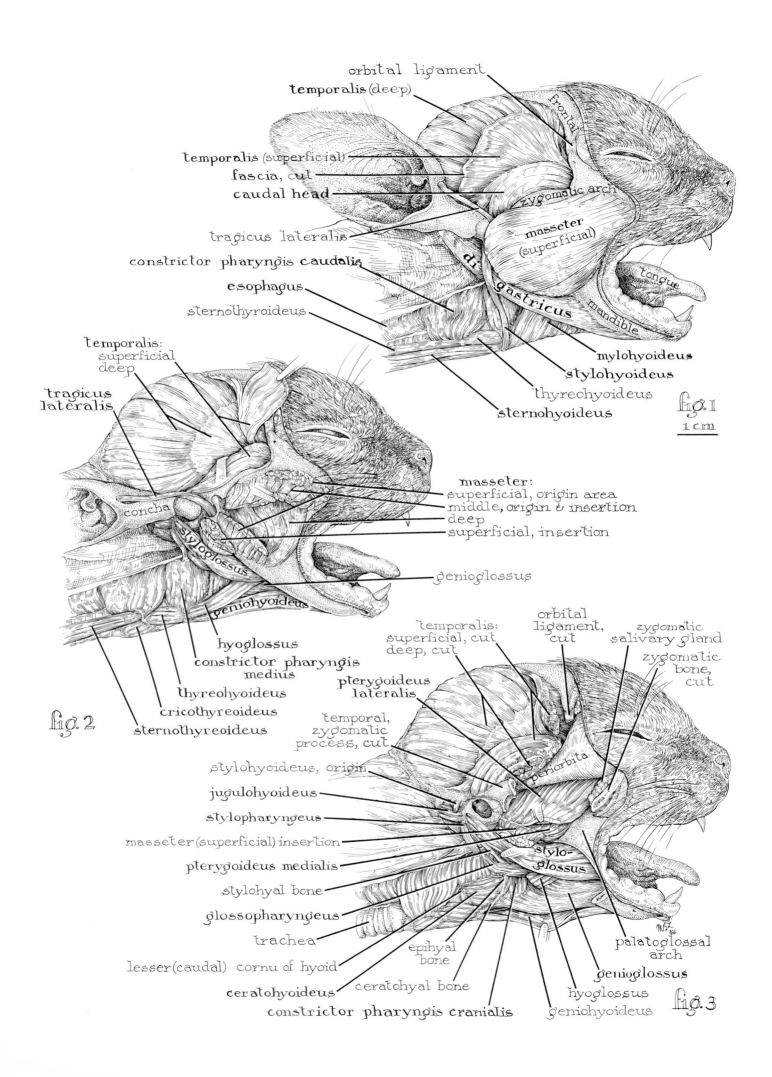

orbital ligament

temporalis (deep)

frontal

temporalis (superficial)

fascia, cut

caudal head

zygomatic arch

masseter (superficial)

tragicus lateralis

constrictor pharyngis caudalis

esophagus

sternothyroideus

digastricus

tongue

mandible

mylohyoideus

stylohyoideus

thyreohyoideus

sternohyoideus

fig. 1

1 cm

temporalis:
superficial
deep

tragicus
lateralis

concha

styloglossus

masseter:
superficial, origin area
middle, origin & insertion
deep
superficial, insertion

genioglossus

geniohyoideus

hyoglossus

constrictor pharyngis
medius

thyreohyoideus

cricothyreoideus

sternothyroideus

fig. 2

temporalis:
superficial, cut
deep, cut

orbital
ligament,
cut

zygomatic
salivary gland

zygomatic
bone,
cut

pterygoideus
lateralis

temporal,
zygomatic
process, cut

periorbita

stylohyoideus, origin

jugulohyoideus

stylopharyngeus

masseter (superficial) insertion

pterygoideus medialis

stylohyal bone

stylo-
glossus

glossopharyngeus

trachea

lesser (caudal) cornu of hyoid

ceratohyoideus

constrictor pharyngis cranialis

epihyal
bone

ceratohyal bone

palatoglossal
arch

genioglossus

hyoglossus

geniohyoideus

fig. 3

The three figures on this plate represent successive stages in the dissection of the deeper muscles of the head and neck. You are likely using the same head on which you studied the superficial muscles of the head and face. For our drawings we used "new" heads and left the skin on anteriorly for the areas with which we are now concerned lie caudad. In the last drawing of the superficial head muscles we had reached the superficial and deep temporal. These drawings go on from there to deeper muscles and to muscles in the throat region. More specifically, these figures deal with the five important muscles of mastication, muscles of the pharynx which are involved in deglutition, and muscle of the hyoid bone and larynx, some of which are involved directly or indirectly in these functions of chewing and swallowing.

The muscles of mastication are the digastricus, temporalis, masseter, and the lateral and medial pterygoideus. The **digastricus** is the important depressor of the mandible, *i.e.*, it opens the jaws. It extends all the way from the caudoventral part of the skull where it originates on the jugular and mastoid processes and the ridge between them. It passes medial to the parotid and mandibular salivary glands and inserts by fleshy attachment to the ventro-medial border of the mandible, some fibers reaching the mandibular symphysis (fig. 1 and Plate 25). In man this muscle has a central tendon and hence cranial and caudal bellies. Only slight indication of this is seen in the cat but its innervation by both the facial and trigeminal nerves suggests its dual origin.

The **temporalis** is the largest and strongest muscle of the head. As seen in Plate 22, figure 2, and in figures 1, 2 and 3 of this plate, it has a thin fascia covering the deep part of the muscle. Muscle fibers adhere to the inside of this fascia and insert to the outside of the coronoid process of the mandible. It constitutes the superficial part of the muscle. A part of the superficial temporal is seen as a semicircular group of fibers extending above the zygomatic arch. The massive deep temporalis originates in the extensive temporal fossa of the skull and inserts over the entire medial surface of the coronoid process of the mandible. It is a powerful elevator of the mandible.

The **masseter muscle,** shown in figures 1 and 2, is composed of three layers. It is covered externally with a strong fascia. All three layers originate from the zygomatic and the zygomatic process of the temporal which constitute the zygomatic arch as seen in figure 2. The superficial and middle layers insert mostly around the ventral border of the masseteric (*coronoid*) fossa of the mandible; the deep layer inserts into the fossa, mostly toward the caudal part, and it mingles with some of the fibers of insertion of the deep temporalis. It is a strong elevator of the mandible.

The **pterygoideus lateralis** and **medialis muscles** are shown in figure 3, medial to the masseter and temporalis muscles which have now been removed. The lateralis arises from the external pterygoid fossa and inserts on the medial surface of the mandible just below the condyloid process. It is an elevator of the jaw. The pterygoideus medialis arises from the internal pterygoid fossa and inserts partly with the external pterygoideus and partly near the angular process of the mandible. It, too, elevates the mandible. Some believe the action of the pterygoideus muscles causes some sidewise movement of the mandible. This appears to be the case at least in man.

The **hyoideus group** of muscles consists of six muscles all of which insert directly or indirectly into the hyoid bone, and most of them cause its movement. It might be well before studying these to refer back to Plate 12, figure 1, to review the parts of the hyoid bone. On figure 1 the following are easily seen: the **stylohyoideus,** raises the hyoid; the **mylohyoideus,** raises the floor of the mouth and pulls the hyoid forward; the **thyreohyoideus,** raises the larynx; and the **sternohyoideus,** draws the hyoid caudad. On figure 2 the mylohyoideus has been removed and the **geniohyoideus** is seen; a muscle which has its origin on the inner surface of the mandible near the symphysis and inserts onto the hyoid pulling it forward. On figure 3, due to removal of superficial structures, the **jugulohyoideus** is seen running from the jugular process over the outer surface of the tympanic bulla to the stylohyoid which it draws backward. The **ceratohyoideus** is also shown here running from the ceratohyal to the caudal cornu of the hyoid.

This plate also shows the three important extrinsic muscles of the tongue, *i.e.*, those which originate outside of the tongue but insert in it (fig. 3). They are the **stylo-glossus** which originates on the mastoid process and the stylohyal; the **hyoglossus** originating on the lateral sides of the body of the hyoid and on the ceratohyal; and the **genioglossus** arising on the inside of the mandibular symphis.

The tongue also has intrinsic muscles which lie entirely within the tongue (im. propria lingua). They are arranged in three sets, longitudinal, transverse and vertical. They are best studied in cross sections of the tongue. With so many muscles the tongue is very versatile in its movements.

A series of muscles is responsible for the movements of the pharynx in the function of swallowing or deglutition. They are all constrictors of the pharynx. The constrictor pharyngis cranialis also draws the pharynx forward. Three of them occur in series in a craniocaudal direction. They are the **constrictor pharyngis cranialis** (*pterygopharyngeus*) (fig. 3), the **constrictor pharyngis medius** (fig. 2), and the constrictor pharyngis caudalis (fig. 1). These all insert into the median dorsal raphe of the pharynx after encircling its walls.

Two other small pharyngeal muscles can be seen in figure 3. They are the **stylopharyngeus,** which forms a band under the middle constrictor and outside of the cranial constrictor of the pharynx and dilates and elevates the pharynx; and the **glossopharyngeus** whose fibers originate from the genioglossus and styloglossus muscles and insert into the median dorsal raphe of the pharynx.

These muscles may seem difficult but from the standpoint of dissection they are much easier than the facial muscles. Many can be separated by using a blunt probe.

A Ventral View of Some Trunk and
Extrinsic Appendicular Muscles
PLATE 25

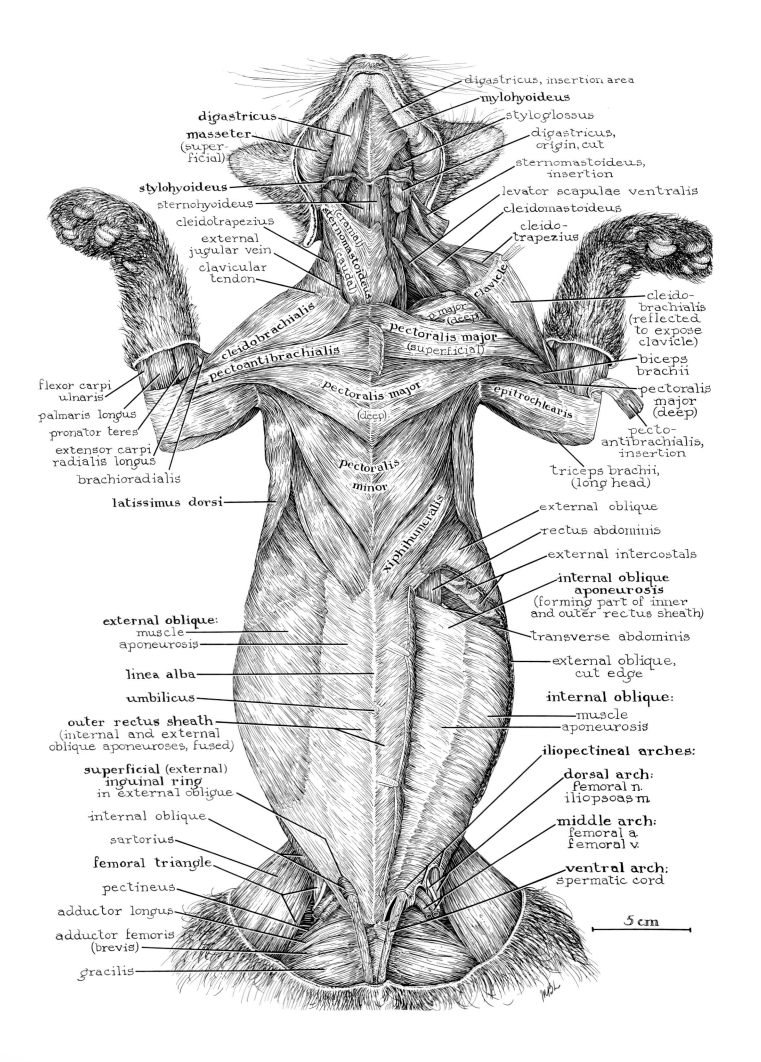

digastricus, insertion area
mylohyoideus
styloglossus
digastricus, origin, cut
sternomastoideus, insertion
levator scapulae ventralis
cleidomastoideus
cleido-trapezius
cleido-brachialis (reflected to expose clavicle)
biceps brachii
pectoralis major (deep)
pecto-antibrachialis, insertion
triceps brachii, (long head)
external oblique
rectus abdominis
external intercostals
internal oblique aponeurosis (forming part of inner and outer rectus sheath)
transverse abdominis
external oblique, cut edge
internal oblique:
muscle
aponeurosis
iliopectineal arches:
dorsal arch: femoral n. iliopsoas m.
middle arch: femoral a. femoral v.
ventral arch: spermatic cord

digastricus
masseter (super-ficial)
stylohyoideus
sternohyoideus
cleidotrapezius
external jugular vein
clavicular tendon
cleidobrachialis
pectoantibrachialis
flexor carpi ulnaris
palmaris longus
pronator teres
extensor carpi radialis longus
brachioradialis
latissimus dorsi
external oblique: muscle aponeurosis
linea alba
umbilicus
outer rectus sheath (internal and external oblique aponeuroses, fused)
superficial (external) inguinal ring in external oblique
internal oblique
sartorius
femoral triangle
pectineus
adductor longus
adductor femoris (brevis)
gracilis

(cranial) sternomastoideus (caudal)
p. major (deep)
pectoralis major (superficial)
clavicle
pectoralis major (deep)
epitrochlearis
pectoralis minor
xiphihumeralis

5 cm

Your cat's muscles can be made to appear almost as clear cut as these if you have carefully removed all of the superficial fascia. I say almost because here the edges of many of the muscles are shown slightly elevated or darkened for the sake of clarity. Also the integumentary muscles are, of course, removed.

If you have done the dissections of the head, you will see a number of familiar terms at the upper end of this drawing; muscles of mastication such as the masseter and digastricus, hyoid muscles as the sternohyoideus, mylohyoideus and stylohyoideus, and a tongue muscle, the styloglossus. Notice particularly the strap-like nature of the stylohyoideus as seen in this view.

All through the illustrations on this system and others there will be this repetition. It is done to help you to learn and to keep important relationships in mind. It is my hope that you will come through this study with a whole animal in mind.

On the animal's left side the dissection goes deeper than on the right. This procedure is regularly followed. It means separating muscles at their **deep fascial** coverings, for the deep fascia holds muscles together like the superficial fascia held the skin to the underlying muscles. This fascia is loose in some areas and makes dissection relatively easy; in others there is less of it or it may be tougher. It does allow muscles to contract without too much involvement of their neighbors. It means too that most of your dissection can be done with your fingers or a blunt probe. The scalpel is used largely in cutting across muscles.

Notice that except in the abdominal and neck region any trunk muscles that there may be are covered over by the extrinsic appendicular muscles. This is particularly true in the thoracic region. Trunk muscles are generally divided into muscles of the cervical, thoracic and lumbar vertebrae, muscles of the ventral and lateral thoracic walls, the diaphragm, muscles of the abdomen and of the tail. Obviously one must know the skeleton to understand the muscles.

Note the position of the clavicle, a bone of the pectoral or shoulder girdle but detached from the rest of the skeleton. Here it is shown on the left and its position is suggested on the right by its tendon. It lies between the **cleidobrachialis** and the **cleidotrapezius** and **cleidomastoideus.** The cleidomastoideus is covered by the **sternomastoideus** on the right side. The **levator scapula ventralis** is seen on the left side. It originates on the atlas and the exoccipital and inserts on the metacromion of the scapula. It pulls the scapula forward.

The most superficial muscle on the ventral thorax is the **pectoantibrachialis** which has its origin on the sternum, at least in part on the manubrium. It inserts on the ulna

and serves to draw the arm mediad, that is, to adduct the arm. On the left side the pectoantibrachialis has been cut and its center removed leaving its origin and insertion. The remaining appendicular muscles in this area are those of the pectoralis group, pectoralis major and minor and the xiphihumeralis which some consider a part of the pectoralis minor. The **pectoralis major** is divided into a narrow **superficial** band and a broader **deep** portion. The superficial portion has its origin on the manubrium and on a raphe in front of it, the deep portion from the manubrium and raphe and the first three divisions of the sternum. Both superficial and deep portions of the pectoralis major insert on the humerus, shown to greater advantage on the next plate.

The **pectoralis minor** originates from the first six divisions of the body of the sternum, sometimes to the xiphoid process. Its insertion is shown on the next plate.

The **xiphihumeralis** has its origin in a median raphe along the xiphoid process and usually by lateral extensions on to the sheath of the rectus abdominis muscle. It passes craniad, narrows into a thin tendon and attaches to the strong fascia called the bicipital arch to which the latissimus dorsi also attaches (See Plate 38, fig. 1). It inserts on the border of the bicipital groove of the humerus.

Note the thin, flat band of muscle on the medial side of the brachium (arm) the **epitrochlearis.** It originates on the outer surface of the ventral border of the latissimus dorsi and inserts on the olecranon process of the ulna. It tends to extend the forearm but is a weak muscle.

Three abdominal muscles of importance are shown here, the external oblique on the intact right side, its fibers extending caudoventrad from its origin, the internal oblique on the left side with its fibers running in a slightly cranioventral direction from its origin, and finally though mostly covered by its rectus sheath, the rectus abdominis runs laterad of the median line with its fibers in a craniocaudal direction.

The **external oblique** is a flat sheet of muscle which originates from the last nine or ten ribs and the lumbodorsal fascia (see Plate 30), and inserts into a broad aponeurosis which attaches to a median raphe in the thoracic region, the linea alba of the mid-abdominal wall to the symphysis pubis and the cranial border of the pubis. From the xiphoid process caudad the aponeurosis of the external oblique attaches to that of the internal oblique and forms the outer sheath of the rectus abdominis. An opening, the **superficial** (*external*) **inguinal ring** (*orifice*) penetrates the caudal border of the aponeurosis to each side of the median line. It is the orifice of the inguinal canal for the passage of the spermatic cord—or in the female, the round ligament.

The **internal oblique** muscle has a complex origin from (1) the lumbar fascia (*aponeurosis*) from the fourth to the seventh vertebrae, (2) the aponeurosis from the ventral half of the **iliac crest,** (3) from the **iliopectineal** (*crural*) **arches.** It inserts by an aponeurosis in the linea alba. Its complex relationship to the rectus muscle and the aponeuroses of the external oblique and transversalis muscles is shown in the next plate.

The **rectus abdominis muscle** originates on the pubic tubercle and inserts on the first and second costal cartilages and the sternum between the first and fourth costal cartilages.

The external and internal oblique and the rectus abdominis muscles are compressors of the abdomen and hence of the abdominal viscera. The external oblique may flex the vertebral column when acting with its mate or bend it laterally when one acts independently of the other.

Finally, note the iliopectineal arches and the structures which pass through these from the abdominal region to the thigh. There will be occasion to study them again later.

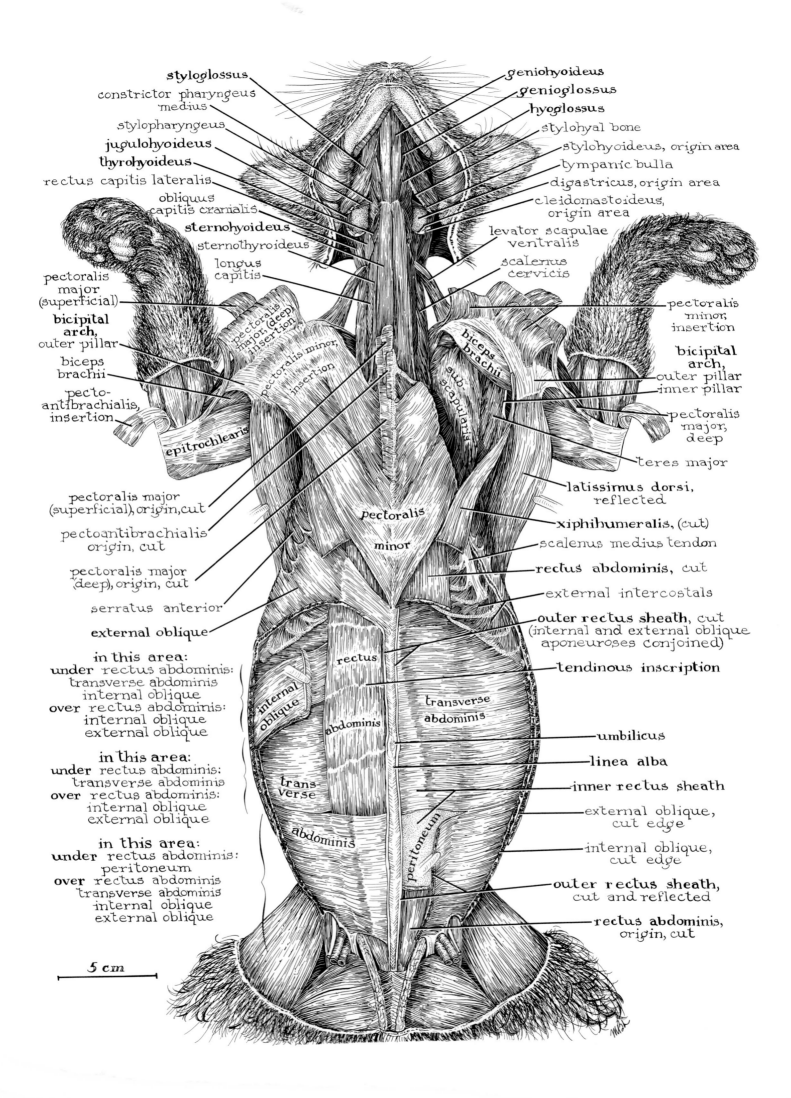

styloglossus

constrictor pharyngeus medius

stylopharyngeus

jugulohyoideus

thyrohyoideus

rectus capitis lateralis

obliquus capitis cranialis

sternohyoideus

sternothyroideus

longus capitis

pectoralis major (superficial)

bicipital arch, outer pillar

biceps brachii

pecto-antibrachialis, insertion

epitrochlearis

pectoralis major (deep), insertion

pectoralis minor, insertion

pectoralis major (superficial), origin, cut

pectoantibrachialis origin, cut

pectoralis major (deep), origin, cut

serratus anterior

external oblique

in this area:
under rectus abdominis:
transverse abdominis
internal oblique
over rectus abdominis:
internal oblique
external oblique

in this area:
under rectus abdominis:
transverse abdominis
over rectus abdominis:
internal oblique
external oblique

in this area:
under rectus abdominis:
peritoneum
over rectus abdominis
transverse abdominis
internal oblique
external oblique

internal oblique

transverse abdominis

rectus abdominis

pectoralis minor

rectus

transverse abdominis

peritoneum

5 cm

geniohyoideus

genioglossus

hyoglossus

stylohyal bone

stylohyoideus, origin area

tympanic bulla

digastricus, origin area

cleidomastoideus, origin area

levator scapulae ventralis

scalenus cervicis

biceps brachii

subscapularis

pectoralis minor, insertion

bicipital arch, outer pillar
inner pillar

pectoralis major, deep

teres major

latissimus dorsi, reflected

xiphihumeralis, (cut)

scalenus medius tendon

rectus abdominis, cut

external intercostals

outer rectus sheath, cut (internal and external oblique aponeuroses conjoined)

tendinous inscription

umbilicus

linea alba

inner rectus sheath

external oblique, cut edge

internal oblique, cut edge

outer rectus sheath, cut and reflected

rectus abdominis, origin, cut

Compare this plate with the last one to see what additional dissection has been done. Note that the sternomastoideus, cleidomastoideus and cleidotrapezius muscles have been removed to give a good view of the sternohyoideus and some of the deep muscles of the neck. The mylohyoideus, stylohyoideus and digastricus muscles have been dissected away in the head region revealing the **geniohyoideus, genioglossus,** and **hyoglossus muscles.** We get our first glimpse of some of the deep head muscles as the **capitis group.**

In the thoracic region the pectoantibrachialis and pectoralis major have been largely removed, but there has been enough left to show their origins and insertions. Indeed one sees the insertions of these muscles better here than in the previous plate. The **latissimus dorsi** is shown in relationship to the **bicipital arch** with which it and the xiphihumeralis become involved at their insertions. Here the arch is seen as a tendinous structure which arches over the **biceps brachii muscle.** Its outer pillar is formed in part from the tendons of the latissimus dorsi and pectoralis minor. Its inner pillar is from the conjoined tendons of the latissimus dorsi and the teres major. We will meet it again in Plate 38, figures 1 and 2.

You see a lot of work represented in this dissection of the abdominal region involving the removal of the very thin external and internal oblique muscles leaving only small parts of them as land marks. In doing this the even thinner **transverse abdominis muscle** has been exposed with its aponeurosis.

A small window has been prepared on the left side by removing a section of the rectus abdominis where it lies directly on the **peritoneum** lining the body cavity.

On the right side, the rectus abdominis is clarified by removing its sheath from a section of its ventral surface. The light lines crossing the muscle are called **tendinous inscriptions** and are indications that this muscle was segmented at some time in its evolutionary history.

The sheath of the rectus abdominis muscle is a very complicated structure. It is formed by the aponeuroses of the other three abdominal muscles and they do not produce a sheath which is the same throughout its length. It has three variations and these are indicated on the left side of the drawing, or to the right of the cat. To pick an example, you may want to know the condition of the sheath at the caudal end. Looking at the listings in that area we find that only the peritoneum lies under (*dorsal*) the rectus abdominis at that point, whereas the aponeuroses of all three of the abdominis muscles are over (*ventral*) it. In its cranial area the aponeuroses of the transverse abdominis and internal oblique are under it and the internal oblique and external oblique are over it. This immediately raises a question. How can the aponeurosis of the internal oblique be in two places—both over and under? Examine it carefully and you will notice that it splits in this area sending one sheet above and one below (Plate 32). Test your own skill in dissection by seeing if you can demonstrate these relationships of the sheath of the rectus abdominis in your cat.

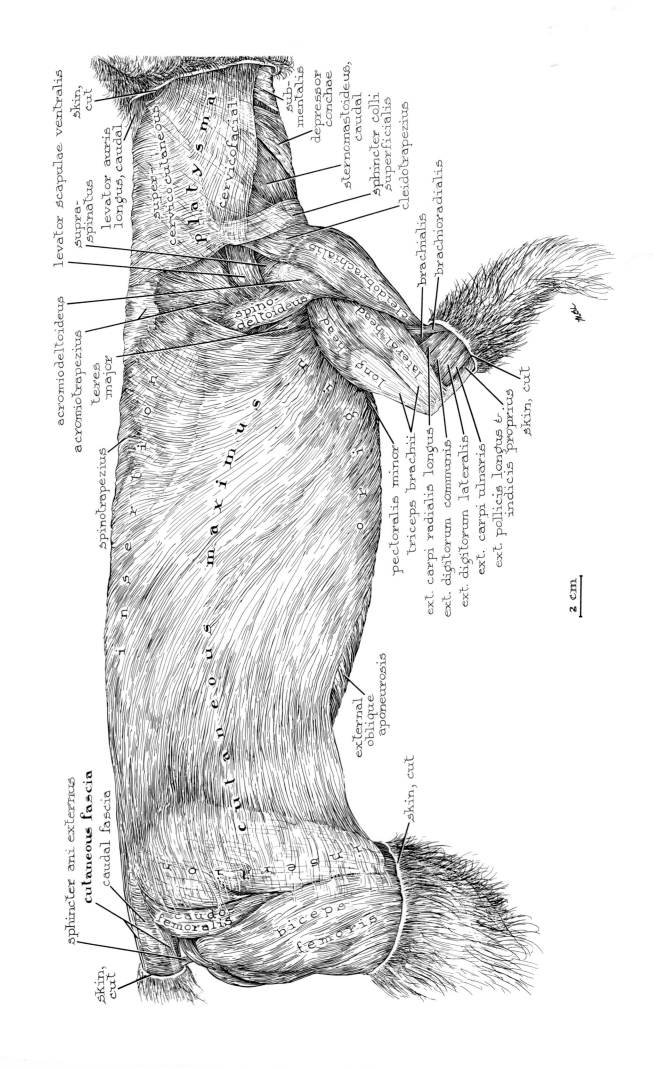

skin, cut

levator scapulae ventralis

super- cervicocutaneous

platysma

cervicofacial

sub- mentalis

depressor conchae

sternomastoideus, caudal

sphincter colli superficialis

cleidotrapezius

acromiodeltoideus

supra- spinatus

levator auris longus, caudal

acromiotrapezius

teres major

spino- deltoideus

cleidobrachialis

brachialis

brachioradialis

spinotrapezius

insertion

long head

lateral head

latissimus dorsi

inferior

pectoralis minor

triceps brachii

ext. carpi radialis longus

ext. digitorum communis

ext. digitorum lateralis

ext. carpi ulnaris

ext. pollicis longus & indicis proprius

skin, cut

cutaneous maximus

external oblique aponeurosis

sphincter ani externus

cutaneous fascia

caudal fascia

insertion

caudo- femoralis

biceps femoris

skin, cut

skin, cut

2 cm

This plate is the first in a series of plates of the lateral views of the musculature of the cat. They start with the one showing the integumentary muscles and each one will take the reader deeper into the musculature of the animal. They will emphasize trunk musculature and muscles which run from the trunk to the pectoral and pelvic girdles which support the limbs. The repetition of muscles on these plates is premeditated as a device for learning and for keeping relationships clear. The descriptions will not repeat to the same degree, but again it will be done where I feel emphasis is needed. No formal course in mammalian anatomy is expected to cover all of the material but it is made available for the sake of completeness, of challenge to the student who wishes to know more and for reference.

The integumentary muscles have their insertions on the skin and are used to shiver the skin, erect the fur, move the vibrissae and give expression to the face. While they are present in many vertebrates, they are best developed in mammals. They are usually partially removed in skinning the cat unless a special effort is made to save them. The two major integumentary muscles are shown in Plate 27, the **cutaneous maximus** or panniculus carnosus and the posterior parts of the **platysma.** The platysma extends on to the head where it has many subdivisions which move the cheeks, ears, lips, eyelids and other parts (Plates 22 to 24).

The **cutaneous maximus** extends over the thoracic and abdominal regions of the body and covers the underlying muscles somewhat like a veil. It forms a thin sheet anteriorly and is reduced posteriorly to strips of muscles which ordinarily adhere to the skin. It has its **origin** from the latissimus dorsi muscle at its ventral cranial end and from the bicipital arch in the axillary region (Plate 26) and from the linea alba and various points ventrally on the thorax. It **inserts** dorsad in the thoracic region with other fibers going caudad to the lumbar and sacral region and onto the dorsal and ventral sides of the tail. This muscle is degenerate in man.

The **platysma** muscle is found on the underside of the skin of the neck and head and will come off when skinning the cat if special care is not taken. At best some is lost with the skin. The head part of the platysma is not shown on this plate but can be seen on Plate 23. It originates in a fascia or the middorsal line from the occiput to the first thoracic vertebra. Some of the most cranial fibers arise beneath the levator auris longus. The fibers pass craniolaterad, the more dorsal ones curving around the ventral side of the ear and reaching the zygomaticus major, a few pass farther craniad to the corrugator supercilii lateralis and may reach the eyelid. The more ventral muscle fibers pass to the side of the face and become involved with the facial muscles, some passing as far forward as the pad which supports the vibrissae and the most ventral ones meeting the fibers of the muscle on the other side just ventrad of the symphysis of the mandible. Except for this point of contact the ventral borders of the muscles of the opposite sides are separated by a wedge-shaped area.

Note that the platysma muscle fibers are interrupted along a line on the side of the neck by an attachment to the skin. This line is between the base of the ear and the center of the cranial border of the scapula. The part of the platysma above this line is called by some authors (Strauss-Durckheim) the supercervicocutaneous, the part below the line of the cervicofacial.

Another band of muscle fibers of irregular occurrence, usually considered a part of the platysma, arises in the fascia in the side of the neck just craniad of the middle of the cranial border of the scapula. It passes caudoventrad to the manubrium crossing the fibers of the other part of the platysma. It is called the sphincter colli superficialis.

Except for the few fibers of the platysma which arise under the levator aurus longus the muscle is subcutaneous.

The other muscles shown on this plate are clearly labeled but will be discussed at a more appropriate time. They are not integumentary muscles.

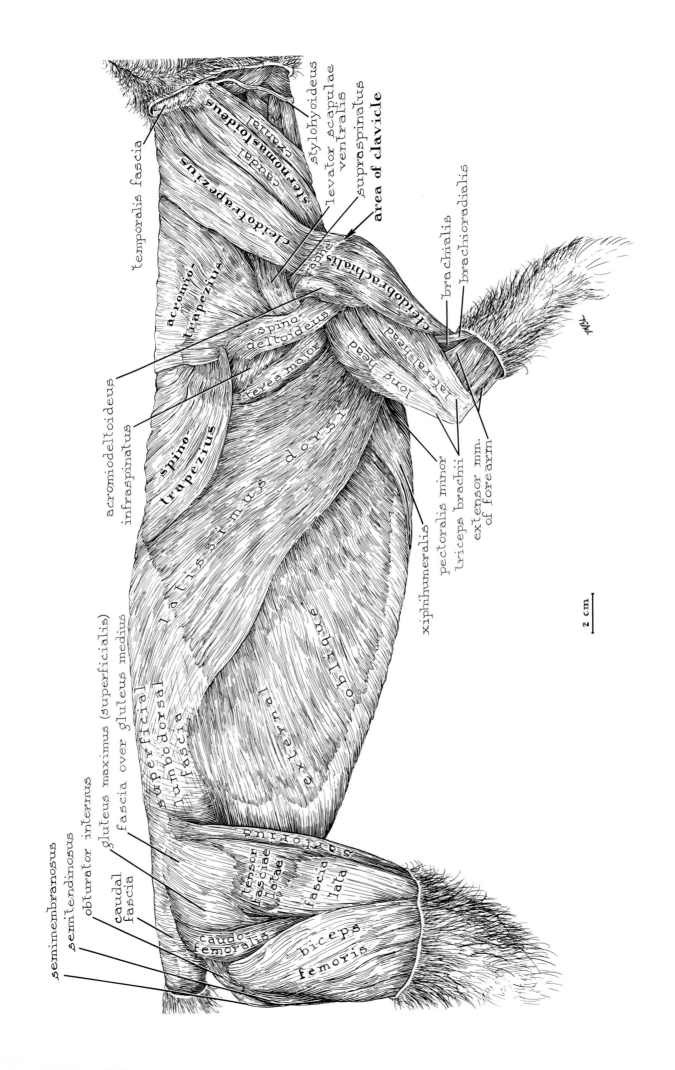

semimembranosus
semitendinosus
obturator internus
gluteus maximus (superficialis)
fascia over gluteus medius
caudal fascia

temporalis fascia
acromiodeltoideus
infraspinatus
spino trapezius
acromio-trapezius
spino deltoideus
teres major

superficial lumbodorsal fascia
latissimus dorsi
external oblique
lateral fascia
tensor fasciae latae
fascia lata
caudo femoralis
biceps femoris
sartorius

sternomastoideus
caudal sternomastoideus
cleidotrapezius
fascia
cleidobrachialis
spino deltoideus
long head
lateral head
teres major

stylohyoideus
levator scapulae ventralis
supraspinatus
area of clavicle
brachialis
brachioradialis

xiphihumeralis
pectoralis minor
triceps brachii
extensor mm. of forearm

2 cm

The superficial fascia and the integumentary muscles have been removed. This leaves the superficial muscles well exposed. One thing should be immediately apparent as one studies this plate. Most of the muscles you see have to do with the movement of the appendicular skeleton. Only two muscles of the trunk are in view, the **external oblique** and the **sternomastoideus.** The external oblique has been described in Plates 25 and 26 and its origin, insertion and relationships will gradually be uncovered again in this series of lateral views of muscles. The sterno-mastoideus is described in Plate 25.

Muscles obscuring the trunk muscles here are mostly those relating to the pectoral girdle and limb which have their origin on the trunk. They are the latissimus dorsi, a flat, triangular muscle that covers a broad area on the abdomen and thorax and the three trapezius muscles, the caudal one overlapping the latissimus dorsi and with the others covering the area from there to the skull.

The **latissimus dorsi muscle** has its origin in the superficial **lumbodorsal fascia** and from the tips of the spinous processes of thoracic and lumbar vertebrae from about the fifth thoracic to the sixth lumbar. It inserts through the bicipital arch into the mediad surface of the shaft of the humerus. It pulls the arm caudodorsad.

The trapezius muscles which together are considered by some to be homologous to the three parts of the human trapezius are the caudal spinotrapezius, the middle acromiotrapezius and the cranial cleidotrapezius.

The **spinotrapezius originates** from the spinous processes and the supraspinous ligaments of most of the thoracic vertebrae from about the fourth to the twelfth. It inserts on the tuberosity of the spine of the scapula and on the fascia of the supraspinatus and infraspinatus muscles as seen on the next plate. It draws the scapula dorsocaudad.

The **acromiotrapezius** lies craniad of the spinotrapezius and its origin extends from the spinous process of the axis to the region of the first to fourth thoracic vertebrae. Most of the tendon of origin is continuous over the midline with the muscle of the other side thus crossing the depression between the vertebral borders of the two scapulae. Cranially the muscle fibers approach closer to the midline narrowing the tendon or even reaching the midline at the cranial border. In this cranial region the tendon attaches by a fascia to the spinous processes of the cervical vertebrae. This muscle inserts on the metacromion and spine of the scapula to the tuberosity and for some distance on the musculotendinous junction of the spinotrapezius muscle. It, with its counterpart on the other side of the body, holds the vertebral borders of the scapulae together or pulls them closer.

The **cleidotrapezius** (*cleidocervicis*) originates from the medial half of the lambdoidal crest and the midline to the caudal end of the spine of the axis. It inserts into the clavicle and the raphe which extends laterally from the clavicle. It draws the scapula craniodorsad.

The **cleidobrachialis** (*cleidodeltoideus*) **muscle** perhaps should not be discussed at this point but since there are some things we want to see under it in the next plate, it is going to be removed. It lies on the cranial surface of the shoulder and appears as a direct continuation of the cleidotrapezius. Indeed some of its fibers are continuous with it and therefore some anatomists designate these two together the brachiocephalicus or the cephalobrachialis. Others consider it homologous to the clavicular portion of the human deltoideus—hence cleidodeltoideus. Its origin is on the clavicle and the raphe lateral to the clavicle. It narrows as it passes distad along the cranial surface of the arm to join the brachialis at its distal end and to insert with it by a flat tendon on the medial surface of the ulna distad of the semilunar notch. Since it is a flexor of the forearm, you wonder why we do not call it the cleidoantibrachialis. Plate 39, figure 2 shows you this insertion. These comments are meant to indicate that decisions about anatomical nomenclature are not always easy and this is part of the reason why we have international congresses on nomenclature—to try to develop a universal language of anatomy.

Although they will be studied in more detail later, you should note the muscles of the hindlimb as they are shown here in relationship to the whole body. The **gluteus muscles** are under the gluteal fascia. The **sartorius** shown on the cranial border of the leg; back of it the **tensor fasciae latae** muscle and its broad fascia lata; caudal to it the large **biceps femoris;** and dorsally between the biceps femoris and the gluteus maximus, the **caudofemoralis.** On the dorsal side around the tail is the **caudal fascia.**

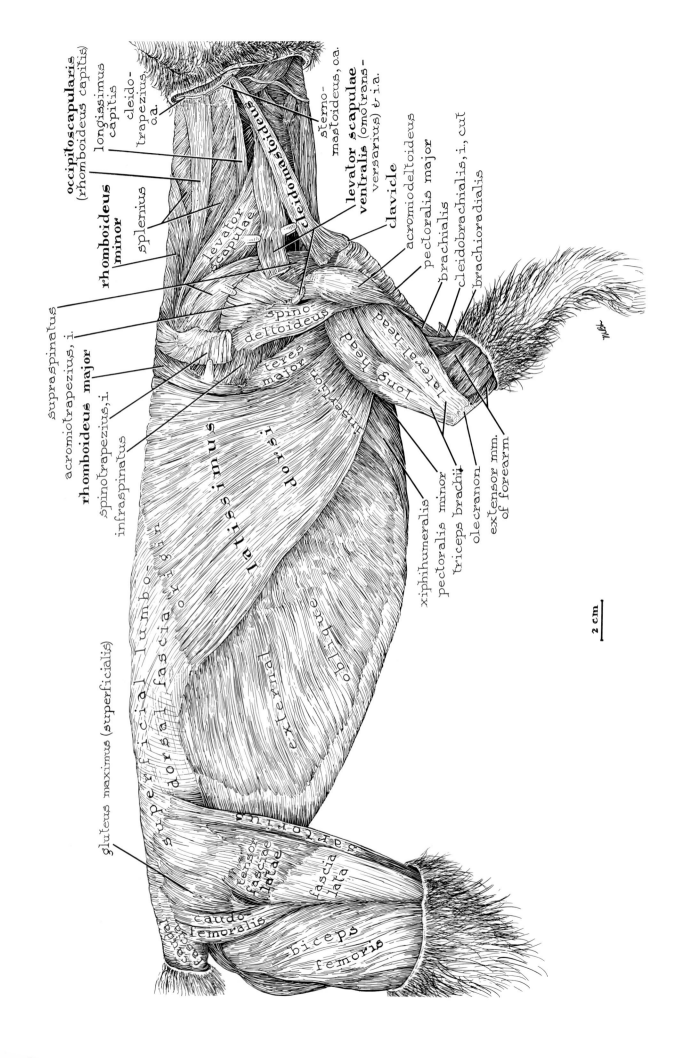

occipitoscapularis (rhomboideus capitis)

rhomboideus minor

longissimus capitis

cleido-trapezius, o.a.

sterno-mastoideus, o.a.

levator scapulae ventralis (omotrans-versarius) & i.a.

clavicle

acromiodeltoideus

pectoralis major

brachialis

cleidobrachialis, i, cut

brachioradialis

splenius

levator scapulae

cleidomastoideus

supraspinatus

acromiotrapezius, i

rhomboideus major

spinotrapezius, i

infraspinatus

spinodeltoideus

teres major

lateral head

long head

pronator teres

latissimus dorsi

xiphihumeralis

pectoralis minor

triceps brachii

olecranon

extensor mm. of forearm

superficial lumbo-dorsal fascia

latissimus

external oblique

gluteus maximus (superficialis)

gluteus medius

tensor fasciae latae

caudo femoralis

fascia lata

biceps femoris

2 cm

This looks much like the last plate but careful examination reveals a difference. With the exception of the latissimus dorsi which is now more fully exposed the muscles discussed with the last plate are for the most part removed, except where it was instructive to leave a bit of their origins and/or insertions. You will see more of this kind of procedure in the plates that follow and you should know that **o.a.** after a muscle name means **origin area** and that an **i.a.** means **insertion area.** Or, there may be just an **o** for **origin** or an **i** for **insertion.**

By the removal of the trapezius group of muscles and the cleidobrachialis, a whole new set of muscles has been made available for examination. If you study this carefully, you can make similar cuts on your cats and arrive at this result, but there will be fat and fascia between the muscle groups which you must remove. You may wonder about those things that look like little clips on some of the muscles. They represent the tips of blunt forceps and they indicate that something, a muscle in this case, is being lifted up or to the side so that you can see underneath, or it may be done to emphasize an origin or insertion.

Notice first that you can now see how far craniad the origin of the **latissimus dorsi** goes on the thorax. Just craniad of it is a narrow band of muscle, the **rhomboideus major,** and almost continuous with it craniad a broader mass, the **rhomboideus minor.** The nomenclature here should trouble you because the minor is larger than the major and probably we should just call the whole thing the **rhomboideus.** Most anatomist do. The reason for this confusion of terms is that there are two distinct muscles in this area in man, the **rhomboideus major** and **minor** and the major is larger than the minor. There has always been an influence from human anatomy carried over into mammalian anatomy generally, because human anatomy was studied in detail before there was much study of other mammals. The **rhomboideus** originates on the supraspinous ligament in the cervical region, from the first four vertebral spinous processes and from interspinous ligaments caudad of this. It inserts to the lateral side of the vertebral border of the scapula at the caudal angle (*the major, Plate 36, fig. 1*); to the medial surface of the vertebral border of the scapula craniad (*the minor, Plate 38, fig. 1*). Its functions are to pull the scapula craniad (*minor*), caudad (*major*) and dorsad (*major and minor*) by use of a part or all parts of the muscle.

The **occipitoscapularis muscle** (*rhomboideus capitis, levator scapulae dorsalis*) is presumably homologous to a part of the human rhomboideus and hence the name rhomboideus capitis sometimes applied to it. It has its origin on the medial half of the lambdoidal ridge beneath the clavotrapezius. It is long, slender and flat and runs

directly caudad and inserts into the cranial angle of the scapula or its tendon may join that of the levator scapulae. It rotates the scapula and draws it forward.

The **cleidomastoideus** has its origin on the mastoid process and its insertion on the clavicle and its lateral raphe. Its action is to pull the clavicle craniad. However, if the clavicle is fixed by action of other muscles, the cleidomastoideus may turn the head and depress the snout. This action is confusing if you define the insertion of a muscle as the end attached to the less stable and the origin to the more stable structure or even if you say that the origin is more proximal, the insertion more distal. The fact is that in cases like this origin and insertion would appear to be interchangeable.

The **levator scapulae** has its origin on the tubercles of the transverse processes of the cervical vertebrae from three through seven and from the ligaments between the tubercles. It inserts on the medial surface of the scapula near the cranial angle. It draws the scapula cranioventrad.

The **levator scapulae ventralis** (*omotransversarius*) is a strap-shaped muscle which originates by two parts, one from the transverse process of the atlas, the other from the ventral surface of the basioccipital bone of the skull. It inserts into the metacromion of the scapula and into the infraspinatus fossa. Its action is to pull the scapula craniad.

Two other muscles should be mentioned here, the spinodeltoideus and the acromiodeltoideus and as was suggested in the discussion of the previous plate maybe a third, the cleidobrachialis (*cleidodeltoideus*). But while the deltoid is a muscle of three parts in man, we are considering it to have only two parts represented in the cat. The **spinodeltoideus** originates on the middle portion of the caudal border of the spine of the scapula and by a tendinous raphe between the spinotrapezius and acromiotrapezius and inserts on the deltoid ridge of the humerus (Plate 37, fig. 1). It is a flexor and outward rotator of the humerus.

The **acromiodeltoideus muscle** originates on the caudal border of the acromion and inserts into the humerus partly with and partly adjacent to the spinodeltoideus. Its action is the same as that of the spinodeltoideus. Refer to Plate 36, figure 2 for a better view of the origins of these deltoideus muscles.

The **teres major muscle** is a thick muscle which lies parallel with the caudal border of the scapula. It has its origin on the upper one third of the caudal border of the scapula and from the fascia of the subscapularis and infraspinatus muscles. It inserts by a tendon common to it and the latissimus dorsi on the medial surface of the shaft of the humerus near the pectoral ridge. It is a rotator and flexor of the humerus.

83

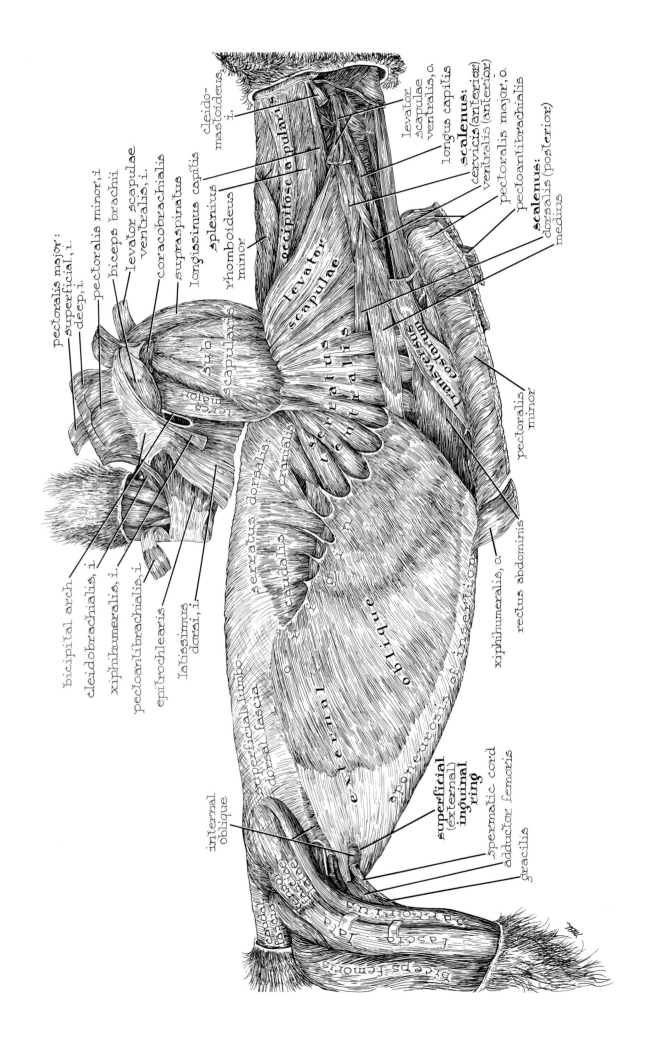

pectoralis major:
superficial, i.
deep, i.
pectoralis minor, i
biceps brachii
levator scapulae
ventralis, i.
coracobrachialis
supraspinatus
longissimus capitis
splenius
rhomboideus minor

cleido-
mastoideus, i.

occipitoscapularis

levator scapulae

serratus ventralis

transversus

bicipital arch
cleidobrachialis, i.
xiphihumeralis, i.
pectoantibrachialis, i.
epitrochlearis
latissimus dorsi, i.

subscapularis
teres major

levator scapulae ventralis, o.
longus capitis
scalenus:
cervicis (anterior)
ventralis (anterior)
pectoralis major, o.
pectoantibrachialis
scalenus:
dorsalis (posterior)
medius

pectoralis minor

serratus dorsalis
cranialis

caudalis

superficial lumbo-
dorsal fascia

rectus abdominis

xiphihumeralis, o.

internal oblique

external oblique

aponeurosis of insertion

superficial (external)
inguinal ring
spermatic cord
adductor femoris
gracilis

caudo-
femoralis
fascia lata

sartorius

biceps femoris

2 cm

The shoulder has been lifted upward in this illustration to show the serratus ventralis and to better advantage the levator scapulae and the origin of the external oblique. The transversus costorum is shown by pulling aside the origins of the pectoralis muscles and the complex scalenus muscle is now quite fully uncovered.

The **external oblique muscle** is shown here in its entirety and its origin should be studied carefully. It arises by tendons from the last nine or ten ribs. These tendons form arches over the slips of the serratus ventralis and serratus dorsalis muscles and to these tendons the muscle fibers are attached. It also takes origin in part from the superficial lumbodorsal fascia. Note that its insertion is not at any point on the ilium as it is in man and there is no inguinal (Poupart's) ligament. The muscles of the hindlimb have been pulled back to show the **superficial** (*external*) **inguinal ring** through which passes the spermatic cord. It penetrates the aponeurosis of the external oblique muscle.

The **serratus ventralis** muscle is easily recognized by its slips or "serrations" each of which originates on a rib, from the first to the ninth or tenth. They converge and join to insert on to the medial surface of the scapula just below the vertebral border and in line cranially with the insertion of the **levator scapulae.** Its function is to support the trunk and to carry it forward and backward. It also aids in inspiration.

The **transversus costarum** is a small rectangular muscle which lies ventrad to the tendon of the rectus abdominis. Its origin is by a tendon from the side of the sternum between the third and sixth ribs. It inserts on the first rib and the lateral portion of its costal cartilage. It draws the sternum forward and may aid in inspiration.

The **scalenus** is a complex muscle composed of many interconnected parts some of which have been considered as separate muscles. It extends from the ventrolateral side of the thorax into the ventral neck region. It is shown on this plate without much dissection of the muscle itself.

Only the fascia has been cleaned away. On the next plate it is further dissected so it may be profitable as you read about it to refer to both plates. The **dorsal portion** has its origin in a very slender flat tendon from the middle of the outer surface of the fourth rib. It extends craniad and emerges from between the second and third slip of the serratus ventralis muscle. The **middle portion** of the scalenus is the longest and has its origin by thin tendons from the ventral ends of ribs six through nine. These all join into a common flat tendon to which the muscle fibers attach. It passes craniad and joins with the other parts of the muscle just craniad of the first rib. One head of the **ventral scalenus** arises from the cartilages of the second and third ribs by tiny tendons which are partly united with the transversus costarum. They join into a narrow muscle band which unites with the other two portions cranial to the first rib. The second part of the ventral scalenus is the **cervical portion** which has a large number of small bundles of muscle fibers which arise from the first rib and the transverse processes of the last six cervical vertebrae and the first thoracic. This portion is shown separated out in the next illustration, Plate 31. The insertion of the scalenus is on the trasverse processes of all of the cervical vertebrae. Its action is to flex the neck when both work together; to bend it sideward when acting unilaterally; and as an aid in inspiration when the neck is fixed.

A few other features of this plate should be noticed although they will be studied and emphasized later. With the scapula and arm lifted into the position it is here you can see the muscles that cover its medial surface—the subscapularis, teres major and at the cranial border, the supraspinatus. Also this shows very well the relationship of the tendon of the xiphihumeralis muscle to the latissimus dorsi and the bicipital arch.

Caudad and dorsad of the serratus ventralis some slips of the serratus dorsalis can be seen. Most of the cranial portion of this muscle is medial to the serratus ventralis.

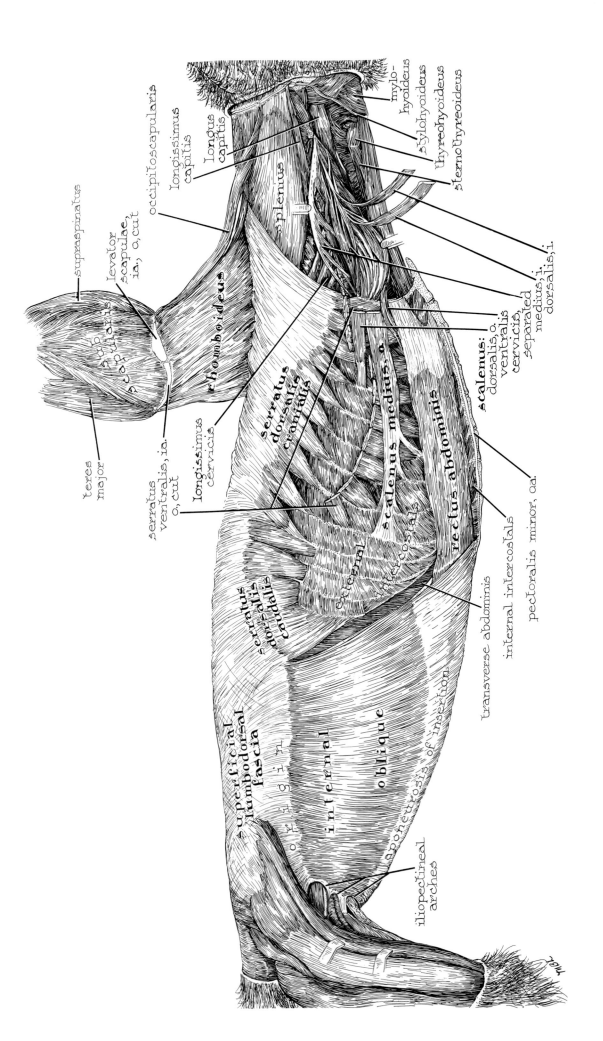

supraspinatus

spinous scapulae

teres major

levator scapulae, ia.) o, cut

serratus ventralis, ia. o, cut

longissimus cervicis

rhomboideus

serratus dorsalis cranialis

occipitoscapularis

longissimus capitis

longus capitis

splenius

mylo- hyoideus

stylohyoideus

thyreohyoideus

sternothyreoideus

separated dorsalis, i. medius, i. dorsalis, i.

scalenus: dorsalis, o. ventralis cervicis,

scalenus medius, a.

rectus abdominis

serratus dorsalis caudalis

sternal

intercostals

transverse abdominis

internal intercostals

pectoralis minor, aa.

origin superficial lumbodorsal fascia

internal

oblique

aponeurosis of insertion

iliopectineal arches

2 cm

In this illustration the external oblique and serratus ventralis have been removed. With the serratus ventralis gone there is nothing left to hold the pectoral girdle and forelimb to the body except the **rhomboideus muscles.** It should be evident by now that the pectoral girdle and forelimb have no articulation with the rest of the skeletal system. In man the only articulation is between the clavicle and the sternum but in the cat the clavicle is diminished in size and "floats" in between muscles. The serratus ventralis and levator scapula act as a cradle in which the trunk is suspended. The rhomboideus and trapezius muscles, the acromiotrapezius in particular, hold the scapulae together dorsally. These are some of the reasons why a cat can jump to the ground from a considerable height, land on the front legs and go uninjured.

The serratus dorsalis cranialis and caudalis muscles are now visible throughout their extent. The **serratus dorsalis cranialis** has its origin in a broad aponeurosis which starts in the median dorsal raphe which extends from the axial spinous process to the spinous process of the tenth thoracic vertebra. This aponeurosis narrows as it extends in a ventral and caudal direction to the fleshy portion of the muscle which inserts by fleshy slips into the outer surfaces of the first nine ribs just ventrad of their angles. Its action is to lift the ribs upward for inspiration.

The **serratus dorsalis caudalis** lies caudad of the serratus dorsalis cranialis. Its origin is in the lumbar spinous processes and the interspinous ligaments by means of an aponeurosis. Its fibers pass cranioventrad to insert by four or five heads into the last four or five ribs. Its action is to draw the last four or five ribs caudally as an aid to expiration.

By removal of the transversus costarum muscle the cranial end of the rectus abdominis muscle is seen with its tendon attaching to the first and second costal cartilages.

The **internal oblique muscle** is shown here in its entirety from its origin on the **superficial lumbodorsal fascia** to its insertion in a ventral aponeurosis which as seen in Plate 26 forms a part of the sheath of the rectus abdominis muscle.

Note that the muscles of the pelvic girdle and hindlimb have been pulled back to reveal the three iliopectineal (*crural*) arches.

Finally this is an opportunity to review the **scalenus** muscle which was quite completely dissected and is well illustrated here. Refer to Plate 30 for the description.

The splenius muscle is partially revealed by lifting up the rhomboideus and occipitoscapularis.

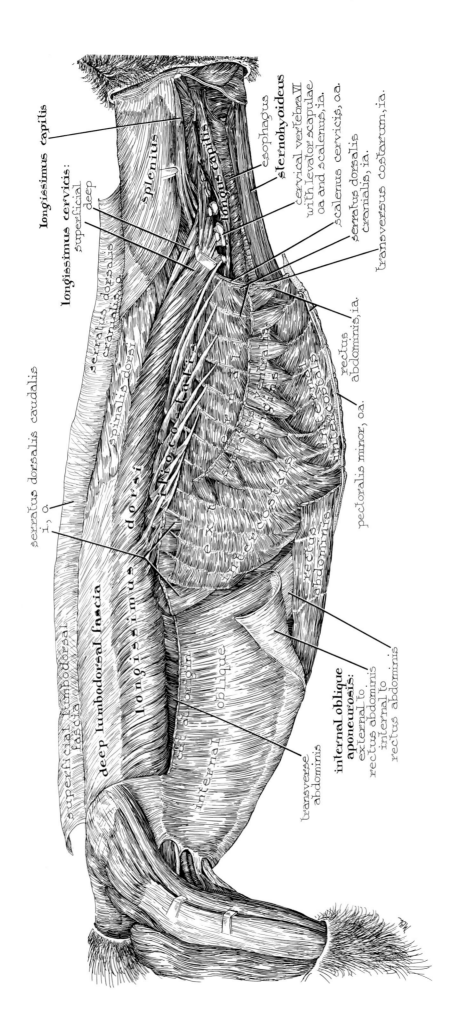

longissimus capitis

longissimus cervicis:
superficial
deep

serratus dorsalis cranialis, a

splenius

esophagus

sternohyoideus

cervical vertebra VI
with levator scapulae
oa and scalenus, ia.

scalenus cervicis, oa.

serratus dorsalis
cranialis, ia.

longus capitis

transversus costarum, ia.

serratus dorsalis caudalis

spinalis dorsi

longissimus dorsi

serratus ventralis

intercostals

external oblique origin

internal oblique

rectus abdominis

intercostales

rectus abdominis, ia.

pectoralis minor, oa.

rectus abdominis

superficial lumbodorsal
fascia

deep lumbodorsal fascia

i, a.

transverse
abdominis

internal oblique
aponeurosis:
external to
rectus abdominis
internal to
rectus abdominis

2 cm

The superficial lumbodorsal fascia and the aponeuroses of the serratus dorsalis cranialis and caudalis have been cut to the right of the midline and lifted to the left uncovering the deep lumbodorsal fascia, the spinalis dorsi and the rest of the splenius muscle. The longissimus dorsi has also been exposed and just below it and above the internal oblique, a small part of the transverse abdominis can be seen. A part of the **aponeurosis of the internal oblique** is shown to demonstrate how, at this point, it divides sending one leaf to the internal surface of the rectus abdominis, the other to the external surface thus contributing to the rectus sheath.

The **longus capitis** is a narrow muscle lying on the ventrolateral surface of the cervical vertebrae and with its fellow on the opposite side forms a trough for the esophagus. It has its origin by five or six heads on the margins of the transverse processes of the second to the sixth vertebrae. It unites with the levator scapulae ventralis at its insertion into the body of the occipital bone medial to the tympanic bulla. It serves to depress the snout.

The **sternohyoideus** with its counterpart lie close together on either side of the median line of the neck. They originate from the costal cartilage of the first rib and from the manubrium of the sternum and insert into the body of the hyoid, the basihyal. The sternothyroideus muscle is very closely associated with the sternohyoideus dorsally and caudally.

With the consideration of the **longissimus dorsi** we are getting an introduction to a complicated group of muscles connected with the vertebral column which are not as fully differentiated into distinct or specific muscles as those of the limbs with which we will be concerned or even with other muscles of the trunk. These begin caudally as a large longitudinal mass of muscle lying between the transverse processes of the vertebrae and the spinous processes. This mass contains fibers which run in various directions and make a variety of connections; to the sacrum, the spinous, transverse, accessory or articular processes of vertebrae; to the vertebral arches, the os coxae and farther craniad where more differentiation of distinct muscles develops to the ribs and to the head. This whole mass, at least the relatively undifferentiated caudal part is often called the **common dorsal extensor** of the vertebral column. Muscles which separate from this group have other functions than extension. The **longissimus dorsi** covered by the deep lumbodorsal fascia is the larger part of this mass. At the caudad end of the thorax this muscle mass gives rise to a lateral muscle part of which becomes connected with the ribs and extends as far forward as the last cervical vertebrae. It is known as the **iliocostalis** as seen on this plate. The rest of the longissimus dorsi continues forward medial to the iliocostalis and into the neck region. Another strip becomes separate from it at about the level of the eighth or ninth thoracic vertebrae and becomes the **spinalis dorsi.** It is an extensor of the vertebral column. In the neck region the poorly separated **longissimus cervicis** is a part of this complex. It is inserted into the cervical transverse processes. And finally the **longissimus capitis** is a well-differentiated cranial extension of the longissimus dorsi. It is a slender muscle lying close against the lateral border of the splenius having its origin by five tendons on the anterior articular processes of cervical vertebrae four through eight. The tendons give rise to muscular slips which in turn unite into a flat belly which inserts by a strong tendon into the mastoid process often closely united with that of the splenius. It is a lateral flexor of the head.

The **splenius** is a flat broad muscle on the side of the neck having its origin from the cervical ligament and cervical fascia and inserts by a broad flat tendon into the entire lambdoidal ridge. The lateral part of the tendon may be fused with that of the longissimus capitis. It is an elevator of the head when it works bilaterally, otherwise it is a lateral flexor of the head.

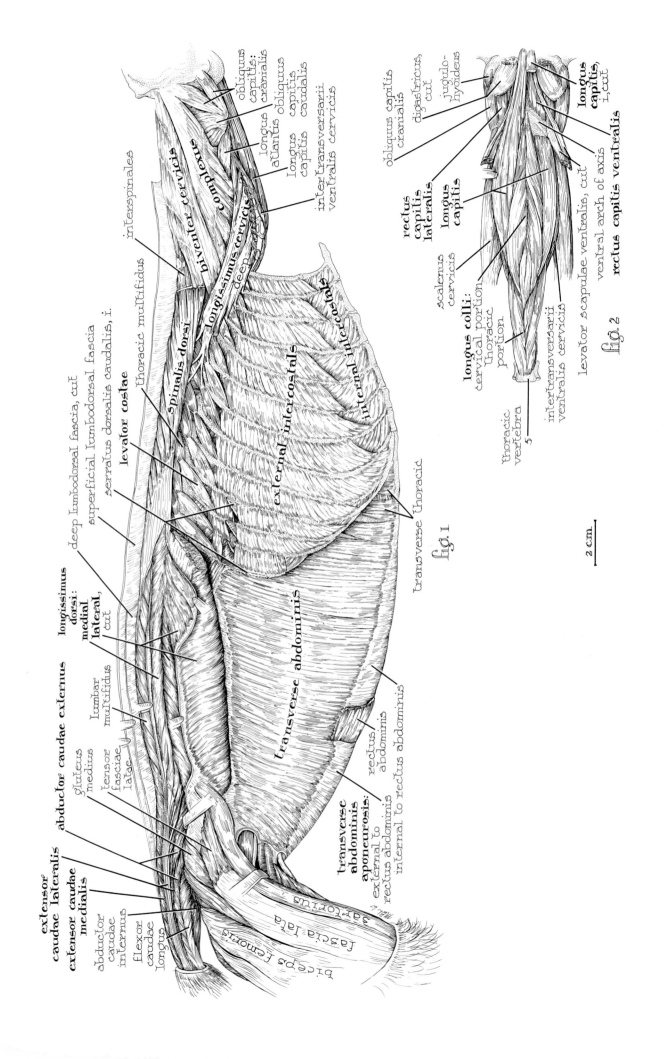

fig. 1

extensor caudae lateralis
extensor caudae medialis
abductor caudae externus
longissimus dorsi: medial / lateral, cut
gluteus medius
tensor fasciae latae
abductor caudae internus
Lumbar multifidus
flexor caudae longus
biceps femoris
fascia lata
sartorius

obliquus capitis: cranialis
obliquus capitis caudalis
longus atlantis
longus capitis
intertransversarii ventralis cervicis
biventer cervicis
complexus
longissimus cervicis
deep
interspinales
thoracic multifidus
Spinalis dorsi
deep lumbodorsal fascia, cut
superficial lumbodorsal fascia
serratus dorsalis caudalis, i.
levator costae
external intercostals
internal intercostals
transverse thoracic
transverse abdominis
rectus abdominis
transverse abdominis aponeurosis: external to rectus abdominis / internal to rectus abdominis

fig. 2

obliquus capitis cranialis
digastricus, cut
jugulo-hyoideus
longus capitis, i, cut
rectus capitis lateralis
longus capitis
scalenus cervicis
longus colli: cervical portion / thoracic portion
thoracic vertebra
5
intertransversarii ventralis cervicis
levator scapulae ventralis, cut
ventral arch of axis
rectus capitis ventralis

2 cm

This plate shows the deepest muscle of the ventral abdominal wall, the **transverse abdominis,** and its relationship to the rectus abdominus. It is an extremely thin muscle whose fibers assume a transverse direction and whose aponeurosis forms a part of the sheath of the rectus abdominis. As illustrated the aponeurosis is on the outside of the rectus muscle caudad but craniad it is internal. The transverse abdominis is continuous craniad with the transverse thoracic which lies inside of the thorax except at its caudal end.

Both external and internal intercostals are shown clearly and one should note the direction of their fibers between the ribs.

In the dorsal cervical region the splenius muscle has been removed to uncover the **biventer cervicis** and **complexus.** These muscles originate by slips from the spinous, cranial articular and transverse processes of cervical and thoracic vertebrae as far back as the third thoracic. They insert into the medial third of the lambdoidal crest of the occipital bone; the biventer medial to the complexus. By their contractions they raise the head and neck or unilaterally flex the head and neck laterally.

The deep **longissimus cervicis** is shown with its slips of origin in the thoracic region separated out. Its insertion is on the cervical transverse processes. It lies in the angle between thoracic and cervical vertebrae in such a relationship as to extend the neck when the muscles of both sides contract. In unilateral action the neck is raised obliquely and turned to one side.

By cutting away most of the thoracic portion of the longissimus dorsi and the iliocostalis a series of twelve spindle-shaped muscles can be seen, the **levator costae.** They have their origins on the transverse processes of the first to the twelfth thoracic vertebrae and insert, after running ventrocaudad to the angle of the rib next caudad from the second to the thirteenth. They are often considered as modified parts of the external intercostals. They pull the ribs craniodorsad and thus aid in inspiration.

In the lumbar and sacral regions the deep lumbar fascia has been cut to reveal the medial and lateral portions of the longissimus dorsi and the multifidis which lies medial to the medial longissimus. Note that the multifidus passes into the tail as the extensor caudi medialis, while the medial part of the longissimus dorsi extends into the tail as the extensor caudi lateralis.

The **extensor caudi medialis** muscles lie to either side of the midline, their medial borders touching one another. They originate on the spinous processes of the sacral and the first caudal vertebrae. They pass caudad to insert on the dorsal surface and the articular processes of the caudal vertebrae. They serve to extend or raise the tail and to some degree move it laterad—lateral flexion.

The **extensor caudi lateralis** lies just laterad of the extensor caudi medialis and is a continuation of the medial portion of the longissimus dorsi. It has its origin on the articular processes of the sacral vertebrae and the transverse processes of the coccygeal vertebrae. Its insertion is by many long, slender tendons on the dorsal surfaces of the caudal vertebrae. The contraction of these muscles extends or raises the tail, while acting unilaterally the tail is flexed laterally. Study these muscles also on Plate 35.

Figure 2 of this plate is a ventral view of the cranial thoracic and cervical regions to detail the muscles of this area. It should be compared to the lateral view of the cervical region in figure 1 and to Plate 34, figure 3 in which some of the same muscles are shown. The **longus colli** shows to special advantage in figure 2. Its thoracic portion extends caudad to thoracic vertebrae five or six and it is made up of three well-defined separate bundles which originate on the ventral surfaces of these vertebrae. They insert into the costal process of the sixth cervical vertebra and possibly on the transverse process of the seventh. The cervical portion is made up of bundles which have their origin on the transverse processes and the sides of the ventral surfaces of the sixth to the third cervical vertebrae and their insertions on the midline of the centra as far forward as the tubercle on the ventral arch of the atlas. The muscles of the two sides coming together on the midline form V-shaped structures with the open end of the V caudad. The last V caudad embraces the cranial part of the thoracic portion of the longus colli. These muscles function to draw the neck downward.

The **longus capitis** muscles lie one on either side of the longus colli and with it form a trough on the ventral side of the neck in which lie the pharynx, esophagus and trachea. The muscles originate on the transverse processes of cervical vertebrae two through six. They unite into a common belly which inserts on the basioccipital between the tympanic bulla and the midline. They draw the neck and head downward; the latter by flexing the atlantooccipital joint.

The left **rectus capitis ventralis** muscle lies dorsad of the longus capitis. Its origin is on the ventral arch of the atlas, its insertion is on the basioccipital caudad of the insertion of the longus capitis. It crosses the atlantooccipital joint which it flexes.

The **rectus capitis lateralis** is seen on figure 2 as a small muscle originating on the ventral surface of the transverse process or "wing" of the atlas. It crosses the atlantooccipital joint to insert in the fossa lateral to the occipital condyle and on the base of the jugular process. It flexes the head laterad.

Finally, take note of the relationships of these muscles to others shown on figure 2.

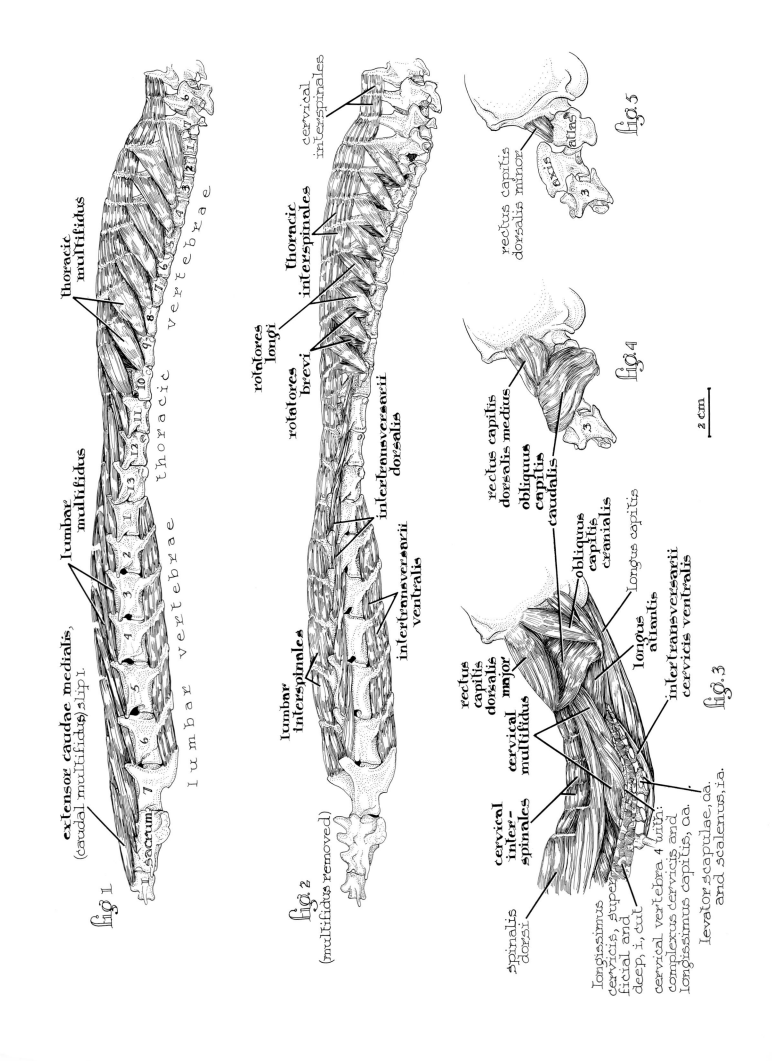

fig. 1

extensor caudae medialis, (caudal multifidus) slip 1.

lumbar multifidus

thoracic multifidus

lumbar vertebrae   thoracic vertebrae

sacrum 7 6 5 4 3 2 1 13 12 11 10 9 8 7 6 5 4 3 2 1

fig. 2
(multifidus removed)

lumbar interspinales

rotatores longi

rotatores brevi

thoracic interspinales

cervical interspinales

intertransversarii dorsalis

intertransversarii ventralis

fig. 3

spinalis dorsi

cervical inter- spinales

rectus capitis dorsalis major

cervical multifidus

obliquus capitis cranialis

rectus capitis dorsalis medius

obliquus capitis caudalis

longus capitis

longus atlantis

intertransversarii cervicis ventralis

longissimus cervicis, superficial and deep, i, cut

cervical vertebra 4 with: complexus cervicis and longissimus capitis, aa.

levator scapulae, aa. and scalenus, i.a.

fig. 4

fig. 5

rectus capitis dorsalis minor

axis atlas 3

2 cm

To uncover the muscles which are illustrated in this plate provides a real challenge to the dissector. Yet, except for the lumbar multifidus in figure 1, the drawings are not idealized, *i.e.* they look pretty much as you actually find them when you have removed the fat and fascia.

Figure 1 emphasizes the lumbar and thoracic portions of the multifidus muscle and its extension into the tail, the extensor caudae medialis or caudal multifidus. The cervical multifidus is shown in figure 3. As a whole the muscle can be said to consist of bundles of fibers originating on the transverse and other lateral processes and parts of the vertebrae and inserting, after passing craniodorsad over one or more vertebrae, on the spinous processes. The muscle is thickest in the **lumbar region** and it is difficult to distinguish separate bundles in the interwoven mass of fibers. They are shown more clearly in figure 1 than you are apt to see them in your cat. The **thoracic multifidus** shows much more separate bands of muscle which are also more distinct and generally pass over only one vertebra between their origin and insertion. Their insertions on the spinous processes are not as near the tips of the processes as is the case with the lumbar multifidus. The **extensor caudae medialis** was described in connection with Plate 33 and is shown again in Plate 35, figure 1. Thoracic and cervical interspinales and intertransversarii are shown in figure 1, but are not labeled.

Figure 2 shows the lumbar, thoracic and cervical interspinales; the dorsal and ventral intertransversarii and the rotatores longi and brevi.

The **interspinales,** as the name indicates, run between the contiguous edges of the spinous processes. They are best developed in the lumbar region but until the lumbar multifidus is removed they are not visible. In the thoracic and cervical regions they are visible above the multifidus. They function as fixators of the vertebral column.

The **intertransversarii** of the lumbar and thoracic region are shown in figure 2. The **dorsal group** is made up of muscle slips which run between the mammillary processes of the seventh lumbar to the twelfth or thirteenth thoracic vertebrae and the accessory processes of the fifth lumbar to the ninth thoracic. The components of the **ventral group** of intertransversarii lie between the transverse processes of the lumbar region. These muscles when contracting unilaterally cause lateral flexion of the vertebral column, when working together they stabilize the column.

The rotatores longi and brevi are confined to the cranial thoracic region. Although Reighard and Jennings, third edition by Elliott, report no "deepest layer of the multifidus" forming "what is sometimes distinguished as the Mm. rotatores," the muscles in figure 2 are found upon removal of the thoracic multifidus. I interpret them to be **Mm. rotatores.**

The **rotatores longi** form a series of eight flat slips of muscle extending between the transverse and spinous processes of two alternate vertebrae. The most caudal slip extends from the transverse process of the tenth thoracic vertebra to the spinous process of the eighth thoracic vertebra. The most cranial muscle slip of the series extends from the transverse process of the third thoracic to the spinous process of the first thoracic vertebra.

The **rotatores brevi** lie deep to the rotatores longi and extend between vertebrae. Eight are seen in figure 2, the first extending between the transverse process of the third thoracic vertebra and the spinous process of the second thoracic vertebra. The eighth one occupies a similar situation between thoracic vertebrae ten and nine. When these muscles act unilaterally, they are, as the name indicates, rotators of the cranial portion of the thoracic vertebral column around the longitudinal axis. When working bilaterally, they cause fixation of the column.

Figures 3 to 5 show some of the skeletal, articular and muscular features which provide the animal with such versatility of movement in the cervical and head regions. Recall the manner in which the first two cervical vertebrae are modified; the dens of the axis over which the ring-like atlas fits and around which it rotates; the atlantooccipital articulation where the rounded condyles of the occipital fit into the concave cranial articulating surfaces of the atlas. Modifications of the interspinales muscles in this area may be the source of the three rectus capitis muscles, the dorsalis major, dorsalis medius and dorsalis minor. These run between atlas, axis and occiput to provide the motive force to move the articulations in various ways. Two obliquus muscles; the obliquus capitis caudalis and cranialis, derived by modification of the multifidus or of the intertransversariis give a greater range of action.

The **rectus capitis dorsalis major** originates on the spine of the axis and inserts on the median part of the occiput. The right and left muscles are joined by a raphe where they meet on the middorsal line. The **rectus capitis dorsalis medius** lies beneath the rectus capitis dorsalis major (fig. 4). It arises cranially on the spine of the axis and its fibers diverge over the atlas to insert on the occipital bone ventrad of the median half of the lambdoidal crest. The **rectus capitis dorsalis minor** (fig. 5) is a short flat muscle lying between the atlas and occiput on the capsule of the atlantooccipital joint. It arises on the cranial border of the dorsal arch of the atlas and inserts on the occipital bone above the foramen magnum and just below the insertion of the dorsalis medius. These three muscles are all extensors of the atlantooccipital joint and by so doing, they lift the snout of the animal.

The **obliquus capitis caudalis** has its origin on the whole lateral surface of the spine of the axis from which its fibers pass obliquely craniolaterad to insert on the lateral margin of transverse process of the atlas (fig. 3 and 4). When acting bilaterally, it fixes the atlantoaxial joint; acting unilaterally it rotates the atlas and hence, the head on the dens of the axis. The **obliquus capitis cranialis** is divided into two parts. It originates on the dorsal surface and lateral border of the transverse process of the atlas. Its fibers pass craniad and diverge to insert into the caudal side of the mastoid process and into a line (*dorsal nuchal line*) parallel with the lambdoidal ridge and ventral to it. It extends the atlantooccipital joint and flexes the head laterally in unilateral action.

The **longus atlantis** (fig. 3) is differentiated from the longissimus dorsi. It originates on the transverse process and the side of the vertebral arch of the third cervical and inserts into the caudolateral angle of the transverse process of the atlas. It extends the neck and turns the head sideways.

The **intertransversarii cervicis ventralis,** as seen here, extends from the transverse process of the first thoracic vertebra to the transverse process of the second cervical touching each of the transverse processes of the intermediate vertebrae. It is closely related to the scalenus lying ventral to it and dorsal to the longus colli.

Note also the **cervical interspinales** and the **cervical multifidus.** In the latter, in contrast to the multiple slips of the thoracic multifidus, the bands of fibers are interconnected to form a fairly well-defined muscle. It has its origin on the articular processes of the last five cervical vertebrae and inserts into the spinous processes of the cervical vertebrae up to the atlas. Its action is to extend the back. The cervical multifidus is sometimes called the semispinales cervicis.

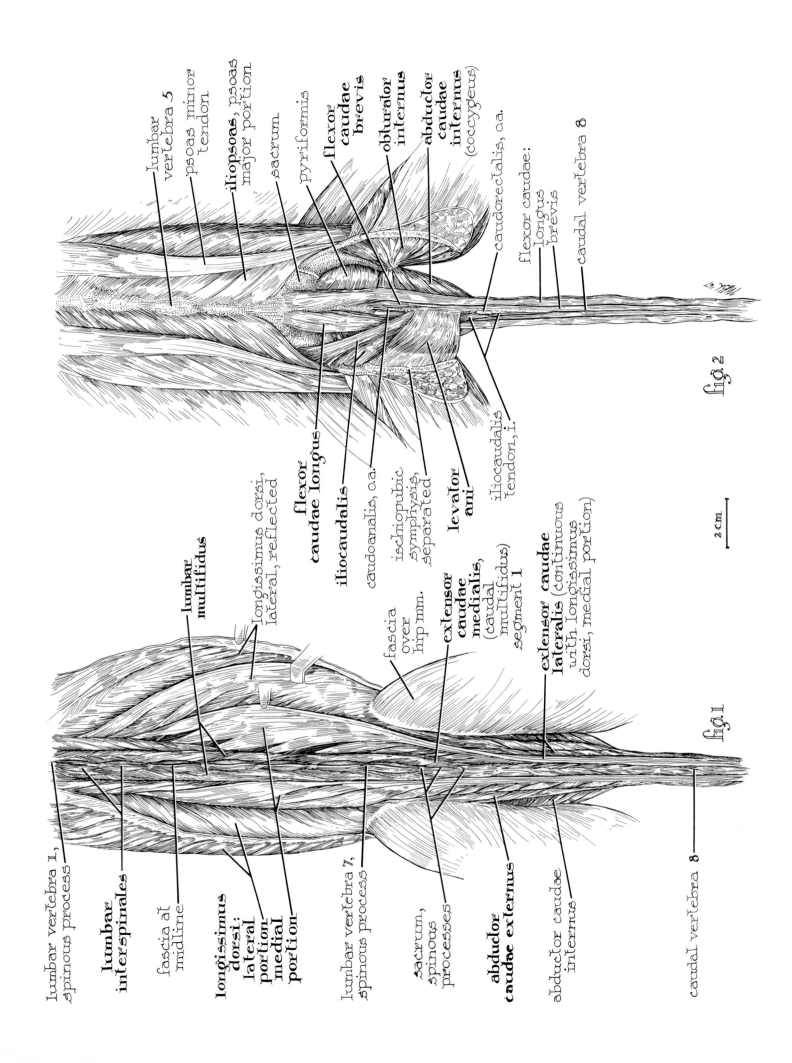

fig. 2

lumbar vertebra 5

psoas minor tendon

iliopsoas, psoas major portion

sacrum

pyriformis

flexor caudae brevis

obturator internus

abductor caudae internus (coccygeus)

caudorectalis, o.a.

flexor caudae: longus brevis

caudal vertebra 8

flexor caudae longus

iliocaudalis

caudoanalis, o.a.

ischiopubic symphysis, separated

levator ani

iliocaudalis tendon, i.

2 cm

fig. 1

lumbar multifidus

longissimus dorsi, lateral, reflected

fascia over hip mm.

extensor caudae medialis, (caudal multifidus) segment 1

extensor caudae lateralis (continuous with longissimus dorsi, medial portion)

lumbar vertebra 1, spinous process

lumbar interspinales

fascia at midline

longissimus dorsi; lateral portion medial portion

lumbar vertebra 7, spinous process

sacrum, spinous processes

abductor caudae externus

abductor caudae internus

caudal vertebra 8

Figure 1 shows the dorsal muscles of the vertebral column from the region of the first lumbar to the eighth caudal vertebrae. The ilia, the dorsal bones of the pelvic girdle appear on each side of the sacrum. They are covered with muscle over which lies a fascia. The muscles illustrated here are some which we have already studied from a lateral view. Their paired arrangement and relationship to each other are better visualized from the dorsal side. The superficial and deep lumbodorsal fasciae have been removed and the **lateral and medial portions** of the **longissimus dorsi** have been separated and spread laterally on the right side. This also reveals to better advantage the **lumbar multifidus.** Note again the **extensor caudae medialis** (*sacrococcygeus dorsalis medialis*), an extension of the multifidus into the tail and the **extensor caudae lateralis** (*sacrococcygeus dorsalis lateralis*), a continuation of the medial portion of the longissimus dorsi. The **abductor caudae externus** is a muscle having its origin from the dorsal surface of the sacrum and from the medial side of the dorsal border of the ilium. It runs caudad lying below the edge of the extensor caudae lateralis and inserts on the transverse processes and lateral surfaces of the caudal vertebrae as far back as the eighth or ninth. It abducts the tail, *i.e.* moves it laterally or away from the median line.

The **abductor caudae internus** (*coccygeus*) seen best on figure 2 lies ventral to the abductor caudae externus and dorsal to the levator ani. It is a large flat muscle with its origin on the spine of the ischium. Its fibers extend dorsomediad and fan out to insert into the transverse processes of the second to the fourth caudal vertebrae. It draws the tail laterally away from the median line of the body and forms a part of the pelvic diaphragm.

Other tail muscles shown in figure 2 are the flexor caudae longus and brevis and the iliocaudalis. The **flexor caudi longus** (*sacrococcygeus ventralis lateralis*) originates on the ventral surface of the last lumbar vertebra, the ventral surface of the sacrum and the transverse processes of the cranial caudal vertebrae. It terminates in long slender tendons which insert along the ventral surface of the tail. It bends the tail downward.

The **flexor caudi brevis** (*sacrococcygeus ventralis medialis*) originates on the ventral surface of the caudal vertebrae from the first to about the eighth. It may attach also to the last sacral vertebra. The insertions of each of its various slips insert into the ventral surface of a caudal vertebra some distance caudad. The muscle terminates at about the level of the tenth caudal vertebra. The two muscles lying in contact on the midline of the tail form a groove for the coccygeal artery.

The **iliocaudalis** relates closely to the levator ani and is sometimes considered to be a part of it. It originates on the ventral half of the medial surface of the ilium. Its fibers after passing caudad insert by a flat tendon into the ventral surface of the caudal vertebrae as far as the seventh or eighth.

By removal of the levator ani, and the iliocaudalis on the animals left side the pyriformis, obturator internus and abductor caudae internus (*coccygeus*) are uncovered. The pyriformis is described in reference to Plate 41; the **obturator internus** in relation to Plate 41; the **abductor caudae internus** above.

Note also in the lower part of this plate the origin area of the caudorectalis and the point of insertion of the tendon of the iliocaudalis. The origin area of the caudoanalis is also shown (See Plate 64, figs. 1 and 2).

At the top of the plate the psoas minor and the **iliopsoas** are shown. They have been considered in relationship to other plates. Their cranial portions are shown in Plates 79 and 80.

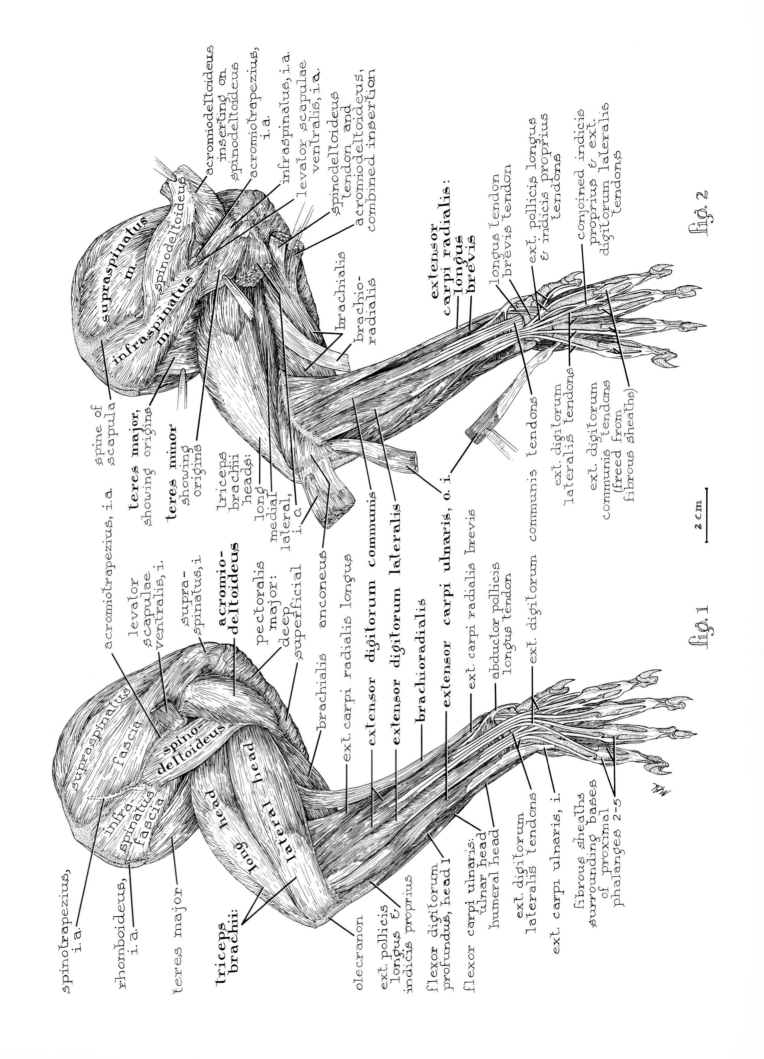

supraspinatus m.

spinodeltoideus

infraspinatus m.

acromiodeltoideus inserting on spinodeltoideus

acromiotrapezius, i.a.

infraspinatus, i.a.

levator scapulae ventralis, i.a.

spinodeltoideus tendon and acromiodeltoideus, combined insertion

brachialis

brachio-radialis

extensor carpi radialis:
longus
brevis

longus tendon
brevis tendon

ext. pollicis longus & indicis proprius tendons

conjoined indicis proprius & ext. digitorum lateralis tendons

fig. 2

spine of scapula

teres major, showing origins

teres minor showing origins

triceps brachii heads:
long
medial
lateral, i.o.

extensor digitorum communis

extensor digitorum lateralis

extensor carpi ulnaris, o. i.

ext. digitorum lateralis tendons

ext. digitorum communis tendons (freed from fibrous sheaths)

fig. 1

2 cm

spinotrapezius, i.a.

rhomboideus, i.a.

acromiotrapezius, i.a.

levator scapulae ventralis, i.

supra-spinatus, i

acromio-deltoideus

pectoralis major:
deep
superficial

brachialis

teres major

ext. carpi radialis longus

anconeus

brachioradialis

extensor carpi ulnaris, o. i.

ext. carpi radialis brevis

abductor pollicis longus tendon

ext. digitorum communis tendons

supraspinatus fascia

infra-spinatus fascia

spino-deltoideus

long head

lateral head

triceps brachii:
long head
lateral head

olecranon

ext. pollicis longus & indicis proprius

flexor digitorum profundus, head 1

flexor carpi ulnaris:
ulnar head
humeral head

ext. digitorum lateralis tendons

ext. carpi ulnaris, i.

fibrous sheaths surrounding bases of proximal phalanges 2-5

The muscles by which the pectoral girdle and forelimb are related to the trunk have been examined. We are now concerned with the muscles which move the limb in relationship to the scapula and those which work entirely within the limb, *i.e.* are intrinsic to the limb.

Figure 1 shows the superficial muscles from the lateral view. The spinodeltoideus and acromiodeltoideus have already been discussed in reference to Plate 29.

The **supraspinatus** and **infraspinatus** muscles originate in the fossae of the lateral surface of the scapula which go by similar names, the supraspinous and infraspinous fossae (fig. 2). The supraspinatus passes by a strong muscular cord over the capsule of the shoulder joint to insert into the great tubercle (*tuberosity*) of the humerus. The infraspinatus sends a flat tendon over the capsule of the shoulder joint into the ventral half of the infraspinous facet on the greater tubercle of the humerus. The supraspinatus is an extensor of the shoulder and advances the limb forward. The infraspinatus is an outward rotator and abductor of the humerus.

The **teres major** has been discussed but its origin is shown very clearly on figure 2. The **teres minor** (fig. 2) is a small muscle originating on the caudal border of the scapula and sometimes attached to the infraspinatus and to the long head of the triceps. It inserts into the greater tubercle of the humerus and works with the infraspinatus in outward rotation of the humerus.

The **triceps brachii** is the largest muscle of the brachium or arm. It has three heads, the lateral, long and medial and the medial in turn has three parts, accessory, intermediate and short. All parts have their origins on various areas of the brachium as indicated in Plates 37 and 39. Only the **long head** has an origin on the glenoid border of the scapula. All heads and parts insert at various points on the olecranon process of the ulna. The triceps is a powerful extensor of the forearm.

In your dissection of the forearm of the cat you will be experiencing an interesting study of fascia. The muscles of the forearm are covered with a strong antebrachial fascia consisting of a superficial layer, a continuation of the general subcutaneous fascia of the arm and a deep layer which is a dense tendinous sheet closely applied to the muscles. Study this complex fascia as carefully as time permits. Note especially that at the wrist the fascia attaches to the longitudinal ridges on the dorsal surface of the head of the radius and thus bridges the intervening grooves, holding the tendons in place. Similarly on the ventral side a transverse ligament forms between the distal ends of the radius and ulna to hold the flexor tendons in place. On the fingers the fascia forms tendon sheaths for the passage of the flexor tendons and strong ring-like annular ligaments form in these sheaths near the base and head of the proximal phalanges to hold the tendons in place.

The muscles emphasized here are the extensor and supinator muscles of the dorsal side of the antebrachium.

The brachioradialis is a ribbon-like muscle lying just beneath the superficial fascia on the lateral side of the brachium and on the ventral and lateral side of the antebrachium. As is indicated by its name it connects the humerus and radius. It is a supinator of the hand.

The **extensor carpi radialis longus and brevis**, the **extensor digitorum communis and lateralis** all have their origins on the lateral supracondyloid ridge of the humerus. The **extensor carpi ulnaris** has its origin on the lateral epicondyle of the humerus. It also receives a small slip from an origin on the ulna. The insertions of these muscles are well indicated by their names as are their actions as extensors of the hand and the extensors of digits; the communis to the phalanges of the second through the fifth digits and the lateralis to the same digits with a conjoined tendon with the indicis proprius. The extensor carpi radialis longus is inserted on the dorsal surface of the second metacarpal; the extensor carpi radialis brevis to the dorsal surface of the base of the third metacarpal (Plate 37, fig. 2). The extensor carpi ulnaris inserts into the tubercle on the ulnar side of the base of the fifth metacarpal.

Note that this plate, as do most, offers a great deal of information not mentioned in the discussion. Many insertion areas of muscles studied are clearly shown here. Also, muscles which are to be discussed on the next plate are shown, where possible, to place them in proper perspective.

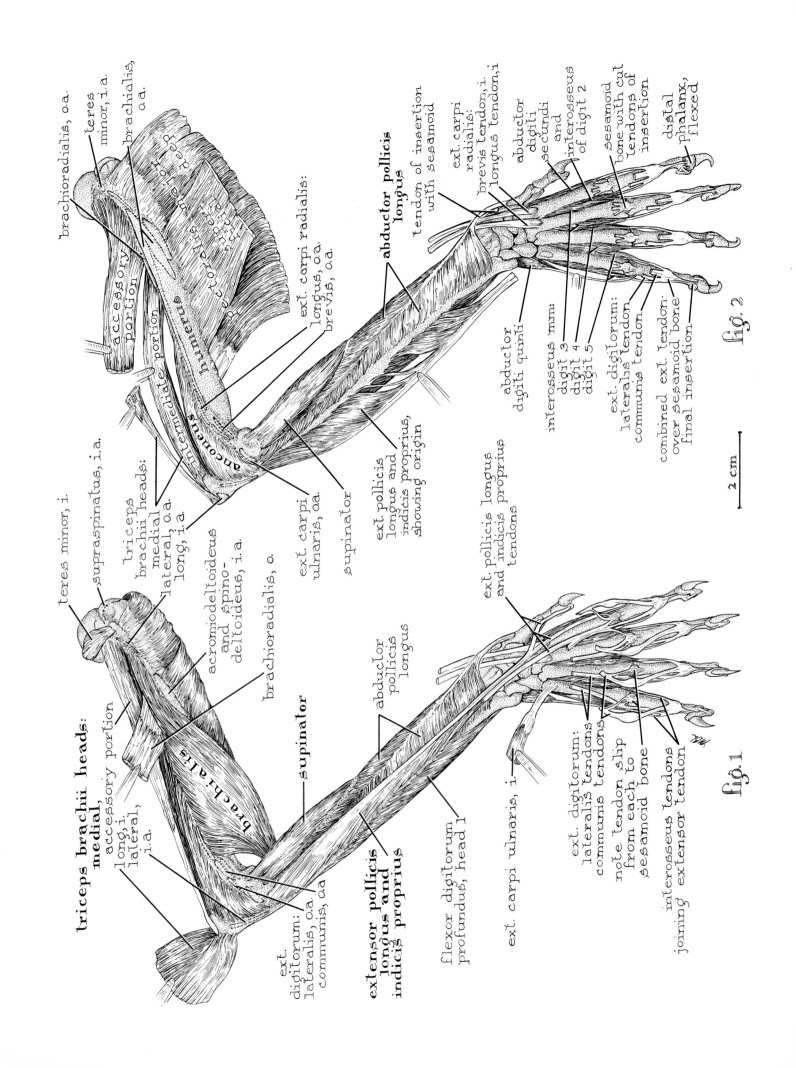

brachioradialis, o.a.

teres minor, i.a.

brachialis, o.a.

biceps brachii

accessory portion

brachialis

intermediate portion

humerus

anconeus

ext. carpi radialis:
longus, o.a.
brevis, o.a.

abductor pollicis longus

tendon of insertion with sesamoid

ext. carpi radialis:
brevis tendon, i
longus tendon, i

abductor digiti secundi and interosseus of digit 2

sesamoid bone with cut tendons of insertion

distal phalanx, flexed

abductor digiti quinti

interosseus mm:
digit 3
digit 4
digit 5

ext. digitorum:
lateralis tendon
communis tendon

combined ext. tendon over sesamoid bone
final insertion

fig. 2

2 cm

teres minor, i.

supraspinatus, i.a.

triceps brachii heads:
medial
lateral, o.a.
long, i.a.

acromiodeltoideus and spino-deltoideus, i.a.

brachioradialis, o.

ext. carpi ulnaris, o.a.

supinator

ext. pollicis longus and indicis proprius, showing origin

ext. pollicis longus and indicis proprius tendons

triceps brachii heads:
medial,
accessory portion
long, i.
lateral,
i.a.

ext. digitorum:
lateralis, o.a.
communis, oa

extensor pollicis longus and indicis proprius

abductor pollicis longus

brachialis

supinator

flexor digitorum profundus, head 1

ext. carpi ulnaris, i.

ext. digitorum:
lateralis tendons
communis tendons

note tendon slip from each to sesamoid bone

interosseus tendons joining extensor tendon

fig. 1

The bellies of the superficial muscles have been dissected away. In most cases, as a result, their points of origin and insertion are much clearer than when the muscles were intact. Note particularly the accessory portion of the **medial head** of the **triceps brachii** on figure 1 and the intermediate portion on figure 2.

The **brachialis** is featured on figure 1 with its V-shaped origin better shown on figure 2. The long limb of the V extends to the proximal margin of the supracondyloid foramen, the short, ventral limb to the middle of the humerus. Muscle fibers take origin only on the limbs of the V, not in between them. The insertion is by a flat tendon which joins the tendon of the cleidobrachialis in a roughened area on the lateral surface of the ulna just distad of the semilunar notch. The brachialis is a flexor of the forearm.

The **anconeus** is a small muscle on the lateral side of the elbow joint (fig. 2). Its origin is on the distal end of the dorsal surface of the humerus sometimes extending on to the lateral epicondyle. It passes over the capsule of the elbow joint to insert on the lateral surface of the ulna from the proximal end of the olecranon process to the distal margin of the semilunar notch. It tenses the capsule of the elbow joint and may rotate the ulna slightly.

The **supinator** is a flat muscle which spirals around the proximal end of the radius having its origin by a short tendon from the lateral side of the annular ligament of the radius and from the radial collateral ligament which runs from the humerus to the radius. It inserts on the dorsal and part of the ventral surface of the proximal two fifths of the radius. It supinates the paw.

The **extensor pollicis longus and indicis proprius** has its origin by short fleshy fibers over three fourths of the lateral surface of the ulna. They converge into a tendon which runs nearly the length of the muscle and parallel to the radius. The tendon divides into three distally, two going to the base of the proximal phalanx of the second digit and one to the pollex. This combined muscle is an extensor of the index finger and the thumb.

The **abductor pollicis longus** (*extensor brevis pollicis or extensor ossis metacarpi pollicis*). Reighard and Jennings call this the extensor brevis pollicis. It appears to fit more closely the description of the abductor pollicis longus both in the dog and in man. This muscle has its origin by fleshy fibers from the lateral surface of most of the length of the ulna and from the dorsal lateral surface of the radius distal to the supinator and from the interosseous membrane between the radius and ulna. The fibers converge to a strong central tendon which runs obliquely over the tendons of the extensor carpi radialis brevis and longus to insert into the radial side of the first metacarpal. There is a sesamoid bone in this tendon at its insertion—the radial sesamoid.

The **interosseus muscles** are on the palmar surface of the paw and will be considered in Plate 39, figure 2. Notice here, however, in figure 1 the connections between the interosseus tendons and those of the tendons of the extensor muscles.

Note also the combined extensor tendons and the final insertion on to the terminal phalanx (fig. 2).

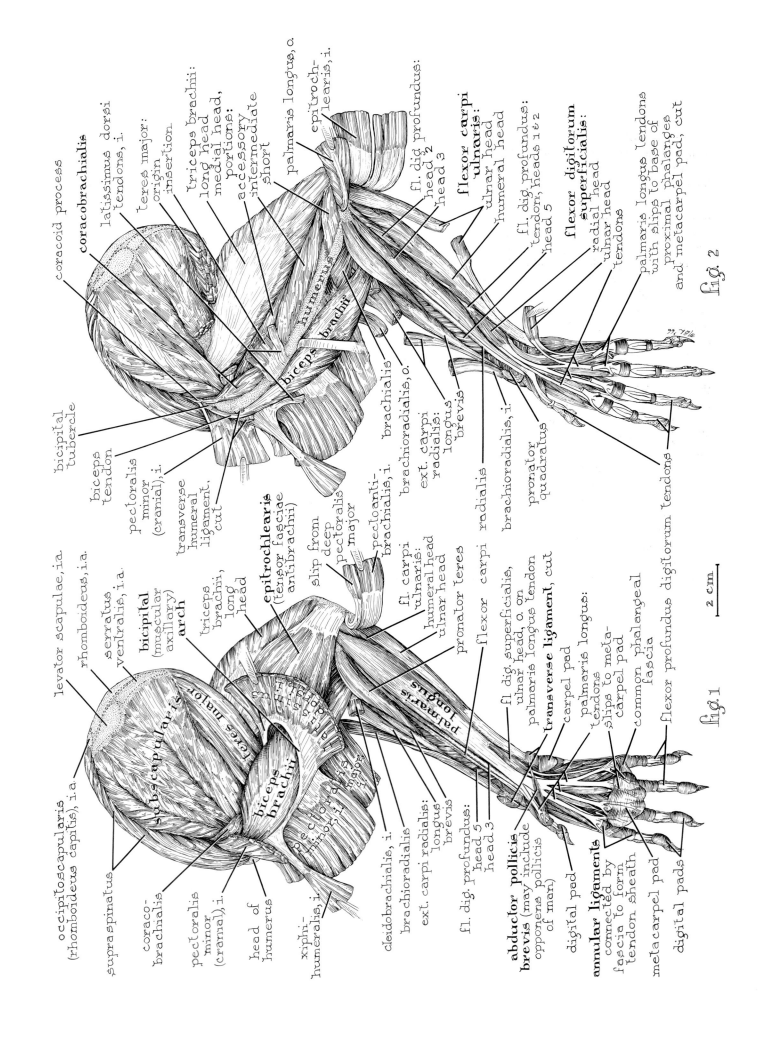

coracoid process

corocobrachialis

latissimus dorsi tendons, i.

teres major: origin insertion

triceps brachii: long head medial head, portions: accessory intermediate short

palmaris longus, o.

epitrochlearis, i.

fl. dig. profundus: head 2 head 3

flexor carpi ulnaris: ulnar head humeral head

fl. dig. profundus: tendon, heads 1 & 2 head 5

flexor digitorum superficialis: radial head ulnar head tendons

palmaris longus tendons with slips to base of proximal phalanges and metacarpel pad, cut

humerus

biceps brachii

brachialis

brachioradialis, o.

ext. carpi radialis: longus brevis

brachioradialis, i.

pronator quadratus

ext. carpi radialis

bicipital tubercle

biceps tendon

pectoralis minor (cranial), i.

transverse humeral ligament, cut

fig. 2

2 cm

occipitoscapularis (rhomboideus capitis), i.a.

levator scapulae, i.a.

rhomboideus, i.a.

serratus ventralis, i.a.

bicipital (muscular axillary) arch

triceps brachii, long head

epitrochlearis (tensor fasciae antibrachii)

slip from deep pectoralis major

pectoanti- brachialis, i.

fl. carpi ulnaris: humeral head ulnar head

pronator teres

flexor carpi radialis

supraspinatus

coraco- brachialis

pectoralis minor (cranial), i.

head of humerus

xiphi- humeralis, i

subscapularis

teres major

biceps brachii

pectoralis minor & major

cleidobrachialis

brachioradialis

ext. carpi radialis: longus brevis

fl. dig. profundus: head 5 head 3

palmaris longus

fl. dig. superficialis, ulnar head, o. on palmaris longus tendon

transverse ligament, cut

carpel pad

palmaris longus: tendons slips to meta- carpel pad

common phalangeal fascia

abductor pollicis brevis (may include opponens pollicis of man)

digital pad

annular ligaments connected by fascia to form tendon sheath

metacarpel pad

digital pads

flexor profundus digitorum tendons

fig. 1

Although we speak of the lateral and medial muscles of the forelimb, one must realize that things are not that definitive. Some lateral muscles will show up in part on the medial views which are now to be studied. Some muscles, the triceps for example, have medial heads which one sees to much better advantage in figure 2 than on previous plates.

The **subscapularis muscle** originates by fleshy fibers to the periosteum over most of the subscapular fossa and to tendinous fibers which partially divide the muscle into two or three parts. Note in figure 1 that a number of insertion areas impinge upon a part of the subscapular fossa—those of the levator scapulae, rhomboideus, serratus ventralis and occipitoscapularis. The teres major at its origin overlaps a part of the subscapularis. In figure 2 notice that the biceps brachii has been pulled aside so you can see the insertion of the teres major and its relationship to the tendon of insertion of the latissimus dorsi. The subscapularis fibers converge to the caudal border of the scapula and insert by a strong, flat tendon into the dorsal border of the lesser tubercle of the humerus.

The **biceps brachii** (figs. 1 and 2) is a distinct fusiform muscle on the cranial surface of the humerus. Its origin is by a strong tendon from the supraglenoid (*bicipital*) tubercle of the scapula within the capsule of the shoulder joint. It lacks the second head from the coracoid process as is found in man. The tendon passes through the capsule of the shoulder joint and then along the intertubercular groove which is closed to form a canal by transverse ligaments. The muscle inserts by a rounded tendon on the radial tuberosity of the radius. It is a flexor and supinator of the hand. Note the relationship of the biceps brachii to the two arms of the **bicipital arch** and also the joining of the xiphihumeralis into one side of the bicipital arch. The small **coracobrachialis** is seen in figure 2. It originates on the coracoid process and inserts on the humerus (Plate 39, fig. 2).

The **epitrochlearis** (*tensor fasciae antebrachii*), discussed previously, is well shown here and in figure 2 is cut and pulled aside to reveal the underlying muscles.

In the antebrachium or forearm a number of muscles of importance are shown. The **palmaris longus** is a flat muscle on the medial surface of the forearm. It originates on the medial epicondyle of the humerus and in the wrist ends in a flat tendon which passes beneath the **transverse ligament** (fig. 1) and divides on the hand into four (*or five*) tendons which diverge to their insertions, all except number one giving off a branch which goes into the trilobed metacarpal pad in the palm. The middle two of these usually have cutaneous branches to the skin of the pad. The tendon is finally inserted into the base of the proximal phalanx except the one to the thumb which divides and inserts into the sesamoid bones at the base of the proximal phalanx. The palmaris longus flexes the proximal phalanx and each digit. The ulnar head of the **flexor digitorum superficialis** (*sublimis*) has its origin on the tendon of the palmaris longus before it divides.

The **abductor pollicis brevis** (*includes opponens pollicis of man*) is a tiny muscle having its origin on the **transverse ligament** and inserting into the base of the proximal phalanx of the pollex. It abducts the pollex or thumb.

Note again on figure 1 the **annular ligaments** which are connected by fascia forming the tendon sheaths. The annular ligaments hold the flexor tendons to the digits.

The **flexor carpi ulnaris** has two heads, one originating on the distal end of the humerus just distad of the medial epicondyle and in common with the second part of the flexor profundus digitorum, the other, the ulnar head, on the lateral surface of the olecranon process. The humeral head is thick; the ulnar head is flat and lies closely against the humeral head. They both insert into the pisiform bone and serve to flex and abduct the forepaw.

The **flexor digitorum superficialis** has an ulnar and a radial part. As stated above, the ulnar part originates from the outer surface of the tendon of the palmaris longus. It gives rise to two slips of muscle whose tendons insert into the base of the middle phalanx of the fourth and fifth digits. The radial part of this muscle originates over the outer surface of a tendon formed by the first and second parts of the flexor profundus. From its two slips tendons go to the second and third digits. There are therefore four tendons of the superficialis muscle. Each of these is perforated by a tendon of the flexor profundus as it passes through the fibrous rings at the base of the proximal phalanx. They then go through the second ring and insert into the base of the middle phalanx. This is a flexor of the proximal and middle digital joints and thereby a flexor of the whole forepaw.

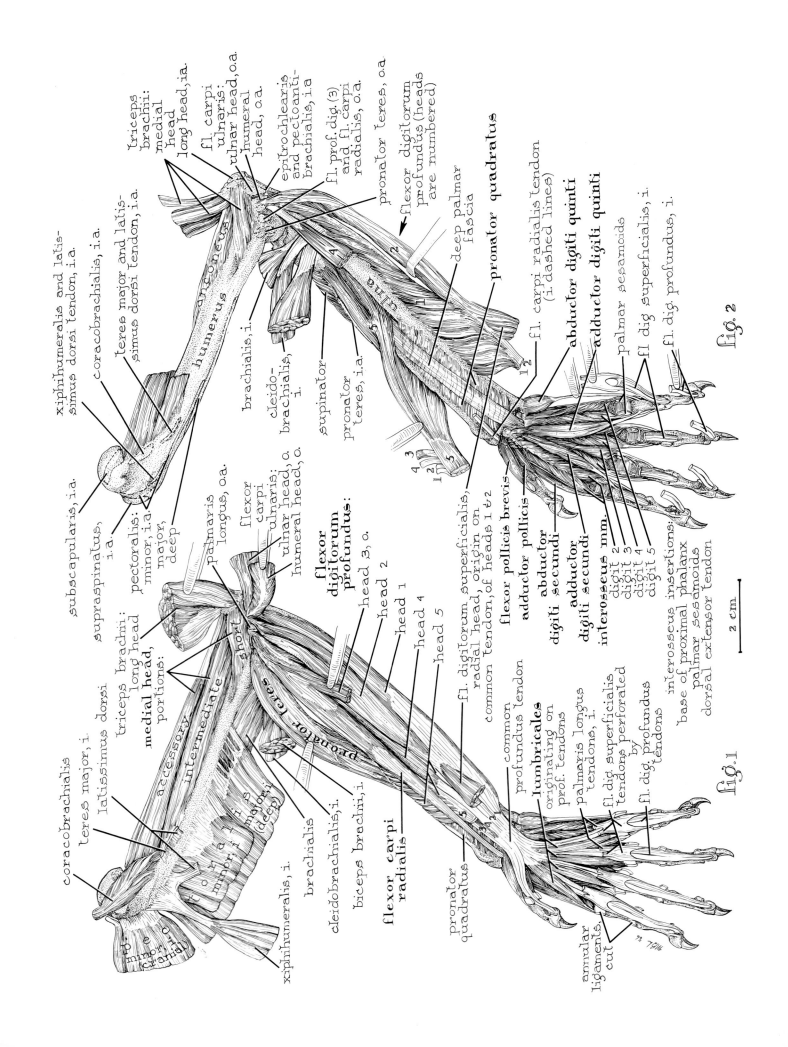

xiphihumeralis and latissimus dorsi tendon, i.a.

coracobrachialis, i.a.

teres major and latissimus dorsi tendon, i.a.

triceps brachii: medial head long head, ia.

fl. carpi ulnaris: ulnar head, o.a. humeral head, o.

epitrochlearis and pectoantibrachialis, ia

pronator teres, o.a.

fl. prof. dig. (3) and fl. carpi radialis, o.a.

flexor digitorum profundus (heads are numbered)

pronator quadratus

deep palmar fascia

fl. carpi radialis tendon (i dashed lines)

abductor digiti quinti

adductor digiti quinti

palmar sesamoids

fl dig superficialis, i.

fl. dig. profundus, i.

anconeus

humerus

brachialis, i.

cleidobrachialis, i.

supinator

pronator teres, i.a.

fig. 2

subscapularis, i.a.

supraspinatus, i.a.

pectoralis: minor, i.a. major, deep

palmaris longus, o.a.

flexor carpi ulnaris: ulnar head, o. humeral head, o.

flexor digitorum profundus:

head 3, o.

head 2

head 1

head 4

head 5

fl. digitorum superficialis, radial head, origin on common tendon, of heads 1 & 2

flexor pollicis brevis

adductor pollicis

abductor digiti secundi

adductor digiti secundi

interosseus mm.

digit 2
digit 3
digit 4
digit 5

interosseus insertions:
base of proximal phalanx
palmar sesamoids
dorsal extensor tendon

coracobrachialis

teres major, i.

latissimus dorsi

triceps brachii: long head medial head portions:

accessory

intermediate

xiphihumeralis, i.

cleidobrachialis, i.

biceps brachii, i.

flexor carpi radialis

pronator teres

pronator quadratus

common profundus tendon

lumbricales originating on prof. tendons

palmaris longus tendons, i.

fl. dig. superficialis tendone perforated by fl. dig. profundus tendons

annular ligaments, cut

fig. 1

2 cm

All that is shown here on the muscles of the humerus is in way of clarification and review. Figure 1 gives the best exposition yet seen of the three parts of the **medial head** of the **triceps brachii**—the **accessory, intermedius** and **short.** By cutting sections out of these three parts, the anconeus (fig. 2) is uncovered. The coracobrachialis muscle is shown over the shoulder joint in figure 1 and in figure 2 it has been removed and only its insertion area (*i. a.*) is indicated. Examine these parts of the two figures in this manner.

The **pronator teres** (fig. 1) originates by a strong tendon from the medial epicondyle of the humerus. It crosses the medial surface of the elbow joint and inserts on the medial border of the proximal half of the radius. It rotates the radius in such a way as to pronate the forepaw. It may also serve as a weak flexor of the elbow.

The **flexor carpi radialis** is a slender muscle with its origin on the medial epicondyle of the humerus. It inserts by a long slender tendon into the bases of the second and third metacarpals after passing through a deep groove which is turned into a canal by overlying tendons. It is a flexor of the forepaw at the carpal joint.

The largest, strongest and deepest of the flexor muscles is the **flexor digitorum profundus.** It has five heads, according to one interpretation, in which the heads are commonly referred to only by number. Another interpretation is that heads two, three and four, since they **all** have their origins on the **medial epicondyle** of the humerus, represent a **humeral head.** Head one, which has its origin from the dorsal half of the medial surface of the ulna for most of its length represents an **ulnar head.** Head five, with its origin from the middle third of the ventral surface of the shaft of the radius, from adjacent parts of the interosseus membrane and from a part of the medial surface of the shaft of the ulna, represents a **radial head.** We choose to follow the "five head" interpretation. In any case, the tendon of insertion of the fifth or radial head is the largest and forms the deep, broad middle part of the common tendon since all the other tendons, one through four, join to it.

The common tendon of the flexor profundus digitorum covers the carpus and much of the metacarpal region and then divides into five tendons, each inserting into the base of the distal phalanx of the digit of the same number. Those tendons of digits two through five perforate the tendons of flexor digitorum superficialis (fig. 1) at the base of the first phalanx of each digit. The tendon of the first digit does not perforate. It is held to the digit by an annular ligament. The tendons of the profundus to digits two through five pass through the tendinous sheaths and are held to the underlying bone by proximal and distal annular ligaments. The profundus is a flexor to all of the digits.

The **lumbricales** (fig. 1) are four small muscles which originate by fleshy fibers from the outer and palmar surface of the profundus tendons, two through five. Flat at their origins, each becomes cylindrical and, curving about the base of one of the four digits, inserts into the radial side of the base of the proximal phalanx. Their action is to draw the digits toward the radial side and to flex the metacarpophalangeal joints.

The **pronator quadratus** is a quadrangular muscle which originates on the distal half of the flexor surface of the ulna and the interosseous membrane (fig. 2). Its fibers pass obliquely and distad to the radial side where they insert into the flexor surface of the radius. The pronator quadratus turns the paw inward.

The first, second, and fifth digits have some special muscles which give them greater versatility of movement than seen in the other digits. The **flexor pollicis brevis** is a small muscle which inserts on the base of the proximal phalanx of the pollex on its ulnar side.

The **adductor pollicis** originates by fleshy fibers from the capitate. It inserts into the base of the proximal phalanx of the pollex on its ulnar side. It pulls the thumb toward the midline of the paw.

The second digit has two special muscles, an **abductor digiti secundi** and an **adductor digiti secundi.** The abductor originates from the radial and ventral side of the base of the second metacarpal and the ventral surface of the trapezium and inserts into the radial side of the proximal phalanx of the second digit and its sesamoid. The adductor originates on the capitate and inserts on the ulnar side of the base of the proximal phalanx of the second digit.

The **abductor digit quinti** originates on the pisiform bone and from the transverse ligament. Its slender tendon inserts into the ulnar side of the proximal phalanx of the fifth digit.

The **adductor digit quinti** originates on the ventral surface of the capitate. It passes distad and toward the ulnar side to insert into the radial surface of the fifth metacarpal and into the base of the proximal phalanx.

The interossei are small muscles which lie on the palmar surfaces of metacarpals two through five. They originate by fleshy fibers from the proximal ends of the metacarpals on the ventral and lateral surfaces. Each divides near the distal end of the metacarpal into two parts which pass onto the sides of the metacarpal. From here they pass to the base of the proximal phalanx and its sesamoids. They also send a tendon dorsad to join the common extensor tendon of the digit. Their function is to flex the metacarpophalangeal joints.

105

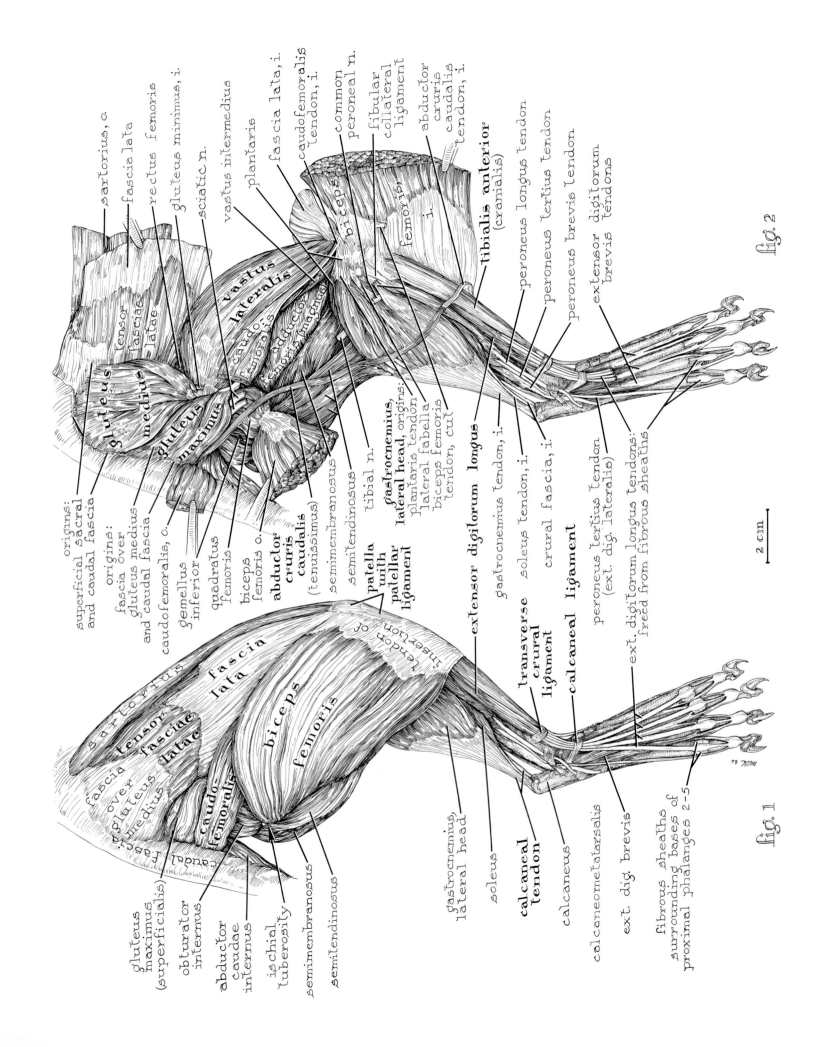

sartorius, o.

fascia lata

rectus femoris

gluteus minimus, i.

vastus intermedius

plantaris

fascia lata, i.

caudofemoralis tendon, i.

common peroneal n.

fibular collateral ligament

abductor cruris caudalis tendon, i.

tibialis anterior (cranialis)

peroneus longus tendon

peroneus tertius tendon

peroneus brevis tendon

extensor digitorum brevis tendons

sciatic n.

tensor fasciae latae

gluteus medius

vastus lateralis

caudofemoralis

adductor femoris (magnus)

biceps femoris, i.

gluteus maximus

origins:
superficial sacral
and caudal fascia

origins:
fascia over
gluteus medius
and caudal fascia

caudofemoralis, o.

gemellus inferior

quadratus femoris

biceps femoris o.

abductor cruris caudalis (tenuissimus)

semimembranosus

semitendinosus

tibial n.

gastrocnemius, lateral head, origins:
plantaris tendon
lateral fabella
biceps femoris
tendon, cut

extensor digitorum longus

gastrocnemius tendon, i.

soleus tendon, i.

crural fascia, i.

calcaneal ligament

peroneus tertius tendon (ext. dig. lateralis)

ext. digitorum longus tendons:
freed from fibrous sheaths

fig. 2

2 cm

gluteus maximus (superficialis)

obturator internus

abductor caudae internus

ischial tuberosity

semimembranosus

semitendinosus

sartorius

tensor fasciae latae

fascia over gluteus medius

caudofemoralis

fascia lata

biceps femoris

tendon of insertion

patella with patellar ligament

transverse crural ligament

calcaneal tendon

caudal fascia

gastrocnemius, lateral head

soleus

calcaneus

calcaneometatarsalis

ext. dig. brevis

fibrous sheaths surrounding bases of proximal phalanges 2-5

fig. 1

'74 TOM

This is the first in a series of six plates in which the muscles of the hindlimb are illustrated from carefully prepared dissections. One wishing to do the dissections using the plates as a guide should experience little difficulty if he does two things: (1) keeps his dissection clean by continuous and careful removal of fascia and (2) studies each plate carefully to see exactly where muscles have been cut. The gluteus muscles are not very clear in figure 1 but after removal of the fascia they appear as in figure 2. If you become concerned about all of the blood vessels and nerves which you are destroying, remember your cat has limbs on the other side which can be used for such studies though you may wish to identify them with your muscle dissection. If so, refer to the appropriate plates.

It might be well to refer back for orientation purposes to Plate 28 where the hindlimb was in place on the body. A part of the tail covered with caudal fascia is shown here to help put the hindlimb in proper perspective.

We have mentioned the gluteal and caudal fascia. Note also the **fascia lata** on the cranial part of the thigh. It covers the vastus lateralis as seen in figure 2 and also dips down caudad of the sartorius and is attached to the vastus medialis as seen in Plate 43, figure 2. Distally it is continuous with the tendon of insertion of the biceps femoris. At its proximal end the fascia lata receives the insertion of the tensor fasciae latae muscle.

The **tensor fasciae latae** muscle has its origin on the ventral border of the ilium and from the fascia covering the ventral border of the gluteus medius and over its cranial half. It covers the proximal third of the lateral cranial surface of the thigh. It inserts into the fascia lata and tenses the side of the thigh and aids in extending the leg.

The **biceps femoris** is a very large flat muscle which covers much of the lateral surface of the thigh. It has its origin from the tuberosity of the ischium by tendinous and fleshy fiber attachment. The fibers diverge and at the knee end in a fascia which inserts into the proximal third of the lateral margin of the tibia and into the lateral margin of the patella. It is an extensor of the thigh and to a small degree an abductor. Note the patella with the **patellar ligament.**

The **caudofemoralis** originates by a flat tendon from the transverse processes of the second and third caudal vertebrae. It inserts into the middle of the lateral border of the patella (fig. 2). It abducts and extends the thigh.

Note the very slender **abductor cruris caudalis** (*tenuissimus*) which originates on the transverse process of the second caudal vertebra and inserts with the biceps fermoris.

The **extensor digitorum longus** is well hidden behind the tibialis anterior (*cranialis*) and even after the fascia is removed the separation is difficult to detect. This muscle originates by a thin tendon from the lateral surface of the lateral epicondyle of the femur. It passes through the capsule of the knee joint and through a groove on the head of the tibia just proximal to the head of the fibula. The belly of the muscle extends the length of the tibia and distally its tendons, along with the tendon of the tibialis anterior, pass through the **transverse crural ligament.** At the ankle joint the tendons of the extensor digitorum longus pass behind the calcaneal ligament. These ligaments hold the tendons to the underlying structures when the muscle contracts. The muscle inserts by four tendons, each tendon being connected with a fibrous sheath surrounding the base of the proximal phalanx and each has a synovial bursa beneath it. Near the distal end of the proximal phalanx the tendon attached on the lateral side to the conjoined tendon of the extensor digitorum brevis and an interosseus muscle and on the medial side to the tendon of an interosseus muscle. The lateral side of the most lateral tendon is not thus attached but is joined by the tendon of the peroneus tertius (Plate 42, fig. 2). The **insertion** of these tendons is on the base of the distal phalanx. At the juncture of the proximal and middle phalanges the tendons pass over fibrous pads.

Going to figure 2 consider the **gluteus maximus** (*superficialis*) **muscle** which in the cat is smaller than the gluteus medius, opposite from the condition in man. It lies between the caudofemoralis and the gluteus medius. Its origin is by fleshy fibers from the tips of the transverse processes of the last sacral and the first caudal vertebrae and from the fascia covering the spinal muscles and the gluteus medius. It inserts by tendon and fleshy fibrous attachments into the caudal side of the greater trochanter. It abducts and extends the thigh (Plate 42, fig. 1).

The **gluteus medius** originates on the dorsal half of the crest of the ilium and the dorsal half of the lateral surface of the ilium and from various fascia in that area. It also attaches to the transverse process of the last sacral and first caudal vertebrae. It inserts on the proximal end of the greater trochanter. It abducts and extends the thigh.

The **gastrocremius** is the muscle which forms the calf of the leg. It arises by two heads, the lateral and medial from the lateral and medial sesamoid bones (*fabellae*) of the femur and from adjacent areas of the femur and the fascia and tendons of that area. The two heads unite opposite the upper third of the tibia and the muscle narrows rapidly into its insertion by the **calcaneal** (*tendon of Achilles*) **tendon** into the proximal end of the calcaneus near its ventral border. This tendon and that of the soleus

107

muscle and the fascia of the shank forms a tubular sheath through which the tendon of the plantaris passes.

The **tibialis anterior** (*cranialis*) is a superficial muscle covering the lateral side of the cranial part of the tibia. Its origin, by fleshy fibers, is from the proximal one sixth of the lateral side of the shaft of the tibia and from the proximal third of the medial side of the shaft and head of the fibula and from the intervening interosseous ligament. It ends distally in a strong tendon which passes beneath the transverse crural ligament. The tendon passes obliquely across the dorsal surface of the foot toward the medial side where it inserts into the outer surface of the first metatarsal. It is a flexor of the foot, an extensor of the second digit and rotates the paw laterally.

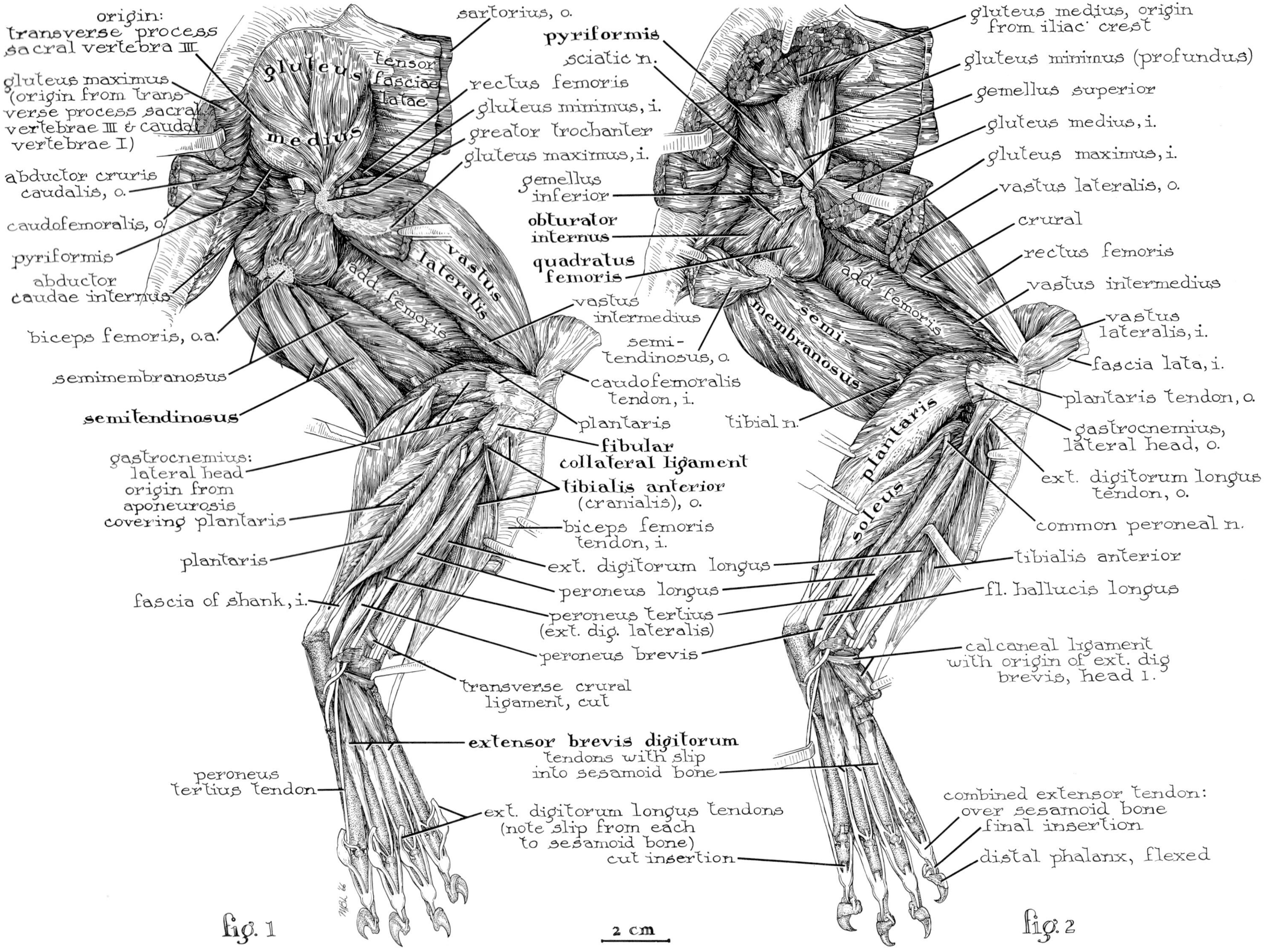

fig. 2

gluteus medius, origin from iliac crest
gluteus minimus (profundus)
gemellus superior
gluteus medius, i.
gluteus maximus, i.
vastus lateralis, o.
crural
rectus femoris
vastus intermedius
vastus lateralis, i.
fascia lata, i.
plantaris tendon, a
gastrocnemius, lateral head, o.
ext. digitorum longus tendon, o.
common peroneal n.
tibialis anterior
fl. hallucis longus
calcaneal ligament with origin of ext. dig. brevis, head I.
combined extensor tendon: over sesamoid bone final insertion
distal phalanx, flexed

add. femoris
semi-membranosus
plantaris
soleus

sartorius, o.
pyriformis
sciatic n.
rectus femoris
gluteus minimus, i.
greater trochanter
gluteus maximus, i.
gemellus inferior
obturator internus
quadratus femoris
vastus intermedius
semi-tendinosus, o.
caudofemoralis tendon, i.
plantaris
tibial n.
fibular collateral ligament
tibialis anterior (cranialis), o.
biceps femoris tendon, i.
ext. digitorum longus
peroneus longus
peroneus tertius (ext. dig. lateralis)
peroneus brevis
transverse crural ligament, cut
extensor brevis digitorum
tendons with slip into sesamoid bone
ext. digitorum longus tendons (note slip from each to sesamoid bone) cut insertion

fig. 1

origin:
transverse process sacral vertebra III
gluteus maximus (origin from transverse process sacral vertebrae III & caudal vertebrae I)
abductor cruris caudalis, o.
caudofemoralis, o.
pyriformis
abductor caudae internus
biceps femoris, o.o.
semimembranosus
semitendinosus
tensor fasciae latae
gluteus medius
vastus lateralis
add. femoris
gastrocnemius: lateral head origin from aponeurosis covering plantaris
plantaris
fascia of shank, i.
peroneus tertius tendon

2 cm

By removing the biceps femoris and the tensor fasciae latae and cutting the caudofemoralis and gluteus maximus and retracting their origins and insertions, additional muscles come to view.

The **semitendinosus** shown here with its belly subdivided, is a muscle of the caudal border of the thigh between the biceps femoris and semimembranosus. It has its origin on the tuberosity of the ischium beneath the origin of the biceps. It passes to the medial side of the tibia to insert into the crest (*cranial border*) of the tibia about 1 centimeter from its proximal end. Its action is to extend the hip and to flex the leg.

The **vastus lateralis** is a part of a large four-parted **quadriceps femoris muscle** which covers the cranial, lateral and medial parts of the femur or thigh. It is a great extensor of the lower leg (*shank*). It inserts into the patella and through it by the ligamentum patellae into the tuberosity of the tibia. Its other three parts are the vastus medialis, Plate 43, figure 2; vastus intermedius, Plate 45, figures 1 and 2; and the rectus femoris, Plate 42, figure 1. The vastus lateralis originates on a triangular area on the caudal and lateral surfaces of the shaft and greater trochanter of the femur. It inserts into the outer surface of the patella near its lateral border.

The **extensor digitorum brevis** lies on the dorsal side of the foot. It has its origin from the distal border of the calcaneal ligament (fig. 2) and from the dorsal surfaces of the proximal ends of the three lateral metatarsals. The muscle ends in three flat tendons which run in the interspaces between the four tendons of the extensor longus digitorum. Each tendon divides into two branches, the lateral branch of each tendon inserts into the cartilaginous plate (*sesamoid bone*) which lies in the metatarsophalangeal joint of the digit on the outer side. "The medial branch joins the lateral side of the extensor longus tendon on the dorsum of the first phalanx" (Reighard and Jennings). It extends the digits. Note these structures on both figures 1 and 2.

Turn now to figure 2. The gluteus medius has been cut and its ends pulled back to reveal the pyriformis, gemellus superior and gluteus minimus (*profundus*).

The **pyriformis** is a triangular muscle with origins from the tips of the transverse processes of the last two sacral and the first caudal vertebrae. It passes through the great sciatic notch and inserts into the greater trochanter just caudal to the insertion of the gemellus superior. It abducts the thigh.

The **quadratus femoris** is a small muscle with its origin on the lateral surface of the ischium near the tuberosity. It inserts into the ventral border of the greater trochanter and about half of the adjacent surface of the lesser trochanter. It is an extensor and outward rotator of the thigh.

The **obturator internus** is a flat, triangular muscle located caudad of the gemellus superior. It originates from the dorsal surface (*medial aspect*) of the ramus of the ischium by a number of small heads. It narrows and passes through the lesser sciatic notch, turns ventrad and ends in a strong tendon inserting into the bottom of the trochanteric fossa of the femur. It is an abductor of the thigh and outward rotator of the hip joint.

The **plantaris** is a large fusiform muscle quite largely covered by the gastrocnemius muscle. It originates by a strong tendon from the lateral border of the patella and by fleshy fibers from the ventral border of the lateral sesamoid. The belly of the muscle is closely united with the lateral head of the gastrocnemius. It ends in a thick tendon which passes through a sheath formed by the tendons of the gastrocnemius and soleus and the shank fascia. It passes next over the grooved end of the calcaneus and onto its ventral surface. It is held in place by two sheets of aponeurosis. The tendon broadens distal to the calcaneus and ends in the flexor brevis digitorum muscle which may be considered a second part of the muscle. It is an extensor of the foot and if truly a part of the flexor digitorum brevis (*superficialis*) a flexor of the digits.

The **soleus** is a flat muscle lying beneath the plantaris. It originates by fleshy fibers from the lateral surface of the head of the fibula and by tendinous fibers from the proximal two-fifths of its ventral border. It inserts as indicated previously in a slender tendon which with the tendon of the gastrocnemius forms the calcaneal tendon.

The gastrocnemius and soleus are sometimes considered a single muscle of three parts like the triceps brachii and called the triceps surae.

This plate contains much more information than that selected for emphasis. It requires careful study.

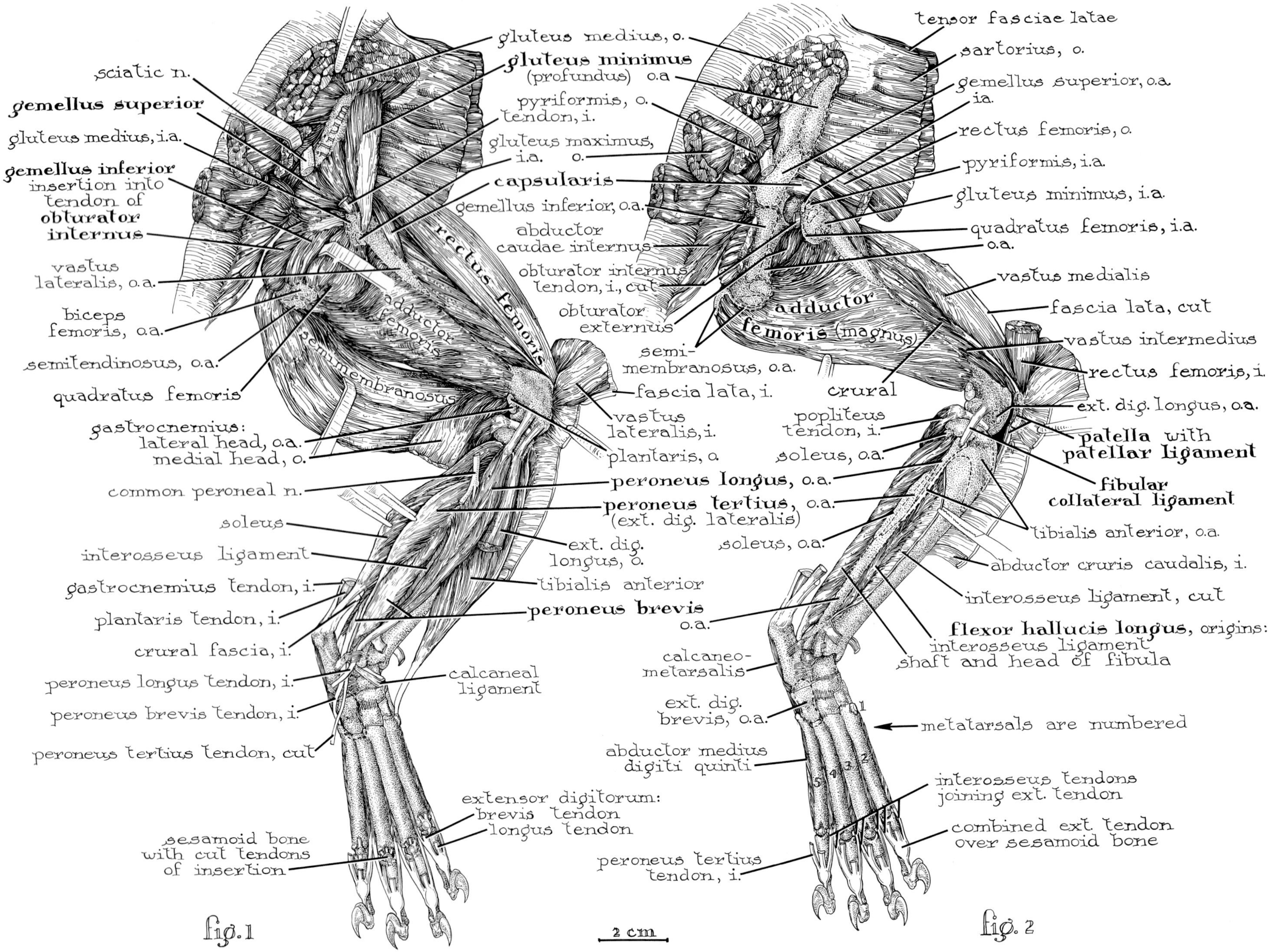

fig. 1

fig. 2

2 cm

tensor fasciae latae
sartorius, o.
gemellus superior, o.a. ia.
rectus femoris, o
pyriformis, ia.
gluteus minimus, i.a.
quadratus femoris, i.a. o.a.
vastus medialis
fascia lata, cut
vastus intermedius
rectus femoris, i
ext. dig. longus, o.a.
patella with
patellar ligament
fibular
collateral ligament
tibialis anterior, o.a.
abductor cruris caudalis, i.
interosseus ligament, cut
flexor hallucis longus, origins:
interosseus ligament
shaft and head of fibula
metatarsals are numbered
interosseus tendons
joining ext. tendon
combined ext. tendon
over sesamoid bone

adductor
femoris (magnus)
crural
semi-
membranosus, o.a.
fascia lata, i.
vastus
lateralis, i.
popliteus
tendon, i.
plantaris, a
soleus, o.a.
peroneus longus, o.a.
peroneus tertius, o.a.
(ext. dig. lateralis)
soleus, o.a.
tibialis anterior
peroneus brevis
o.a.
calcaneo-
metatarsalis
ext. dig.
brevis, o.a.
abductor medius
digiti quinti
peroneus tertius
tendon, i.

gluteus medius, o.
gluteus minimus
(profundus) o.a
pyriformis, o.
tendon, i.
gluteus maximus,
i.a. o.
capsularis
gemellus inferior, o.a.
abductor
caudae internus
obturator internus
tendon, i., cut
obturator
externus

sciatic n.
gemellus superior
gluteus medius, i.a.
gemellus inferior
insertion into
tendon of
obturator
internus
vastus
lateralis, o.a.
biceps
femoris, o.
semitendinosus, o.a.
quadratus femoris
gastrocnemius:
lateral head, o.a.
medial head, o.
common peroneal n.
interosseus ligament
gastrocnemius tendon, i.
plantaris tendon, i.
crural fascia, i.
peroneus longus tendon, i.
peroneus brevis tendon, i.
peroneus tertius tendon, cut

rectus femoris
adductor
femoris
semimembranosus
soleus
extensor digitorum:
brevis tendon
longus tendon
calcaneal
ligament
sesamoid bone
with cut tendons
of insertion

The belly of the pyriformis has been removed in figure 1, giving a better view of the gemellus superior and the quadratus femoris has been pulled back revealing more of the gemellus inferior. The semitendinosus has been removed to show better the semimembranosus. The tibialis anterior has been pulled to the side and the extensor digitorum longus cut to give a good view of the peroneus muscles.

The **gemellus superior** is closely joined to the gemellus inferior caudally and to the gluteus minimus (*profundus*) cranially. Its origin is by fleshy fibers on the dorsal border of the ilium and ischium. Its fibers converge into a strong tendon which inserts into an area dorsad of the tip of the greater trochanter (fig. 2). It rotates and abducts the femur.

The **gemellus inferior** is just caudad of the gemellus superior, and has its origin on the dorsal one-half of the whole lateral surface of the ischium. Its insertion is into the inner surface of the tendon of the obturator internus and a few fibers into the capsule of the hip joint. It is an abductor of the thigh. Some consider it a separate head of the obturator internus.

The **gluteus minimus** (*produndus*) is the third of the gluteus group of muscles. It is long and triangular and lies with its caudal border against the cranial border of the gemellus superior. It originates from the ventral half of the ilium and tapers to an insertion at the base of the dorsal surface of the greater trochanter on its lateral side. It rotates the femur outward.

The **capsularis** (*gluteus quartus*) **muscle** is a narrow bundle of muscle fibers which runs obliquely beneath the gluteus minimus and gemellus superior. It originates on the ilium and inserts by fleshy fibers on the middle line of the cranial surface of the femur, distad of the greater trochanter. It flexes the thigh.

The **rectus femoris** is the large round central part of the quadriceps femoris. It has its origin by a strong tendon in a triangular area craniad of the acetabulum on the lateral surface of the ilium. It inserts into the oblique area on the outer surface of the patella in connection with the vastus lateralis.

The **peroneus longus** is the largest of three peroneus muscles which have their origin on the fibula and insert on the foot. The **longus** originates on the lateral surface of the head of the fibula and from the proximal half of the shaft of the fibula. At the level of the lower third of the fibula it becomes tendenous, the tendon passing through the groove on the lateral surface of the lateral malleolus which is converted into a canal by a transverse ligament. "The tendon then passes through the groove on the peroneal tubercle of the calcaneus, then turns onto

the sole of the foot and passes through the peroneal groove on the cuboid bone. It then turns mediad and passes through the groove between the ventral process of the lateral cuneiform and metatarsal until it reaches the first metatarsal. The entire groove is converted into a canal by the overlying ligaments" (Reighard and Jennings). Its insertion is into the base of the first and fifth metatarsal but with branches to the outer sides of the remaining metatarsals. It is a flexor of the foot.

The **peroneus tertius** lies beneath the peroneus longus. It originates from the lateral surface of the fibula below the longus. The belly of the muscle is short and ends in a slender tendon that passes through the groove on the ventral border of the lateral malleolus. The groove is converted into a canal by a transverse ligament. After leaving the canal the tendon runs along the lateral margin of the foot. It is united to the sesamoid bone at the base of the proximal phalanx of the fifth digit by a band of tissue. It passes then to the lateral border of the extensor tendon of the fifth digit as it passes to the middle phalanx. It extends and abducts the fifth digit and flexes the foot.

The **peroneus brevis** lies beneath the other two peroneus muscles taking its origin by fleshy fibers from the distal half of the fibula. At the lateral malleolus it ends in a tendon which passes through the canal on the ventral surface of the malleolus. It is surrounded by a synovial bursa in the canal. Emerging on to the foot, it passes over the dorsal surface of the calcaneus on its lateral border. It inserts into the tubercle on the lateral side of the fifth metatarsal. It is an **extensor** of the foot. The **peroneus muscles are sometimes called fibularis.**

On figure 2, note particularly the origin areas (o.a.) indicated on the os coxa and also the cut ends of muscles studied indicating their general areas of origin and insertion.

The **adductor femoris** (*magnus*) is well shown in figure 2. Its origin is by muscle fibers along the whole length of the ischiopubic bones to the side of the symphysis and from the tendon of origin of the gracilis muscles. It inserts into the whole ventral (*caudal*) surface of the shaft of the femur. It is a powerful extensor of the thigh and an adductor.

Note the **patella** and the **patellar ligament** and also the **fibular collateral ligament.**

The **flexor hallucis longus** is mentioned here to call your attention to this lateral view of its origin by fleshy fibers from the whole interosseus ligament, the shaft and head of the fibula and from the ventral surface of the tibia. It will be discussed at greater length in reference to Plate 44.

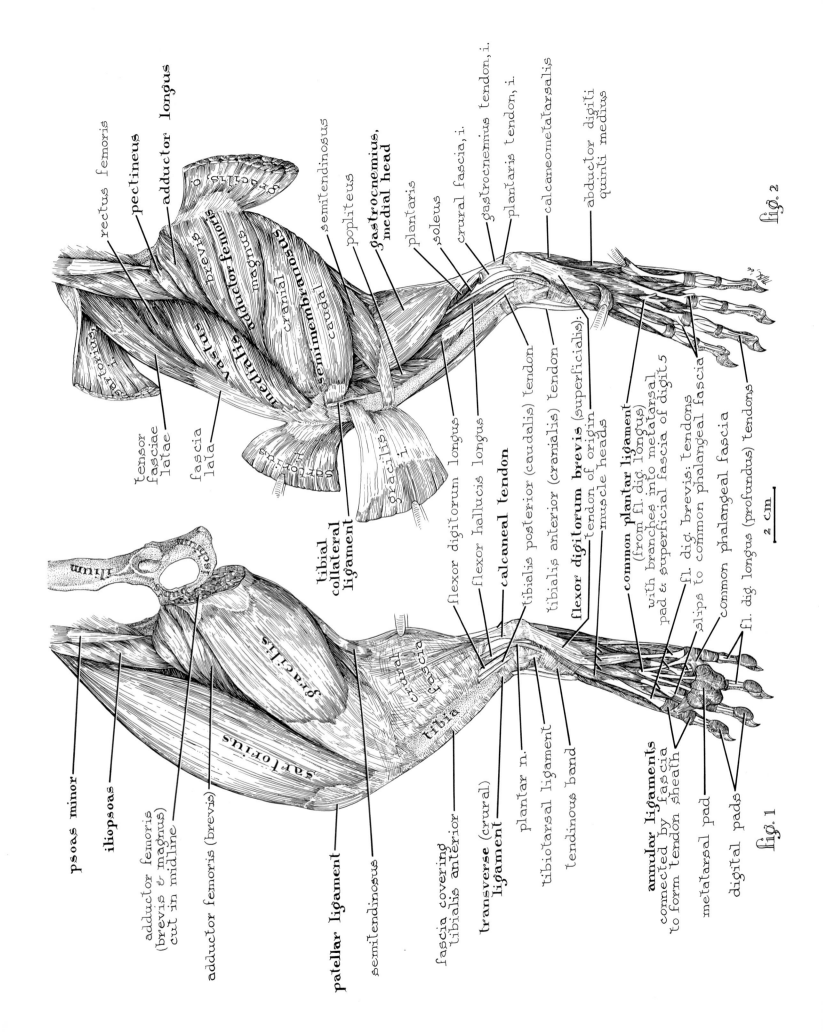

rectus femoris

pectineus

adductor longus

gracilis o.

semitendinosus

popliteus

gastrocnemius, medial head

plantaris

soleus

crural fascia, i.

gastrocnemius tendon, i.

plantaris tendon, i.

calcaneometatarsalis

abductor digiti quinti medius

vastus medialis

adductor femoris brevis

adductor magnus

semimembranosus

caudal

sartorius i.

gracilis, i.

ilium

ischium

tuber ischii

gracilis

crural fascia

tibia

sartorius

tensor fasciae latae

fascia lata

tibial collateral ligament

flexor digitorum longus

flexor hallucis longus

calcaneal tendon

tibialis posterior (caudalis) tendon

tibialis anterior (cranialis) tendon

flexor digitorum brevis (superficialis): tendon of origin muscle heads

common plantar ligament (from fl. dig. longus) with branches into metatarsal pad & superficial fascia of digit 5

fl. dig. brevis: tendons slips to common phalangeal fascia

common phalangeal fascia

fl. dig. longus (profundus) tendons

fig. 2

2 cm

psoas minor

iliopsoas

adductor femoris (brevis & magnus) cut in midline

adductor femoris (brevis)

patellar ligament

semitendinosus

fascia covering tibialis anterior

transverse (crural) ligament

plantar n.

tibiotarsal ligament

tendinous band

annular ligaments connected by fascia to form tendon sheath

metatarsal pad

digital pads

fig. 1

Figure 1 shows the muscles of the thigh with the fascia removed. The **crural fascia** has been left on the leg (*shank*) and the **patellar ligament** is shown. The **transverse crural ligament, tibiotarsal ligament and annular ligaments** and the tendon sheaths between the latter should be observed. Note the metatarsal and digital pads on the hindpaw. At the top of the drawing the psoas minor and iliopsoas muscles appear. They are muscles of the ventral side of the vertebral column and will be shown in later drawings in their entirety. The os coxa is placed in figure 1 for reference.

The **sartorius muscle** is a broad band of muscle which runs down the medial surface of the thigh and extends onto the cranial border. Its origin is from the cranial half of the crest of the ilium and the medial half of the cranial border. Its insertion is on the medial border of the proximal end of the tibia, the medial epicondyle and the patella and adjacent fasciae. It flexes the thigh and possibly the shank.

The **gracilis** is a subcutaneous muscle on the ventral half of the medial surface of the thigh. It is flat and thin and has its origin by a tendon from the caudal three-fourths of the pelvic symphysis. The tendon may be common to the two sides for part of its attachment and is also closely related to the origin of the adductor femoris. It inserts by a thin aponeurosis, in part continuous with that of the sartorius, into the medial surface of the tibia near its proximal end. It adducts the thigh and extends the hip joint.

The **flexor digitorum brevis** (*superficialis*) is visible on the plantar surface of the hindpaw in figures 1 and 2. As seen in figure 2 it is a continuation of the plantaris tendon. It has a broad tendon of origin. It divides into four slips distally which diverge to the four toes. Each ends in a flat tendon at the distal end of the metatarsal and is wrapped around the tendon of the deep flexor which perforates it and divides it into two tendons distad of the first annular ligament. The two halves of the tendon unite beneath the perforating tendon and they pass together with the perforating tendon through the two **annular ligaments.** The two annular ligaments are attached by fascia so as to form a tendon sheath between them which is lined by synovial membrane. Before the tendons passed into the proximal annular ligament the two middle tendons unite— each with the tendon of the corresponding lumbrical muscle. Each also gives off a branch to the common phalangeal fascia. Their function is to flex the middle phalanges of the digits.

Figure 2 with the sartorius and gracilis muscles cut and pulled aside at their origin and insertion ends and the crural fascia removed uncovers a number of other muscles some of which we have described previously, others we have not. All of them are seen from a different point of view. Note the **tibial collateral ligament** between the femur and tibia. It is one of the supporting and binding ligaments of the knee joint. Note also the **medial head of the gastrocnemius** which we discussed with the lateral head—also the plantaris and soleus.

The **vastus medialis,** another component of the quadriceps femoris is seen here on the medial side of the femur partly overlapped by the fascia lata. Its origin is on the shaft of the femur in the region of the linea aspera. It inserts into the medial border of the patella and the ligamentum patellae.

The **adductor longus** is a thin muscle which covers the dorsal half of the medial surface of the adductor femoris. It originates by fleshy fibers from the cranial border of the pubis and inserts into the linea aspera of the femur. This muscle forms a part of the boundary of a depression among the muscles on the proximal medial part of the leg called the iliopectineal depression which contains the femoral artery and vein and saphenous nerve. The medial edge of the adductor longus is subcutaneous; its lateral edge is against the pectineus.

The **adductor femoris** (*magnus and brevis*) lies between the semimembranosus and the adductor longus and is the most massive adductor of the hip joint. It arises along the whole length of the pelvic symphysis and adjacent parts of the ramus of the ischium. It inserts along nearly the whole length of the lateral lip of the linea aspera on the caudal surface of the femur.

The **pectineus muscle** originates by muscular fibers from the cranial border of the pubis. It passes over the iliopsoas muscle and has its insertion on the shaft of the femur just distad of the lesser trochanter and between the adductor femoris and the vastus medialis. It adducts the thigh.

The **semimembranosus** is a large mass of muscle consisting of caudal and cranial portions, the cranial portion bordering the adductor femoris (*magnus*). It lies medial to the semitendinosus and lateral to the gracilis. Its origin is by short tendon fibers from the caudal border of the tuberosity and ramus of the ischium. It inserts into the medial surface of the femur on the epicondyle and into the adjacent medial surface of the tibia behind the tibial collateral ligament. It is an extensor of the thigh and a flexor of the shank.

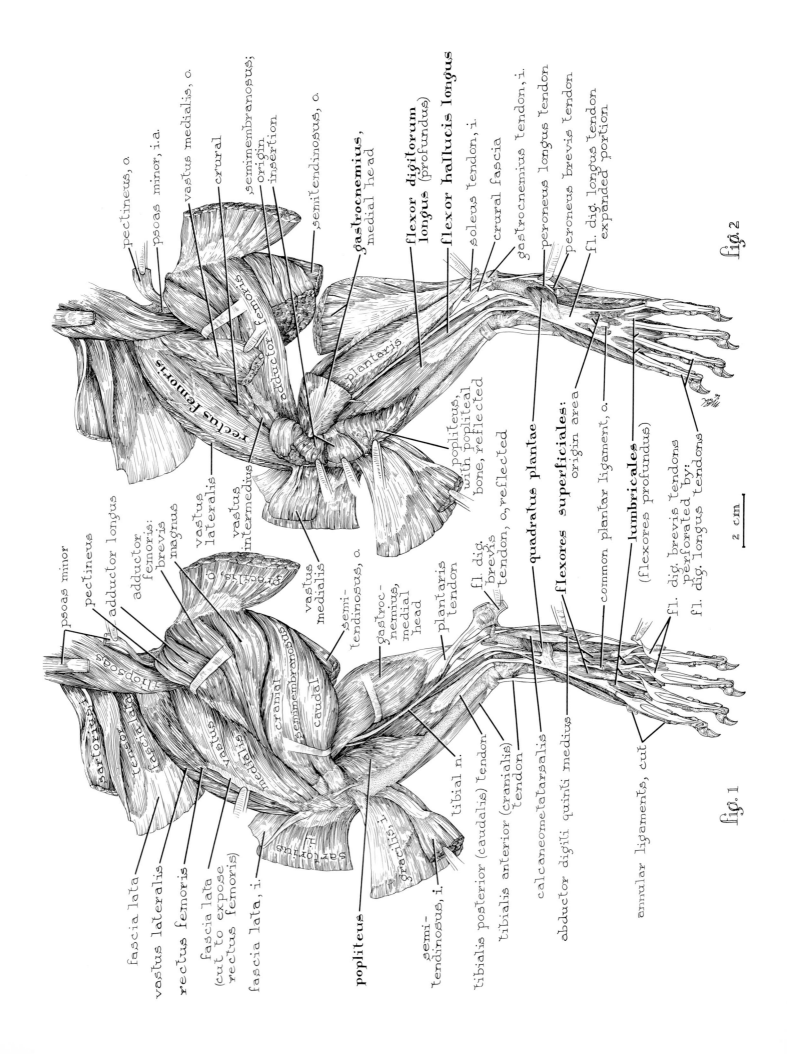

pectineus, o.

psoas minor, i.a.

vastus medialis, o.

crural

semimembranosus; origin insertion

semitendinosus, o.

gastrocnemius, medial head

flexor digitorum longus (profundus)

flexor hallucis longus

soleus tendon, i.

crural fascia

gastrocnemius tendon, i.

peroneus longus tendon

peroneus brevis tendon

fl. dig. longus tendon expanded portion

adductor femoris

rectus femoris

plantaris

popliteus, with popliteal bone, reflected

fl. dig. brevis tendon, o., reflected

quadratus plantae

flexores superficiales: origin area

common plantar ligament, o.

lumbricales (flexores profundus)

fl. dig. brevis tendons perforated by: fl. dig. longus tendons

fig. 2

fascia lata

vastus lateralis

rectus femoris

fascia lata (cut to expose rectus femoris)

fascia lata, i.

psoas minor

pectineus

adductor longus

adductor femoris: brevis magnus

vastus lateralis

vastus intermedius

vastus medialis

iliopsoas

sartorius

tensor fasciae latae

fascia lata

vastus medialis

sartorius

gracilis, i.

semimembranosus: cranial caudal

semitendinosus, o.

gastrocnemius, medial head

plantaris tendon

popliteus

semitendinosus, i.

tibial n.

tibialis posterior (caudalis) tendon

tibialis anterior (cranialis) tendon

calcaneometatarsalis

abductor digiti quinti medius

fl. dig. brevis tendon, o.

annular ligaments, cut

fig. 1

2 cm

Figure 1 reviews for us the muscles in the region of the thigh. Note the relationship of the iliopsoas and the tendon of the psoas minor to the pectineus. The fascia lata has been cut to expose the rectus femoris.

In figure 2 in this same region the rectus femoris is further exposed by cutting the vastus medialis and pulling it back from its origin. Also exposed by this cut is the vastus intermedius, the fourth component of the quadriceps femoris, and a small crural muscle which I consider a part of the vastus medialis. Strauss-Durckheim call it the crural.

The **popliteus** is a triangular muscle having its origin by a long tendon on the plantar surface of the lateral epicondyle of the femur. Its tendon contains a sesamoid bone, the popliteal bone. The tendon invaginates the joint capsule and passes under the fibular collateral ligament. The muscle passes obliquely and medially over the popliteal space to the medial side of the tibia where it inserts on the proximal third of the tibia's ventral and medial surface. It flexes the knee joint and rotates the leg inward.

Before going to a consideration of the flexor digitorum longus (profundus) it might be well to consider some of the interpretations given to the flexor muscles of the digits. In speaking of the flexor digitorum brevis it was indicated that the term superficialis was used by some authorities instead of brevis and that the term included also the muscle discussed earlier in this book as the plantaris (Miller, Christensen and Evans, 1964). By this interpretation which seems a reasonable one, the flexor digitorum superficialis would have two fleshy portions, the large plantaris and the small digital mass on the plantar surface of the foot.

The cat has a reduced first digit (hallex) and the tendon of the flexor hallucis longus has become joined with that of the flexor digitorum longus. The head corresponding to the flexor hallucis longus is larger than that of the flexor digitorum longus. On the plantar surface of the foot the small tendon of the flexor digitorum longus joins the larger tendon of the flexor digitorum hallucis to form the deep plantar tendon. It is common to look upon this whole complex as the two-headed flexor digitorum profundus (Miller, Christensen and Evans, 1964). In this book the flexor digitorum longus tendon is considered to extend to the distal phalanges of the digits.

The **flexor hallucis longus** lies against the caudal surface of the tibia and fibula. It originates on the proximal three-fifths of caudal surface of the fibula, from the tibia distad of the lateral oblique line to within 1 centimeter of the distal end of the shaft and from the whole interosseus membrane. The muscle is large and fusiform and ends in a strong flat tendon which passes over the groove on

the talus and over the groove on the sustentaculum tali. These two grooves are converted into a canal by fibrous transverse ligaments and the canal is lined with synovial fluid—a tendon sheath. Upon emerging from this canal the tendon broadens and receives on its medial side the tendon of insertion of the flexor digitorum longus.

The **flexor digitorum longus** is a flat, short muscle lying lateral to the popliteus and medial to the flexor hallucis longus. It is narrow at its origin where it arises from the head of the fibula, the caudal surface of the tibia over its proximal half and from an aponeurosis between the tibia and the tibialis caudalis. At the distal end of the lower third of the tibia it tapers into a narrow tendon which runs along the caudomedial border of the femur with the even thinner tendon of the tibialis caudalis. Together these tendons pass through the groove of the medial malleolus with the tibialis caudalis tendon lying craniad. They are enclosed in a tendon sheath. The tendon of the flexor digitorum longus now joins that of the flexor hallucis longus. This common tendon gives origin, on its outer surface, to the lumbricales and to the common plantar ligament. The common tendon continues to broaden until it reaches the middle of the metatarsals where it divides into four tendons which insert on the distal phalanges of the digits. To review their relationship to the flexor digitorum brevis (superficialis) to the tendon sheaths and annular ligaments return to the description of the last plate, page 115.

The **lumbricales** muscles, six in number, may be divided into groups of three. The three larger ones, the **flexores superficialis,** have their origin from the outer surface of the expanded portion of the tendon of the flexor digitorum longus and insert by slender tendons which unite with the tendons of the flexor digitorum brevis at their entrance to the first annular ligament and which pass to the three lateral digits (fig. 1). The one to the fifth (*lateral*) digit may be absent. The other three lumbricales, the **flexores profundus,** have their origins in the longus tendon in the three intervals between the four divisions of that tendon. They insert by very slender tendons into the medial sides of the third, fourth and fifth digits. They move the third, fourth and fifth digits in a medial direction.

The **quadratus plantae** is sometimes called the plantar head of the flexor digitorum longus. It is a thin flat muscle which has its origin on the dorsal part of the lateral surface of the cuboid and calcaneus. It inserts by a flat tendon which lies between the flexor digitorum brevis (superficialis) and flexor digitorum longus (profundus) tendons, into the medial surface of the flexor digitorum longus. It supports the longus (profundus) tendon, helping to hold it in position.

117

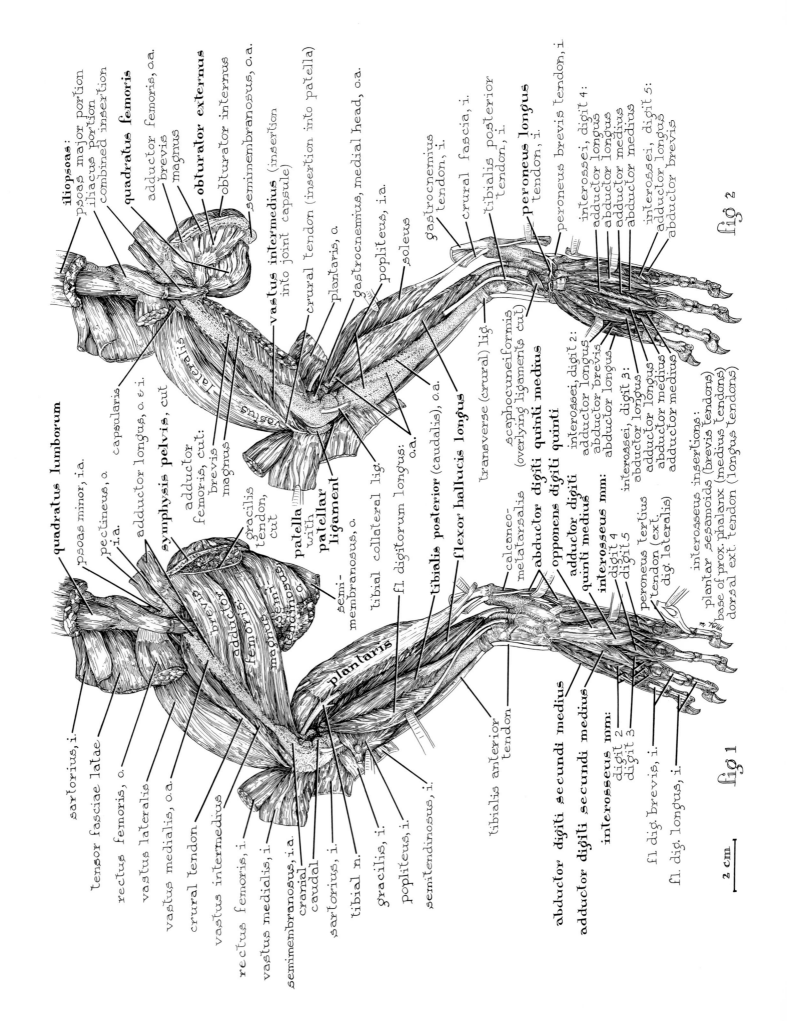

iliopsoas:
psoas major portion
iliacus portion
combined insertion

quadratus femoris

adductor femoris, o.a.
brevis
magnus

obturator externus

obturator internus

semimembranosus, o.a.

vastus intermedius (insertion
into joint capsule)

crural tendon (insertion into patella)

plantaris, o.

gastrocnemius, medial head, o.a.

popliteus, i.a.

soleus

gastrocnemius
tendon, i.

crural fascia, i.

tibialis posterior
tendon, i.

peroneus longus
tendon, i.

peroneus brevis tendon, i

interossei, digit 4:
adductor longus
abductor longus
adductor medius
abductor medius

interossei, digit 5:
adductor longus
abductor brevis

fig 2

quadratus lumborum

psoas minor, i.a.

pectineus, o
i.a.

capsularis

adductor longus, o. e.i.

symphysis pelvis, cut

adductor
femoris, cut:
brevis
magnus

gracilis
tendon,
cut

patella
with
patellar
ligament

semi-
membranosus, o.

tibial collateral lig.

fl digitorum longus:
o.a.

tibialis posterior (caudalis), o.a.

flexor hallucis longus

transverse (crural) lig.

scaphocuneiformis
(overlying ligaments cut)

calcaneo-
metatarsalis

abductor digiti quinti medius

opponens digiti quinti

adductor digiti
quinti medius

interosseus mm:
digit 4
digit 5

interossei, digit 2:
adductor longus
abductor brevis
abductor longus

interossei, digit 3:
abductor longus
adductor longus
abductor medius
adductor medius

peroneus tertius
tendon (ext.
dig. lateralis)

interosseus insertions:
plantar sesamoids (brevis tendons)
base of prox. phalanx (medius tendons)
dorsal ext. tendon (longus tendons)

sartorius, i.

tensor fasciae latae

rectus femoris, o.

vastus lateralis

vastus medialis, o.a.

crural tendon

vastus intermedius

rectus femoris, i

vastus medialis, i.

semimembranosus, i.a.
cranial
caudal

sartorius, i.

tibial n.

gracilis, i.

popliteus, i.

semitendinosus, i.

abductor digiti secundi medius

adductor digiti secundi medius

interosseus mm:
digit 2
digit 3

fl dig. brevis, i.

fl. dig. longus, i.

tibialis anterior
tendon

2 cm

fig 1

The thigh muscles have been dissected further to show the very deep muscles on the femur. It is best to study these two figures together and make close comparisons. Figure 2 represents our final dissection and if you have followed conscientiously through the complete series on the hindlimb, you should appreciate more its complexity, understand better the diversity in its movements and know something also of its limitations.

The crural muscle and its tendon of insertion into the patella is clearly shown. The **vastus intermedius'** insertion into the knee joint capsule is clear. The **quadratus lumborum** described in detail in reference to Plate 80 on page 223 is shown here for the first time in the series in order to tie it in with the muscles which are close to its insertion. It is a muscle of the group found ventral to the vertebral column. The **iliopsoas** has been cut away caudally to reveal the quadratus lumborum and at the same time to distinguish clearly between the psoas major and iliacus which are so close caudally that they are called by the combined name of iliopsoas. Their combined insertion is shown in figure 2.

The **symphysis pelvis** (*ischiopubic symphysis*) is shown cut and in figure 2 the **obturator externus** is shown with the obturator internus covering their respective sides of the obturator foramen. The **quadratus femoris** described as one of the muscles in the lateral series is shown.

The **obturator externus** is a triangular muscle which lies under the adductor femoris whose origin area (o. a.) is shown here. It arises by muscular fibers from the medial lip of the obturator foramen and from both dorsal and ventral surfaces of the rami of the pubis and ischium. Its fibers converge to a strong flat tendon which inserts into the proximal part of the trochanteric fossa. It is a flexor and outward rotator of the thigh.

On the leg or shank, the **tibialis posterior** (*caudalis*) is shown by pulling back the flexor digitorum longus (fig. 1). On figure 2 the tibialis posterior is removed and its origin area is outlined. It is a flat muscle lying between the flexor digitorum longus and flexor hallucis longus and originating in part on the caudal surface of the tibia as shown here and also on nearly the whole medial surface of the head of the fibula and from the outer surface of the flexor hallucis longus. The muscle ends in a slender tendon at about the half-way point on the tibia and passes parallel to the tendon of the flexor digitorum longus and with it through the dorsal groove on the medial surface of the distal end of the tibia and is described in connection with that muscle. The tendon then passes onto the plantar surface of the foot, passes through a groove on the ventral surface of the scaphoid bone and then divides and inserts into the outer tuberosity of the scaphoid and onto the

proximal end of the ventral surface of the middle cuneiform. It is an extensor of the foot.

The **flexor hallucis longus** is shown here through most of its extent. The insertion area for the medial head of the gastrocnemius is shown in part. The **plantaris** (*flexor digitorum brevis superficialis*) is shown more completely than in previous plates.

A number of muscles of the sole of the foot are revealed by removal of the broad tendons of the flexor digitorum superficialis and profundus. Three of these involve the fifth digit.

The **abductor digiti quinti medius** (fig. 1) originates on the ventral surface of the calcaneus and from the fifth metatarsal. It inserts by a thin tendon into the lateral side of the base of the proximal phalanx of the fifth digit.

The **adductor digiti quinti medius** originates with the adductor digiti secundi medius from the middle of the ligament covering the peroneal canal. It inserts on the medial side of the base of the proximal phalanx of the fifth digit.

The **opponens digiti quinti** originate from the middle of the ligament which covers the peroneal canal and insert on the medial side of the shaft of the fifth metatarsal.

The following muscles are members of the **interosseus muscle group.** Only two will receive special mention in this discussion. All of them are detailed by careful listing on the illustration.

The **adductor digiti secundi medius** is the fifth part of the interosseus of the second digit with its origin from the middle of the ligament covering the peroneal canal, along with the adductor of the fifth digit. The two muscles diverge and it inserts into the outer side of the base of the first phalanx of the second digit.

The **abductor digiti secundi medius** (*intermedius*) has its origin from the ventral process of the lateral cuneiform with four other parts of the interossei, three of which become abductors. This one inserts into the medial side of the base of the proximal phalanx (fig. 1). The other abductors, brevis and longus can be seen on figure 2. Brevis inserts into the medial sesamoid of the metatarsophalangeal joint; longus into the medial side of the tendon of the extensor digitorum communis near the distal end of the proximal phalanx. The fourth portion inserts into the lateral side of the extensor tendon near the distal end of the proximal phalanx and is therefore an adductor longus digiti secundi.

The interossei of the third, fourth and fifth digits are essentially alike. They form the adductories and abductories digiti tertii and quarti.

Two small tarsal muscles are shown; the calcaneometatarsalis shown on figure 1; the scaphocuneiformis on figure 2.

If you have done all of the dissection which has been described and illustrated on the muscular system, you may need a new specimen with which to continue your study. In most courses in anatomy where cats are used, the dissection of muscles is limited and the same specimen can be used for the study of the celom and viscera. If a new specimen is used, the skin can be left on and it will help to protect the specimen from drying out.

To open the celom, a short transverse incision should be made just craniad of the pubic symphysis large enough so you can insert a finger. Lift up the body wall from the underlying viscera and make a longitudinal incision with scissors or scalpel about one-half inch to the left of the **linea alba** (*median line*) and terminating at the level of the xiphoid process of the sternum. At this point it will be necessary to cut through the costal cartilage to the left side of the sternum in order to continue the incision craniad. The diaphragm also is present in this area. Study Plate 46 before you make this cut to get an idea of the relationship of structures. Great care must be taken not to damage the mediastinal pleura which attaches along the median line of the internal surface of the sternum. When you reach the first rib, you will be cutting bone or the rib may be separated easily at the sternocostal junction. Also at the first rib you leave the thoracic cavity and are in the neck region. Your incision may be extended for about 4 centimeters into the neck but be careful to lift the skin before you cut to avoid injury to the underlying structures.

Additional incisions should now be made to enable you to open the cavities more widely. An incision along the right side of the linea alba starting at the pubis can be carried craniad to the point where it is craniad of the membrane supporting the urinary bladder—the median unbilical ligament. At that point the incision can be carried to the left to join the left incision. Next cut the body wall laterad from the area of the pubic symphysis and also in the area just behind the diaphragm following the line of the diaphragm's attachment to the body wall. These walls can now be spread broadening your view of the viscera.

Craniad of the diaphragm make a cut along the right side of the sternum similar to that on the left and join it craniad of the sternum with the left incision. Now cut the side walls of the thorax laterad just craniad of the diaphragm and spreading each wall to the side, nick each of the ribs near the vertebral column using your scalpel. A little pressure applied to each rib individually will cause it to break at the nicked point. When this is done on each side, the thoracic walls can be further spread. It may help to cut laterally along the caudal surface of the first rib

but be very cautious in this area that you do not damage the adjacent structures.

By these procedures you have opened the **thoracic cavity** craniad of the diaphragm and the **abdominal** (*peritoneal*) **cavity** caudad of the diaphragm. You have preserved the median supporting ligament of the urinary bladder. In the thoracic region you have opened the two **pleural cavities**—one on each side of the **mediastinal pleura.** You can see that these pleural cavities before dissection were completely closed to the outside and separated from each other by the mediastinal pleura.

Though it is not clear as a result of gross examination all of the internal surfaces of the body wall and the cranial and caudal surfaces of the diaphragm are covered with a very thin transparent tissue composed of flat cells, one layer thick, the **mesothelium.** It is supported underneath by a basement membrane of connective tissue and the two together constitute a **serous membrane.** Beneath the serous membrane in the walls of the thoracic cavity is a sheet of connective tissue, the **fascia endothoracica.** Similarly in the peritoneal or abdominal cavity there is under the serous membrane a connective tissue layer, the **transversalis fascia** (*endoabdominal fascia*).

Serous membranes not only line the walls of the celom where they are called **parietal serous membranes,** but they are reflected over the organs which constitute the **viscera** thus giving the organs some support and serving as passageways and therefore as support for blood vessels, nerves and lymphatics which supply these organs. These reflections of the serous membranes on to the organs are composed of a layer of serous membrane on each side. They are double walled. They are known collectively and generally as mesenteries but they are given more specific designations by anatomists. Hence, the term **mesentery** is restricted to those membranes which connect the **hollow organs** of the digestive system to the body wall; **ligaments** apply to the double serous membranes which connect other viscera to the body walls like the broad ligament of the uterus or the falciform ligament of the liver; **omenta** are double serous membranes connecting organs like the lesser omentum between the stomach and the liver. The serous membranes on the surfaces of the viscera are called **visceral serous membranes.**

Anatomists are as specific in the naming of parietal and visceral serous membranes as they are in naming the mesenteries. On the walls of the thoracic cavities, since the lungs are found there, we have the **parietal pleura** and where it is reflected onto the lung dorsally, the **pulmonary ligament** and on the lung itself, the **visceral pleura** (Plates 46 and 55). The parietal pleura is also broken down in our nomenclature to the **costal pleura** on the part of the thoracic wall supported by the ribs; the **diaphragmatic pleura** on the diaphragm; and the **mediastinal pleura** where the pleura from the two sides form the walls of a central thoracic area, the **mediastinum.** Similarly, below the diaphragm we have the **parietal peritoneum** on the walls of the cavity and **visceral peritoneum** on each organ. The mesenteries, omenta and ligaments have specific names often reflecting the names of the organs they support, such as mesocolon.

The same kind of naming procedure applies to the various parts of the celom. Thus we have the **pleural cavities** between the parietal and visceral pleura and the **peritoneal cavity** between the **parietal** and **visceral peritoneum.**

To complete the dissection of the celom, remove the mediastinal pleurae being careful not to damage the contents of the **mediastinum.** The largest organ in the mediastinum is the heart and notice that it is enclosed in its own sac. The outer fibrous covering of the sac is the **fibrous pericardium** and is comparable to the fibrous layer under the parietal pleura, the fascia endothoracica. Cut it open and you see the heart inside but here as with other organs the heart is covered by a visceral serous membrane, the **visceral pericardium,** and the fibrous coat is covered internally with a **parietal pericardium.** The visceral and parietal pericardia are continuous at the base of the heart. The space between them, a part of the celom, is the **pericardial cavity.**

In the living animal the serous membranes, visceral and parietal, lie one against the other so that the pleual and pericardial cavities that we speak about are only **potential spaces.** There is only a thin film of serous fluid on their surfaces so that they can move freely one against the other. The same thing is generally true about the peritoneal cavity though the situation becomes more complicated there because of the variety of organs and the functions they perform. If parietal and serous membranes do become joined due to inflammation, for example, the condition is referred to as **adhesions** and they may interfere with the normal functioning of the organs.

It is common practice to refer to the various organs as being in the celom; the heart in the pericardial cavity and the lungs in the pleural cavities, for example. A sophisticated anatomist knows that this is not strictly the case; it only appears to be so. The pleural cavity is between the parietal and visceral pleura—the lung is not there—just a small amount of fluid. A knowledge of descriptive embryology helps you understand these things but lacking that perhaps the simple schematic figure below (Fig. 6) will help you to see how it comes about.

We ordinarily say that these are closed cavities but you will see one important exception in your studies. In the female cat there is a direct passage from the peritoneal cavity to the outside through the reproductive organs.

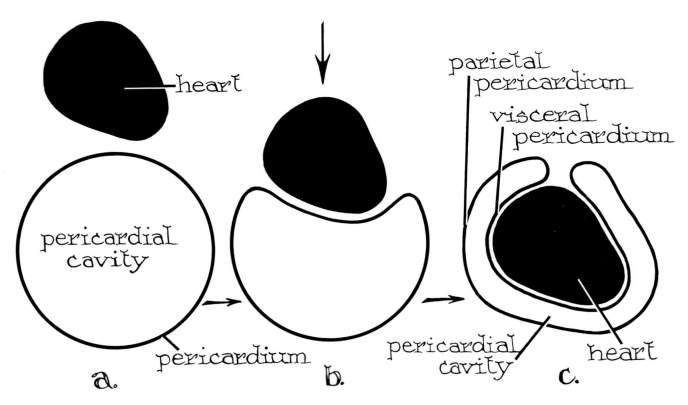

Fig. 6. *Schematic representation to show how the heart develops its relationship to the pericardium and pericardial cavity. a, heart and pericardium separated; b, heart encroaches upon pericardium; c, heart and visceral and parietal pericardia in normal relationship (pericardial cavity exaggerated in size). (Crouch,* Functional Human Anatomy, *Lea & Febiger.)*

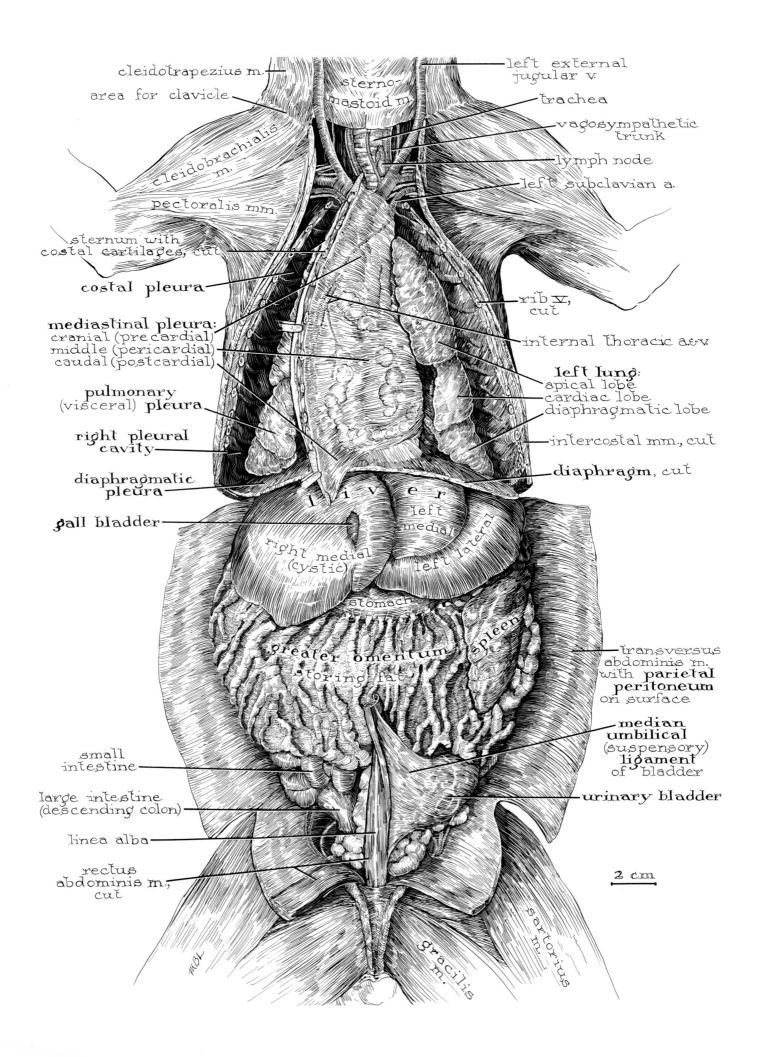

cleidotrapezius m.

area for clavicle

cleidobrachialis m.

pectoralis mm.

sternum with costal cartilages, cut

costal pleura

mediastinal pleura:
cranial (precardial)
middle (pericardial)
caudal (postcardial)

pulmonary (visceral) pleura

right pleural cavity

diaphragmatic pleura

gall bladder

small intestine

large intestine (descending colon)

linea alba

rectus abdominis m., cut

sterno-mastoid m.

left external jugular v.

trachea

vagosympathetic trunk

lymph node

left subclavian a.

rib V, cut

internal thoracic a.&v.

left lung:
apical lobe
cardiac lobe
diaphragmatic lobe

intercostal mm., cut

diaphragm, cut

liver
left medial
right medial (cystic)
left lateral

stomach

spleen

greater omentum storing fat

transversus abdominis m. with parietal peritoneum on surface

median umbilical (suspensory) ligament of bladder

urinary bladder

gracilis m.

sartorius m.

2 cm

This plate and the following one provide good views of the celomic cavities and viscera as seen from the ventral aspect. They should be studied for general relationships of organs to each other and to the supporting serous membranes, *i.e.* the mesenteries, omenta and ligaments. All structures shown will be presented later in greater detail.

In this plate the lateral thoracic walls have been partially removed but the sternum is left and pulled to one side to show how the **mediastinal pleura** divides the thorax into two separate **pleural cavities.** The heart and its pericardial membranes are enclosed between the two halves of the mediastinal pleura in the **mediastinum.** Also other organs are found in the mediastinum as will be seen in the next illustration, Plate 47.

Below the **diaphragm** the body wall has been cut ventrally and spread laterally to show the organs of the **peritoneal cavity** in a relatively undisturbed condition. The **liver** occupies most of the area caudad of the diaphragm. The stomach is largely hidden dorsal to the left lobes of the liver. The **spleen** can be seen on the left side but the remaining abdominal organs are nearly covered by the **greater omentum** which hangs like an apron from the greater curvature of the stomach. At its distal, free end it lies dorsal to the **urinary bladder.** Note that columns of fat run in the greater omentum following the pathways of small arteries. This is one of the major storehouses of fat in an obese specimen. Between the columns of fat the omentum is so thin that it is transparent. Some segments

of the small intestine and colon are seen in the right caudal part of the peritoneal cavity.

A part of the caudal midventral wall made up of the **linea alba** and portions of the **rectus abdominis muscles** are left in order to show the **median umbilical** (*suspensory*) **ligament** of the urinary bladder which attaches to the midventral line internally. The **urinary bladder** rests in the left caudal part of the peritoneal cavity.

The **spermatic cords** (not labeled) lie between the hind legs in the ischiopubic region. Craniad of the thoracic cavity in the lower neck region the trachea, the vagosympathetic trunk and some blood vessels are exposed.

The **pleural cavities** are well shown here. The right and left **pleural cavities** are completely separated by the **mediastinal pleurae** which also form between their layers the mediastinum described on page 127, Plate 47. The serous mediastinal pleura continues on to the walls of the rib cage as the **costal pleura** and on to the cranial surface of the diaphragm as the **diaphragmatic pleura.** These three components constitute the **parietal pleura.** The parietal pleura is in turn reflected at the roots of the lungs on to the lung surface as the **visceral** (*pulmonary*) **pleura.** These serous pleurae are all adherent to the underlying structures. Between the parietal and visceral pleurae is the **pleural cavity** which in the living subject is only a potential space.

The serous pleurae and the pleural cavities show the same relationship to the lungs as do the serous pericardia and pericardial cavity to the heart.

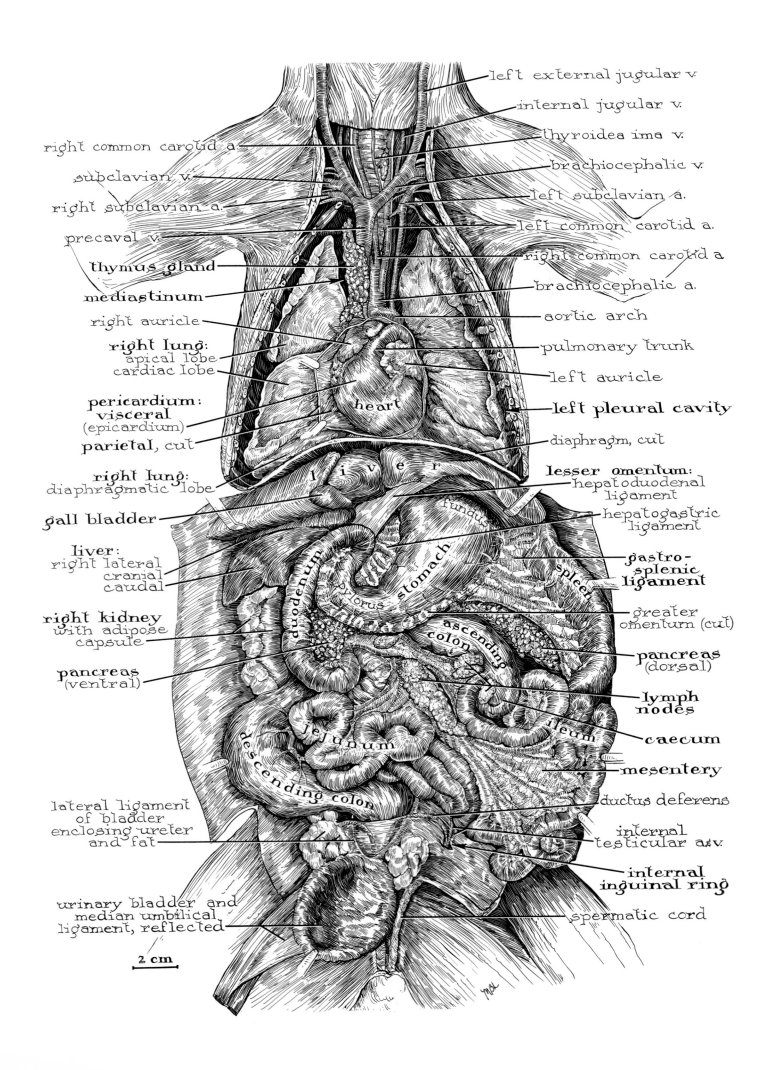

left external jugular v.

internal jugular v.

thyroidea ima v.

brachiocephalic v.

left subclavian a.

left common carotid a.

right common carotid a.

brachiocephalic a.

aortic arch

pulmonary trunk

left auricle

left pleural cavity

diaphragm, cut

lesser omentum:
hepatoduodenal ligament

hepatogastric ligament

gastro-splenic ligament

greater omentum (cut)

pancreas (dorsal)

lymph nodes

caecum

mesentery

ductus deferens

internal testicular a.&v.

internal inguinal ring

spermatic cord

right common carotid a.

subclavian v.

right subclavian a.

precaval v.

thymus gland

mediastinum

right auricle

right lung:
apical lobe
cardiac lobe

pericardium:
visceral
(epicardium)

parietal, cut

right lung:
diaphragmatic lobe

gall bladder

liver:
right lateral
cranial
caudal

right kidney
with adipose capsule

pancreas (ventral)

lateral ligament of bladder enclosing ureter and fat

urinary bladder and median umbilical ligament, reflected

l i v e r

fundus

stomach

pylorus

duodenum

ascending colon

spleen

ileum

jejunum

descending colon

heart

2 cm

The mediastinal pleura and sternum have been removed revealing the contents of the mediastinum. The ventral part of the **parietal pericardium** has been cut away and spread exposing the heart. The greater omentum has been largely removed except at its attachment to the greater curvature of the stomach and to the spleen. The viscera have been spread out to better reveal all structures. The urinary bladder has been pulled caudally and to the right to show its **lateral ligaments, ureters,** and the **ductus deferens.** A portion of the right medial lobe of the liver has been cut away and the remaining portion elevated to show the **cranial and caudal parts of the right lateral liver lobe.**

The **mediastinum,** revealed here by removal of its lateral walls, the mediastinal pleura, is divided into a **cranial** part anterior to the heart, a **middle** part containing the heart and a **caudal** part caudad of the heart. It may also be divided for purposes of description into **dorsal** and **ventral** portions by a frontal plane passing through the roots of the lungs. The cranial part contains the **thymus,** major blood vessels as the aortic arch, precava, and brachiocephalic artery, parts of the trachea and esophagus and thoracic duct. Besides the heart the middle part contains the descending aorta, parts of the trachea, esophagus, and thoracic duct, the bronchi and related blood vessels. The caudal part contains parts of the descending aorta, the esophagus, thoracic duct and vagus nerve. The heart, much of the thymus and the internal thoracic arteries and veins are among the organs in the ventral mediastinum. The esophagus, thoracic ducts and most of the descending thoracic aorta are in the dorsal mediastinum. (See Plates 46 and 78.)

The heart lies in a slightly oblique position in the middle mediastinum with its apex to the left of the median plane. The apex of the heart is also directed slightly ventrad, the base therefore slightly dorsad. The heart extends from about the fifth to the ninth ribs.

The serous pericardial membranes and cavity were described on page 121 in the discussion of Plates 46 to 48.

The serous parietal pericardium lines the inside of a **fibrous pericardium** and cannot be separated from it. The fibrous pericardium forms a complete sac enclosing the heart and the serous pericardium and the pericardial cavity. It attaches only at the base of the heart where it is continuous into the fibrous layers of the major blood vessels of the heart.

The alimentary tract, its supporting structures and related glands are shown in the peritoneal cavity. The peritoneal cavity, like the pericardial and pleural cavities, is lined with a serous membrane, the **parietal peritoneum** which is reflected on to the surfaces of the various organs forming double-walled serous membranes known as mesenteries, omenta and ligaments. On the organs the serous membrane is called the **visceral peritoneum.**

The **lesser omentum,** a part of the ventral mesentery of the embryo is shown between the lesser curvature of the stomach and the cranial part of the duodenum and the liver. It is divided into hepatoduodenal and hepatogastric ligaments. The **gastrosplenic ligament,** connecting the stomach and spleen, the greater omentum and the **mesentery** of the small intestine, much twisted or coiled as is the small intestine, are all parts of the dorsal mesentery. Other parts of the dorsal and ventral mesenteries are seen in later plates.

The sequence of organs of the digestive system can be followed: the stomach, duodenum, jejunum, ileum, ileocolic juncture, cecum, colon and rectum. The rectum is not seen on this plate though the descending colon can be seen leading to it. The liver and pancreas are glands of the digestive system. These organs of the digestive system lack the symmetry of other organs largely because of the necessity of fitting a long diversified tube into a short body. The small intestine alone is about three times that of the body of the cat. The colon in the specimen from which this drawing was made was in an atypical position. The ascending colon is normally on the right side, the descending colon on the left.

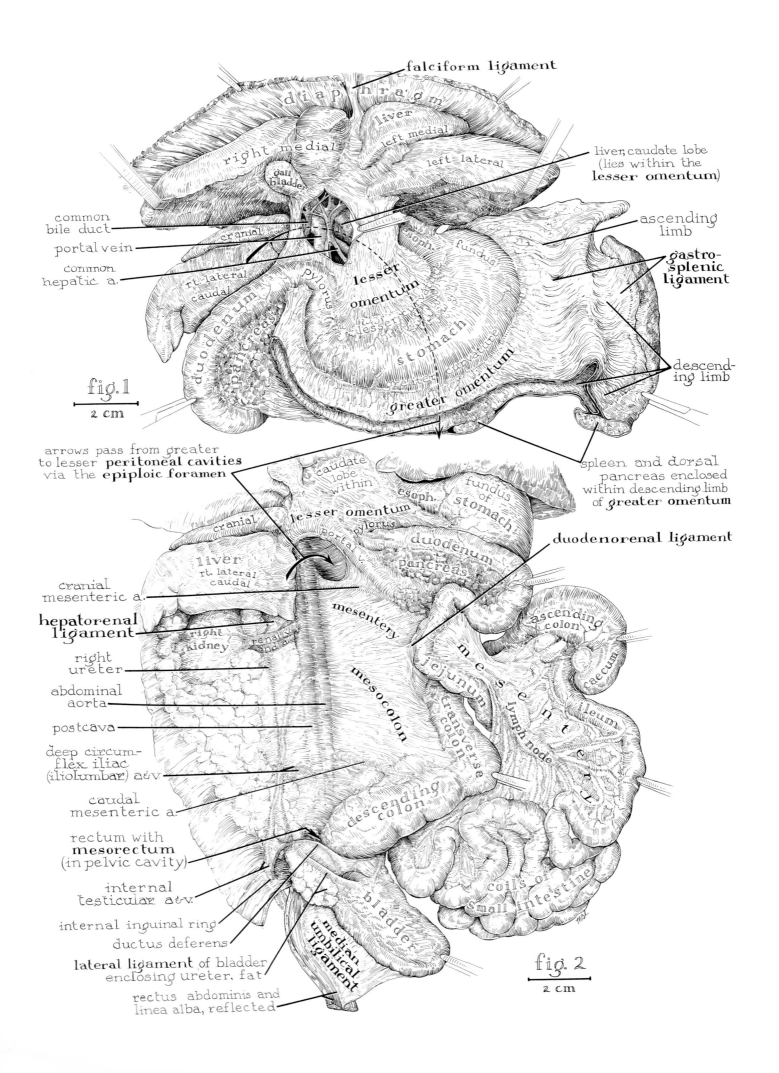

falciform ligament

diaphragm

liver
right medial
left medial
left lateral

gall bladder

liver, caudate lobe
(lies within the
lesser omentum)

common
bile duct

portal vein

common
hepatic a.

cranial

esoph.

fundus

ascending
limb

**gastro-
splenic
ligament**

rt. lateral
caudal

pylorus

duodenum

pancreas

lesser
omentum

lesser curvature

stomach

greater curvature

descend-
ing limb

greater omentum

fig. 1

2 cm

arrows pass from greater
to lesser **peritoneal cavities**
via the **epiploic foramen**

spleen and dorsal
pancreas enclosed
within descending limb
of **greater omentum**

caudate
lobe
within

esoph.

fundus
of
stomach

cranial

lesser omentum

pylorus

portal v.

duodenum

pancreas

duodenorenal ligament

liver
rt. lateral
caudal

cranial
mesenteric a.

mesentery

ascending
colon

**hepatorenal
ligament**

right
kidney

renal a. and v.

m e s e n t e r y

jejunum

caecum

lymph node

ileum

right
ureter

mesocolon

transverse colon

abdominal
aorta

postcava

deep circum-
flex iliac
(iliolumbar) a. & v.

caudal
mesenteric a.

descending
colon

coils of
small intestine

rectum with
mesorectum
(in pelvic cavity)

internal
testicular a. & v.

internal inguinal ring

ductus deferens

lateral ligament of bladder
enclosing ureter, fat

rectus abdominis and
linea alba, reflected

bladder

**median
umbilical
ligament**

fig. 2

2 cm

This plate shows more clearly and in greater detail some of the features of the mesenteries, omenta and ligaments which were seen in the preceding plate. Figure 1 emphasizes the relationships around the stomach, the greater and lesser omenta, the liver, spleen, pancreas and duodenum. Figure 2 deals mostly with structures lower down in the peritoneal cavity. The stomach, small and large intestines and pancreas are all pulled to the left side to uncover the dorsal wall of the peritoneal cavity and to disclose the **epiploic foramen** which leads into the cavity of the **greater omentum,** the omental bursa or **lesser peritoneal cavity.**

In Figure 1 an opening has been cut in the ventral wall of the **lesser omentum** which connects the lesser curvature of the stomach and a part of the duodenum with the liver. Through the opening can be seen the structures which pass through or lie within this omentum, namely the common bile duct, hepatic artery, portal vein and the caudate lobe of the liver. The broken-line arrow suggests the position of the epiploic foramen leading into the omental bursa. The cut edges of the **greater omentum** are seen attached to the greater curvature of the stomach. Since the greater omentum was derived by an elongation of the mesogastrium each of its walls has two serous layers like any other mesentery. The walls of the greater omentum are called the ventral or **ascending limb** and the dorsal or **descending limb.** The omental bursa between these two double walls is only a potential cavity and is closed normally except at the **epiploic foramen** through which it communicates with the **greater peritoneal cavity.** The

epiploic foramen shows best in figure 2 where its boundaries can be observed.

Returning again to figure 1 note that the spleen and the left end of the pancreas lie between the two layers of the descending limb of the greater omentum. That supporting the spleen and connecting it to the stomach is the **gastrosplenic ligament** or **omentum.** Cranially the **falciform ligament** extends from the diaphragm to the liver.

Figure 2 shows clearly the **median umbilical** and **lateral ligaments** of the bladder. The **mesorectum** is not visible but is continuous with the **mesocolon** which, beyond the colon craniad, becomes the **mesentery** of the small intestine. The mesorectum and mesocolon have broad attachments to the middorsal body wall but the mesentery has a very limited one near the caudal end of the kidneys, even though its intestinal border runs all the way from the ileum to the duodenum. The duodenal mesentery is drawn out at its caudal end into a fold, the **duodenorenal ligament** which connects the duodenum and kidney. Finally, note the short **hepatorenal ligament** which attaches the caudal division of the right lateral lobe of the liver to the kidney.

Some of the viscera are without mesenteries, ligaments or omenta but lie behind the parietal peritoneum of the body wall. They are said to be **retroperitoneal.** Examples are the kidneys, ureters, and blood vessels such as the abdominal aorta and the postcava as seen in figure 2.

The digestive and respiratory systems have close anatomical relationships in their cranial portions. At the caudal end of the pharynx they become separate and remain so through the rest of their parts. If we look at the phylogeny or the ontogeny of these two systems, we see even a closer relationship than is exhibited in adult mammals.

In most fishes the nasal cavities do not communicate with the oral cavity and do not form a functional organ of the respiratory system. They serve rather an olfactory function. In the Dipnoans (*lung fishes*) and in the amphibians the nasal cavities open into the oral cavity and they add to the olfactory function that of respiration. In the Reptiles the internal openings of the nasal cavities (*internal nares*) are shifted caudad by the development of the hard palate. In birds and mammals a soft palate is added behind the hard palate and the internal nares are shifted still farther caudad to open into the pharynx rather than into the oral cavity. Air then passes directly from the outside into the pharynx and the mouth serves only the digestive system. However, in respiratory infections when the nasal passageways become partially or completely occluded, the animal can resort to mouth breathing.

In fish the pharynx is perforated by gill slits, usually six pairs, which are situated between the skeletal gill arches which support the pharynx. Between the internal and external gill slits is the gill pouch and between the gill pouches the branchial bars which bear the internal or external gills which are used for respiratory purposes. The pharynx is now an essential organ of respiration as well as a food carrier. While amphibians have gill slits and internal and external gills in larval stages, gills (*external*) are found only in the adults of some urodeles (*tailed amphibians*). Gill slits are transitory in the development of reptiles, birds and mammals but gills are not found at any stage in the ontogeny of these animals.

The modifications of the pharynx are very great as it changes over during phylogeny from the essential respiratory organ of fishes to the relatively simple air and food passageway of higher vertebrates. The tympanic (*middle ear*) cavity and auditory tube represent an outgrowth of the first gill pouch, while the endodermal linings of other gill pouches persist as certain glands such as the thymus, parathyroids, tonsils and epithelial bodies. The thyroid gland develops as an outgrowth of the floor of the pharynx between the second gill arches.

Behind the last gill slits the **lungs** develop as an evagination of the midline of the pharyngeal floor. The single evagination elongates and divides to give rise to the two lungs. As lengthening continues in higher vertebrates, such as birds and mammals, the median evagination becomes the trachea and at its caudal end its divisions are the principle bronchi which lead into the lungs.

The Digestive System

The digestive system is the oldest of the body. Embryologically it is initiated at the time of gastrulation. Its beginnings are marked phylogenetically with the gastrovascular cavity and its entodermal lining as seen in the "gastrula-like" Coelenterates such as Hydra.

Basically, the digestive system, at least from the esophagus to the anus, is a tube within a tube—an inner tube (Fig. 7). Actually since the tube has become so much longer than the body which houses it, it has become very much coiled, particularly the small intestine. While this obscures the pattern indicated in Figure 7, it does not change it. Material can pass from mouth to anus with no membranous barriers to be crossed in traveling the whole route. The lumina (*insides*) of its organs are in direct communication with the outside world and are themselves, in a very real sense, a part of the external environment though each has its own peculiar ecology. To gain access to the internal environment by whatever route—digestive, respiratory, integumentary or excretory—material must pass through living membranous barriers—the linings of organs or the integument itself. Therefore, food, before it is available to the body, must be digested to enable it to pass through the walls of the stomach or intestine (*absorption*) to reach the blood or lymph components of the internal environment of the body.

The **organs of digestion** are those which carry out the functions of **ingestion, digestion, absorption** and **egestion.** Ingestion is the taking in of food at the mouth. Digestion refers to that series of complex mechanical and chemical processes which reduce the food from large molecules to smaller, simpler or soluble molecules capable of being absorbed. Absorption is the passing of food through the walls of the digestive organs to reach the bood and lymph. For adequate absorption, the digestive system must provide extensive surfaces. These are provided by lengthening the tube and by the development of villi and other structures on the absorbing surfaces. Egestion is the elimination of indigestible and other materials at the caudal end of the digestive tube, through the anus. Another problem which the digestive system must meet is that of moving the food mass through the organs at a pace commensurate with efficient digestion and absorption.

The organs which carry out these functions constitute the digestive tube and accessory organs. Those forming the tube are from the cranial to the caudal end, the mouth, pharynx, esophagus, stomach, small intestine, large intes-

131

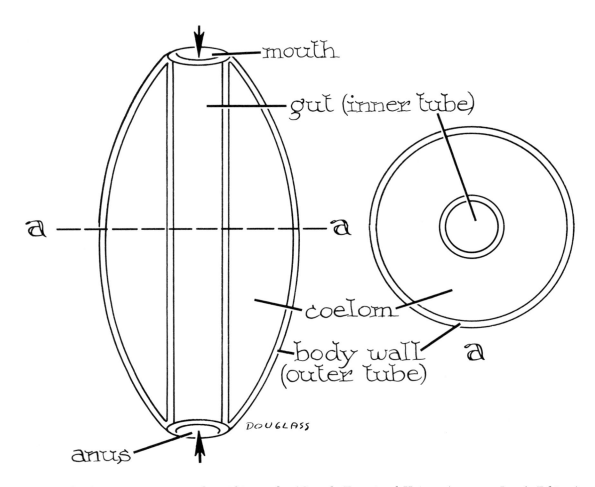

Fig. 7. *The digestive system—a tube within a tube.* (Crouch, Functional Human Anatomy, *Lea & Febiger.*)

tine and anus (Plate 47). That portion caudal to the pharynx, *i.e.* from esophagus to anus is often called the **alimentary canal.** The accessory organs are the teeth, tongue, salivary glands, liver, gallbladder and pancreas. The wall of the digestive tube is lined throughout with mucous membrane which is richly supplied with secretory cells and intrinsic glands such as gastric, duodenal, intestinal and mucous.

The **mouth** (os), the first organ of the digestive system, in a restricted sense would include only the opening between the lips. More commonly it is considered to be the oral cavity which extends from the lips to the pharynx. It is a narrow cavity cranially, widens out in the central region and as far back as the last teeth where it narrows again to form the **isthmus of the fauces.** It is divided into the **vestibule** (*vestibulum oris*) and the **oral cavity proper** (*cavum oris proprium*). (Plates 49 and 55)

The **vestibule** lies between the teeth and gums internally, and the lips and cheeks externally. It opens to the outside anteriorly through a U-shaped slit, the oral fissure between the lips. When the mouth is closed, the vestibule communicates with the oral cavity proper by means of interdental spaces and through the spaces between the lower canines and premolars the **diastema** (Plate 49,

fig. 2). When the mouth is open, the vestibule is greatly reduced and the oral cavity proper is much enlarged.

The **lips** are thick folds of skin which are hairy externally and covered internally with mucous membrane. Inside of the lips on the median line are membranous folds which tie the lips closely to the jaws, the **frenula.** The lower lip has another frenulum on each side between the canine and premolar. A deep external groove marks the median line of the upper lip. It is called the plithrum and marks the point of union of the two halves of the upper lip. Caudad the two lips join each other forming the commissure of the lips and here unite with the cheeks. The muscles of the lips can be seen in Plates 22 and 23. The orbicularis oris makes up a large part of the lips inside of the skin.

The **cheeks** are small and thin and extend from the lips to the ramus of the mandible. They are covered with hair externally and with smooth, but folded mucous membrane internally. The space between the cheeks and the mandible is sometimes called the **buccal cavity.** It is very small in the cat. The ducts of the parotid, molar and zygomatic (*infraorbital*) salivary glands open on the inner surface of the cheeks. A few buccal glands are usually present.

The **oral cavity proper** is bounded dorsally by the hard palate craniad and the soft palate caudad. They separate

132

the nasal fossae and the nasopharynx above from the oral cavity and oral pharynx below. The **hard palate** is supported by the horizontal laminae of the premaxillae (*incisive*) and maxillae and the palatine processes of the palatine bones. The mucosa covering the oral side of the hard palate is elevated into seven or eight curved transverse folds which are concave caudad (Plate 55).

Craniad of the first transverse ridge of the hard palate is a rounded papilla, the **incisive papilla.** To each side of the papilla is an incisive or **nasopalatine canal** which leaves the oral cavity and leads caudodorsad into the floor of the nasal fossae.

The **soft palate** is rather long and narrow and like a curtain separates the caudal part of the nasal cavities and the nasal pharynx from the mouth. It attaches cranial to the caudal border of the hard palate, to the ventral border of the perpendicular plates of the palatine and to the pterygoid processes and hamuli of the sphenoid. Caudad it ends in a free arched border in the region of the epiglottis (Plate 55). It, with the tongue, form the walls of the narrow passage between the oral cavity proper and the oral pharynx. This passage is called the **isthmus of the fauces** (*isthmus faucium*). From the sides of the soft palate caudad two folds of the mucosa pass ventrad. The cranial one goes to the side of the tongue, the caudal one to the floor of the pharynx. These are called **pillars of the fauces** or palatoglossal and palatopharyngeal arches. Between them is a shallow fossa in which is found the **palatine tonsil.**

The floor of the mouth is formed mainly by the tongue which extends as far back as the oral pharynx and the hyoid bone. The latter is sometimes called the bone of the tongue. On the lower surface of the tongue the loose mucosa forms a median vertical **lingual frenulum** which runs to the floor of the mouth below the tongue (Plate 49).

The sides of the oral cavity proper are formed by the teeth and the gums, the latter covering the alveolar borders of the mandible, maxillae and premaxillae.

The ducts of the mandibular and sublingual glands open through a papilla which lies to the side of the median line at the cranial end of the floor of the mouth. The opening of the mandibular duct is on the lateral side of the papilla, that of the sublingual duct on the medial side.

The **tongue** is mostly a mass of interwoven bundles of muscle fibers with a surface mucosa. It can be viewed to better advantage by cutting through the skin and masseter muscle and if necessary the ramus of the mandible to open the mouth wider. The tongue is widest in the middle and tapers slightly at both the base and the apex. It is divided into three parts. The caudal third is the **root of the tongue** through which the extrinsic muscles enter except for a part of the genioglossus which enters through the frenulum; the **body of the tongue,** the central portion; the **apex of the tongue,** the cranial end or tip of the tongue. Its surfaces are called the **dorsum, venter,** and **margins of the tongue.** The margin is the junction of the dorsum and venter.

The tongue is covered with a cornified stratified squamous epithelium which is closely applied to the tongue on the dorsum and margins. On the ventral side the epithelium is less cornified and loose giving rise to the fold mentioned above, the lingual frenulum.

The dorsum and margins of the tongue are provided with **papillae** which are projections of the tongue surface made up of the corium or dermis and the lamina propria covered externally with epithelium (Plate 111, fig. 2). The papillae are of five types—filiform, fungiform, foliate, vallate (*circumvallate*), and conical. The **filiform papillae** are horny, many of them have spines and point caudad. They are most abundant in the central portion of the free end of the tongue. The **fungiform papillae** are small mushroom-shaped structures. They are scattered over the surface of the tongue concentrating along the edge and in a central area caudad of the filiform papillae. At the back of the tongue at each edge is a row of larger and flattened papillae which some consider as a modification of the fungiform papillae but are considered here as **foliate papillae.** The **vallate papillae** are placed in two rows of two or three papillae each which come together in the form of a V at the base of the tongue. They are circular and have a trench-like depression or moat around them. The conical papillae are on the root of the tongue and are similar to the filiform. All papillae, except the conical and filiform, have **taste buds** embedded in them. Collectively they give a characteristic texture to the tongue quite evident when a cat licks your skin.

The **intrinsic muscles of the tongue** run in three directions and therefore are called the longitudinal, transverse and vertical muscles. They attach to the integument of the tongue and account, along with the extrinsic muscles, for its great versatility of movement. The extrinsic muscles, the genioglossus, hyoglossus and styloglossus have been described and are illustrated in Plate 24, figure 3.

The tongue functions in the chewing and swallowing of the food and contains the receptors, the taste buds, for the sense of taste.

The salivary glands are the major digestive glands which open into the mouth. They are the parotid, mandibular, sublingual, zygomatic and molar. They are illustrated on Plate 50 and described on page 143.

The cat is born with no erupted teeth. It has during its lifetime two sets of teeth, the first deciduous or temporary followed by the permanent teeth. This condition is called **diphyodent** (*two sets*). The **deciduous teeth** number 26 and consist of 12 incisors, 4 canines and 10 molars. These are replaced later by the **permanent teeth** of which there are 30; 12 incisors, 4 canines, 10 premolars and 4 molars. The number and type of teeth in mammals is usually indicated by a tooth formula as follows: $\frac{3\text{–}1\text{–}3\text{–}1}{3\text{–}1\text{–}2\text{–}1}$ in which the numbers above the line indicate the teeth in one half of the upper jaw and the numbers below the line, in one half of the lower jaw. The first number indicates the incisors, the second the canines, the third the

133

premolars and the last the molars. In the kitten's skull with 26 deciduous teeth the formula would be $\frac{3\text{-}1\text{-}0\text{-}3}{3\text{-}1\text{-}0\text{-}2}$ which means 3 incisors, 1 canine, 0 premolars in one half of the upper and lower jaws and 3 molars above and 2 below (Plates 3 to 5).

When all of the teeth in an animal are alike as in some fishes and Amphibians, they are said to be **homodont.** When the teeth vary in form as has been seen in the cat, they are called **heterodont.** Some animals have no capacity to renew teeth and are designated **monophyodont,** those which have two sets in a lifetime as in most mammals, are **diphyodont,** and those with a continuous succession of teeth as in fishes, amphibians and reptiles are **polyphyodont.** Teeth ankylosed on the side of a bone are called **pleurodont,** those fastened to the summit or crest of the bone are **acrodont** and those placed in sockets are **thecodont.** The mammals and some reptiles fall into the last category.

A mammalian tooth when cut in longitudinal section shows the following component parts (Fig. 8). The crown is the shiny upper part of the tooth which in the living

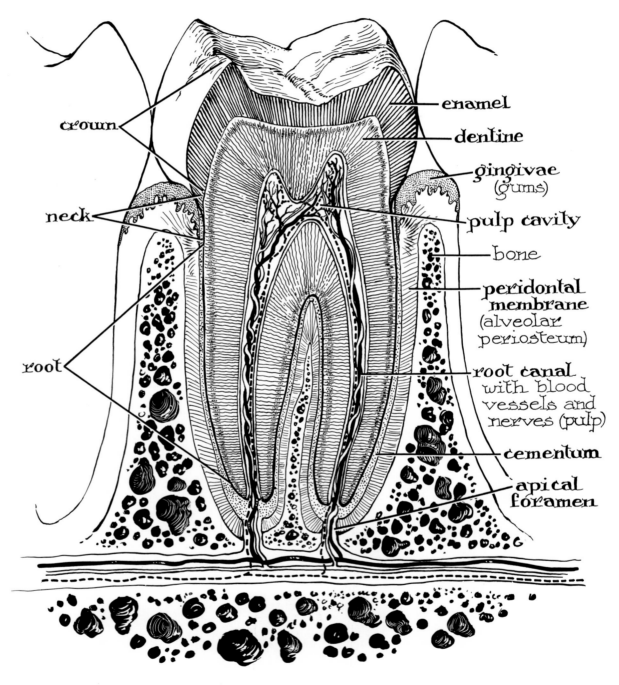

Fig. 8. *Longitudinal section of a molar tooth of man shown in its alveolus.* (*Crouch,* Functional Human Anatomy, *Lea & Febiger.*)

134

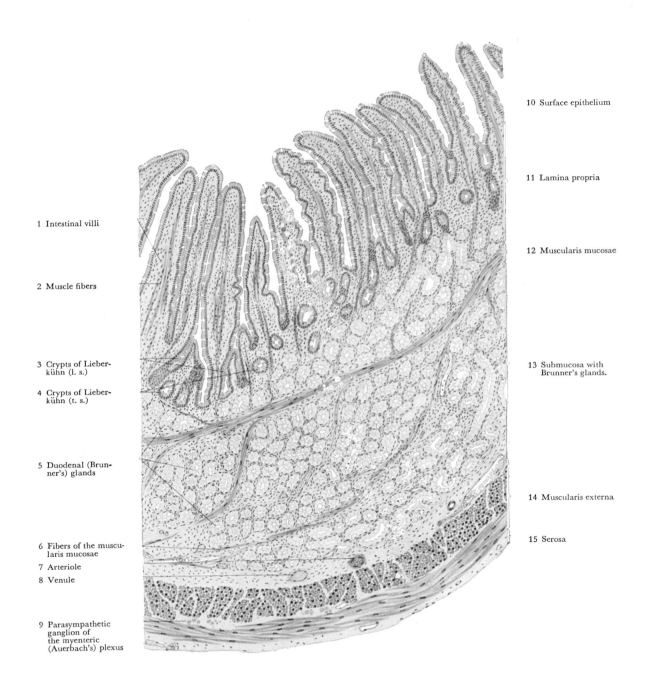

1 Intestinal villi

2 Muscle fibers

3 Crypts of Lieber-
kühn (l. s.)

4 Crypts of Lieber-
kühn (t. s.)

5 Duodenal (Brun-
ner's) glands

6 Fibers of the muscu-
laris mucosae

7 Arteriole

8 Venule

9 Parasympathetic
ganglion of
the myenteric
(Auerbach's) plexus

10 Surface epithelium

11 Lamina propria

12 Muscularis mucosae

13 Submucosa with
Brunner's glands.

14 Muscularis externa

15 Serosa

Fig. 9. *Small Intestine: Duodenum (longitudinal section). Stain: hematoxylin-eosin. 50×. (di Fiore, Atlas of Human Histology,* 3rd Ed., *Lea & Febiger.*)

subject projects above the gums; the **root,** the dull lower part which inserts into the jaw bones; the **dentin,** the bone-like material making up most of the tooth; the **enamel,** the shiny hard covering over the dentin of the crown of the tooth; the **cement,** a thin layer of bony substance which covers the dentin of the root and holds the tooth along with a peridontal ligament in the socket; and finally, the **pulp cavity,** the internal space in the tooth which is filled with **pulp** composed of blood vessels, nerves and connective tissue.

The **pharynx** lies dorsad and caudad of the soft palate and the oral cavity. It extends caudad to a position dorsad of the larynx where it joins the esophagus. A median

section of the head and neck of the cat demonstrates these relationships very well or by cutting through the skin, masseter muscle and the ramus of the mandible on each side will enable one to view the pharynx. By cutting along the median line of the soft palate it can be spread to show the **nasopharynx.** At the cranial end of the nasopharynx note the two choanae (*internal nares*) which mark the end of the nasal passageways. Just caudad of the choanae on the lateral walls of the nasopharynx note the openings into the **auditory tubes** (*Eustachian*) which lead into the middle ear cavity. Immediately behind the oral cavity is the isthmus of the fauces which opens into the **oropharynx** and below the oropharynx the pharynx extends dorsad of

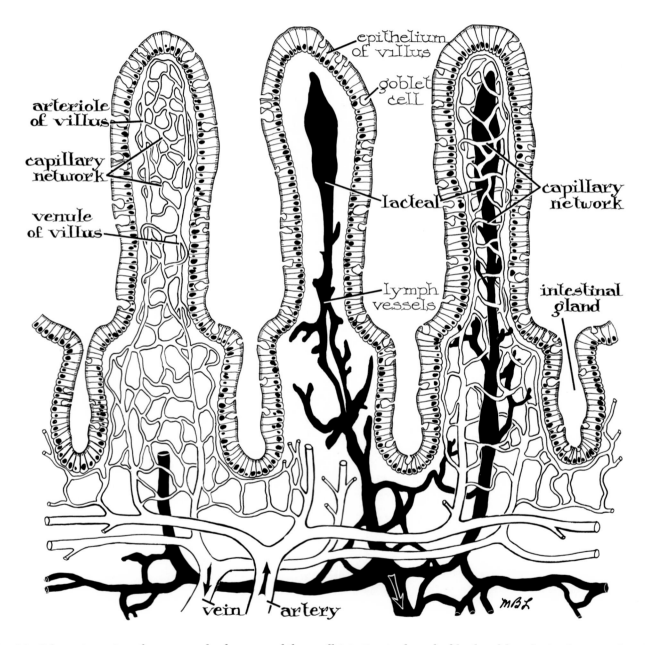

Fig. 10. *Schematic section of mucosa and submucosa of the small intestine to show the blood and lymph circulation in the villi. The villus to the left shows only the blood vessels; the villus in the center only the lymph vessel; the one on the right has both, which is the normal condition for all villi. (Crouch,* Functional Human Anatomy, *Lea & Febiger.)*

136

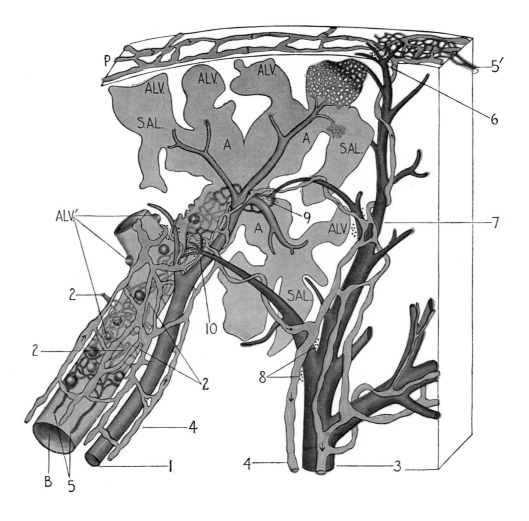

Fig. 11. *General scheme of a primary lobule, showing the subdivisions of* (B) *a respiratory bronchiole into two alveolar ducts; and the atria* (A), *alveolar sacs* (S.AL.) *of one of these ducts. ALV', alveoli scattered along the bronchioles; P, pleura, 1, pulmonary artery, dividing into smaller radicles for each atrium, one of which terminates in a capillary plexus on the wall of an alveolus; 2, its branches to the respiratory bronchiole and alveolar duct; 3, pulmonary vein with its tributaries from the pleura 6, capillary plexus of alveolus, and wall of the atrium 9 and alveolar duct 10; 4, lymphatics; dotted areas at 7, 8, 9 and 10, indicating areas of lymphoid tissue; 5, bronchial artery terminating in a plexus on the wall of the bronchiole; 5', bronchial artery terminating in pleura.* (Miller, The Lung, *courtesy of Charles C Thomas.*)

the larynx as the **laryngopharynx** which joins the esophagus. The pharynx serves in the conduct of air from the choanae to the glottis which opens into the larynx. The oropharynx and laryngopharynges also carry food. Note that in the oropharynx the food and air streams cross over the air going into the ventral larynx the food into the dorsal esophagus. A complex swallowing mechanism facilitates this process. Sometimes it does not work effectively and the cat "chokes" on food which gets into the larynx. This reflex mechanism causes the return of the food to the pharynx.

The remainder of the digestive system is described in connection with Plates 51 to 53 and will not be repeated here. It should be noted, however, that all of these organs have a common pattern of structure in their walls, though they vary in details. Figure 9 demonstrates this common pattern and it is outlined below.

Tunica mucosa (mucous membrane)	Epithelium Lamina propria (connective tissue)
Tela submucosa (connective tissue)	Muscularis mucosae (smooth muscle)
Tunica muscularis (externus) (smooth muscle)	Circular Longitudinal
Tunica serosa	Lamina propria (connective tissue) mesothelium (simple squamous epithelum)

A few of the variations from this common pattern as seen in specific organs are as follows:

The **epithelum** of the tunica mucosa is stratified squamous epithelium in the esophagus, simple columnar from the stomach to the rectum becoming stratified squamous again in the anal canal (Plate 53, figs. 2 and 3).

The **tela submucosa** contains **esophageal glands** in the esophagus, **duodenal** (*Brunner's*) **glands** in the duodenum (Fig. 9). It is without glands in the other organs, their microscopic glands being confined to the muscosa (Plate 53, fig. 3).

The **tunica muscularis** has obliquely arranged muscle fibers in the stomach in addition to the circular and longitudinal fibers.

The small intestine has some special structures besides its great length to increase the epithelial surface for purposes of more efficient absorption. These are the intestinal villi with their blood and lymph capillaries (Fig. 10 and Plate 53, fig. 3). The villi are so small and so numerous as to give a velvet-like appearance to the lining of the small intestine.

The Respiratory System

The **respiratory system** serves a variety of functions the primary one being to provide the mechanism necessary to exchange the respiratory gases oxygen and carbon dioxide between the external and internal environment. Oxygen is taken into the blood; carbon dioxide is released into the air

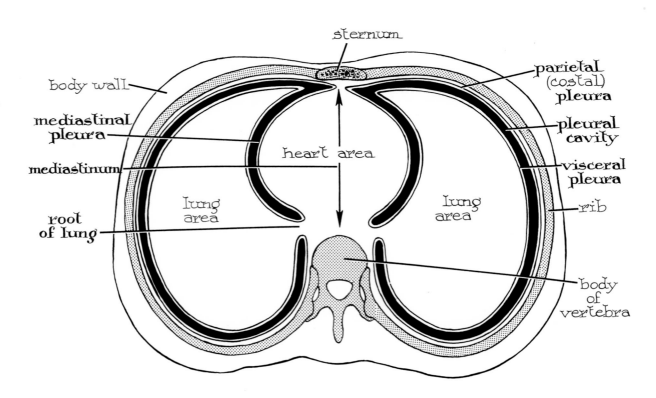

Fig. 12. *Schematic cross-section of the thorax to show the mediastinum, pleural membranes, and pleural cavities. (Crouch, Functional Human Anatomy, Lea & Febiger.)*

138

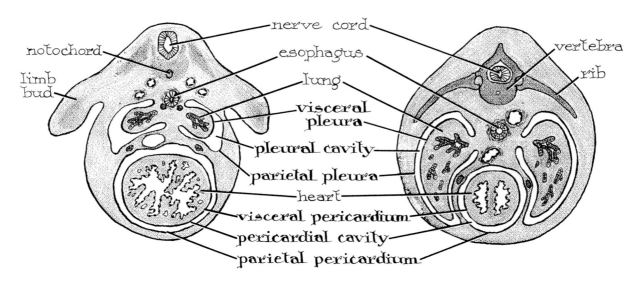

Fig. 13. *Scheme showing growth and development of lungs and pleural cavities in relationship to the heart and pericardial cavity.* (*Modified from Patten.*)

of the lungs. Since the lungs perform this primary function of gaseous exchange they are often referred to as the essential organs of respiration. The other organs of the system, the nasal passageways, pharynx, larynx, trachea and bronchi and the diaphragm and other respiratory muscles serve to maintain a flow of air through the lungs and are sometimes called accessory organs of respiration. They maintain **breathing** or pulmonary ventilation.

Other functions of the respiratory system are voice production and dissipation of heat and moisture. Since the receptors for the sense of smell are in the upper reaches of the nasal passageways, this system is involved in olfaction.

The nasal passages have a complex bony framework as shown in Plates 6 and 49. The air passing through these complex passageways which in life are covered with a highly vascular and glandular mucosa is to some degree

warmed, cleaned and moistened. It passes through the **choanae** into the **nasal pharynx** described above. Passing through the remainder of the pharynx it enters the glottis of the larynx and passes down the trachea to the bronchi and hence to the lungs. (Plate 55 and description). The minute structure of the lung is shown in Figure 11.

Review also at this time the relations of the lungs to their serous or pleural membranes and to the pleural cavities (Plates 46 and 47). Figure 12 shows schematically a cross section through the thorax in which heart, lungs and celomic relationships are clearly indicated. Figure 13 suggests schematically how the lungs are derived and how they develop their relationships to pleural cavities and to the heart and pericardial cavity. In the left hand figure they are pushing out from the lung bud into the mesenchyme of the thorax. In the right figure they have attained their position to either side of the heart.

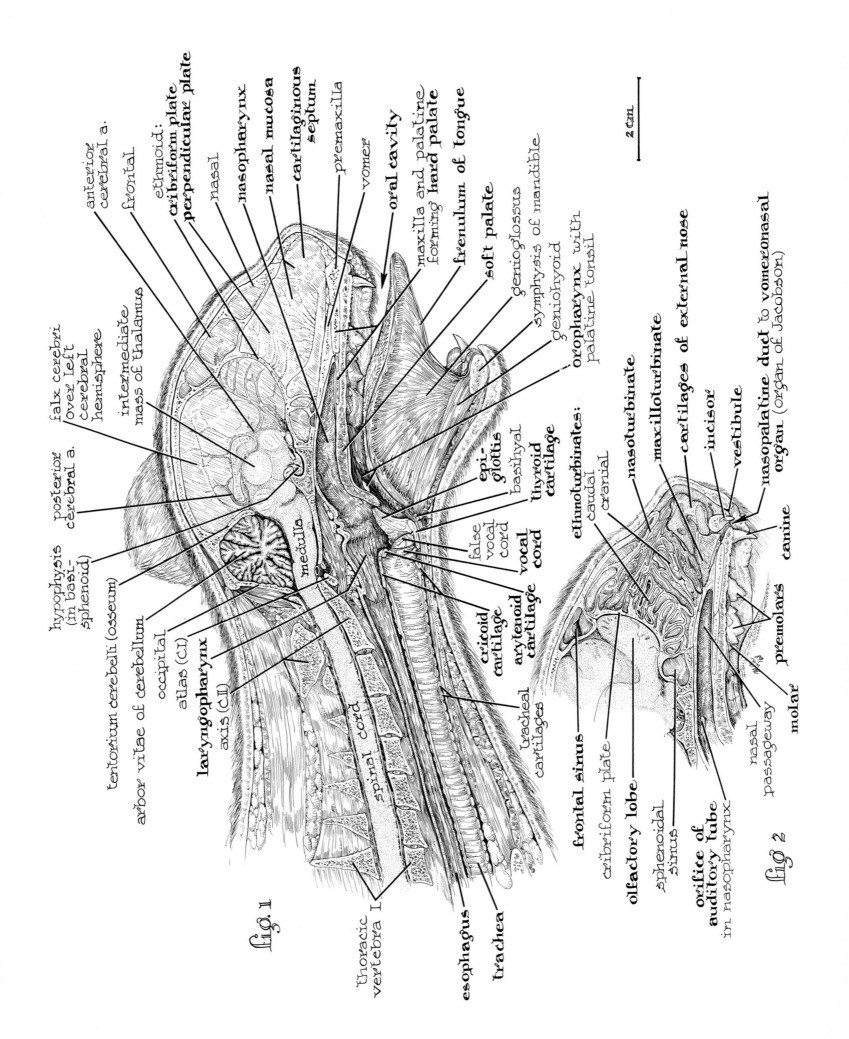

fig. 1

anterior
cerebral a.

ethmoid:
cribriform plate
perpendicular plate

nasal

nasopharynx

nasal mucosa

cartilaginous
septum

premaxilla

vomer

oral cavity

maxilla and palatine
forming hard palate

frenulum of tongue

soft palate

genioglossus

symphysis of mandible

geniohyoid

oropharynx with
palatine tonsil

frontal

falx cerebri
over left
cerebral
hemisphere

intermediate
mass of thalamus

posterior
cerebral a.

hypophysis
(in basi-
sphenoid)

tentorium cerebelli (osseum)

arbor vitae of cerebellum

occipital

atlas (CI)

laryngopharynx

axis (CII)

medulla

epi-
glottis

basihyal

thyroid
cartilage

false
vocal
cord

vocal
cord

cricoid
cartilage

arytenoid
cartilage

spinal cord

tracheal
cartilages

thoracic
vertebra I

esophagus

trachea

2 cm

fig. 2

frontal sinus

cribriform plate

olfactory lobe

sphenoidal
sinus

orifice of
auditory tube
in nasopharynx

ethmoturbinates:
caudal
cranial

nasoturbinate

maxilloturbinate

cartilages of external nose

incisor

vestibule

nasopalatine duct to vomeronasal
organ (organ of Jacobson)

canine

premolars

molar

nasal
passageway

Organs of the Digestive and Respiratory Systems of the Cat in the Head and Neck Regions

PLATE 49

In this illustration, figure 1 is a drawing of the head and neck region as seen in the median (midsagittal) section. Figure 2 is a portion of nasal and mouth regions as seen in parasagittal section. Their purpose is mainly to clarify some of the relationships of digestive and respiratory structures to each other and to other features of head and neck anatomy. It should be emphasized that at the cranial or rostral end the respiratory structures are dorsal to those of the digestive system. This relationship is reversed in the pharyngeal region so that the esophagus of the digestive system comes to lie dorsad to the trachea of the respiratory system.

Since most of the structures in this plate have been or will be presented in greater detail in other plates, our treatment of them here will be general, except as certain basic relationships will be stressed. For example, the various turbinates are more critically analyzed in figure 2 of Plate 6, while the overall relationships of structures in the head region are better demonstrated in this plate.

Since figure 1 is a median section, the **perpendicular plate** of the ethmoid and the **cartilaginous septum** of the nose are shown intact. The **cribriform plate** of the ethmoid is seen forming the caudal wall of the nasal cavities and as you should recall, allowing the passage of olfactory fibers. In the floor of the nasal cavities are seen the premaxilla, maxilla and palatines forming the **hard palate** and separating nasal and mouth cavities. Above the hard palate is the nasal passageway which is continuous caudad with the **nasopharynx** which lies dorsal to the soft palate. The **soft palate** is a musculomembranous curtain which is a continuation caudad of the hard palate and it separates the nasopharynx from the mouth cavity. The **orifices of the auditory tubes**, two longitudinal slits, lie on each lateral wall of the nasopharynx (fig. 2). They lead into the auditory tubes which connect the nasopharynx with the tympanic (*middle ear*) cavity. Behind the soft palate is the pharynx proper which is often divided into an oropharynx continuous with the oral cavity and a laryngopharynx dorsal to the larynx. In figure 1 the **oropharynx** is indicated as including the palatine tonsil which lies in the pillars of the fauces in a tonsilar fossa. This means of identifying the cranial limit of the oropharynx is used in human anatomy. The caudal limit of the oropharynx is the ba-

sihyal shown in figure 1. The **laryngopharynx** extends from the basihyal to the caudal end of the cricoid cartilage and leads into the **esophagus** and into the larynx and trachea. It is at this point that we see the crossing over (*chiasma*) of the food and air passageways. The air enters the glottis (*cut open here*) and passing the false vocal cords and (*true*) **vocal cords** is in the larynx from which it moves into the **trachea**. Note the various cartilages of the larynx, the **epiglottis, thyroid, arytenoid,** and **cricoid** and caudad of the larynx the tracheal cartilages which guard against the collapse of the trachea.

Return now to the mouth in figures 1 and 2. Also review the general discussions, pages 132 through 134, on the digestive and respiratory system. The mouth cavity is divided into the **vestibule** (fig. 2) which lies between the teeth and the lips and cheeks. The mouth cavity proper or oral cavity lies inside of teeth and jaws and is bounded dorsally by the hard palate and by the soft palate and pillars of the fauces caudally. The tongue forms most of the floor of the mouth. Ventrad of its free edges the mucosa forms a prominent median vertical fold which unites the tongue with the floor of the mouth rostrally. This fold is called the **frenulum of the tongue.** Notice that the genioglossus muscle passes through the frenulum into the tongue proper and that the geniohyoid muscle is shown extending from the basihyal to the symphysis of the mandible.

In figure 2 note the **nasopalatine duct leading to the vomeronasal organ** (*organ of Jacobson*), a structure innervated by terminal, olfactory and trigeminal branches and probably contributing to the function of olfaction and possibly taste.

Figure 2 also reviews for us the types of teeth as seen here in the left side of the upper jaw-the maxilla and premaxilla. Note also the **frontal sinus** and the **olfactory lobe** of the brain against the cribriform plate. The sphenoidal sinus is also shown.

Returning now to figure 1 review the relationships of the spinal cord and brain to the vertebral column and skull bones forming the cranial cavity. Note particularly the falx cerebri covering over the left cerebral hemisphere and the tentorium cerebelli (*oseum*) between the cerebrum and cerebellum. The hypophysis is seen in the hypophyseal fossa of the sphenoid bone. The arbor vitae of the cerebellum is sectioned medially.

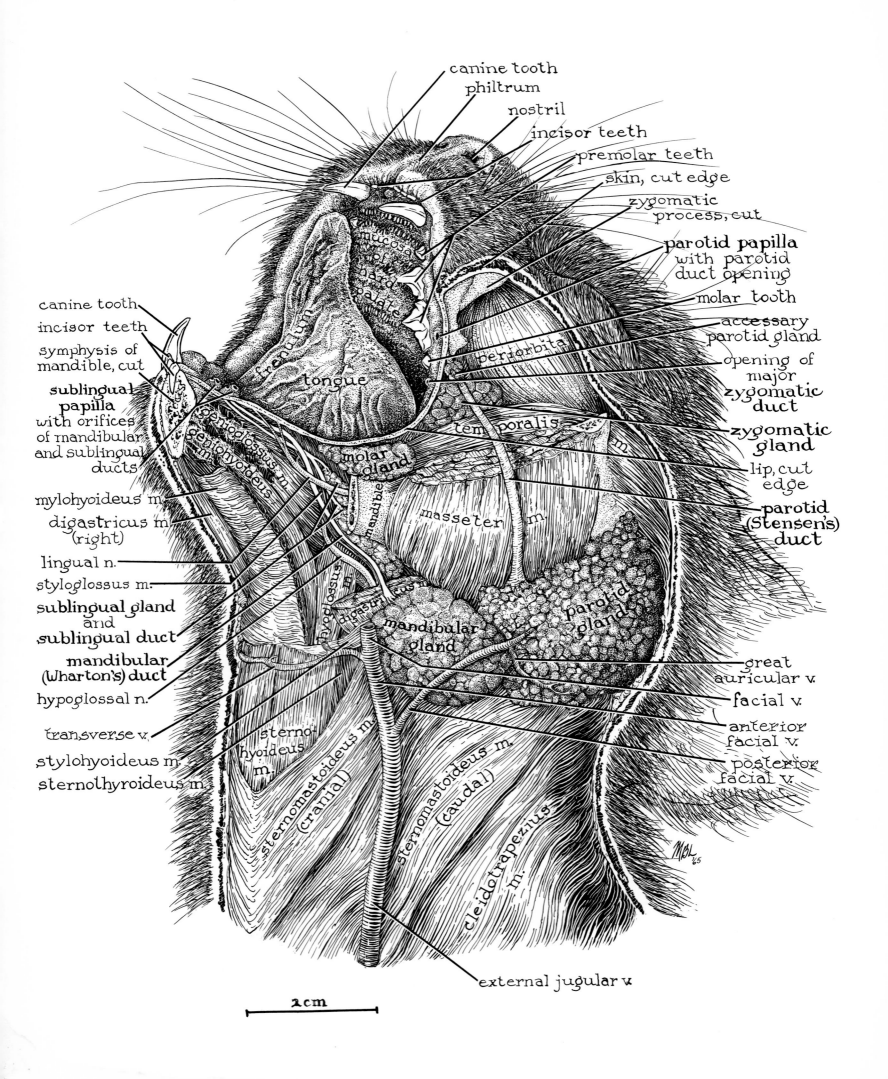

canine tooth

philtrum

nostril

incisor teeth

premolar teeth

skin, cut edge

zygomatic process, cut

parotid papilla with parotid duct opening

molar tooth

accessary parotid gland

opening of major **zygomatic duct**

zygomatic gland

lip, cut edge

parotid (Stensen's) duct

great auricular v.

facial v.

anterior facial v.

posterior facial v.

external jugular v.

mucosa of hard palate

periorbita

temporalis m.

masseter m.

parotid gland

mandibular gland

frenulum

tongue

molar gland

mandible

hyoglossus m.

digastricus m.

sternomastoideus m. (cranial)

sternomastoideus m. (caudal)

cleidotrapezius m.

canine tooth

incisor teeth

symphysis of mandible, cut

sublingual papilla with orifices of mandibular and sublingual ducts

mylohyoideus m.

digastricus m. (right)

lingual n.

styloglossus m.

sublingual gland and **sublingual duct**

mandibular (Wharton's) duct

hypoglossal n.

transverse v.

stylohyoideus m.

sternothyroideus m.

genioglossus m.

geniohyoideus m.

sterno-hyoideus m.

2 cm

MBL '65

The salivary glands, the liver and the pancreas, all lie outside of the digestive tube but empty their secretions into it where they aid the digestive processes. They are extrinsic glands. We should be reminded that there are many kinds of microscopic glands which lie within the walls of the various digestive organs and empty secretions into these organs. They are intrinsic glands.

This plate shows also some of the features of the mouth, the first organ of the digestive system. The teeth and tongue should remind us that digestion of food involves mechanical means as well as chemical.

There are five pairs of **salivary glands** in the cat all derived embryologically from the epithelium of the developing mouth and maintaining their connection with it in the adult by means of their ducts. They make an important contribution to the saliva of the mouth, but there are many microscopic glands in the mucosa of the mouth which also contribute. The salivary glands are the parotid, madibular, sublingual, molar and zygomatic (*orbital*). The molar is absent in some mammals like man and dog. All of the glands have a finely lobulated appearance as seen in the illustration. The ducts of each are also shown except that of the molar gland.

The **parotid gland,** the largest of the five, lies ventrad of the ear over the basal portion of the auricular cartilage. Its cranial border slightly overlaps the caudal border of the masseter muscle. Its caudal border overlaps the sternomastoideus and cleitotrapezius (*cleidocervicalis*) muscles. The posterior facial vein tunnels into the gland near its ventral border. It also borders on the mandibular gland at this point. The **parotid** (*Stensen's*) **duct,** made up of branches from inside of the gland, leaves its cranial ventral angle and passes in the superficial fascia over the external face of the masseter muscle. It turns inward at the cranial border of the masseter m. penetrating the cheek to enter the vestibule where it travels as a white ridge on the mucosa of the cheek opening in a small papilla opposite the most prominent cusp of the last premolar tooth. Some accessory parotid gland tissue can be seen at the cranial end of the duct just in front of the zygomatic gland.

The **mandibular gland** lies caudad of the masseter muscle and ventrad of the parotid gland. It is crossed near its caudal border by the posterior facial vein and the anterior facial vein passes along to its ventral border. The **mandibular** (*Wharton's*) **duct** leaves the medial surface of the gland and runs craniomedially in close association with the sublingual gland. It passes behind the mylohyoideus and digastricus muscles and against the outer surface of the styloglossus and then continues craniad against the oral mucosa and parallel to the mandible to open at the apex of the **sublingual papilla** lateral to the median line and the frenulum of the tongue in the cranial part of the floor of the oral cavity. It is accompanied by the duct of the sublingual gland.

The **sublingual gland** fits against the mandibular gland and appears to be an extension of it. It then stretches along the mandibular duct narrowing as it goes craniad. The sublingual duct follows the mandibular duct running at first dorsad of it and then mediad. It opens on the medial side of the apex of the sublingual papilla. In some cases the mandibular and sublingual glands open through separate papillae. There may also be separate openings of parts of the sublingual gland directly through the oral mucosa independent of the sublingual duct.

The **molar gland** lies between the mucosa of the lower lip and the orbicularis oris muscle at the cranial border of the masseter muscle. It is quite flat and extends forward from a broad base to a narrow apex to a point between the first premolar and the canine. Its several ducts open directly through the mucosa into the oral cavity.

The **zygomatic gland** (*orbital or infraorbital*) is found only in the dog and cat among the domestic animals. It is usually ovoid or pyramidal in shape and lies mediad of the zygomatic arch in the lower part of the orbit and in contact with ventral part of the periorbita. Ventrally it rests against the mucosa of the mouth just caudad of the molar teeth. Its **major duct** leaves the ventral end of the gland and enters the oral cavity through its mucosa about 3 millimeters caudad of the molar tooth. Additional **minor ducts** may open caudad of the major duct.

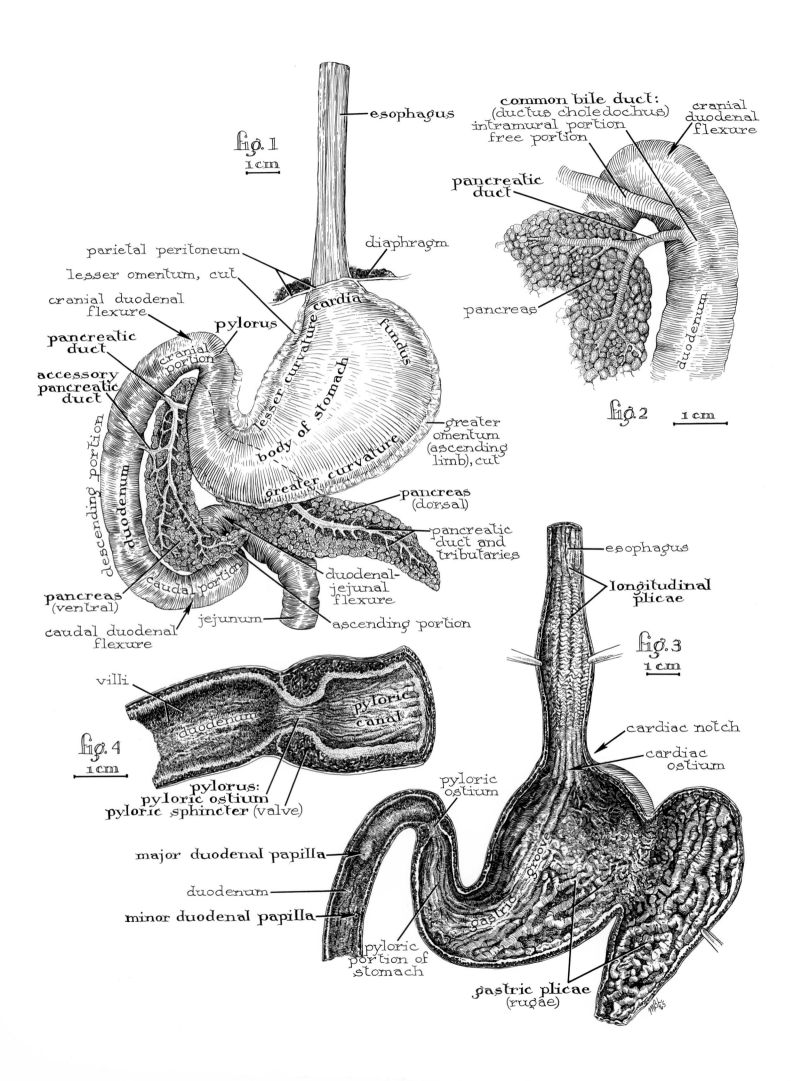

fig. 1
1cm

esophagus

diaphragm

parietal peritoneum

lesser omentum, cut

cranial duodenal flexure

pylorus

pancreatic duct

cranial portion

accessory pancreatic duct

cardia

lesser curvature

fundus

body of stomach

descending portion

duodenum

greater curvature

greater omentum (ascending limb), cut

pancreas (ventral)

caudal portion

caudal duodenal flexure

jejunum

duodenal-jejunal flexure

ascending portion

pancreas (dorsal)

pancreatic duct and tributaries

common bile duct: (ductus choledochus) intramural portion free portion

cranial duodenal flexure

pancreatic duct

pancreas

duodenum

fig. 2   1 cm

esophagus

longitudinal plicae

fig. 3
1cm

cardiac notch

cardiac ostium

villi

duodenum

pyloric canal

fig. 4
1cm

pylorus: pyloric ostium pyloric sphincter (valve)

major duodenal papilla

duodenum

minor duodenal papilla

pyloric ostium

pyloric portion of stomach

gastric groove

gastric plicae (rugae)

This plate deals with external and internal features of four connected organs of the upper alimentary tract, the esophagus, stomach, duodenum and the pancreas with its system of ducts. The common bile duct from the liver is shown in figure 2 in relationship to the pancreatic duct.

Figure 1 shows the external anatomy of these four organs. Only the lower thoracic and the abdominal portions of the **esophagus** are shown. Its cervical portion is continuous with the laryngeal pharynx above. It is a dorsoventrally flattened tube when empty. It is about 1 centimeter wide and lies in the median plane except between the thyroid glands and the bifurcation of the trachea into primary bronchi where it swings to the left and lies laterodorsad of the trachea. Its upper cervical portion lies dorsal to the trachea and ventral to the longus colli muscles. Its lower thoracic portion lies ventral to the aorta and it pierces the diaphragm about 2 centimeters from the dorsal wall. Its attachment to the diaphragm is loose, allowing longitudinal movement. Since in the thorax it passes through the dorsal mediastinum, it has no outer serous coat. Its short abdominal portion joins the stomach at the cardia and, like the stomach, has a serous coat—the serosa or visceral peritoneum. The right side of the esophagus continues straight with the lesser curvature of the stomach but at a considerable angle on the left side with the greater curvature of the stomach, the cardiac notch. Internally, the lining of the esophagus is thrown into a series of longitudinal folds which in the lower thoracic region have a scale-like surface (fig. 3). These folds make the esophagus distensible to accommodate a bolus of food.

The **stomach,** the largest dilatation of the digestive system, lies in a more or less transverse position with more of its bulk to the left of the median plane than to the right. It fits into an extensive concavity on the caudal surface of the liver and is in contact at its dorsal surface with the dorsal part of the diaphragm. Its ventral surface is largely covered by the liver except when the stomach is distended with food (Plate 46) when it touches the abdominal wall.

The entrance to the stomach is called the **cardia,** its outlet the **pylorus.** The major divisions of the stomach are the cardiac portion, fundus, body and pyloric portion. It has **greater** and **lesser curvatures** and dorsal and ventral surfaces.

In figure 3 the opened stomach shows the folds on its inner surface, the **gastric plicae.** The height of the plicae depends on the amount of distension of the stomach. They are highest in an empty stomach. The plicae run generally longitudinally but are quite tortuous. They start in the pyloric end and run along the greater curvature and through the body of the stomach and fade out in the fundus. Along the side of the lesser curvature the plicae

are less tortuous and less crowded and run directly from the cardiac to the pyloric ostia. They constitute the most direct route through the stomach, the gastric groove.

While no true sphincter valve develops at the cardiac ostium, a pronounced thickening of the circular muscles in the pylorus produces a distinct **pyloric sphincter** at the **pyloric ostium** which controls the passage of the acid chyme into the duodenum (fig. 4).

The **duodenum** is the first and shortest part of the small intestine. It makes a sharp angle with the pylorus, the **cranial duodenal flexure,** and thereby extends caudad and slightly to the right. It then extends almost directly caudad for 8 or 10 centimeters caudad of the pylorus, the descending portion, where it again makes a U-shaped turn, the **caudal duodenal flexure,** extending thus cranially and to the left for 4 or 5 centimeters, the **ascending portion,** where, without definite external demarcation, it joins the jejunum at the **duodenal-jejunal flexure.** Its total length is about 14 to 16 centimeters.

The duodenum receives ducts from two large glands, the liver and the pancreas (fig. 2). On the dorsal internal wall of the duodenum about 3 centimeters from the pylorus there appears a **major duodenal papilla** at the apex of which is an opening from the space inside the papilla called the **ampulla** (*of Vater*) (fig. 3). The **common bile duct** from the liver and the **pancreatic duct** empty into this ampulla. About 2 centimeters caudoventrad of the major duodenal papilla is a smaller papilla, the **minor duodenal papilla,** which receives the **accessory pancreatic duct.**

Figure 2 shows the pancreatic duct and common bile duct as they reach the duodenum externally on the dorsal side. The common bile duct after contacting the duodenum, runs a short distance within the substance of the wall before it enters the ampulla. This part of the duct is called the **intramural portion** as opposed to the free duct. The pancreatic duct enters the same ampulla so that these two ducts have a common opening into the duodenum.

The **pancreas** (fig. 1) is about 12 centimeters long and varies in width from 1 to 2 centimeters. At about its middle it is bent so that one part forms an acute angle to the other. The right half fits into the inside curve of the duodenum and on its other side parallels the right side of the greater curvature of the stomach. It lies within the duodenal omentum.

Figure 1 shows some of the substance of the pancreas removed to uncover the pancreatic ducts and their tributaries. These form a common duct at about the angle of the gland which enters the ampulla. The **accessory duct** forms by the union of small branches, some of which anastomose with those of the pancreatic duct.

145

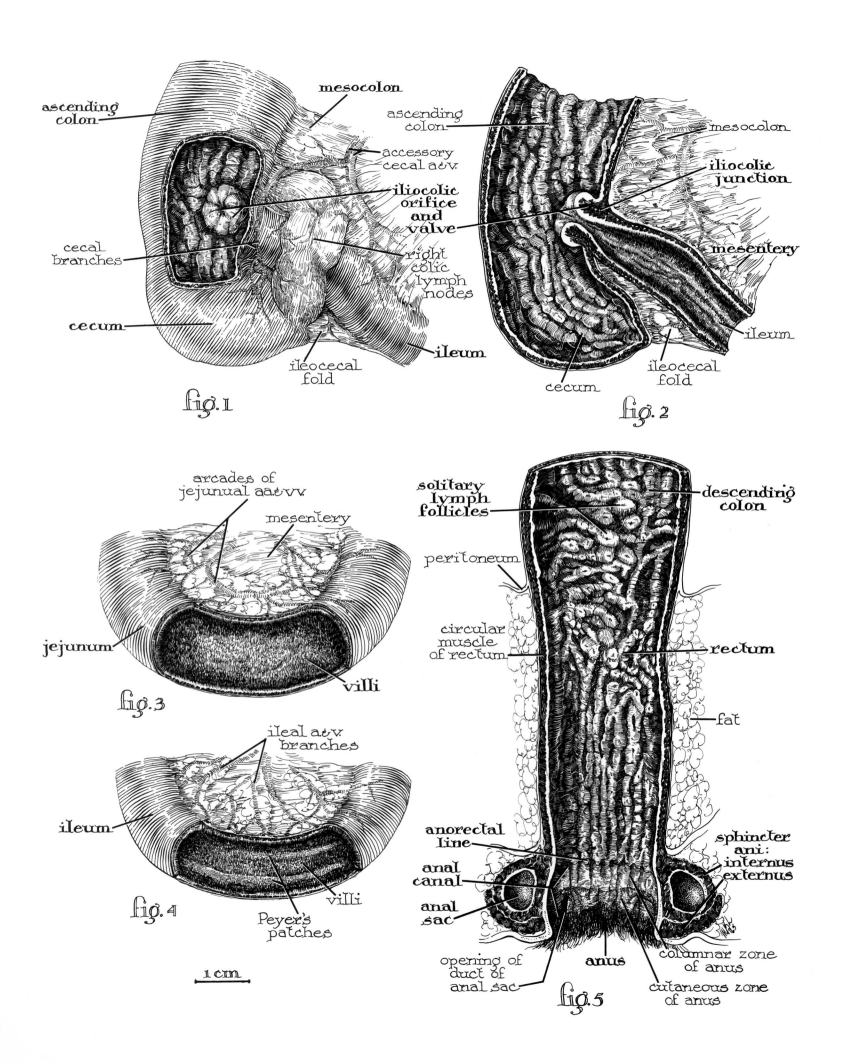

fig. 1

ascending colon

mesocolon

accessory cecal a.&v.

iliocolic orifice and valve

right colic lymph nodes

cecal branches

cecum

ileocecal fold

ileum

fig. 2

ascending colon

mesocolon

iliocolic junction

mesentery

ileum

cecum

ileocecal fold

fig. 3

arcades of jejunual aa.&vv.

mesentery

jejunum

villi

fig. 4

ileal a.&v. branches

ileum

villi

Peyer's patches

1 cm

fig. 5

solitary lymph follicles

descending colon

peritoneum

circular muscle of rectum

rectum

fat

anorectal line

anal canal

anal sac

opening of duct of anal sac

anus

sphincter ani: internus externus

columnar zone of anus

cutaneous zone of anus

Review the relationships of these organs and their supporting structures by referring back to Plate 48. The jejunum and ileum compose the bulk of the small intestine. The jejunum begins at the duodenojejunal flexure and the ileum ends at the ileocolic orifice. While the duodenum is relatively fixed, the jejunum is the most mobile part of the alimentary tract. No clearcut macroscopic or microscopic differences can be found to distinguish between the jejunum and the ileum. Some believe Galen applied the term jejunum because it is usually found to be emptier than the rest of the intestine.

The large intestine consists of the cecum, colon, rectum and anal canal. It begins at the ileocolic orifice and ends at the anus. It is a short and relatively unspecialized tube two to three times the diameter of the small intestine. The large intestine of the cat resembles that of man more than it does that of other domestic mammals except the dog. It has no haustra, tenia, vermiform process, epiploic appendices, or sigmoid colon.

Figures 1 and 2 are external and internal views of the **ileocolic junction,** the **ileocolic valve** and the **cecum.** The opening of the ileum into the colon is on its side about 1½ centimeters from its end. This leaves a blind pouch, the **cecum,** which is cone-like at its terminus. There is no vermiform process or appendix. The cecum is attached to the ileum by the **ileocecal fold.** Internally the cecum, as well as the colon, show longitudinal and sometimes circular folding of their mucosae. These folds or "plicae" are readily effaced by distension and they are not ordinarily considered the same as the plicae of the stomach. Indeed the colon is usually described as being smooth internally, lacking even the villi found in the small intestine. It does have numerous solitary lymph nodules as does the cecum.

The **ileocolic valve** or sphincter should be examined. As shown in figures 1 and 2 it consists of a part of the ileum which pushes a short distance into the colon. Its circular muscle layer is thickened to form the sphincter muscle.

The **rectum** begins at about the level of the pelvic inlet and ends at the beginning of the anal canal (fig. 5). It has longitudinal folds and numerous solitary lymph nodules. Its lower portion is without a serous peritoneum as it passes through the body wall.

The **anal canal** is a centimeter or less in extent and is a specialized part of the alimentary tract (fig. 5). It is surrounded by smooth and striated muscle forming respectively the **internal** and **external anal** sphincters. Spherical **anal sacs** are found one on each side of the anal canal each being about the size of a pea. Each empties by a small duct into the anal canal. They are of considerable clinical importance since they sometimes become enlarged due to accumulation of secretion or they may become abscessed causing constipation. They may rupture to the outside of the anus causing fistulas.

The mucosa of the anal canals is usually divided into three zones: the outer most, the cutaneous zone; the next, the intermediate zone; and the innermost, the columnar zone. The cutaneous zone is divided into two parts, an outer and inner. The outer, since it is a hairless area outside the anus, does not properly belong to the anal canal. The inner cutaneous zone is only a few millimeters wide and is the area into which the anal sacs open. Circumanal glands lie beneath the cutaneous zone and their ducts open through it. The intermediate zone is no more than a millimeter wide and is a sharp edged, scalloped fold which completely encircles the anal canal. It is often called the anocutaneous line. Like the cutaneous zone its surface consists of stratified squamous epithelium. The columnar zone contains longitudinal and oblique ridges called the anal columns. The zone is about 7 millimeters wide. It terminates cranially at the **anorectal line.** Caudally the columns are united by the anocutaneous line forming anal sinuses. Anal glands also empty into the anal canal.

Figure 3 is an opened section of the jejunum to show the velvety surface of the mucosa caused by the presence of intestinal villi. Figure 4 is a similar section of the ileum to show villi and aggregated lymph nodes or Peyer's patches.

The blood supply to some of the structures is also shown.

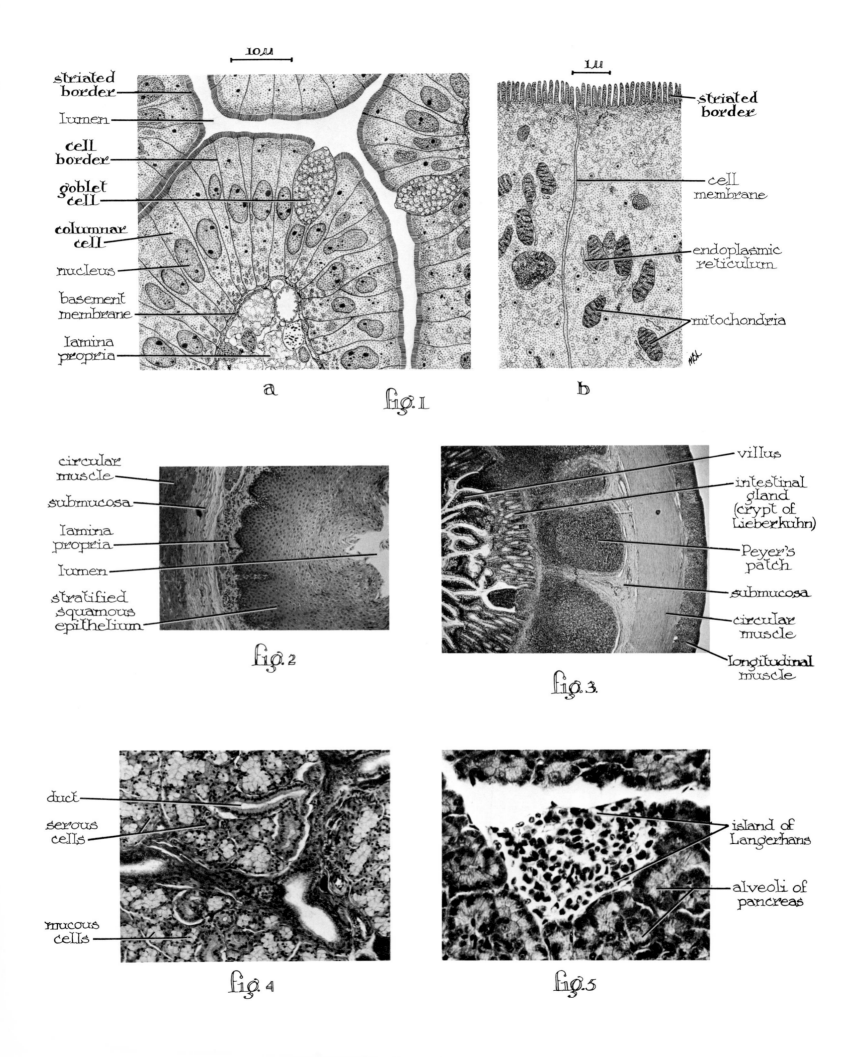

striated border

lumen

cell border

goblet cell

columnar cell

nucleus

basement membrane

lamina propria

10μ

a

fig. 1

1μ

striated border

cell membrane

endoplasmic reticulum

mitochondria

b

circular muscle

submucosa

lamina propria

lumen

stratified squamous epithelium

fig. 2

villus

intestinal gland (crypt of Lieberkühn)

Peyer's patch

submucosa

circular muscle

longitudinal muscle

fig. 3.

duct

serous cells

mucous cells

fig. 4

island of Langerhans

alveoli of pancreas

fig. 5

Microscopic Anatomy of Some Digestive Organs
PLATE 53

(Crouch, Functional Human Anatomy, Lea & Febiger)

Figure 1 is a drawing modified from electron micrographs from Rhodin, **An Atlas of Ultrastructure.** Figure 1a is a cross section of intestinal villi from the mucosa of the jejunum. It is magnified about 2500 times. It shows parts of four villi with the lumen in white. Note that two of the villi are so closely opposed that their striated borders almost touch. The surface cells are **columnar** with oval nuclei, most of them centrally placed in the cells. The cells taper slightly toward their bases and rest on a basement membrane which in turn rests on the lamina propria. The free surface of the cell, next to the lumen, is provided with a **striated border.** Interspersed among the columnar cells are some **goblet cells** which secrete mucous.

Figure 1b shows the distal portion of two adjacent columnar cells of the duodenum magnified 15,000 times. The **striated border** is seen here as closely placed finger-like projections emerging from the free surface of the cell. The cytoplasm of the striated border is more dense than that of the rest of the apical portion of the cell and it may contain fibrillar material which serves as a cytoskeleton to give some rigidity to the cell processes. The apical part of the cell contains mitochondria, lysosomes (*not labeled*), and endoplasmic reticulum.

Figure 2 is a cross section of a part of the wall of the **esophagus** magnified about 100 times. Note especially the mucosa with its thick stratified squamous epithelium next to the lumen and inside of the epithelium the lamina propria. The submucosa and circular muscle layers are also shown.

Compare figure 2 with figure 3 which is a cross section of the **ileum** magnified about 48 times. Note that the mucosa has a covering of simple columnar epithelium (see fig. 1) and is provided with numerous villi and intestinal glands. The submucosa is marked by the presence of masses of lymphoid tissue which constitute the Peyer's patches. Circular and longitudinal muscle layers are clearly shown and though not evident, the outer surface is covered by a serous membrane (*visceral peritoneum*).

Figure 4 is a photomicrograph of a section of the mandibular salivary gland which through its duct opens into the mouth. It is a compound tubuloalveolar gland consisting of glandular tissue or parenchyma and of supporting connective tissue elements. Note that the glandular tissue of the alveoli contains clear mucous cells and granular serous (*albuminous*) cells. The mucous cells produce a viscid secretion made up almost entirely of mucin. The serous or albuminous cells produce a watery secretion which lacks mucus but contains salts, proteins and the enzyme ptyalin. The parotid salivary gland has only serous type cells, whereas both are found in the mandibular and sublingual salivary glands.

Figure 5 is a photomicrograph of a section of the pancreas, a digestive gland which opens into the duodenum. It shows the secreting alveoli which empty their products into the duct system of the pancreas as seen in Plate 51, figures 1 and 2. Also shown is an islet or island of Langerhans which belongs to the endocrine system—the glands of internal secretion. The island of Langerhans has no ducts but empties its secretion, insulin, into the blood stream. It has marked effects upon sugar metabolism in the body.

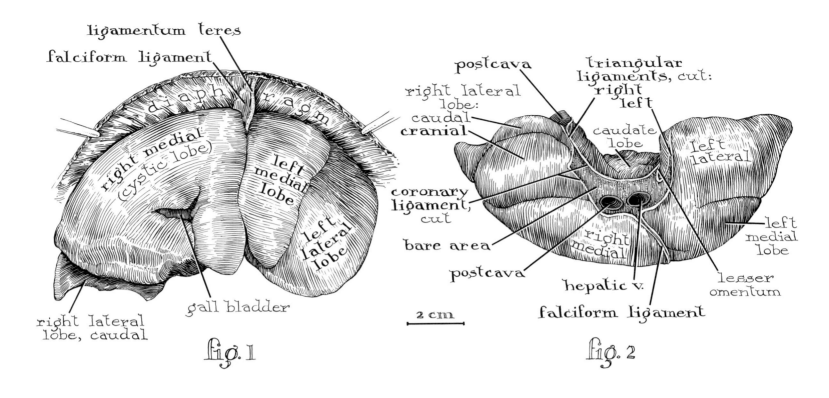

ligamentum teres

falciform ligament

diaphragm

right medial (cystic lobe)

left medial lobe

left lateral lobe

right lateral lobe, caudal

gall bladder

fig. 1

postcava

right lateral lobe: caudal cranial

triangular ligaments, cut: right left

caudate lobe

left lateral

coronary ligament, cut

right medial

bare area

postcava

hepatic v.

left medial lobe

lesser omentum

falciform ligament

2 cm

fig. 2

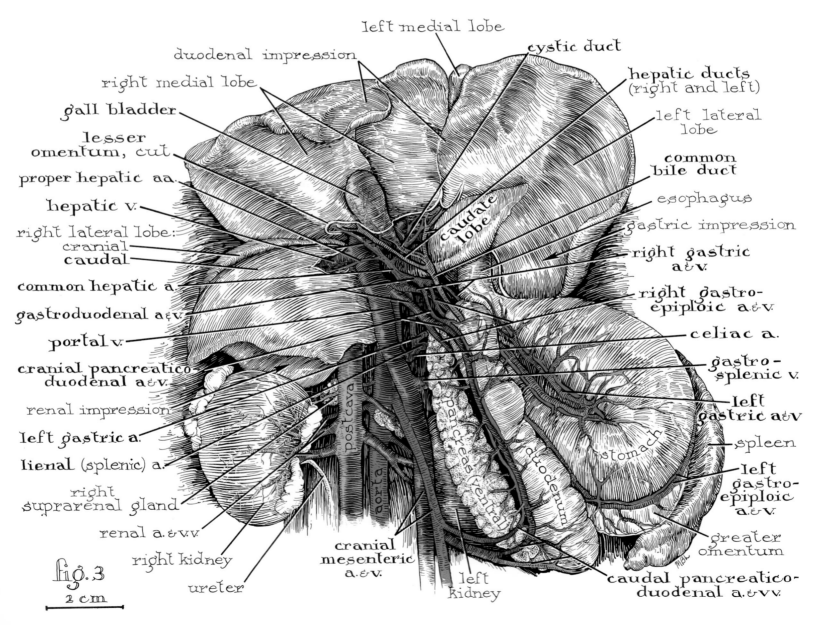

left medial lobe

duodenal impression

right medial lobe

gall bladder

lesser omentum, cut

proper hepatic aa.

hepatic v.

right lateral lobe: cranial caudal

common hepatic a.

gastroduodenal a.&v.

portal v.

cranial pancreatico-duodenal a.&v.

renal impression

left gastric a.

lienal (splenic) a.

right suprarenal gland

renal a.&vv.

right kidney

ureter

cranial mesenteric a.&v.

left kidney

cystic duct

hepatic ducts (right and left)

left lateral lobe

common bile duct

esophagus

gastric impression

right gastric a.&v.

right gastro-epiploic a.&v.

celiac a.

gastro-splenic v.

left gastric a.&v.

spleen

left gastro-epiploic a.&v.

greater omentum

caudate lobe

postcava

aorta

pancreas (ventral)

duodenum

stomach

caudal pancreatico-duodenal a.&vv.

fig. 3

2 cm

The liver not only serves the digestive system with which its complex duct system is connected through the common bile duct, but plays important roles in protein, fat and carbohydrate metabolism; makes and stores vitamin A; elaborates agents involved in the clotting of blood, such as fibrinogen and prothrombin and heparin which prevents clot formation. It contains phagocytic cells, is a blood resevoir, a detoxifying agent, a producer of blood cells in the embryo, a destroyer of hemoglobin from dead red blood cells and stores iron and copper. It is the largest gland in the body.

Figures 1 and 2 present two views of this important organ and figure 3 shows its external duct system and its special relationship to the blood-vascular system—the portal circulation. The arteries are red; portal vein and its tributaries are purple; other veins are blue; and the gallbladder and duct system are green.

Figure 1 is a ventral view of the liver showing its lobed arrangement, the collapsed **gallbladder** in the **right median** or cystic lobe and the **falciform ligament** coming down from the diaphragm. In the caudal margin of the falciform ligament is the **ligamentum teres** (*round ligament*) which is homologous to the umbilical vein of the fetus.

Figure 2 is a view of the cranial surface of the liver showing again the falciform ligament but, in addition, the **coronary** and **right** and **left triangular ligaments** which have been cut free from the diaphragm to which they secure the liver. Notice especially that the area within the cut coronary and triangular ligaments is free of serous membrane (*visceral peritoneum*) and referred to as the **bare area** of the liver. The postcava and an hepatic vein are shown in cross section.

Figure 3 shows the liver lobes spread out to give a better view of the **gallbladder** and the **caudate lobe.** Also it enables one to see the **hepatic ducts** which drain bile from the liver lobes, and the **cystic duct** through which the bile backs up into the gallbladder when the **common bile duct** is closed at its distal end at the duodenal papilla. When the bile reaches the gallbladder, it is concentrated by the absorption of water into the blood stream. When the duodenal papilla opens, bile then flows into the duodenum both from the gallbladder and from the hepatic ducts.

Figure 3 also shows the veins of the portal system which collect blood from the stomach, duodenum, pancreas and the small and large intestines. This blood is carried to the portal vein and goes directly into the complex circulation in the liver. It contains absorbed food substances like amino acids, glucose, inorganic salts, water, etc. The portal system is distinctive in that veins carry the blood into the sinusoids (capillaries) of the liver and it is picked up by hepatic veins which carry it into the postcava. It is a system of capillaries between veins, whereas in other parts of the body capillaries lie between arteries and veins.

Notice, too, that the liver receives arterial blood through its **hepatic artery** and its branches, as does any other organ. Many of these vessels will be shown again in the plates detailing the blood supply of the viscera.

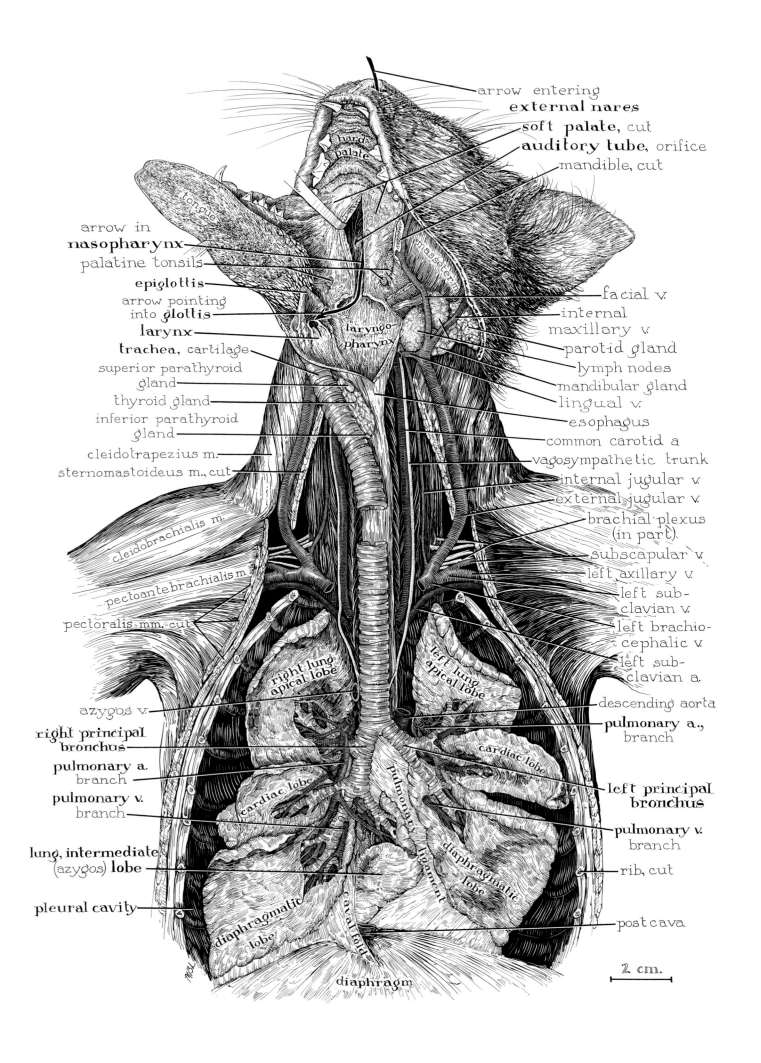

arrow entering
external nares
soft palate, cut
auditory tube, orifice
mandible, cut

hard palate

tongue

masseter

arrow in
nasopharynx
palatine tonsils
epiglottis
arrow pointing
into **glottis**
larynx
trachea, cartilage
superior parathyroid
gland
thyroid gland
inferior parathyroid
gland
cleidotrapezius m.
sternomastoideus m., cut

laryngo-
pharynx

facial v.
internal
maxillary v.
parotid gland
lymph nodes
mandibular gland
lingual v.
esophagus
common carotid a.
vagosympathetic trunk
internal jugular v.
external jugular v.
brachial plexus
(in part)
subscapular v.
left axillary v.
left sub-
clavian v.
left brachio-
cephalic v.
left sub-
clavian a.

cleidobrachialis m.

pectoantebrachialis m.

pectoralis mm., cut

right lung
apical lobe

left lung
apical lobe

azygos v.
right principal
bronchus
pulmonary a.
branch
pulmonary v.
branch

cardiac lobe

cardiac lobe

pulmonary ligament

diaphragmatic
lobe

descending aorta
pulmonary a.,
branch
left principal
bronchus
pulmonary v.
branch

lung, intermediate
(azygos) **lobe**

rib, cut

pleural cavity

diaphragmatic
lobe

caval fold

post cava

2 cm.

diaphragm

This illustration emphasizes the organs of the respiratory system: the nose, pharynx, larynx, trachea, principal bronchi, lobar bronchi and lungs. It also demonstrates the close relationship between digestive and respiratory systems. They share a common pharynx, and the mouth can and does on some occasions serve for inspiration and expiration of air. Embryologically speaking, the trachea and lungs are outgrowths of the primary digestive tube.

In this illustration, the gross anatomy of the entire **respiratory system** is shown though not in equal detail for all organs. It is designed to give continuity to the system and to show relationships of respiratory organs to other structures. One mandible has been cut and in part removed and the tongue, larynx and part of the trachea are displaced to the right to reveal the laryngopharynx. The soft palate has been cut on the medial line and opened to show one of the auditory tubes. A section of the trachea has been removed to show the esophagus lying dorsad. The heart, thymus and major blood vessels have been removed to uncover the roots of the lungs, the bronchi and other structures. The arteries are red except the pulmonary artery which is blue; the pulmonary veins are red, the others blue.

The **nose**, considered here in the broad sense, includes an **external nose**, consisting of the movable and flexible cartilages and the fixed bony structures, and an **internal nose** or nasal cavity with its complicated mucus-lined passageways. The external nose can best be understood by reference to Plates 6 and 9 on the skeletal system. The internal nose or nasal cavity is also made clear by the same plates if one can imagine all of the passageways lined by a soft, moist and vascular mucous membrane which helps to cleanse, warm and moisten the inspired air.

The air enters the external nares, passes through the nasal cavities and enters the **nasal pharynx** as indicated by the arrow. The arrow can be followed back through the **oral pharynx** and **laryngeal pharynx** to the point where it enters the glottis of the larynx.

The **larynx** (Plate 49) or voice box consists of a set of complicated cartilages, ligaments and muscles and internally the true and false vocal cords. It passes air into the **trachea** which is a straight tube lying ventrad of the esophagus and dividing at its caudal end at the level of the fifth or sixth rib into two **principal bronchi**. It passes through the neck region and into the thoracic cavity through its inlet. On either side of the trachea lie the common carotid arteries and the vagosympathetic trunks. The trachea is lined with a mucosa with ciliated epithelium and many mucous glands and a connective tissue framework externally which contains supporting C-shaped cartilages. The C-shaped cartilages are placed with the open part of the C caudad, where the free ends are connected by muscle and connective tissue. These tracheal cartilages insure that the trachea will not collapse thereby preventing the possibility of the suffocation of the animal. Its glandular and ciliated mucosa serves as a device for keeping foreign material out of the delicate lung structures. The foreign bodies get caught in the mucous secretion of the glands and the cilia sweep the whole mass upward into the pharynx.

Laterad of the cranial end of the trachea and larynx are the paired lobes of the thyroid glands and embedded in these are the small parathyroid glands. These are endocrine glands; the secretions of the thyroid gland contributing to the maintenance of a proper metabolic balance and the parathyroids to the regulation of calcium metabolism.

The **principal bronchi** lead into the right and left lungs at their roots along with the pulmonary arteries and veins. In the lungs the principal bronchi each divide into two branches, the **cranial** and **caudal lobar bronchi.** The **right cranial lobar bronchus** supplies only the **apical lobe** of the lung. The right caudal lobar bronchus gives rise to three large **segmental** (*tertiary*) **bronchi,** one to each of the remaining lobes of the lung, the **cardiac, intermediale,** and **diaphragmatic lobes.** The **left cranial lobar bronchus** divides into two **segmental bronchi,** one each to the apical and cardiac lobes, while the **caudal lobar bronchus** becomes the **segmental bronchus** of the diaphragmatic lobe. The cartilaginous "rings" are found on all of these bronchi and continue also on the smaller subdivisions of this "respiratory tree" until they reach about 1 millimeter in diameter.

The bronchi ultimately lead into bronchioles, bronchioles into alveolar ducts, alveolar ducts into alveolar sacs and these into alveoli—the terminal units or "leaves" of the "respiratory tree."

Plate 56 shows more clearly the relationships of bronchi and pulmonary arteries and veins and should be consulted at this time.

This plate shows clearly the **pulmonary ligament** of the left lung which is a connection between the visceral and parietal pleura. One occurs also on the right lung. Note also that on the right side there is a special fold of pleura called the **caval fold** which lies next to the postcava and also encloses, in a kind of medial compartment, the intermediate lobe of the right lung.

The mediastinal pleura should be reviewed by referring to Plate 46 where it is shown and described. It divides the thorax into right and left compartments and forms the mediastinum.

The lungs contain also in their system of bronchioles

smooth muscle and elastic tissue. Because the lungs lie in closed cavities and their air ventilation system opens to the outside and is subject to atmospheric pressure, when the thorax expands in response to contractions of the respiratory muscles, such as the diaphragm, the lungs also expand and air rushes in. When the respiratory muscles relax and the thoracic cavity reduces in size, and because also of the elastic tissue in the lungs, air is passed out of the lungs—or may be forced by contractions of the abdominal muscles which compress the viscera and put pressure on the diaphragm further reducing the size of the thorax. The lungs, however, are passive in all of this "pulmonary ventilation" or breathing.

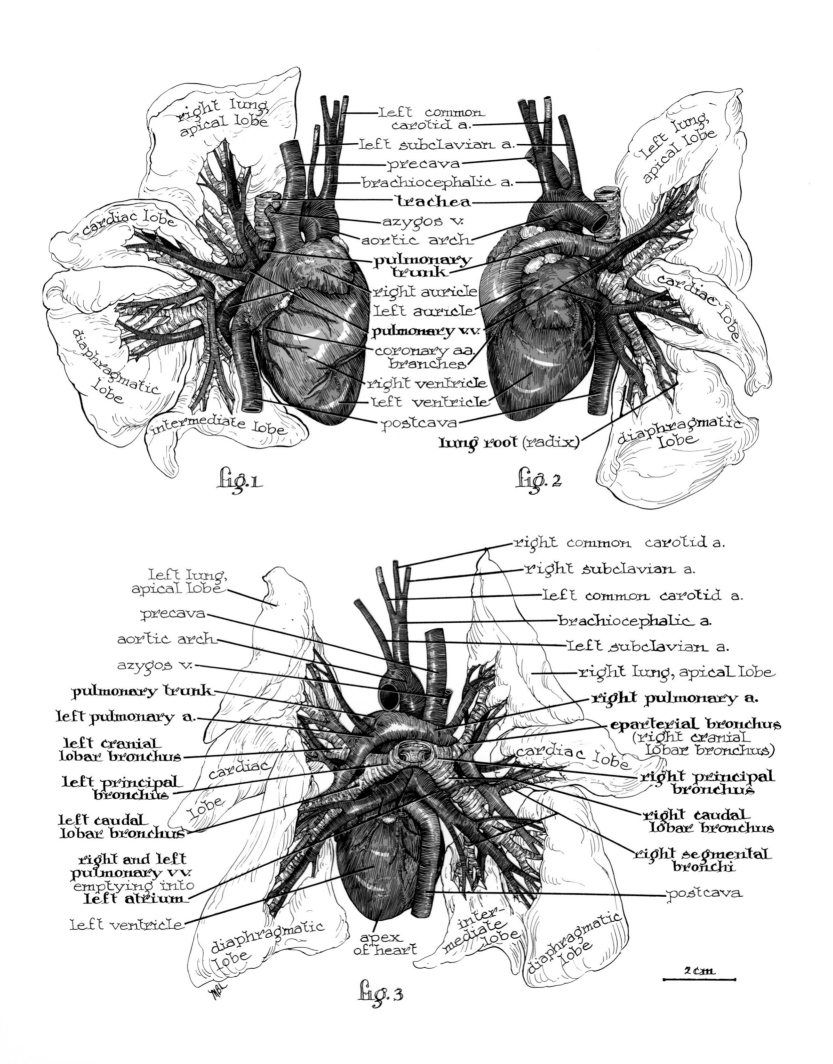

fig. 1

right lung, apical lobe

cardiac lobe

diaphragmatic lobe

intermediate lobe

Left common carotid a.
Left subclavian a.
precava
brachiocephalic a.
trachea
azygos v.
aortic arch
pulmonary trunk
right auricle
left auricle
pulmonary v.v.
coronary aa. branches
right ventricle
left ventricle
postcava

fig. 2

Left lung, apical lobe

cardiac lobe

diaphragmatic lobe

lung root (radix)

Left lung, apical lobe
precava
aortic arch
azygos v.
pulmonary trunk
left pulmonary a.
left cranial lobar bronchus
left principal bronchus
left caudal lobar bronchus
right and left pulmonary v.v. emptying into Left atrium
left ventricle

cardiac lobe

diaphragmatic lobe

apex of heart

inter-mediate lobe

diaphragmatic lobe

right common carotid a.
right subclavian a.
Left common carotid a.
brachiocephalic a.
Left subclavian a.
right lung, apical lobe
right pulmonary a.
eparterial bronchus (right cranial lobar bronchus)
right principal bronchus
right caudal lobar bronchus
right segmental bronchi
postcava

fig. 3

2 cm

The relationships of the heart and lungs in situ in the thoracic cavity are shown in Plates 46, 47, and 55. This Plate 56 deals with the intricate relationships between the arteries, veins and bronchi of the lungs.

Figures 1 and 2 show right and left sides of the heart respectively and the accompanying lung. Figure 3 is a dorsal view of the heart and lungs with vessels and bronchi. Arteries, veins, and bronchi enter or leave the lungs at their **roots** (*radices*). Arteries are red except the pulmonary artery which is blue; the veins are blue except the pulmonary veins which are red.

Recall that the lungs are completely separated by the mediastinal pleura and that the heart with its pericardium lies between the mediastinal pleura where it is a part of the mediastinum (Plate 46). It is only where the trachea divides into **right and left principal bronchi** to supply the lungs that they have any connection (fig. 3).

The principal bronchi each divide at first into two main branches, the **lobar** (*secondary*) **bronchi** (fig. 3). The right cranial lobar bronchus lies dorsad of the pulmonary artery and is known as the **eparterial bronchus.** All the other bronchi lie ventrad of pulmonary arteries and are hyparterial. The right cranial lobar bronchus supplies only the apical lobe of the lung and hence has no other large branches but has many small branches within the lobe. The right caudal lobar bronchus gives rise to three large **segmental** (*tertiary*) **bronchi** one to each of the remaining lobes of the lung, the **cardiac, diaphragmatic** and **intermediate lobes.**

The **left cranial lobar bronchus** divides into two segmental branches, one each to the apical and cardiac lobes of the lung, while the caudal lobar bronchus becomes the segmental bronchus of the diaphragmatic lobe of the left lung and subdivides only within that lobe. Segmental bronchi and the lung tissue they ventilate constitute bronchopulmonary segments. It should be noticed that the lungs are quite separate except as they are connected by the bronchi and connective tissue.

As indicated in all three figures on this plate, the trachea and the principal lobar and segmental bronchi are all supported by cartilaginous elements as are their further subdivisions in the lungs, the bronchioles. They keep the tubes from collapsing. The bronchioles are also provided with spiral bands of smooth muscle.

The **pulmonary trunk** is continuous with the cranial end of the conus arteriosus (Plate 70, fig. 1). It passes craniodorsad and slightly to the left for about 1 or $2\frac{1}{2}$ centimeters and then divides into right and left branches. The left branch passes ventrad of the descending thoracic aorta to reach the left lung dorsad of the left cranial lobar bronchus (fig. 3). The right branch passes under the aortic arch and ventral to the right cranial lobar bronchus and reaches the lung at about the junction of the cranial lobe with the rest of the lung. Here it divides to serve the four lobes.

The **pulmonary veins** enter the left atrium usually in three groups. The first group comes from the apical and cardiac lobes of the right lung; the second from the corresponding lobes of the left side and the third from the remaining lobes of both sides. Figure 2 shows a variation from this pattern since the branch from the cardiac lobe joins the one from the diaphragmatic lobe rather than joining the branch from the apical lobe. Within the lobes, the veins follow closely the pattern of the bronchi.

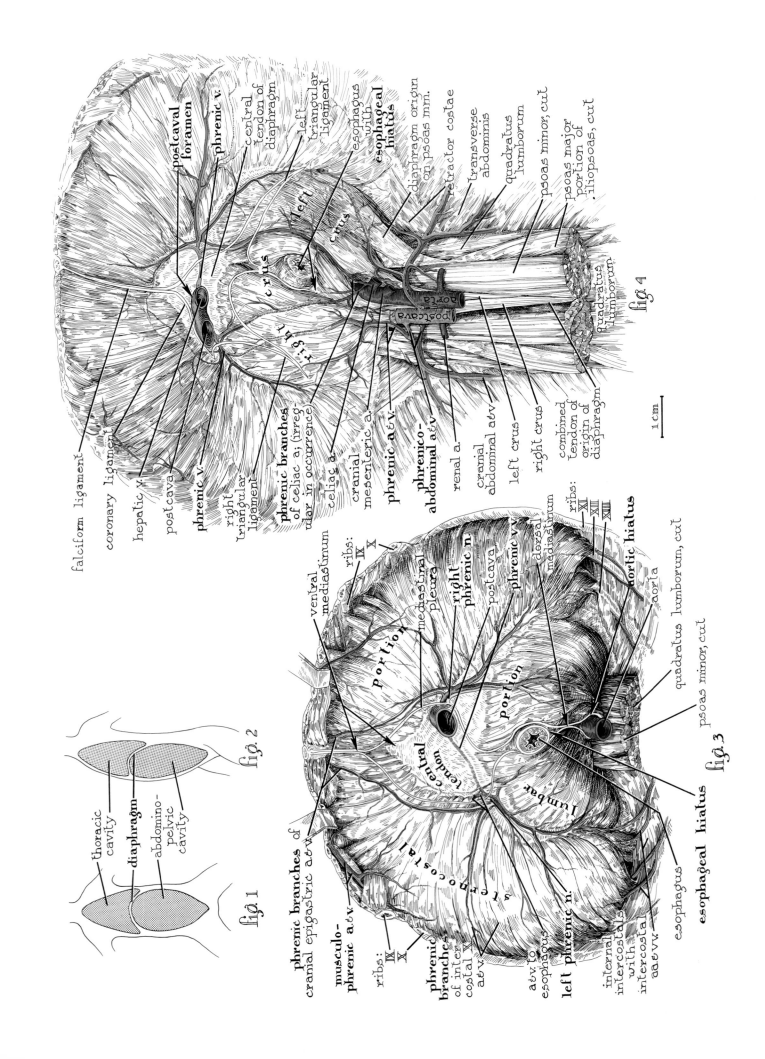

fig.1

thoracic cavity

diaphragm

abdomino-pelvic cavity

fig.2

fig.3

phrenic branches of cranial epigastric a.&v.

musculo-phrenic a.&v.

ribs: IX X

phrenic branches of intercostal a.&v.

a.&v. to esophagus

left phrenic n.

internal intercostals with intercostal a.&v.v.

esophagus

esophageal hiatus

ventral mediastinum

ribs: IX X

mediastinal pleura

right phrenic n.

postcava

Portion

Portion

central tendon

sterno-costal

Portion

lumbar

phrenic v.

dorsal mediastinum

ribs: XI XII XIII

aortic hiatus

aorta

quadratus lumborum, cut

psoas minor, cut

fig.4

falciform ligament

coronary ligament

hepatic v.

postcava

phrenic v.

right triangular ligament

phrenic branches of celiac a; (irregular in occurrence)

celiac a.

cranial mesenteric a.

phrenic a.&v.

phrenico-abdominal a.&v.

renal a.

cranial abdominal a.&v.

left crus

right crus

combined tendon of origin of diaphragm

quadratus lumborum

postcaval foramen

phrenic v.

central tendon of diaphragm

left triangular ligament

esophagus with **esophageal hiatus**

diaphragm origin on psoas mm.

retractor costae

transverse abdominis

quadratus lumborum

psoas minor, cut

psoas major portion of iliopsoas, cut

left crus

right crus

left

aorta

postcava

aorta

1 cm

The **diaphragm** is a musculotendinous dome-shaped wall which separates the thoracic from the peritoneal (*abdominal*) cavities. Figures 1 and 2 show schematically the diaphragm doming into the thoracic cavity. Notice in figure 2 that the thoracic cavity extends farther caudad dorsal to the diaphragm than it does ventrally. The cranial surface of the diaphragm is covered over most of its area with the diaphragmatic pleura; the caudal surface, except in the central area, by parietal peritoneum.

The peripheral part of the diaphragm is **muscular** and takes origin from the xiphoid process of the sternum, the last five ribs, the lumbar vertebrae and at one point on the psoas muscle. Because of these attachments, the muscular portion of the diaphragm is divided into **sternal, costal** and **lumbar portions.** We often combine the sternal and costal and call them the **sternocostal portion.** These muscular portions radiate into a central tendon into which they insert.

The **central tendon** as described by Reighard and Jennings, revised by Elliott (1934), is "thin and irregularly crescent-shaped, with the convexity ventrad and the horns of the crescent prolonged as two tendinous bands which end in two triangular membranous portions of the diaphragm, one on each side of the spinal column." (See fig. 3.)

There are three principal openings in the diaphragm to allow the passage of certain structures between the thoracic and peritoneal cavities (fig. 3). They are the **aortic hiatus,** the most dorsal, for the passage of the aorta, azygos vein and thoracic duct; the **esophageal hiatus,** ventral to the aortic hiatus and in the lumbar portion, for the passage of the esophagus, vagus nerves and some small esophageal blood vessels; the **postcaval foramen,** in the central tendon, for the postcava and some branches of the **right phrenic nerve.**

The **function** of the diaphragm is to retract the central tendon thus increasing the volume and lowering the pressure in the closed thoracic cavity. In this circumstance, air comes in through the open passageways of the respiratory system to enter the lung for purposes of external respiration, making oxygen available to the blood and removing carbon dioxide from it.

Figures 3 and 4 have a great amount of detail drawn into them which should be studied. Figure 3 shows clearly the serrations by which the costal portion of the diaphragm attaches to ribs 11, 12, and 13. The internal intercostal muscles between these ribs are also shown along with their intercostal arteries and veins. Medially, the psoas minor and quadratus lumborum are shown extending

into the thoracic region. The **sternal** portion of the sternocostal muscle is seen in the ventral sternal region and, at this point, the **ventral mediastinum** is seen delimited by the mediastinal pleura which spreads out in the central region of the diaphragm and closes in again dorsally around the esophagus and aorta-the **dorsal mediastinum.** The **left phrenic nerve** reaches the diaphragm having traveled in a fold of the mediastinal pleura much of the way from its origin from the fifth and sixth cervical nerves. The **right phrenic nerve** is seen close to the postcava which it follows near its termination in the diaphragm. A number of phrenic veins and arteries are seen in figure 3: the **phrenic branches** of intercostal artery and vein number ten on the left; the **musculophrenic artery and vein** which are main (*terminal*) branches of the internal thoracic artery and vein as seen on Plate 79, figure 1; the **phrenic branches** of the cranial epigastric artery and vein; and **phrenic veins** coming into the postcava. Arteries and veins to the esophagus are also seen near the left phrenic nerve.

Figure 4 shows the caudal surface of the diaphragm. Here we see the origin of the **lumbar portion** of the diaphragm. It arises as a combined tendon on the ventral surfaces of the bodies of the second, third and fourth lumbar vertebrae, between the psoas and quadratus lumborum muscles. This tendon then diverges into two, the right one much stronger than the left, and from them arise muscle fibers constituting the **right and left crura** of the diaphragm. The **aortic hiatus** lies between the right and left crura; the **esophageal hiatus** is within the larger **right crus.** The left crus receives a small contribution from the psoas muscle.

In this region of the origin of the crura, we see coming from the abdominal aorta, the celiac, cranial mesenteric and phrenicoabdominal arteries. The celiac, in some cases, sends **phrenic arteries** into the diaphragm as shown here and the phrenicoabdominal artery regularly has a **phrenic branch.** The **phrenicoabdominal artery** and phrenic branch are accompanied by veins which bear the same names.

Recall that the liver lies against this caudal surface of the diaphragm and is attached to it by falciform, coronary and right and left triangular ligaments. The cut edges of these ligaments are shown in figure 4. The area enclosed by the coronary and triangular ligaments has no peritoneum and is called a bare area corresponding to a similar area on the cranial face of the liver. Refer to Plate 54, figure 2. Note also in figure 4 that a hepatic vein from the liver empties into the postcava just before the postcava passes through the **postcaval foramen.**

159

The structures and functions of the urinary and genital systems have had such a close relationship during vertebrate evolution (*phylogeny*) that they are frequently considered together (Fig. 14). This same close relationship is seen in the embryological development of the systems (Plate 69). A full separation of the two is not attained in the males of any vertebrates. It is accomplished only in the females of adult primates. Our understanding of one system is therefore dependent upon a knowledge of the other. In plates 58 and 63 structures of both systems are shown in proper relationship. In plates 59, 60, 61 and 65 the organs of each system are separated out for more critical study.

THE URINARY SYSTEM

Definition and Functions. The urinary system is often called the **excretory system.** Its functions, while they center around the excretion of the waste products of metabolism, are more varied and more important than that alone suggests. By excreting excesses of water and of

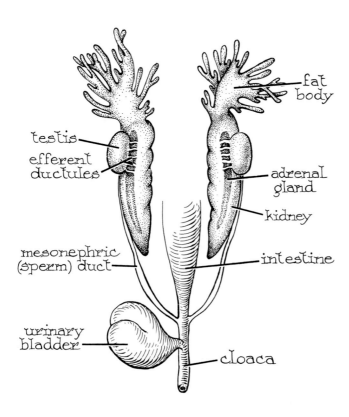

Fig. 14. *Urogenital system of male frog showing close relationship of the urinary and genital organs.* (*Crouch,* Functional Human Anatomy, *Lea & Febiger.*)

inorganic salts the osmotic relationships in the body are maintained at an optimal level. Also the hydrogen ion concentration (*acid-base balance*) of the internal environment is kept in equilibrium. Since it is the **kidneys** that perform these important, complex and selective actions, they are often called the **essential organs** of excretion. The remaining organs of the urinary system, the **ureters, urinary bladder** and **urethra,** may be referred to as the **accessory organs** since they serve the kidneys by transporting, storing and eliminating the urine. Theirs is an easier task than that of the kidneys and they are correspondingly simpler in structure. In the function of excretion materials must cross living membranes which separate the blood in capillaries from the lumina of the excretory tubules; in elimination there are no membranes to cross. It involves only the evacuation of hollow organs.

THE REPRODUCTIVE SYSTEM

Definition and Functions. The reproductive or genital system is unique among body systems in the kind of functions it performs. While the others serve to support, protect, maintain, organize, and coordinate the individual, the reproductive system is concerned with the perpetuation and organization of the species. Since reproduction in mammals is sexual, an intimate and complex relationship between individuals is involved—more intimate and perhaps as complex as any aspect of living.

The individual, in its relatively short life time, adjusts to its environment through inate or learned mechanisms. Its reactions to change in the internal or external environment are rapid physiological or behavioral responses. Its survival depends on the capacity to compete with the physical and biological pressures and demands of the environment. A large part of individual behavior is directed toward perpetuating the species. This involves (1) producing sufficient numbers and variety of offspring, (2) placing them in an environment in which they can survive and (3) providing care for them during development.

The "life" of a species is a long one. Hereditary mechanisms are such that a species remains constant over long periods. It is conservative. Yet, the same mechanism accounts for change, but, except in rare cases, change involving many hundreds or thousands of years.

Coming back now to the individual animal and its reproductive system we find that in mammals it consists of the essential organs; **ovaries** in the female, **testes** in the male. These produce the **gametes—eggs** or **sperms** which together are the "seed" for the next generation. These organs also produce hormones and hence are **endocrine glands.** These endocrine glands are in turn under the

161

influence of other endocrine glands—ultimately the pituitary. They have far ranging influence not only in the life of the individual animal, but directly or indirectly influence the physiology and behavior of individuals to bring them together to mate and to provide for the continuation of their kind.

In the male the other organs of the reproductive system have to do with the passage of the sperms and their introduction into the female reproductive tract. They must also provide a proper environment for the sperm through secretions of various glands so that the sperms remain viable. These organs are the scrotum, epididymis, ductus deferens, urethra and penis. The prostate and bulbourethral (*Cowper's*) glands provide the necessary secretions for transport and viability of the sperms. The scrotum and penis are external organs; the others are internal.

In the female, besides the ovaries, the other organs serve to transport the eggs, or if the eggs are fertilized, the horns of the uterus house the embryos and fetuses during their period of development of about 63 days. The fetuses are delivered (*parturition*) through the body of the uterus, the vagina and urogenital sinus. A clitoris, homologous to the penis, occurs in the ventral floor of the urogenital sinus.

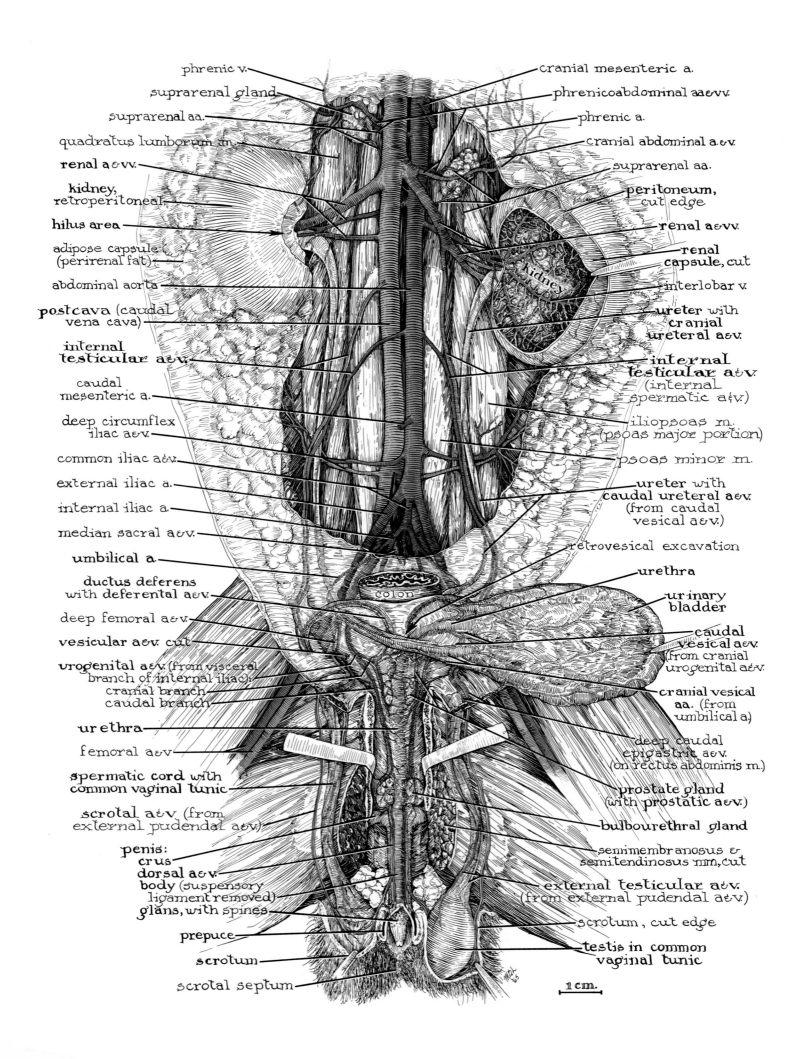

phrenic v.

suprarenal gland

suprarenal aa.

quadratus lumborum m.

renal a & vv.

kidney, retroperitoneal

hilus area

adipose capsule (perirenal fat)

abdominal aorta

postcava (caudal vena cava)

internal testicular a & v.

caudal mesenteric a.

deep circumflex iliac a & v.

common iliac a & v.

external iliac a.

internal iliac a.

median sacral a & v.

umbilical a.

ductus deferens with deferental a & v.

deep femoral a & v.

vesicular a & v. cut

urogenital a & v. (from visceral branch of internal iliac): cranial branch caudal branch

urethra

femoral a & v.

spermatic cord with common vaginal tunic

scrotal a & v. (from external pudendal a & v.)

penis: crus dorsal a & v. body (suspensory ligament removed) glans, with spines

prepuce

scrotum

scrotal septum

cranial mesenteric a.

phrenicoabdominal aa & vv.

phrenic a.

cranial abdominal a. & v.

suprarenal aa.

peritoneum, cut edge

renal a & vv.

renal capsule, cut

interlobar v.

ureter with cranial ureteral a & v.

internal testicular a & v. (internal spermatic a & v.)

iliopsoas m. (psoas major portion)

psoas minor m.

ureter with caudal ureteral a & v. (from caudal vesical a & v.)

retrovesical excavation

urethra

urinary bladder

caudal vesical a & v. (from cranial urogenital a & v.

cranial vesical aa. (from umbilical a.)

deep caudal epigastric a & v. (on rectus abdominis m.)

prostate gland (with prostatic a & v.)

bulbourethral gland

semimembranosus & semitendinosus mm, cut

external testicular a & v. (from external pudendal a & v.)

scrotum, cut edge

testis in common vaginal tunic

colon

kidney

1 cm.

Since this plate is to be followed by others which detail the individual urogenital structures shown here, this description will be limited to basic relationships of organs. The parietal peritoneum has been removed from a part of the dorsal wall to show better the underlying organs.

The **kidneys** are paired, compact organs having the typical kidney bean shape. They lie to either side of the abdominal aorta and postcava in the region between the third and fifth lumbar vertebrae. The right kidney lies 1 or 2 centimeters craniad of the left one-about 1 centimeter in this specimen. Only the ventral surfaces of the kidneys are covered with peritoneum and, hence, they are **retroperitoneal** in position. Both kidneys are embedded in an adipose capsule and are loosely fixed in position by renal fascia. The kidneys of the cat are said to be more loosely attached than in the dog.

Within the above investments each kidney has a strong, thin **fibrous capsule,** shown cut away, in part, on the left kidney. It dips inward at the hilus on the medial border of the kidney, where it lines an area called the **renal sinus** shown in Plate 59, figure 2. It also covers the walls of the renal arteries, veins and nerves and forms a covering for the renal pelvis.

The **renal arteries** are shown branching from the abdominal aorta to enter the hilus of the kidney. The renal veins leave the kidney at the hilus to join the postcava. Also the ureter, the duct of the kidney pelvis, leaves the hilus of the kidney and passes caudad to the urinary bladder. The ureters are musculofibrous tubes which pass caudad dorsal to the parietal peritoneum and ventral to the psoas muscles and to the deep circumflex iliac arteries and veins. Nearing the urinary bladder they cross dorsal to the ductus deferens which loops around them and, in the lateral ligament of the bladder (see Plate 61), reach the dorsolateral part of the neck of the bladder which they enter by passing obliquely through the bladder wall.

The **urinary bladder** is a musculomembranous, pear-shaped organ which lies in the lower abdominal cavity between the ventral body wall and the descending colon. It has been pulled to the left side in this illustration. It varies in size, shape and position depending on the amount of urine it contains. Caudad, the bladder has a long neck which passes dorsad of the ischiopubic symphysis into the pelvic cavity.

The urinary bladder is covered with peritoneum and is held in place by its neck and by three peritoneal folds, the **median umbilical** and the **lateral ligaments.** Within the folds of these lateral ligaments pass blood vessels, the ureters and ductus deferens (Plate 61). Between the urinary bladder and the colon and rectum is a deep space called the retrovesical excavation.

Since the urethra differs so much between the male and female, it will be discussed with the plates where those systems are detailed—Plates 61 and 65. It is enough to say here that in the male the urethra is relatively long and complicated as it passes through the penis.

The male reproductive system pictured here shows the **testis** in the integumental scrotum which has been opened. A scrotal septum separates the two testes within the scrotum. From the testis the spermatic cord runs craniad close to the median line and ventral to the ischiopubic symphysis and enters the external inguinal ring, passes through the inguinal canal and enters the abdominal cavity at the internal inguinal ring (Plate 61, fig. 3). At this point, the **internal testicular arteries** and **veins** and the **ductus deferens,** which were carried in the cord, go their separate ways. The **internal testicular arteries** can be seen traveling craniad, lying behind the peritoneum to join the abdominal aorta of which they are branches. The **right** and **left internal testicular veins** differ in their termination. They are asymmetrical. The left one enters the renal vein, the right one the postcava. The **ductus deferens** loops around the ureter as indicated above and turns caudad to enter the prostate gland.

The **prostate gland** lies around the beginning of the urethra, the prostatic urethra. Caudad of the prostate can be seen the **bulbourethral glands** and next, caudad, the **penis** consisting of **crus, body** and the **glans** with its **spines.** The glans is covered with the **prepuce.**

The blood supply to most of the urinary and reproductive organs in this area is shown but need not be described since they are named mostly for the organs they supply or drain.

The **suprarenal** (*adrenal*) **glands** are shown well in this plate along with their blood supply and should be studied carefully. They lie craniad of the kidneys and are asymmetrical in their position.

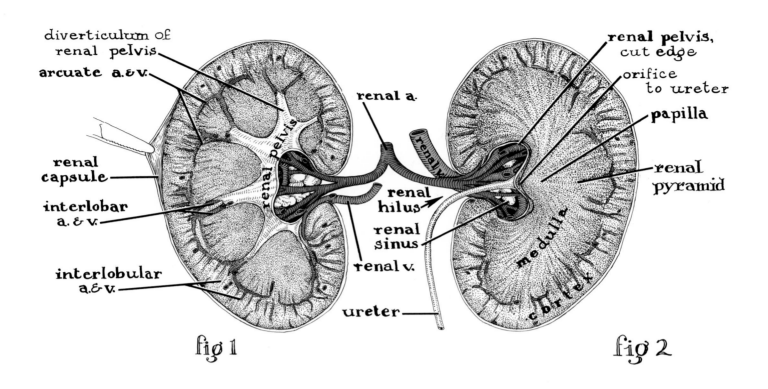

fig 1

diverticulum of renal pelvis

arcuate a.&v.

renal capsule

interlobar a.&v.

interlobular a.&v.

renal pelvis

renal a.

renal hilus

renal sinus

renal v.

ureter

fig 2

renal pelvis, cut edge

orifice to ureter

papilla

renal pyramid

medulla

cortex

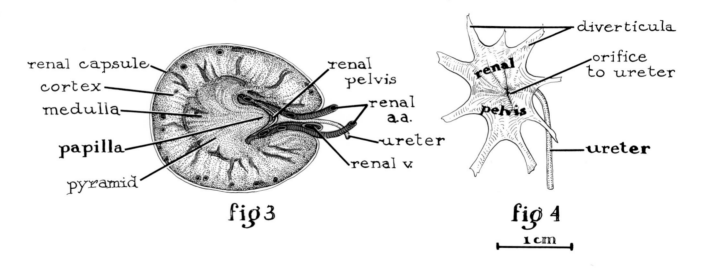

fig 3

renal capsule

cortex

medulla

papilla

pyramid

renal pelvis

renal a.a.

ureter

renal v.

fig 4

diverticula

orifice to ureter

renal pelvis

ureter

1 cm

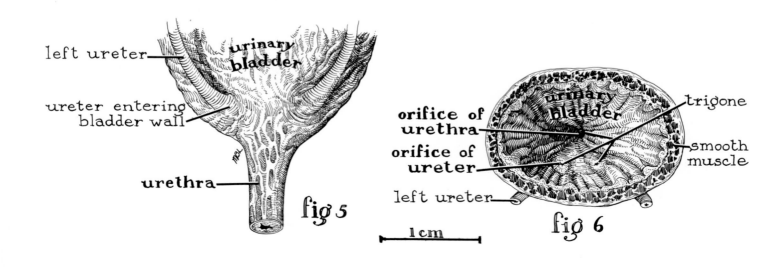

fig 5

left ureter

urinary bladder

ureter entering bladder wall

urethra

fig 6

orifice of urethra

orifice of ureter

left ureter

urinary bladder

trigone

smooth muscle

1 cm

This plate details some of the general anatomy of the urinary part of the urogenital system presented on Plates 58 and 63.

Figures 1 and 2 are **dorsal** views of the **kidneys;** the right one (fig. 2) cut in mid-frontal section, the left one (fig. 1) also sectioned in the frontal plane but at a deeper level. Figures 1 and 2 both show the **renal sinuses** with their openings at the hilus and with the renal arteries entering, and the renal veins and ureter leaving. The sinus also contains the expanded end of the pelvis whose cut edge can be seen in the right kidney. Fat fills the remaining space in the sinus. The right kidney (fig. 2) shows the common **renal papilla** and the **renal pyramid** of the **medulla.** The renal papilla opens into the pelvis and has on its intrapelvic surface a variable number of orifices which are the openings of the **papillary ducts.** The cortex constitutes the peripheral portion of the kidney substance and lies immediately beneath the renal fibrous capsule. It is made up of the renal corpuscles and convuluted tubules in contrast to the medulla with its collecting tubules and papillary ducts as will be seen in the next plate. Figure 1 also shows small bits of the arcuate, interlobar, and interlobular arteries and veins which are also detailed in the next plate.

Figure 3 is a kidney cut in cross section in the region of the renal sinus and renal pelvis. It shows again the structures discussed above, the renal papilla being particularly prominent.

Figure 4 is the renal pelvis and the cranial end of the ureter dissected out of the renal sinus. It shows the central orifice into the ureter and the branch-like **diverticula** which reach out as is shown in figure 1 and embrace the curved pyramid which is a part of the medulla of the kidney.

Figure 5 is the caudal end of the dorsal surface of the **urinary bladder** showing the points of entry of the ureters and the "neck of the bladder" or **urethra.** Reighard and Jennings as revised by Elliott (1934) calls the part labeled here the urethra, the neck of the bladder, and considers the urethra in the male as starting at the prostate and in the female at the urogenital sinus which makes for a very long bladder neck. I am inclined to use the term membranous urethra for the so-called neck of the bladder which means that the urethra has two membranous portions, the other below the prostate.

Figure 6 is the opened floor of the urinary bladder showing the orifices of the ureters and urethra which are positioned in such a way as to mark the corners of a triangle. This triangular area is called the **trigone.**

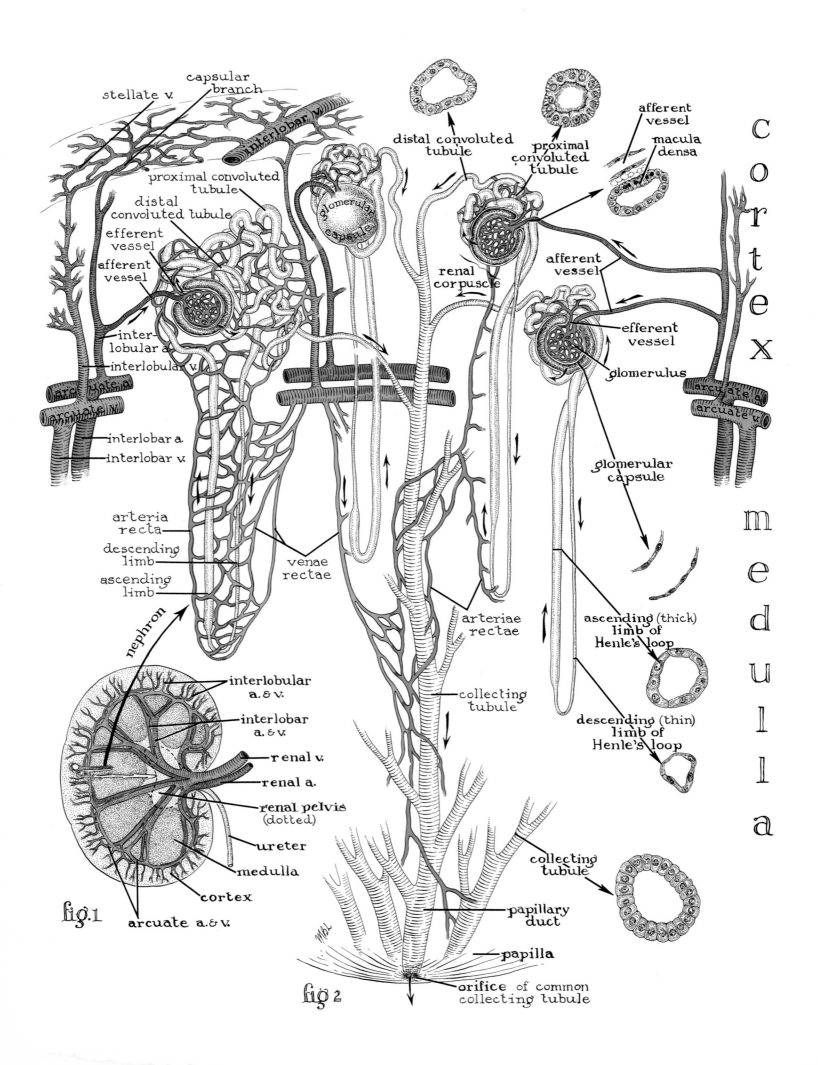

stellate v.

capsular branch

interlobar a.

distal convoluted tubule

proximal convoluted tubule

afferent vessel

macula densa

proximal convoluted tubule

distal convoluted tubule

efferent vessel

afferent vessel

glomerular capsule

afferent vessel

renal corpuscle

afferent vessel

efferent vessel

glomerulus

cortex

inter-lobular a.

interlobular v.

arcuate a.

arcuate v.

interlobar a.

interlobar v.

glomerular capsule

arcuate a.

arcuate v.

arteria recta

descending limb

ascending limb

venae rectae

ascending (thick) limb of Henle's loop

descending (thin) limb of Henle's loop

nephron

arteriae rectae

collecting tubule

medulla

interlobular a. & v.

interlobar a. & v.

renal v.

renal a.

renal pelvis (dotted)

ureter

medulla

cortex

collecting tubule

fig.1

arcuate a. & v.

papillary duct

papilla

fig 2

orifice of common collecting tubule

Schematic Drawing of Kidney Tubules and their
Relations to the Blood Supply
PLATE 60

In this plate, figure 2, I am deviating from the use of plates dealing only with gross anatomy to show the structural and functional unit of the kidney, the kidney tubule or nephron.

In figure 1 there is repeated some of the anatomy already described in Plate 59. Shown in addition, however, are the branches of the **renal artery** and **vein** inside of the kidney substance. These have pertinence in attempting to understand the kidney tubule and its vascular relationships. Note that **interlobar arteries** and **veins** branch from the renal vessels and pass between the **secondary pyramids.** These in turn give rise to **arcuate vessels** at the corticomedullary junction and from these in turn **interlobular arteries** and **veins** penetrate the cortex giving off **afferent vessels** (see fig. 2) to the glomeruli and to the cortex substance and fibrous capsule. Notice especially that the arcuate arteries do not form complete arches over the secondary pyramids, while the arcuate veins do.

Figure 1 also shows a **nephron,** enlarged, of course, to indicate the relationships of its various parts to the gross regions of the kidney. This is shown also in figure 2, with the **cortex** lying above the arcuate arteries and veins, the **medulla** below. In the upper left hand corner of figure 2 a piece of the fibrous capsule on the surface of the kidney is shown with branches of interlobular artery supplying it and its surface well covered with **stellate veins** which drain into the interlobular veins which extend on to its surface.

The remaining description relates to figure 2. A **nephron** starts in the **cortex** by means of a **glomerular capsule** which is the proximal end of the tubule. It is a blind tube with its end enlarged and pushed in on itself like pushing one side of a soft rubber ball in towards its other side. The glomerular capsule is therefore a double layered structure. It is often called Bowman's capsule. Each layer of its double wall is composed of a single layer of flat or squamous cells as shown on the right side of the illustration. The glomerular capsule encloses a knot of capillaries, the **glomerulus,** coming from one of the **afferent vessels** (*an arteriole*) which is a branch from the interlobular artery. The glomerulus and the glomerular capsule together constitute a **renal** (*malpighian*) **corpuscle.** It is here that the tubule collects a filtrate of the blood plasma.

The glomerular capsule leads by a short neck into the **proximal convoluted tubule,** shown in the upper right part of the plate. It leads a tortuous course in the cortex and constitutes much of the cortical substance. Its walls are composed of a single layer of cuboidal or pyramidal cells with large oval nuclei and on the free surface of the cells, a brush border. It finally runs to the medullary region and

tapers down to continue into the U-shaped **loop of Henle,** specifically into its **descending thin limb** where the epithelium changes abruptly from the pyramidal or cuboidal cells to flattened (*squamous*) cells. This is seen in the lower right side of the plate. The proximal convoluted tubule plays an important role in urine formation by reabsorbing much of the filtrate produced by the renal corpuscle.

The **descending thin limb** of Henle's loop pushes into the medulla and then makes an abrupt turn and as an ascending limb returns to the cortex of the kidney. The change from the descending thin limb to the **ascending** (*thick*) **limb** is also quite abrupt as the **squamous cells** are replaced by **cuboidal cells.** The ascending limb passes into the cortex and returns to the renal corpuscle of its nephron and attaches itself to its afferent arteriole. That side of the tube in contact with the afferent arteriole forms an area of taller cells with crowded nuclei called the **macula densa.** Its function is not known with any degree of certainty. It is seen in the upper right hand corner of the plate.

The tubule next passes into the **distal convoluted tubule** which has a somewhat lower **cuboidal epithelium** than the proximal convoluted tubule and a larger lumen. The cells may have microvilli on their free surfaces. One is seen at the top of the plate. Each distal convoluted tubule is continued by a short **collecting tubule** still in the cortex of the kidney. As the collecting tubule runs into the medulla of the kidney it is quite straight and receives no further connection until it reaches the inner zone of the medulla, when it is joined at vertical angles by several similar tubules which form a **papillary duct** which opens on the apex of a **papilla.** These collecting tubules have at first a low simple columnar epithelium which gradually becomes higher as the collecting tubules become larger and is of maximum height in the papillary ducts (*ducts of Bellini*).

The filtrate produced from the plasma of the blood at the renal corpuscle has been partly reabsorbed, some materials have been added to it; it has become much concentrated and arrives in the pelvis of the kidney as urine. This is possible because coming from each glomerulus there is an **efferent arteriole** which forms capillary networks around the proximal and distal convoluted tubules and the loop of Henle. It is between the blood of these capillaries and the filtrate in the renal tubules that the changes are made which produce the urine from a plasma filtrate. **Arteria recta** are formed also from the efferent arterioles which send vessels toward the medulla and form capillary networks around the collecting tubules and the lower part of Henle's loop. **Venae rectae** return this blood to the arcuate and interlobar veins.

169

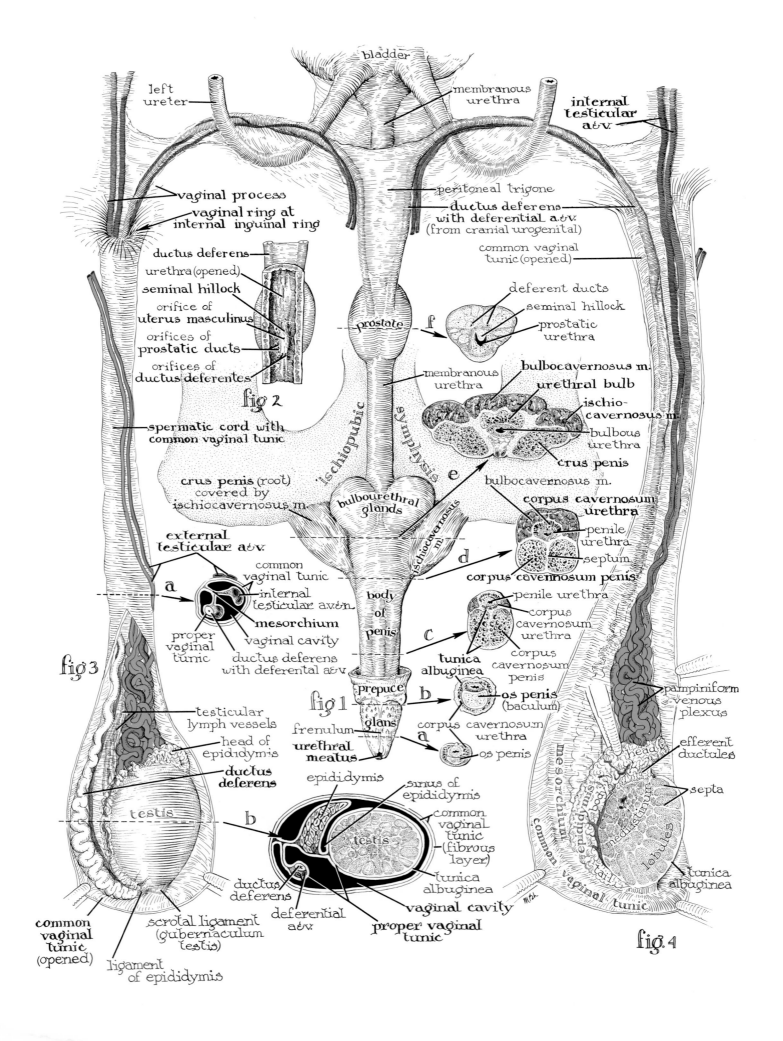

bladder

left
ureter

membranous
urethra

internal
testicular
a.&v.

vaginal process

vaginal ring at
internal inguinal ring

ductus deferens

urethra (opened)

seminal hillock

orifice of
uterus masculinus

orifices of
prostatic ducts

orifices of
ductus deferentes

fig 2

spermatic cord with
common vaginal tunic

crus penis (root)
covered by
ischiocavernosus m.

external
testicular a.&v.

peritoneal trigone

ductus deferens
with deferential a.&v.
(from cranial urogenital)

common vaginal
tunic (opened)

deferent ducts

seminal hillock

prostatic
urethra

prostate   f

membranous
urethra

bulbocavernosus m.

urethral bulb

ischio-
cavernosus m.

bulbous
urethra

crus penis

bulbocavernosus m.

corpus cavernosum
urethra

penile
urethra

septum

corpus cavernosum penis

e

ischiopubic

symphysis

bulbourethral
glands

ischiocavernosus
m.

d

common
vaginal tunic

internal
testicular a.&n.

mesorchium

vaginal cavity

ductus deferens
with deferental a.&v.

proper
vaginal
tunic

a

body
of
penis

penile urethra

corpus
cavernosum
urethra

corpus
cavernosum
penis

c

tunica
albuginea

fig 3

testicular
lymph vessels

head of
epididymis

ductus
deferens

testis

prepuce

glans

fig 1

frenulum

urethral
meatus

b

os penis
(baculum)

corpus cavernosum
urethra

os penis

a

penile
urethra

pampiniform
venous
plexus

efferent
ductules

epididymis

sinus of
epididymis

common
vaginal
tunic
(fibrous
layer)

testis

tunica
albuginea

vaginal cavity

proper vaginal
tunic

b

ductus
deferens

deferential
a.&v.

mesorchium

common vaginal tunic

head

body

tail

mediastinum

lobules

septa

tunica
albuginea

fig. 4

common
vaginal
tunic
(opened)

scrotal ligament
(gubernaculum
testis)

ligament
of epididymis

The male genital organs are illustrated here, mostly in dorsal view, and as removed from their normal position and relationships in the animal. The urinary bladder, urethra and the ureters, each with a **ductus deferens** looping around it, serve as landmarks as do the ventral parts of the pelvic girdle with the ischiopubic symphysis lying ventral to the membranous urethra. Compare this with the system shown in position in the animal and seen in ventral view in Plate 58.

Figure 1 is primarily a study of the **penis** seen here in ventral view, *i.e.* the urethral surface. Its prepuce, a fold of skin which covers the glans of the penis, has been folded back. The **glans,** thus shown, is seen to be covered on the surface by sharp, recurved horny papillae or **spines.** A frenulum along its midventral surface connects it to the prepuce. The **urethral meatus** and opening is seen on the free end of the glans of the penis. Moving craniad we see the body of the penis and next the root or **crus penis** covered by the **ischiocavernosus muscle.** The bulbourethral glands which open into the root of the penis lie at the cranial end of the penis. Notice that each part of the penis is shown in cross section in figure 1a, b, c, d and e. The exact location of each cross section is indicated by a dotted line across the penis and an arrow to the appropriate drawing. Sections a and b are through the glans and show the cavernous urethra running through the **corpus cavernosum** (*spongiosum*) **urethra** and lying within the groved surface of the **baculum** or **os penis.** The glans is soft, the surface consisting of squamous epithelium. Section c is through the body of the penis. The **corpus cavernosum penis** takes the place of the baculum. It, like the corpus cavernosum urethra, is composed of open spongy tissue which becomes filled with blood to erect the penis. The corpus cavernosum penis becomes divided into two parts by a septum as seen best in section d taken at the proximal end of the body of the penis. The **bulbocavernosus muscle** has appeared on the ventral side. It becomes more expansive in the next section e through the region of the root of the penis and the **urethral bulb.** This section cuts through the **crura of the penis** (*crus penis*) covered by the **ischiocavernosus muscles,** here shown on just one side. The ischiocavernosus muscle has its origin on the caudal border of the ramus of the ischium and it inserts into the whole outer surface of the crus penis. The bulbocavernosus muscles cover the ventral surface of the penis from the bulb to the proximal end of the lateral surface of each corpus cavernosum penis to the distal third of the body of the penis. Some of their fibers originate from the external anal sphincter.

The **prostate gland** is shown at the cranial end of the membranous urethra. The **prostatic urethra** passes through this gland and receives ducts from it as also seen in cross section f. Note the U-shape of this part of the urethra. The **ductus deferentes** can be seen on the dorsal surface of the prostate and in the section f entering through the wall of the prostate.

Additional structures involved in the prostatic urethra are seen in figure 2 as are some already mentioned. The ventral wall has been removed to show on the dorsal wall a raised area, the **seminal hillock,** in the top of which there is a small opening leading into a tube that extends dorsally into the prostate. This is called the **uterus masculinus** (*prostatic utericle*) and is homologous to the caudal part of the vagina of the female.

Lying between the two converging deferent ducts at the top of figure 1 is a fold of peritoneum called the peritoneal trigone.

To understand figures 3 and 4 which show the testes, their duct system and blood supply, we must realize that the testes, like the ovaries, have their origin embryologically on the dorsal wall of the abdominal cavity in close relationship to the kidneys. During development they migrate caudad into the scrotum into which there has moved a pouch of the peritoneum called the **vaginal process** (see fig. 3). They bring with them their ducts and blood vessels. They develop in the scrotal sac the same relationship to the celom (**vaginal cavity**) and its peritoneal lining and to their mesenteries as do other organs. Their surface layer is a visceral epithelium.

In figure 3 the testis and epididymis have been removed from the scrotum. Their **common vaginal tunic** has been cut and removed in part. This tunic is lined on the inside by the **parietal peritoneum** outside of which is a **fibrous layer** the transversalis fascia, which is the same layer as is found under the peritoneum in the abdominal cavity. Attached to the medial dorsal part of the testis cranially is the head of the epididymis which passes around the cranial end of the testis to the lateral dorsal side, the body of the epididymis, and then caudally as the tail of the epididymis to the caudal pole of the testis. These parts are labeled on figure 4. From the tail of the epididymis the **ductus deferens,** at first much coiled, runs craniad and straightens out as it passes into the spermatic cord. The spermatic cord, like the testis, is covered with the same common vaginal tunic. Cross sections (a) and (b) will help you to understand these relationships. Section (a), a cross section of the **spermatic cord,** for example, shows the common vaginal tunic with its lining of parietal peritoneum (outer layer of proper vaginal tunic) and an outer fibrous layer as seen also around the testis in section (b). The ducts deferens and its blood vessels, the internal testicular arteries, veins and nerves are suspended in the vaginal cavity by a thin double membrane, the **mesorchium** or mesentery of these structures including, as seen in (b) the testis and epididymis. The two layers of the mesorchium split and surround these organs forming their visceral peritoneum, the inner layer of the **proper vaginal tunic.** Between the layers of the proper vaginal tunic is the **vaginal cavity** (See section a). On the outside of the spermatic cord are shown the external testicular arteries and veins.

The spermatic cord is covered ventrally by the skin as it passes craniad. At the abdominal wall it enters the external inguinal ring (Plate 58), passes through the inguinal canal and enters at the internal inguinal ring. Here the arteries, veins, nerves and the ductus deferens separate. The ductus deferens, as we have noted, loops over the ureter and turns caudad to join the urethra in the prostate gland.

Figure 4 shows the right testis sectioned to reveal the internal structure. It is enclosed in a dense, white fibrous capsule, the tunica albuginia, and, outside of this, the inner layer of the proper vaginal tunic. Running lengthwise in the center of the testis is a fibrous cord, the mediastinum. At the attachment of the epididymis to the testis on the dorsomedial border, the tunica albuginia sends connective tissue partitions, the **septa,** into the testis which converge centrally on the mediastinum. Between these septa are the lobes of the testis which contain coiled seminiferous tubules in the walls of which are sperm producing cells and supporting and nourishing cells. Sperm formed in the tubules move toward the epididymal attachment to the testis, pass through straight tubules formed by the union of seminiferous tubules and, in turn, into a network of tubules, the rete testes. From the rete testis, efferent ductules, seen in figure 4, lead into the epididymis. The sperm pass then into the ductus deferens.

Finally, notice the complex **pampiniform plexus** which is a part of the venous system from the testis and epididymis.

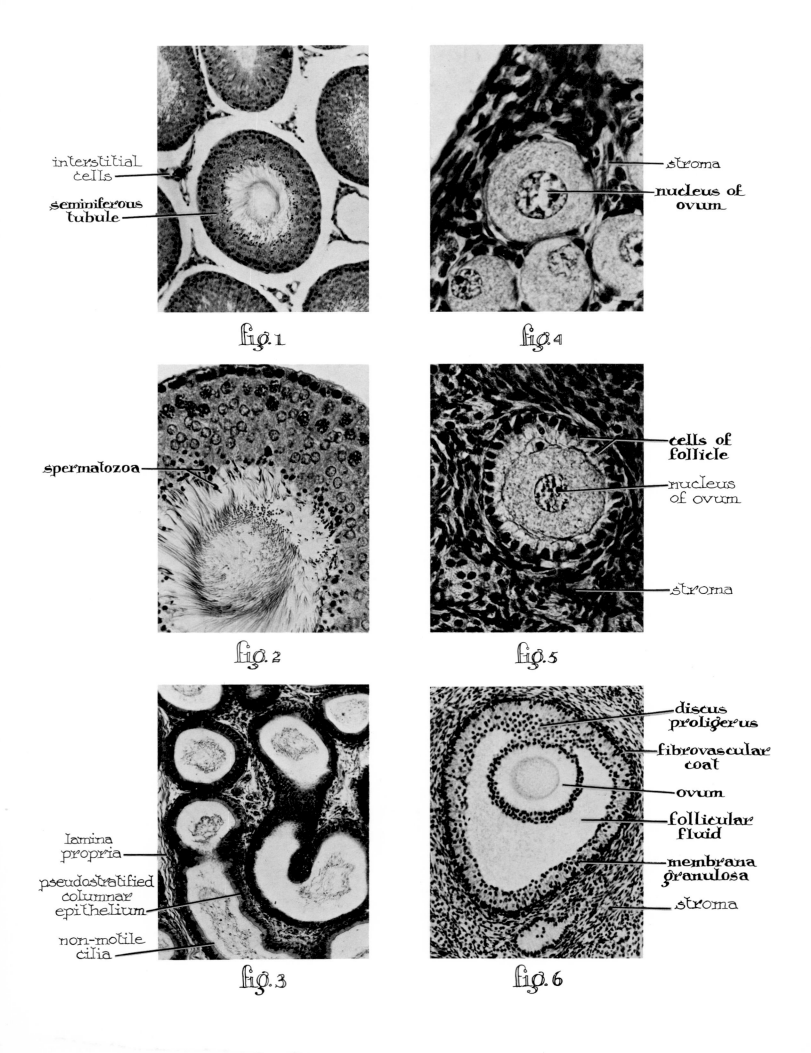

interstitial cells

seminiferous tubule

fig. 1

spermatozoa

fig. 2

lamina propria

pseudostratified columnar epithelium

non-motile cilia

fig. 3

stroma

nucleus of ovum

fig. 4

cells of follicle

nucleus of ovum

stroma

fig. 5

discus proligerus

fibrovascular coat

ovum

follicular fluid

membrana granulosa

stroma

fig. 6

This plate shows a number of photomicrographs of parts of the testis and ovary and the sperms and eggs which they produce. Figure 1 is a section of a number of **seminiferous tubules** of the testis of a rat magnified about 180 times. Among the tubules are patches of interstitial cells which are endocrine in function producing the male sex hormone or testosterone. Within the walls of the seminiferous tubules various stages in the process of meiosis are shown. Meiosis is a process by which **spermatozoa** (*sperm cells*) are produced from the spermatogonia of the seminiferous tubules. These sperm cells, as a result of meiosis, have a chromosome number of one half of that of the parent cells and their hereditary potential has been established. Figure 2 is a photomicrograph of one seminiferous tubule magnified much higher, about 720 times, and showing more clearly the various stages in meiosis. **Spermatozoa** are seen clearly in the lumen of the tubule.

Figure 3 is a cross section of the **epididymis** of a rat. It is the principal storehouse for sperms and it adds an essential secretion to the fluid in which the spermatozoa are activated and stored. It is lined with tall pseudostratified columnar epithelium with non-motile cilia. The epithelium rests on a distinct basement membrane. The fibrous lamina propria contains a thin layer of smooth muscle. Contraction of this muscle during ejaculation forces the spermatozoa into the ductus deferens.

Figures 4 to 6 are photomicrographs which show in sequence the growth of **follicles** of the ovary of the cat. These follicles house the **ova** or eggs and during follicular growth the ova undergoes meiosis, as did the sperm, and with the same result that the mature egg has half the chromosome number of the body cells and has established its own hereditary potential. When the sperm fertilizes an egg and adds its complement of chromosomes to those of the sperm, the chromosome number of the species is again attained and the hereditary material of sperm and egg—representing male and female parents—determine the hereditary make up of the new individual.

Figure 4 (*magnification 700×*) shows a number of primary follicles located in the **stroma** (*supporting tissue*) of a cat's ovary. In the new born **human** female it is estimated that there may be as many as 400,000 of these primary follicles in both ovaries. Yet only about 400 of these mature during the reproductive life of a woman. The others, with their ova, degenerate at various stages in their development. A small ring of flattened follicular cells can be seen surrounding the ovum. Note the large nucleus of the ovum.

Figure 5 (360×) is a photomicrograph of a **growing follicle** and the contained ovum. There have been changes in the ovum and its nucleus. The ovum is now larger and a refractile, dark staining cell membrane, the zona pellucida, (*not labeled*), appears around it. A polar body, indicating meiotic activity, appears at the upper side of the ovum. The follicle cells, now columnar rather than squamous, are numerous around the ovum and form at first a single layer. Later, through mitotic proliferation, the follicle grows, the simple epithelium becomes stratified and the follicle becomes oval with the ovum eccentric in position. With further growth of the follicle spaces appear among the follicular cells and these spaces fill with a clear liquor folliculi or **follicular fluid.** Finally, these spaces coalesce to form a single larger fluid-filled cavity. This final stage in the growing follicle is the **vesicular** or **Graafian follicle** as seen in figure 6 (*magnified only 180×*). The ovum here is far to one side in a mound of cells, the **discus proligerus** (*cumulus oophorus*). The follicular cavity is large surrounded by a stratified epithelium, the **membrana granulosa.** Outside of this is a **fibrovascular coat** and outside of it, in turn, the stroma or connective tissue of the ovary. A more careful analysis of the structure of the vesicular follicle would reveal a basement membrane between the follicle cells, the membrana granulosa and the fibrovascular coat. The latter can also be shown to consist of an inner loose connective tissue and vascular layer the theca interna and an outer more dense fibrous layer the theca externa, which is continuous into the stroma of the ovary.

As the vesicular follicle continues to enlarge by addition of cells and secretion of more fluid, it comes to be close to the ovary surface and to bulge from it. Finally, as secretion of fluid continues and cell proliferation stops the pressure mounts within the follicle, the bulging area thins and finally ruptures. Fluid seeps into the celom around the ovary and finally the ovum is extruded, a process called **ovulation.** Active movements of the fimbria of the uterine tube, which "grasp" the ovary, and the movement of cilia presumably cause the ovum to enter the ostium of the tube (Plate 63). Fertilization may take place when the ovum is in the celom. More often it happens in the uterine tube.

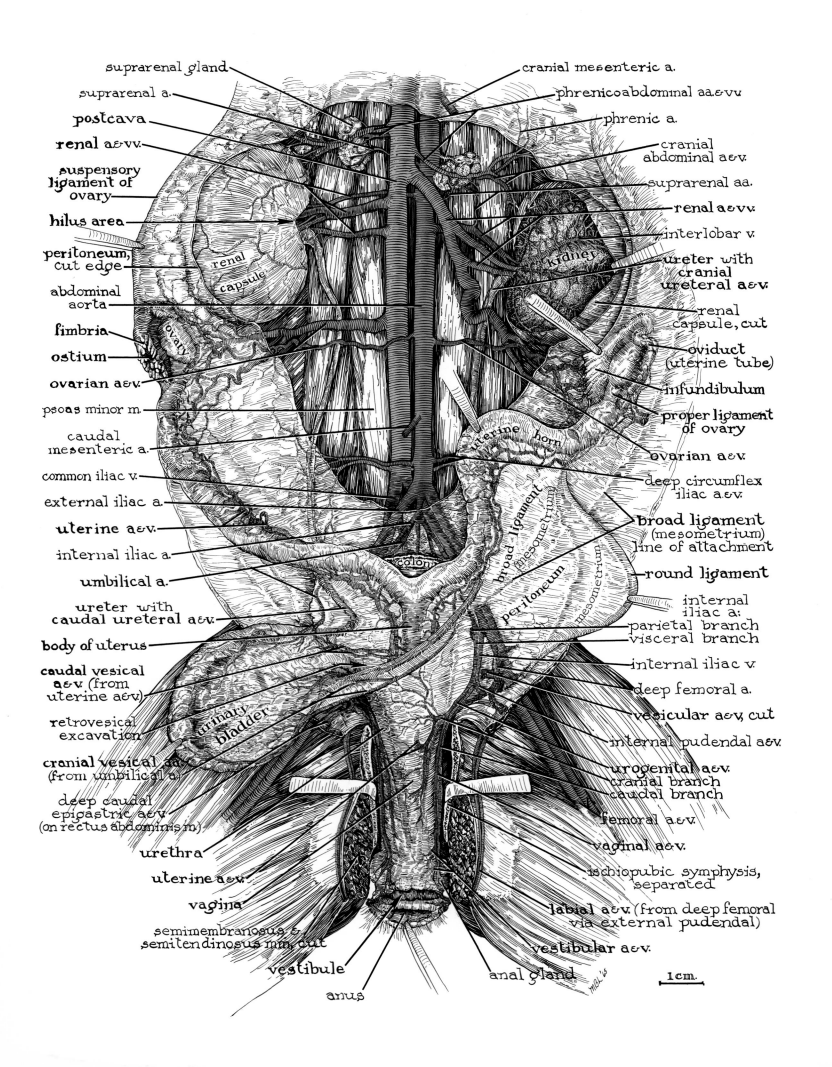

suprarenal gland

suprarenal a.

postcava

renal a.&v.

suspensory ligament of ovary

hilus area

peritoneum, cut edge

abdominal aorta

fimbria

ostium

ovarian a.&v.

psoas minor m.

caudal mesenteric a.

common iliac v.

external iliac a.

uterine a.&v.

internal iliac a.

umbilical a.

ureter with caudal ureteral a.&v.

body of uterus

caudal vesical a.&v. (from uterine a.&v.)

retrovesical excavation

cranial vesical aa. (from umbilical a.)

deep caudal epigastric a.&v. (on rectus abdominis m.)

urethra

uterine a.&v.

vagina

semimembranosus & semitendinosus mm. cut

vestibule

anus

renal capsule

ovary

cranial mesenteric a.

phrenicoabdominal aa.&vv.

phrenic a.

cranial abdominal a.&v.

suprarenal aa.

renal a.&vv.

interlobar v.

kidney

ureter with cranial ureteral a.&v.

renal capsule, cut

oviduct (uterine tube)

infundibulum

proper ligament of ovary

ovarian a.&v.

deep circumflex iliac a.&v.

uterine horn

broad ligament (mesometrium) line of attachment

broad ligament (mesometrium)

round ligament

peritoneum

mesometrium

internal iliac a.

parietal branch

visceral branch

internal iliac v.

deep femoral a.

vesicular a.&v., cut

internal pudendal a.&v.

urogenital a.&v.

cranial branch

caudal branch

femoral a.&v.

vaginal a.&v.

ischiopubic symphysis, separated

labial a.&v. (from deep femoral via external pudendal)

vestibular a.&v.

anal gland

colon

urinary bladder

1 cm.

Since the urinary system has been discussed in connection with the male urogenital system (Plate 58), that part will be omitted here. Again notice the suprarenal glands and their blood supply and the asymmetry in the genital veins, here called **ovarian veins.**

The female reproductive system consists of the ovaries, oviducts, uterus, vagina, urogenital sinus and vulva. The **ovaries** lie caudad of the kidney and on a line with them. They are small oval bodies about 1 centimeter in length. The left one is usually slightly more caudad than the right one in keeping with the position of the kidneys. Notice that the left ovary is supported by a part of the broad ligament, the **mesovarium** (Plate 65, fig. 1). A narrow tube, the **oviduct** (*uterine tube*), with a funnel-shaped **infundibulum** to the left, encircles the ovary. On the right side the open end of the infundibulum, the **ostium,** is seen with its mucus finger-like folds the **fimbria.** Eggs coming from the ovary must pass into the celom and then enter the ostium of the infundibulum to reach the oviduct.

The **oviduct** ascends along the infundibular side of the ovary, bends over its cranial end and descends along the opposite side, usually undergoing a few convolutions. At the caudal end of the ovary it joins the horn of the uterus. The ovary, at its caudal end, is held to the horn of the uterus by the **proper** (*ovarian*) **ligament** of the ovary, a part of the broad ligament which goes forward to the middle of the last one or two ribs.

The horn of the uterus is supported by a part of the broad ligament known as the **mesometrium** as shown on the left side. There is also a **round ligament,** a thickening in the edge of a peritoneal fold which derives from the broad ligament. The round ligament extends from the cranial end of the uterine horn and runs caudad to the internal inguinal ring and usually passes through the inguinal canal along with its peritoneal covering (*the processus vaginalis*) and terminates subcutaneously in or near the vulva. The suspensory, the proper and the round ligaments all contain smooth muscle which allows for stretching of these parts during pregnancy. The round ligament is homologous to the gubernaculum of the male.

The right and left horns of the uterus join medially to form the **body of the uterus.** The bladder has been pulled to the right side to reveal the body of the uterus. The body of the uterus lies ventral to the colon and rectum and dorsal to the bladder. It extends into the cranial end of the pelvis where it projects into the vagina forming a prominent papilla, the cervix, shown in Plate 65. The vagina extends dorsad of the symphysis of the pelvis to just craniad of the caudal border of the ischiatic symphysis where it receives the urethra from the urinary bladder. The remainder of the tube from this point is the **vestibule** or **urogenital sinus** and is homologous with the urethra of the male. It extends to the external opening which lies just ventrad of the anus. The external opening of the urogenital sinus forms the vulva. These structures, shown on this plate only in terms of their external form, are presented in greater detail in Plate 65, figures 1 to 6.

In this plate as in that of the male system the blood supply is well illustrated and the names of the vessels conform mostly to the names of the organs they supply or drain. Study them carefully and refer also to Plates 67 and 80.

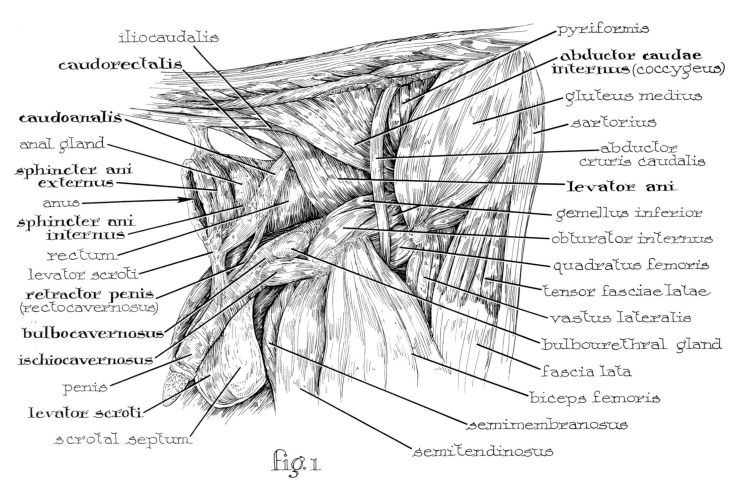

iliocaudalis

caudorectalis

caudoanalis

anal gland

sphincter ani externus

anus

sphincter ani internus

rectum

levator scroti

retractor penis (rectocavernosus)

bulbocavernosus

ischiocavernosus

penis

levator scroti

scrotal septum

pyriformis

abductor caudae internus (coccygeus)

gluteus medius

sartorius

abductor cruris caudalis

levator ani

gemellus inferior

obturator internus

quadratus femoris

tensor fasciae latae

vastus lateralis

bulbourethral gland

fascia lata

biceps femoris

semimembranosus

semitendinosus

fig. 1

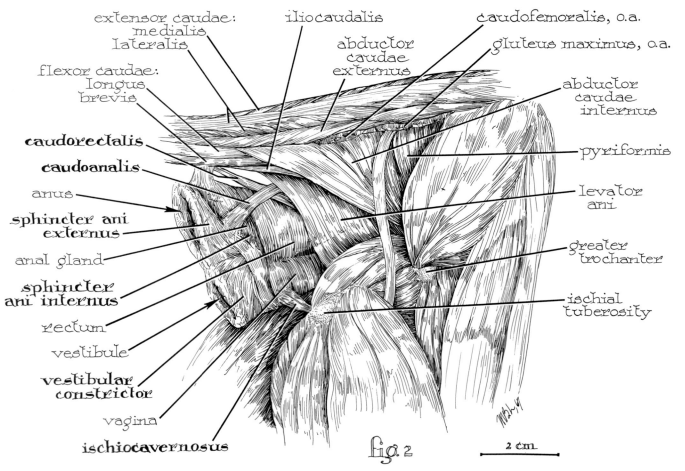

extensor caudae: medialis lateralis

iliocaudalis

abductor caudae externus

caudofemoralis, o.a.

gluteus maximus, o.a.

flexor caudae: longus brevis

caudorectalis

caudoanalis

anus

sphincter ani externus

anal gland

sphincter ani internus

rectum

vestibule

vestibular constrictor

vagina

ischiocavernosus

abductor caudae internus

pyriformis

levator ani

greater trochanter

ischial tuberosity

fig. 2

2 cm

This plate shows the terminal organs of the urogenital and digestive systems and some of the related musculature. The area between the anus and the urogenital openings is the **perineum.** A number of the muscles of the hindlimb are also shown for orientation purposes. The dissection of this area requires skill and patience for many of the muscles are thin and heavily covered with fat and fascia.

Examine figure 1, the male pelvis, carefully. Note especially the position of the penis projecting in a caudoventrad direction and how the organs of the urogenital and digestive system are supported from the vertebral column by the various muscles. The **abductor caudae internus** (*coccygeus*) and the **levator ani** form a vertical **pelvic diaphragm** which closes off the pelvic cavity. The rectum and urethra pass through this diaphragm. The abductor caudae internus has been described in reference to Plate 35. The levator ani originates on the pelvic symphysis and inserts into the midventral line of the bodies of the third to the fifth caudal vertebrae. Note the close relationship of this muscle dorsally to the iliocaudalis.

The **sphincter ani externus** (figs. 1 and 2) is composed of striated muscle fibers and is involved ventrally with the levator scroti (*female-levator vulvae*). The external sphincter is held to the integument of the dorsum of the tail at the level of the fifth caudal vertebra by a fascia containing a few muscle fibers. This divides around the sides of the tail joining ventrally and continuing ventrad as a median structure which separates around the anus forming the external anal sphincter—a muscle about 5 millimeters wide. Below the anus the fascia and a few muscle fibers continue to the scrotum as the levator scroti. Some of the fibers of the sphincter ani externus pass on to the anal pouch where, with fibers from the sphincter ani internus, they form the constrictors of the anal pouch.

The **sphincter ani internus** is a continuation of the circular muscle layer of the rectum and anal canal. It forms a band, broad above and narrow below, around the anus.

The sphincter ani muscles close the anus but allow it to open for passage of feces.

Note the **caudorectalis** muscle which arises from the ventral surface of the sixth and seventh caudal vertebrae. It is a narrow band, unpaired at first, but as it passes cranioventrad it divides into two portions which extend laterad over the sides of the rectum. It serves to draw the rectum caudad and to aid in the process of defecation.

The **retractor penis** (*rectocavernosus*) is a small bundle of fibers which arises in two halves from the ventral side of the sphincter ani internus. These two parts join into a single muscle which joins the ventral surface of the penis and passing caudad is inserted into the corpus cavernosum. It pulls the penis craniodorsad.

The **caudoanalis** muscle lies between the levator ani and the caudorectalis muscles. Dorsad it originates on the ventral surface of the second and third caudal vertebrae. Extending caudoventrad it inserts into the sphincter ani muscles or neighboring fascia. Associated with the caudoanalis and just craniad of it one usually sees the caudocavernosus not shown in this illustration.

The ischiocavernosus and bulbocavernosus muscles have been described in relationship to the study of the male reproductive organs, Plate 58.

Note the similarities and differences between the muscles of the male and female reproductive systems. The similarities far outweigh the differences which is not surprising when one recalls that in development all individuals go through a common or indifferent stage from which the male and female systems later differentiate.

The **vestibular constrictor** (*rectovaginalis*) constricts the vestibule and aids in copulation by tightening behind the glans of the penis. The vestibular constrictor arises in the sides of the sphincter ani internus from which it passes ventrocaudad to insert into the ventral surface of the urogenital sinus caudad of the insertion of the ischiocavernosus.

179

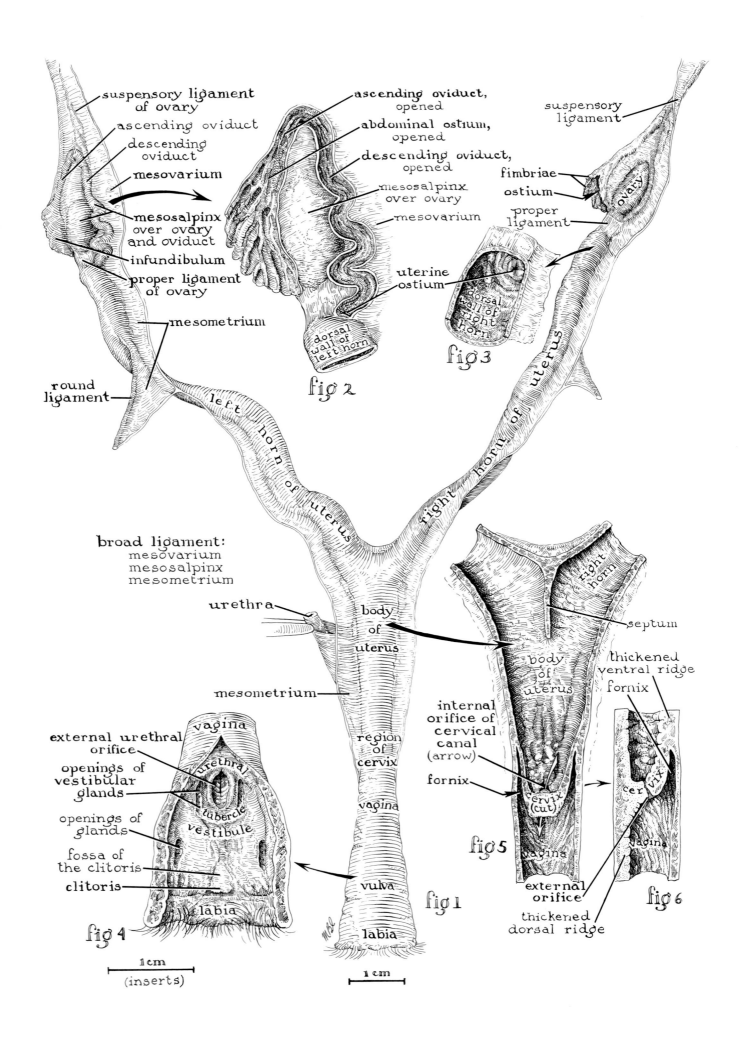

suspensory ligament of ovary

ascending oviduct

descending oviduct

mesovarium

mesosalpinx over ovary and oviduct

infundibulum

proper ligament of ovary

mesometrium

round ligament

ascending oviduct, opened

abdominal ostium, opened

descending oviduct, opened

mesosalpinx over ovary

mesovarium

uterine ostium

dorsal wall of left horn

fig 2

suspensory ligament

fimbriae

ostium

ovary

proper ligament

dorsal wall of right horn

fig 3

left horn of uterus

right horn of uterus

broad ligament:
mesovarium
mesosalpinx
mesometrium

urethra

body of uterus

mesometrium

region of cervix

vagina

vulva

labia

external urethral orifice

openings of vestibular glands

openings of glands

fossa of the clitoris

clitoris

vagina

urethral tubercle

vestibule

labia

fig 4

right horn

septum

body of uterus

internal orifice of cervical canal (arrow)

fornix

cervix (cut)

vagina

external orifice

fig 5

thickened ventral ridge

fornix

cervix

vagina

thickened dorsal ridge

fig 6

fig 1

1 cm
(inserts)

1 cm

In figure 1 the female genital system is shown intact as removed from its normal relationships in the body. Its connection with the urinary system through the urethra is also indicated. Figures 2 and 3 show in a more detailed fashion some of the more important relationships between the ovary, the oviducts and horn of the uterus. Figure 4 shows the external urethral orifice and figures 5 and 6 illustrate the relationship between the vagina and body of the uterus.

The usual sequence in reproduction in the female is that the egg breaks through the surface of the ovary, enters the celom and then makes its way into the oviduct by way of the abdominal ostium. It passes into the horn of the uterus and if fertilized develops there. The horns lead into the body of the uterus; the body of the uterus by way of the cervix leads into the vagina and finally the vagina leads into the terminal portion of the tube, the vestibule which receives the orifice of the urethra and is therefore sometimes called the urogenital sinus. External genital organs are the clitoris and the labia at the opening of the vestibule. Vestibule, labia and clitoris make up the **vulva.**

Figures 1 and 2 may be used to review some of the supporting structures of the ovary, oviduct, horns and body of the uterus. The **broad ligament,** the largest of the supporting structures consists of the **mesovarium** supporting the ovary and extending over the ovary and oviduct as the **mesosalpinx** and the **mesometrium** which supports the horns and body of the uterus. In addition the ovary is supported by a **suspensory ligament** cranially and a **proper ligament** caudally which ties it to the cranial portion of the horn of the uterus. Finally a **round ligament,** enclosed in a fold of peritoneum arising from one layer of the wall of the mesometrium, continues from the proper ligament to the abdominal wall at the site of the internal inguinal ring. It passes through the inguinal canal, the external inguinal ring and terminates near the vulva.

Figure 2 is an enlarged drawing of the left ovary and its oviduct and the cranial end of the horn. The oviduct and infundibulum are sectioned to show internal structure. The **oviduct** (*uterian tube, Fallopian tube*) has at the left side of the ovary an expanded funnel-shaped **infundibulum** equipped with a number of finger-like ridges which project from the edge of the infundibulum as the fimbria. The infundibulum leads into the **abdominal ostium** of the oviduct. The oviduct ascends to the cranial pole of the ovary, turns around it and pursues a wavy course caudal along the medial side of the ovary. It attaches to the proper ligament of the ovary caudally and opens by the **uterine ostium** into the left horn of the uterus. Figure 3 shows the cranial end of the right horn of the uterus with the ventral wall removed to show more clearly the papilla through which the oviduct enters the uterine horn.

Figure 4 depicts the lower end of the **vagina** where it joins the **vestibule** at the entrance of the urethra (See fig. 1). At the entrance of the urethra there is a **urethral tubercle** in the ventral wall of the vestibule. The urethra opens from the center of the tubercle and to either side of it are the openings of **vestibular glands.** In the ventral wall of the vestibule is the **fossa of the clitoris** and the **clitoris,** very tiny, projecting from it. Examined microscopically the clitoris can be shown to have roots (*crura*), a body and glans. These parts are homologous to the parts with the same names in the penis of the male. The labia are skin folds which are homologous to the scrotum in the male.

Figure 5 shows cranially the septum which partially divides the body of the uterus internally. Caudally the narrowed **cervix** of the uterus projects into the cranial end of the vagina leaving a space, the **fornix,** between the cervical and vaginal wall. The cervix has an **internal orifice** cranially and an **external orifice** caudally as shown in figures 5 and 6. Between internal and external cervical orifices is a **cervical canal.**

Figure 6 is a median section of the uterus and vagina which, with figure 5, gives a truer picture of the cervix and fornix. The ventral wall of the uterus has a thickened ridge and the dorsal wall of the vagina has a similar one. The cervix lies between these thickened areas in an oblique position. The fornix is most prominent ventral and lateral to the cervix.

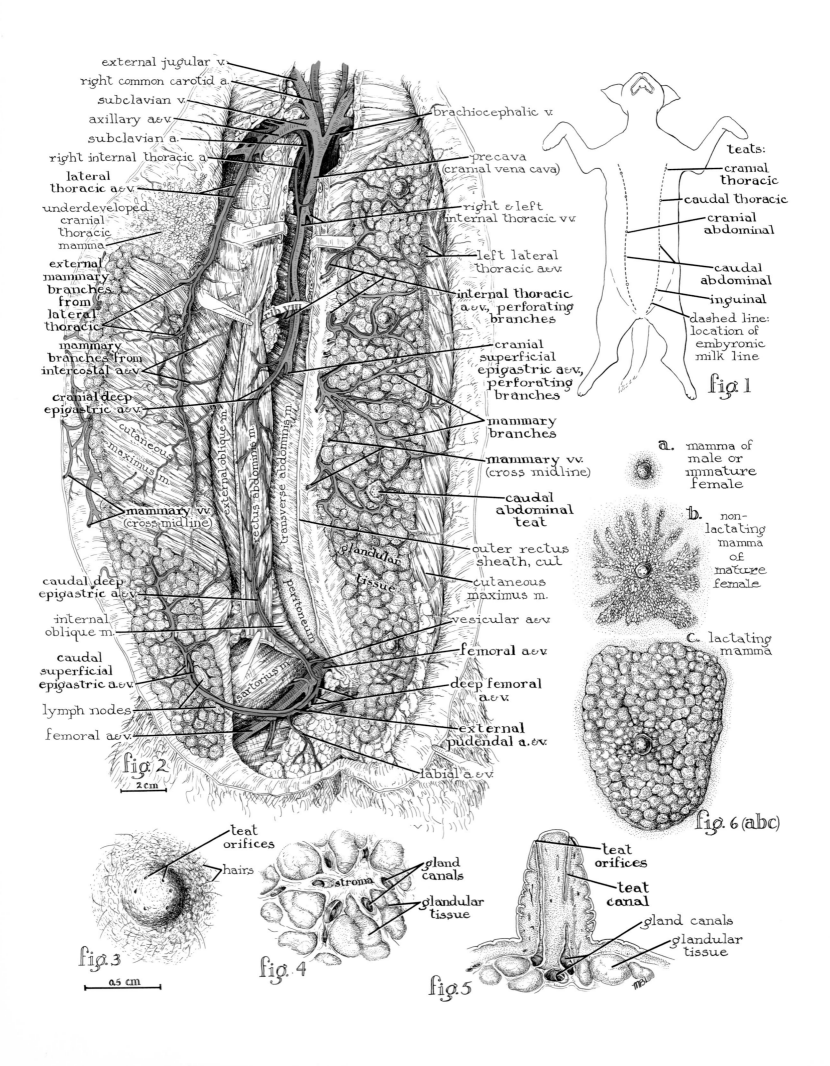

external jugular v.
right common carotid a.
subclavian v.
axillary a.&v.
subclavian a.
right internal thoracic a.
lateral thoracic a.&v.
underdeveloped cranial thoracic mamma
external mammary branches from lateral thoracic
mammary branches from intercostal a.&v.
cranial deep epigastric a.&v.
cutaneous maximus m.
mammary vv. (cross midline)
caudal deep epigastric a.&v.
internal oblique m.
caudal superficial epigastric a.&v.
lymph nodes
femoral a.&v.

brachiocephalic v.
precava (cranial vena cava)
right & left internal thoracic vv.
left lateral thoracic a.&v.
internal thoracic a.&v., perforating branches
cranial superficial epigastric a.&v., perforating branches
mammary branches
mammary vv. (cross midline)
caudal abdominal teat
outer rectus sheath, cut
cutaneous maximus m.
vesicular a.&v.
femoral a.&v.
deep femoral a.&v.
external pudendal a.&v.
labial a.&v.

external oblique m.
rectus abdominis m.
transverse abdominis m.
peritoneum
glandular tissue
sartorius m.

fig. 2
2 cm

teats:
cranial thoracic
caudal thoracic
cranial abdominal
caudal abdominal
inguinal
dashed line: location of embyronic milk line

fig. 1

a. mamma of male or immature female

b. non-lactating mamma of mature female

c. lactating mamma

fig. 6 (abc)

teat orifices
hairs

fig. 3
0.5 cm

gland canals
glandular tissue
stroma

fig. 4

teat orifices
teat canal
gland canals
glandular tissue

fig. 5

The Mammary Gland—Its Structure, Relationships and Blood Supply
PLATE 66

This plate is a study of the structure and relationships of all the mammary glands of a lactating female. They are modified cutaneous (*skin*) glands of the apocrine type. Figure 1 is a schematic representation of the ventral side of the cat with dotted lines to represent the embryonic milk line along which the mammary glands with their teats develop. On the right side five teats are found, while on the left only four are present. Probably the cranial abdominal one is absent as seen in the same cat used for dissection in figure 2. While five pairs is the most common number of teats in the cat, fewer may occur and also additional ones may appear along the milk line. This situation is also true in man and other mammals.

Figures 3, 4 and 5 show something of the structure of one mamma in a lactating animal. Figure 3 is a surface view showing a teat with **teat orifices,** the naked character of the teat and the hair which grows in close to its base. Figure 4 is a cross section of a gland showing the enlarged **glandular cells,** the **gland canals** and the supporting tissue or **stroma.** Figure 5 shows a single teat canal and teat orifice with gland canals and glandular tissue below.

Figure 6 pictures in (a) a surface view of a mamma and teat in a male or immature female; in (b) a nonlactating mamma in an adult female and in (c) a lactating mamma. Figure 2 shows how in a lactating female the mammae all grow together into a continuous mass running most of the length of the ventral side of the trunk.

Figure 2 emphasizes the relationship of the mammary glands to the skin and to their blood supply. The glands lie in the subcutaneous tissue with the teats and the gland tissue itself a product of the epidermis of the skin. They are a good example of an organ or structure which anatomically belongs to one system, the skin or integument, and physiologically belong to another system, the reproductive. They lie outside of the skin muscles, the cutaneous maximus, as indicated on the right side of figure 2.

The skin on the left side of figure 2 has been removed except on and around the teats and the mammary gland has been left intact with its surface blood supply exposed. The skin and cutaneous muscle on the right body wall has been cut and pulled to the side leaving the mammary gland intact. Caudally, the dorsal side of the gland, with its blood vessels, is exposed. The right, outer sheath of the rectus abdominis muscle has been cut and the rectus muscle turned to the side to show its under surface and its blood supply and the peritoneum and transverse abdominis muscle which lie beneath it. Craniad the ventral thoracic wall on the right side has been cut and pulled aside and the internal thoracic arteries and veins freed from it and exposed.

The blood supply of the mammary glands is now quite largely exposed and can be identified. The **internal thoracic artery** sends **perforating branches** through the ventral thoracic wall to supply primarily the **thoracic mammae** (See fig. 1). The **thoracic mammae** get an additional blood supply from perforating branches from the **intercostal artery** and from the **lateral thoracic artery.** The latter vessel can be seen on the right side of the thorax. It is a branch of the axillary artery. Just in back of the eighth rib, after giving off a musculophrenic artery, the internal thoracic artery becomes the **cranial epigastric artery.** This artery now divides sending a **cranial deep epigastric artery** into the rectus abdominis muscle and a **cranial superficial epigastric artery** through the body wall to supply mostly the **cranial abdominal mamma.** Arising from the **deep femoral artery** is the **caudal deep epigastric artery** which passes forward in the rectus abdominis and in the region of the unbilicus anastomoses with the cranial deep epigastric artery. Also from the deep femoral artery there arises an **external pudendal artery** which after giving off branches in the perineal region continues as the **caudal superficial epigastric artery** to the dorsal side of the mammary gland supplying the **inguinal** and **caudal abdominal mammae** and anastomosing with the cranial superficial epigastric artery.

The veins of the mammae of the cat follow closely the arteries described above. It should be especially noticed that some of the mammary veins cross the midline. Ordinarily this does not happen with paired vessels.

The extensive blood supply as seen on this plate is evident only in lactating females.

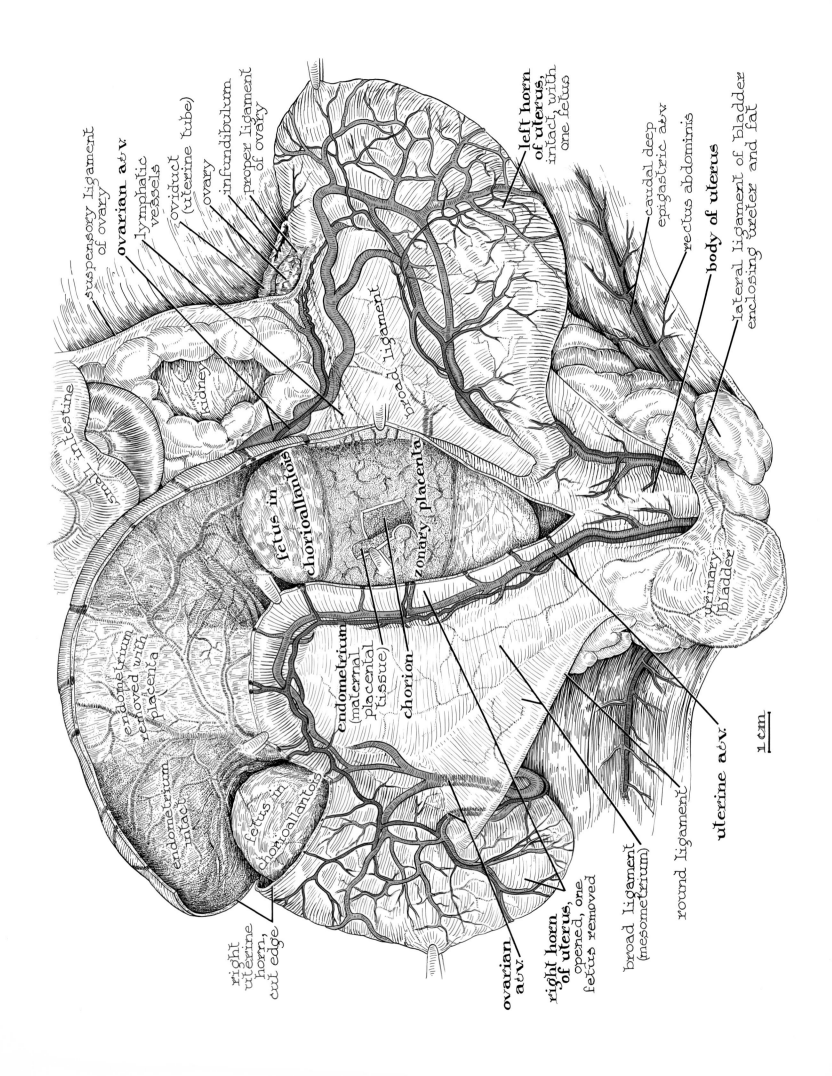

suspensory ligament of ovary

ovarian a&v.

lymphatic vessels

oviduct (uterine tube)

ovary

infundibulum

proper ligament of ovary

left horn of uterus, intact with one fetus

caudal deep epigastric a&v.

rectus abdominis

body of uterus

lateral ligament of bladder enclosing ureter and fat

large intestine

small intestine

kidney

broad ligament

fetus in chorioallantois

endometrium removed with placenta

zonary placenta

endometrium (maternal placental tissue)

chorion

urinary bladder

endometrium intact

fetus in chorioallantois

1 cm

right uterine horn, cut edge

ovarian a&v.

right horn of uterus, opened, one fetus removed

broad ligament (mesometrium)

round ligament

uterine a&v.

While it is not a part of the purpose of this book to go into the study of embryological development, it does seem appropriate to include some information on the fetus and its relationship to maternal tissue—specifically the fetal membranes and placenta. The fetal membranes are the amnion, chorion, allantois and yolk sac. The **placenta** is a structure derived from maternal and fetal tissues through which the fetus can derive oxygen and nourishment from maternal blood and through which the fetus can return to the mother's blood the waste products of its metabolism. It is also a source of hormones. By this device the first 63 days—the intrauterine days—of a cat's life are possible. It is the fetal membranes and placenta which constitute the "afterbirth". It is commonly eaten by the female as a "nourishing morsel" stimulating hormone production and uterine contraction after birth.

Among the cats acquired for a laboratory course in mammalian anatomy there are usually a number of pregnant females. They are always and universally a focus of interest—for both the old and the young among the students. This illustration and the one that follows should answer many of the questions that are always asked. Shown here is a gravid uterus. The **left horn** is unopened and contains one fetus. The **right horn** of the uterus contained three fetuses. It was cut open and one fetus removed for study and illustration as seen on the next plate; one is shown in position with its zonary placenta and chorioallantois in normal position; the third was only partially exposed. Review the internal organs of the urogenital system which are clearly shown; the ovary and its ligaments, the oviduct or uterine tube with the infundibulum, the ligaments of the right and left horns of the uterus, and the urinary bladder and its lateral ligaments. Note the body of the uterus.

The **zonary placenta** is one which is composed of a complete band of thickened vascular tissue. It is the type found not only in the cat but in the dog, seal and other mammals. Some mammals, like the raccoon, have an incomplete band. A diffuse type of placenta with chorionic villi scattered over the whole chorion is seen in the pig; a cotyledonous type with scattered, large cotyledons is seen in the cow; a discoidal or disk-like type is found in bats, rodents and man. All true placentas function in essentially the same way.

By examining closely the fetus with its intact fetal membranes and placenta in this illustration you can see the relationship of maternal and fetal tissue. The **endometrium,** the lining of the uterus and its horns, forms the outer part of the placenta. Here a flap has been cut in the endometrium to show the **chorion** below. Where the fetus was removed, the endometrium was removed also. The intact endometrium is seen adjacent to the area where the placenta attached. The **chorioallantois** shown here is, as the name suggests, a combination of two fetal membranes the outer chorion and the inner allantois. This relationship and the relationship of these fetal membranes to the amnion, the innermost fetal membrane, is shown better in the next plate.

To fulfill its functions the placenta is, of course, a highly vascular structure. Examine now the blood supply to the uterus and ovary. They are supplied by the **uterine arteries** and the **ovarian arteries,** respectively. The uterine arteries particulary are much enlarged in the gravid uterus. Compare them with the same arteries seen in Plate 63. Note particularly the **uterine** and **ovarian veins.** They are especially large, sometimes tortuous and anastomose freely in the uterine wall. Thus the fetuses are provided with an adequate circulation of blood. Note that we are considering at this point maternal blood to the uterus and placenta. In the next plate the source of the fetal blood to the placenta is shown and described.

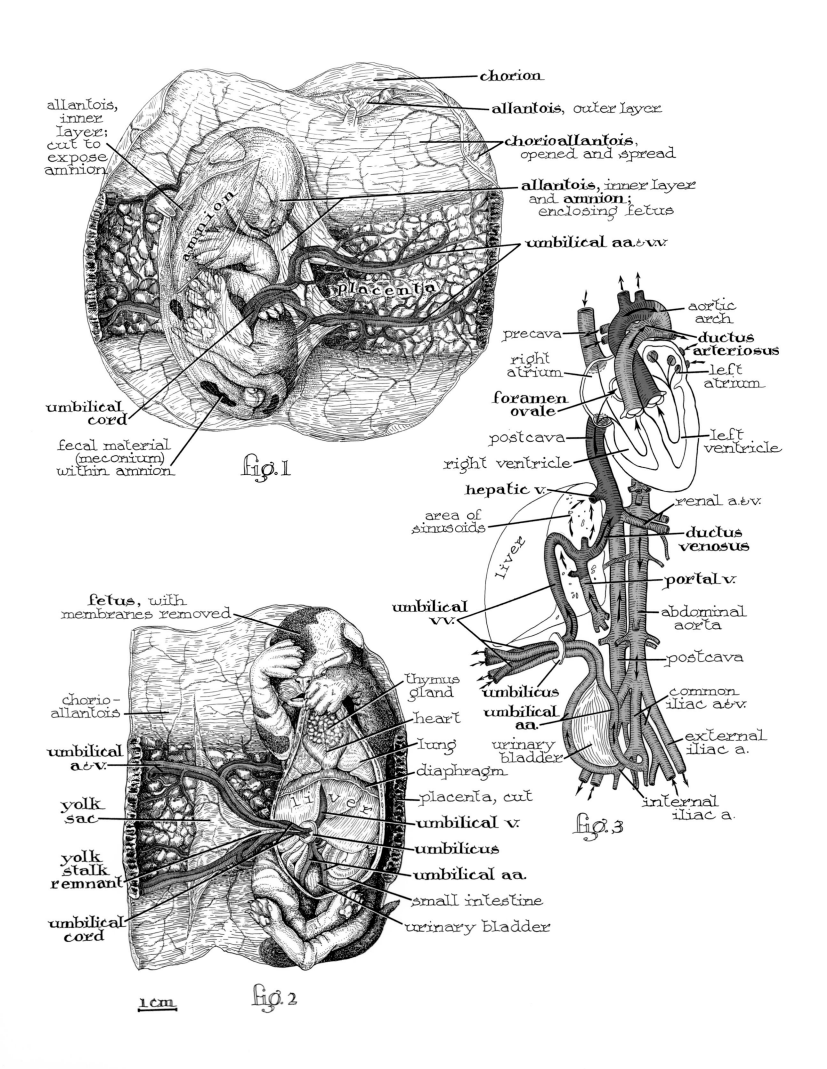

chorion

allantois, outer layer

chorioallantois, opened and spread

allantois, inner layer and amnion; enclosing fetus

umbilical aa.&vv.

allantois, inner layer; cut to expose amnion

amnion

placenta

umbilical cord

fecal material (meconium) within amnion

fig. 1

fetus, with membranes removed

chorio-allantois

umbilical a.&v.

yolk sac

yolk stalk remnant

umbilical cord

thymus gland

heart

lung

diaphragm

placenta, cut

umbilical v.

umbilicus

umbilical aa.

small intestine

urinary bladder

liver

fig. 2

1 cm

aortic arch

precava

right atrium

foramen ovale

postcava

right ventricle

hepatic v.

area of sinusoids

umbilical vv.

umbilicus

umbilical aa.

urinary bladder

liver

ductus arteriosus

left atrium

left ventricle

renal a.&v.

ductus venosus

portal v.

abdominal aorta

postcava

common iliac a.&v.

external iliac a.

internal iliac a.

fig. 3

Figure 1 shows the fetus with the chorioallantois and placenta cut, opened and spread. The inner layer of the allantois is cut to uncover the thin amnion within. The umbilical cord is revealed. Note that there is some fecal material within the amnion. It is called meconium.

Since we do not go into embryological development in this book, it is difficult to explain fully the constitution of the fetal membranes and how they came to be related as we see them here. The more interested student will wish to refer to a good embryology book. The **chorion** (fig. 1) is the outermost of the fetal membranes. It is double in the histological sense that its outer surface is from the ectoderm germ layer, its inner surface is extraembryonic mesoderm. The **amnion** is the innermost fetal membrane and it too is double; its innermost surface is ectodermal and its outer surface is extraembryonic mesoderm. The allantois, since it forms as a diverticulum of the hindgut wall, has a lining or inner surface of endoderm and an outer surface of extraembryonic mesoderm. The allantois, essentially a sac, pushes out between the amnion and chorion so that its outer surface of extraembryonic mesoderm becomes applied to the amnion (*inward*) on the one hand and the chorion (*outward*) on the other. Therefore, when you remove an embryo in its fetal sac and cut that sac and open it, you are cutting through the chorion and allantois or the **chorioallantois** as seen in figure 1. The embryo is now enclosed only in a sac composed of the allantois on the outside and the amnion inside. In this figure this relationship is shown by making a slit in the allantois to show the amnion underneath, or inside. Figure 1 makes it apparent that it is the chorioallantois which forms the fetal portion of the placenta, especially the chorion. From the latter villi are produced into which project fetal blood vessels. These villi appear widespread initially over the chorion but gradually are lost except in the area of the zonary placenta. Here by pushing into the now highly vascular endometrium of the maternal uterus a close relationship is established between fetal and maternal blood. Here exchanges of nutrient materials, oxygen, carbon dioxide and other wastes take place.

In figure 1 the endometrium is not shown on the placenta. It can be seen by referring back to the previous plate.

Note also in figures 1 to 3 the **umbilical arteries,** carrying carbon dioxide and other wastes, coursing out through and branching in the **placenta** and the **umbilical veins** which carry blood, now laden with oxygen and nutrients, back into the fetus. Where these vessels run together along with the yolk stalk and the allantoic stalk, we have **umbilical cord** which joins the body wall at the **umbilicus.** The umbilical cord is covered by the amnionic tissues.

The **yolk stalk** mentioned above is the proximal portion of the **yolk sac** which is considered one of the fetal membranes (fig. 2). The yolk sac is a diverticulum of the midgut region. In those animals whose embryos and fetuses are dependent on stored food, the yolk sac is large—as in birds, reptiles and primitive mammals. In those mammals, such as the cat, and most other mammals (*the placentals*) a placenta is formed and the yolk sac remains small. The first blood vessels form within its walls.

Figure 2 also shows the fetus with the amnion and inner wall of the allantois removed. The ventral body wall has been cut away, except at the **umbilicus,** and the **umbilical arteries** can be seen as they pass from the fetus, through the umbilicus and **umbilical cord** to the placenta. The **umbilical veins** can be seen coming from the placenta, passing through umbilical cord and umbilicus to enter the fetus.

Figure 3 is a schematic drawing to show the fetal circulation. The oxygen content of the blood in the various vessels of the fetus is indicated as follows. Those vessels with the blood richest in oxygen are indicated in red; those poorest in oxygen by blue. The vessels with the next to the richest oxygen content are shown in reddish purple; those with next to the poorest in oxygen content by bluish purple. Blood enters the fetus through the **umbilical veins** which go to the liver. In the liver branches of the umbilical veins may empty into the **portal vein,** may empty directly into the liver substance or may join directly the postcava by way of a **ductus venosus.** That blood which enters the liver directly or by way of the portal vein is picked up by the **hepatic veins** and carried into the postcava. The postcava enters the right atrium of the heart and here its blood is shunted by the valve of the postcava through the **foramen ovale** of the interatrial wall into the left atrium where it is mixed with a small amount of blood from the lungs. The blood is then passed into the left ventricle and out through the aorta where most of it is sent to the head and upper limbs—only a small amount passing into the descending aorta and to the body as a whole. Recall that this blood has come in part from the placenta where it has taken on oxygen and nutrients and where it has lost its waste products, in part from the lower limbs and in part from the viscera and lungs. It is a mixed blood and yet the "richest" and as is usual among vertebrates the head receives the richest blood. Returning from the head and upper limbs through the precava the blood enters the right atrium and goes directly to the right ventricle and is pumped into the pulmonary artery. Since the lungs are not active during fetal life, only a small amount of blood goes to them. Most of it is shunted across the **ductus arteriosus** to the aorta. The aorta distributes a part of this blood to the viscera and lower limbs; the bulk of it being

187

returned to the placenta by way of the umbilical arteries.

It is well now to try to relate fetal circulation to that of the adult by outlining some of the changes which take place at birth. When the lungs become functional in respiration at the time of birth, the pulmonary circulation is opened up. The ductus arteriosus gradually occludes and becomes the ligamentum arteriosum. The foramen ovale closes, though not completely at first, and a double circulation in the heart is established. The foramen ovale becomes the fossa ovalis. The parts of the internal iliac (*hypogastric*) arteries which we have called **umbilical arteries** gradually become obliterated to become fibrous cords, the **lateral umbilical ligaments.** The **umbilical vein** no longer needed, becomes the **ligamentum teres;** the **ductus venosus,** the **ligamentum venosum** or it may remain in part as an hepatic vein.

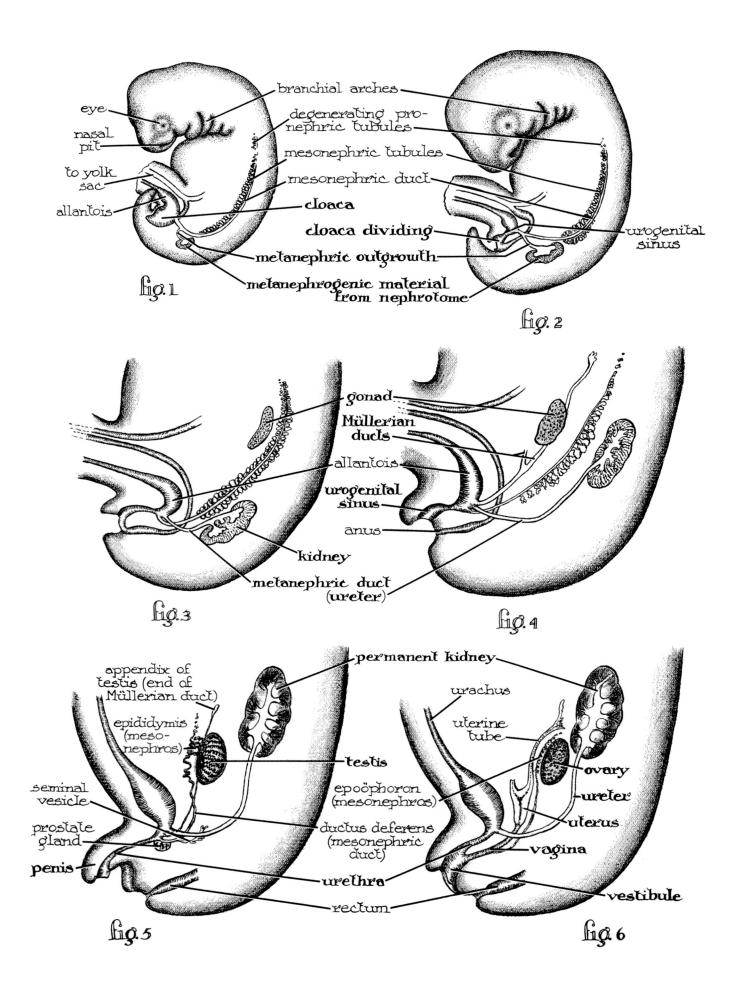

eye

nasal pit

to yolk sac

allantois

branchial arches

degenerating pro-nephric tubules

mesonephric tubules

mesonephric duct

cloaca

cloaca dividing

metanephric outgrowth

metanephrogenic material from nephrotome

fig. 1

urogenital sinus

fig. 2

gonad

Müllerian ducts

allantois

urogenital sinus

anus

kidney

metanephric duct (ureter)

fig. 3

fig. 4

appendix of testis (end of Müllerian duct)

epididymis (meso-nephros)

seminal vesicle

prostate gland

penis

permanent kidney

testis

epoöphoron (mesonephros)

ductus deferens (mesonephric duct)

urethra

rectum

fig. 5

urachus

uterine tube

ovary

ureter

uterus

vagina

vestibule

fig. 6

(Crouch, Functional Human Anatomy, Lea & Febiger)

The closeness of the urinary and genital systems in ontogeny and phylogeny has been emphasized. This illustration shows schematically sections of human embryos demonstrating the formation of the human kidney (*metanephros*) and its ducts, the division of the cloaca, the formation of the urinary bladder and their relationship to the male and female reproductive organs. The same sequence of events occurs in the development of the cat or of most mammals. Most of the organs discussed here are derived primarily from the mesoderm germ layer of the embryo with minor contributions from the ectoderm and entoderm.

Mammals (*also reptiles and birds*) produce in their embryological development a linear sequence of three kidneys—the **pronephros, mesonephros,** and **metanephros.** Only the last remains as the functional kidney of the adult. This is essentially a recapitulation of the evolution or phylogeny of vertebrate kidneys. The pronephros—sometimes called the "head kidney"—is the earliest and the simplest and is functional in the adults of only a few of the early chordates, such as the hagfish, and in the developmental stages of fish and amphibians. The adult kidney of fishes and amphibians is the mesonephros or "middle kidney". It in turn gives way in reptiles, birds, and mammals to the metanephros or "hind kidney".

All three of these kidneys are made up of uriniferous tubules which have a common origin in the urogenital ridge which lies to either side of the middorsal line of the embryo. These tubules lead into a common longitudinal excretory duct which runs caudad and empties into the cloaca—a chamber common to urinary, genital and digestive systems. While in most exocrine or duct glands the glands represent an outgrowth of their ducts, the uriniferous tubules of the kidneys develop independently and secondarily join with their ducts.

In figure 1 the **pronephros** has developed and is already undergoing degeneration of its tubules. From these tubules came, however, the beginning of the pronephric duct growing toward the cloaca. In this figure the mesonephros has begun to develop and its tubules have joined the pronephric duct. Note also that there is, in figure 1, an outgrowth near the caudal end of the mesonephric duct called the **metanephric outgrowth.** This outgrowth has joined with some **metanephric material** from the nephrotome (*urogenital fold*)—specifically some metanephric tubules. The relationship of the three types of kidneys are already established at this early age—about 6 weeks in the human embryo.

Note that the mesonephric tubules are very numerous compared with pronephric tubules. Also observe the relationship of the **cloaca** to the mesonephric duct, the

allantois and the stalk of the yolk sac and the digestive tube.

Figure 2 shows further degeneration of the pronephros; the mesonephros remains the dominant kidney; the metanephros has enlarged and the cloaca is beginning to divide into two parts, one associated with the urinary system and called a urogenital sinus, the other, the rectum of the digestive system.

In figure 3 a **gonad** has appeared, destined to become either a testis or an ovary but not yet differentiated at this early, indifferent (*sexually*) stage. The cloaca is further divided; the metanephric and mesonephric ducts have separated, at their caudal ends.

Figure 4 shows an additional structure, the **Müllerian or oviducts** which are part of the female reproductive system. Their cranial ends open into the celom as the ostium (*not labeled*). Notice also that in this figure the cloaca, as such, is lost in that it has been divided into a distinct urogenital sinus and a rectum each opening separately on the body surface. The metanephric duct, now called a **ureter,** has elongated; the metanephros has enlarged; the mesonephros is undergoing degeneration; the mesonephric ducts are still present. At the figure 4 stage the embryo is still indifferent sexually even though a Müllerian duct is present. As far as any gross structures are concerned it could develop into a male or a female. Yet, as is well known, the direction of its differentiation, that is, toward male or female, was determined at the time of fertilization of the egg from which this embryo developed. Figure 5 indicates what takes place when a male animal is developed; figure 6 is the condition in the female.

Study figure 5 carefully and note the following changes which have taken place as compared to figure 4. The gonad has differentiated into a **testis.** The mesonephros has largely disappeared except for a small portion whose tubules become associated with the tubules of the testis to form the epididymis. The mesonephric duct, no longer needed as a urinary duct, becomes the ductus deferens for carrying sperms to the urogenital sinus, now called the **urethra.** The urethra passes through a developing organ the phallus which becomes the **penis.** A prostate gland develops around the area where the ductus deferens joins the urethra and in man, though not in the cat, seminal vesicles develop as outgrowths of the ductus deferens. The Müllerian ducts are degenerating but will leave a few remnants in the adult male such as the appendix of the testis. The ureter empties into the neck of the bladder which was derived from the stalk of the allantois (see fig. 4). The metanephros or **permanent kidney** is again shown.

Figure 6 is drawn to show what happens as the female urogenital system develops from the indifferent stage. Most obvious is the growth and differentiation of the Müllerian or oviduct to produce the **uterus, uterine tube** and **vagina** and the gonad to produce an **ovary.** Since the mesonephros and mesonephric (*Wolffian*) duct are not needed, they degenerate but remnants remain such as the epoöphoron.

The vagina and the urethra have separate openings into a **vestibule.** The urachus, seen in both male and female as a continuation craniad of the allantois (*bladder*) becomes the median umbilical ligament. The structure forming the penis in the male (fig. 5), the phallus (*not labeled*) forms a clitoris, homologous to the penis, in the female. The clitoris is not labeled in figure 6 though it is shown.

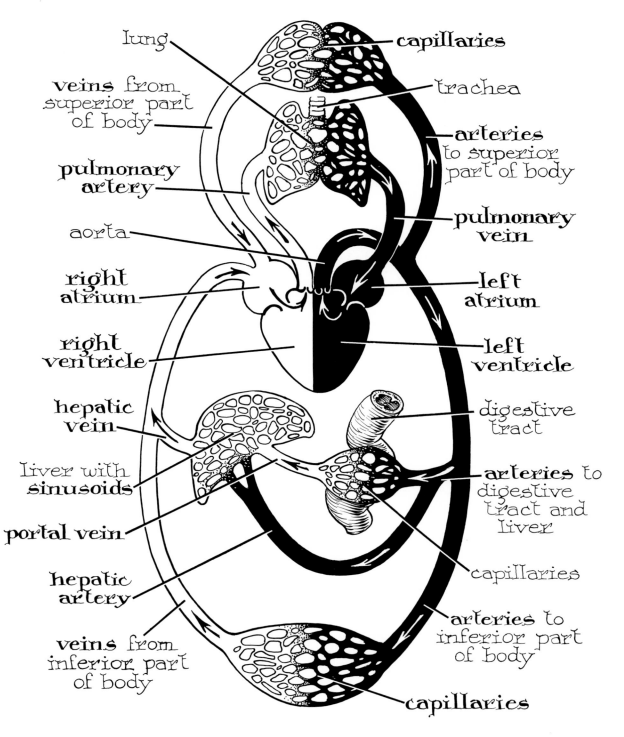

lung

capillaries

veins from superior part of body

trachea

pulmonary artery

arteries to superior part of body

aorta

pulmonary vein

right atrium

left atrium

right ventricle

left ventricle

hepatic vein

digestive tract

liver with sinusoids

arteries to digestive tract and liver

portal vein

hepatic artery

capillaries

veins from inferior part of body

arteries to inferior part of body

capillaries

Fig. 15. *Schematic representation of the double and closed system of circulation. Oxygenated blood is shown in black; the non-oxygenated blood in white. The arrows indicate the direction of flow of blood. (Crouch,* Functional Human Anatomy, *Lea & Febiger.)*

Circulatory systems appear to have had their genesis when the **mesoderm** evolved as a third germ layer and by its differentiation added to the complexity and mass of organisms. This resulted in large numbers of cells losing their functional relationship with the external environment from which oxygen and nourishment were obtained and into which their waste products were excreted. Their survival was possible because from this same mesoderm there evolved a **system of transportation** which could pick up needed materials at points of entry and deliver the wastes of metabolism to points of excretion. It made possible for each living cell of the body an environment of tissue fluid containing the life-giving materials and providing a medium also for the excretion of wastes. This is the **circulatory system.**

The circulatory system varied in form, extent and efficiency in different kinds of animals through evolutionary time and in turn was a controlling factor in the size and activity of organisms. It became involved in functions of organisms other than that of transportation. Among these are **protection, temperature regulation,** and the **control** and **coordination** of certain functions by transportation of hormones and other agents. Indeed, all parts of the body are dependent upon it.

The circulatory system of mammals may be divided into two parts, the **cardiovascular** (*blood vascular*) and the **lymphatic divisions** (*lymph vascular*). The latter is treated by some as a separate system, but its structure and functional relationships are in all vertebrates close to the blood vascular. There are some functions which the lymphatic division alone provides. Its lymph nodes serve as filters which aid in preventing the spread of infection and also are centers for the production of lymphocytes—a special type of blood cell (Plate 94, fig. 1). The lymphatic capillaries called lacteals in the villi of the intestinal wall serve in the absorption of fatty acids and glycerine.

The circulating tissues of this system are blood, tissue fluid and lymph. The blood is confined normally to the heart and the system of vessels associated with that organ— the **blood vessels**—arteries, arterioles, capillaries, venules and veins. The **lymph** similarly is found within the vessels and nodes of the lymphatic division, while the **tissue fluid** occupies intercellular spaces.

In mammals, as in birds, the cardiovascular circulatory system is a closed and a double system (Fig. 15). It is **closed** in the sense that the blood as such is confined to the heart and its system of vessels. Whole blood escapes into the tissues to mingle with tissue fluid only when the walls of the vessels are damaged by mechanical means or by disease processes. A blow to the skin—around the eye, for example—will rupture the capillaries, blood will escape into the tissue spaces and the area will darken— the "black eye" of common terminology.

While the closed system is not limited to mammals and birds among the vertebrates, the double system is. The **double system** is one in which the heart is completely divided into right and left portions. There is a pulmonary circuit which returns oxygenated blood to the heart and a systemic circuit which supplies the body as a whole, including the lungs, with this oxygenated blood. Fish, on the other hand, have single systems of circulation where the heart has but one atrium and one ventricle and the blood passes from the heart to the ventral aorta, then to the gills for oxygenation and from there to the dorsal aorta for circulation to the body as a whole. Since blood pressure is reduced by passage through the gills, the system operates at low pressure levels relative to that of the double system where blood coming back to the heart from the lungs is given another boost in pressure as the left ventricle ejects it into the aorta. Some fish and all amphibians and reptiles have systems which are in transition, that is, intermediate between the single and double types. It is interesting too, that birds and mammals, in their embryology, recapitulate many of the stages seen in these transitional forms. It is for this reason that the sciences of comparative anatomy and embryology are so essential to an understanding of the gross anatomy of adult animals and indeed many of the anomalies which are frequently found.

The heart and the various kinds of blood and lymphatic vessels are illustrated on Plates 70 to 72 and 94, and described on the pages facing the illustrations. Figure 16 below shows the valvular structures in veins. Similar valves are found in lymphatic vessels.

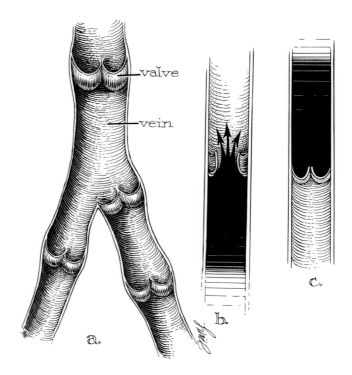

Fig. 16. a, *veins opened to show structure and position of valves.* b, *valve opened as blood flows through in direction indicated by the arrows.* c, *valve closed as blood fills cusps and backflow of blood is prevented.*

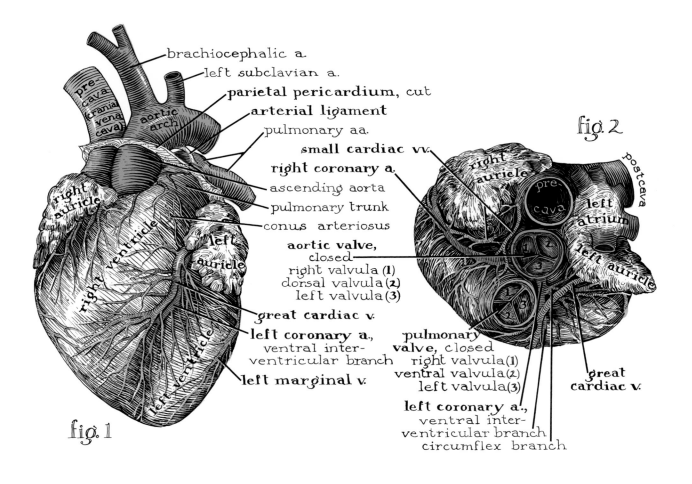

fig. 2

brachiocephalic a.

left subclavian a.

parietal pericardium, cut

arterial ligament

pulmonary aa.

small cardiac vv.

right coronary a.

ascending aorta

pulmonary trunk

conus arteriosus

aortic valve,
closed
right valvula (1)
dorsal valvula (2)
left valvula (3)

great cardiac v.

left coronary a.,
ventral inter-
ventricular branch

left marginal v.

pre-
cava
(cranial
vena
cava)

aortic
arch

right
auricle

right
ventricle

left
ventricle

left
auricle

right
auricle

pre-
cava

postcava

**left
atrium**

left auricle

pulmonary
valve, closed
right valvula(1)
ventral valvula(2)
left valvula(3)

**great
cardiac v.**

left coronary a.,
ventral inter-
ventricular branch
circumflex branch

fig. 1

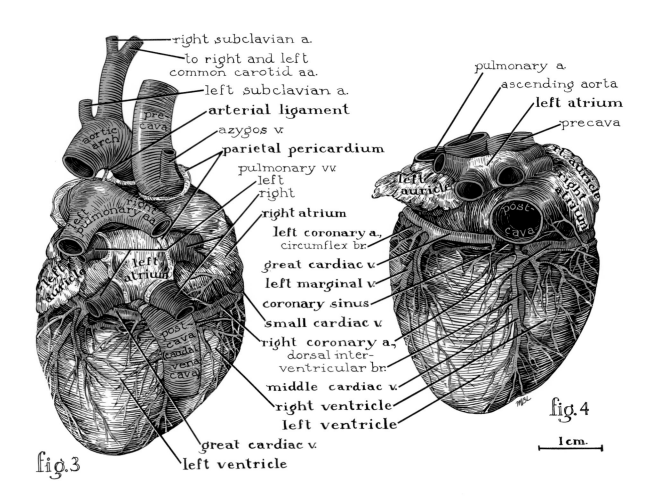

right subclavian a.

to right and left
common carotid aa.

left subclavian a.

arterial ligament

azygos v.

parietal pericardium

pulmonary vv
left
right

right atrium

left coronary a.,
circumflex br.

great cardiac v.

left marginal v.

coronary sinus

small cardiac v.

right coronary a.,
dorsal inter-
ventricular br.

middle cardiac v.

right ventricle

left ventricle

pulmonary a.

ascending aorta

left atrium

precava

aortic
arch

pre-
cava

left
right
pulmonary aa.

left
auricle

left
atrium

post-
cava
(caudal
vena
cava)

left
auricle

right
atrium

post-
cava

great cardiac v.

left ventricle

fig. 3

fig. 4

1 cm.

In this plate the four-chambered character of the heart is shown as reflected in surface features as seen from four different views. The appendages of the atria, the right and left auricles; the major blood vessels to and from the heart; and the vessels of the heart itself, the coronary circuit, are featured. The parietal pericardium has been removed by cutting it around the base of the heart as indicated on figures 1 and 3. The transparent visceral pericardium or epicardium is intact on the heart surface.

Figure 1, a **ventral view,** shows the heart as you would see it by removing the ventral wall of the thorax and the **parietal pericardium.** The bulk of this surface is the wall of the **right ventricle** and its **conus arteriosus** and **pulmonary trunk** which leads into the pulmonary arteries. The surface of the right ventricle is separated from that of the **left ventricle** by an **interventricular groove** occupied by the **great cardiac vein** and the ventral interventricular branch of the **left coronary artery.** This groove marks the position of the **interventricular septum** separating right and left ventricles internally. This groove starts on the caudodorsal side of the conus arteriosus under the left auricle and runs obliquely toward the right of the apex of the heart which it does not quite reach. The **apex** of the heart is thus a part of the left ventricle.

Connecting the aortic arch and pulmonary trunk is the **arterial ligament** which is the remnant of the **ductus arteriosus** of the fetus which served as a functional connection between these two vessels. Thus until this closes after birth there is not complete double circulation. The arterial ligament is shown again in figure 3.

The **right auricle** overlapping the right ventricle; the **precava** entering the heart; the ascending aorta coming from behind the conus arteriosus and leading to the aortic arch and its branches are also shown.

Figure 2 shows the cranial surface of the heart with its major vessels cut off close. **Aortic** and **pulmonary** (*semilunar*) **valves** are shown each with three valvulae. These valves prevent the return of blood to the ventricles at the end of ventricular contraction (*systole*). The **precava** and **postcava** are shown entering the **right atrium** behind the expansive right auricle; the pulmonary veins (not labeled on fig. 2) are shown entering the **left atrium** (See fig. 3). Note especially the **right and left coronary arteries** leaving the ascending aorta just cranial to the aortic valve. The right coronary artery and the circumflex branch of the left coronary artery travel in the **coronary** (*atrioventricular*) **groove** along with the great cardiac vein.

Figures 3 and 4 are dorsal views of the heart and its vessels. In figure 4 the main vessels are cut back and the heart tipped up at its apex to provide a better view of the coronary groove and its contents. This groove is the outward evidence of the **atrioventricular septum** separating the atria and ventricles (See Plate 70). It encircles the heart except cranioventrally where the conus arteriosus interrupts it. There is much fat around the coronary vessels in life which obscures the groove. The vessels of the coronary groove receive from or provide branches for the atrial and ventricular walls of the heart. The right coronary artery sends its dorsal interventricular branch into the dorsal interventricular groove and the middle cardiac vein follows this groove into the **coronary sinus,** best seen in figure 4. The coronary sinus empties into the right atrium (Plate 71, fig. 1). Some vessels, like the small **cardiac veins,** may empty directly into the postcava or into the right atrium.

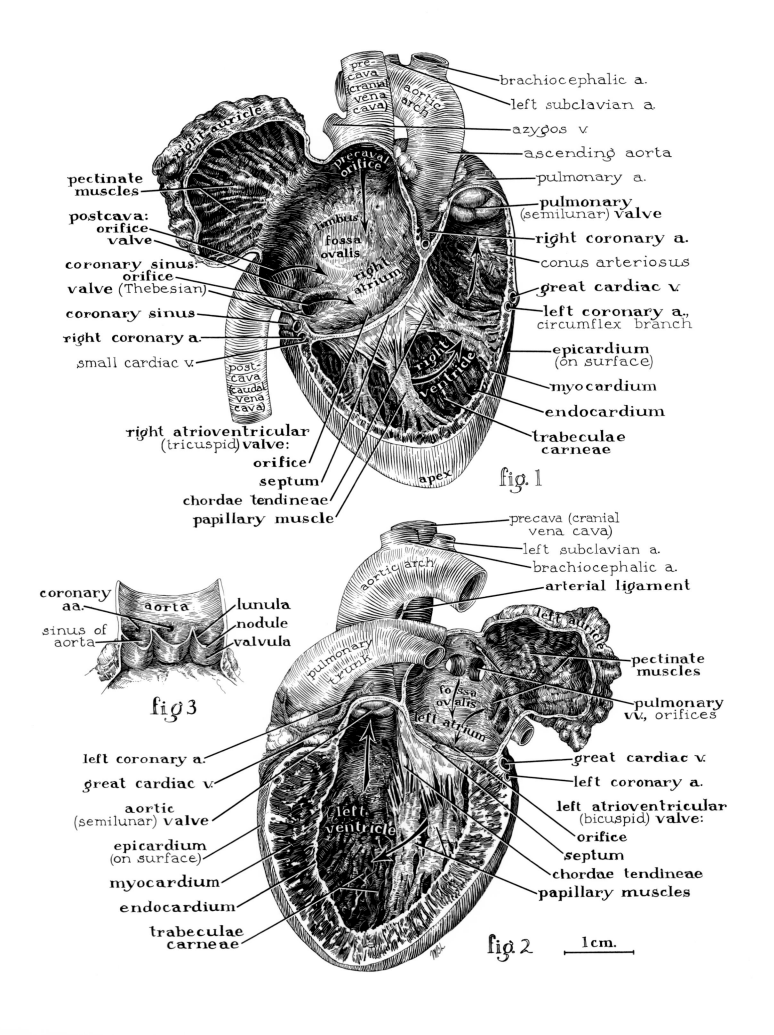

pre-cava (cranial vena cava)

aortic arch

brachiocephalic a.

left subclavian a.

azygos v.

ascending aorta

pulmonary a.

pulmonary (semilunar) valve

right coronary a.

conus arteriosus

great cardiac v.

left coronary a., circumflex branch

epicardium (on surface)

myocardium

endocardium

trabeculae carneae

right auricle

precaval orifice

umbus

fossa ovalis

right atrium

right ventricle

apex

pectinate muscles

postcava: orifice valve

coronary sinus: orifice valve (Thebesian)

coronary sinus

right coronary a.

small cardiac v.

postcava (caudal vena cava)

right atrioventricular (tricuspid) valve:

orifice

septum

chordae tendineae

papillary muscle

fig. 1

coronary aa.

aorta

lunula

nodule

valvula

sinus of aorta

fig 3

precava (cranial vena cava)

left subclavian a.

brachiocephalic a.

arterial ligament

aortic arch

left auricle

pulmonary trunk

pectinate muscles

fossa ovalis

left atrium

pulmonary vv., orifices

left coronary a.

great cardiac v.

aortic (semilunar) valve

epicardium (on surface)

myocardium

endocardium

trabeculae carneae

left ventricle

great cardiac v.

left coronary a.

left atrioventricular (bicuspid) valve:

orifice

septum

chordae tendineae

papillary muscles

fig. 2

1 cm.

The major features of the internal anatomy of the heart chambers and their walls and valves are shown in this plate. Figure 1 shows the opened right side of the heart, figure 2 the left and figure 3 a detail of a semilunar valve. Arrows indicate the direction of blood flow. The atria are the receiving chambers of the heart; the right one for the **systemic circuit** which includes the coronary circulation in the heart wall; the left one for the **pulmonary circuit.** The atria pass the blood through their respective atrioventricular ostia into the right and left ventricles which are the dispensing chambers of the heart. The right ventricle pumps the blood to the lungs through the pulmonary arteries hence feeding into the pulmonary circuit; the left ventricle pumps blood into the aorta which carries it to all parts of the body, the systemic circuit.

There is considerable difference in the thickness of the walls of the right and left ventricles, in the myocardial layer, reflecting the difference in work required to maintain flow and pressure in the long systemic circuit as compared to the short pulmonary circuit. The thickness of right and left atrial walls is not significantly different. The myocardial layer is thin since little work is required to move the blood from the atria to the ventricles. The type of muscle tissue in the heart wall is cardiac (Plate 21, fig. 1).

Three layers constitute the heart wall: endocardium, myocardium and epicardium, or visceral pericardium as shown clearly in figure 2. The **endocardium** consists of a thin free surface of **endothelium** which lines all the heart chambers, covers the valve surfaces and is continued as the innermost tissue of the blood vessels. It is supported underneath by fibrous tissue and some smooth muscle which is thicker in the valve cusps. The **myocardium** or cardiac muscle layer is described above. The **epicardium** or **visceral pericardium** is a thin, transparent **serous membrane** consisting of a surface layer, the **mesothelium,** underlaid by a stroma containing some elastic fibers. It is continuous at the base of the heart with the parietal serous pericardium which is reflected back over the heart to form a closed sac. The space between the visceral and parietal pericardium is called the pericardial cavity and contains a small amount of pericardial fluid.

The **right atrium** (fig. 1) has three main orifices bringing blood into the chamber; a cranial one from the precava and a caudal one from the postcava; and close to the opening of the postcava is the small orifice of the **coronary sinus,** guarded craniad by the valve of the **coronary sinus** (*valve of Thebesius*). A fourth opening, the **atrioventricular orifice** leads into the right ventricle and is guarded by an **atrioventricular valve** (*tricuspid*) which will be described with the right ventricle. In the wall which separates right and left atria, the **interatrial septum,** a small circular thin area, can be seen which is the **fossa ovalis.** It is deepened some on the cranial side by a ridge, the **limbus** of the fossa ovalis. In fetal life the fossa ovalis is open and hence called the **foramen ovale.** This allows a flow of blood between the atria and makes the cardiovascular system an incomplete double system. When this foramen fails to close after birth, the kitten may die as a result of the mixing of oxygenated and nonoxygenated blood.

From the right and cranial side of the atrium the **right auricle** extends ventrally as a blind ear-like appendage. Externally it lies at the right of the base of the aorta and precava (fig. 1). Its internal lining is strengthened by freely branching and interlacing muscular bands, the **pectinate muscles.** Pectinate muscles are found to a lesser extent in the atrium, mostly near the opening into the auricle.

The **left atrium** (fig. 2) lies at the left caudodorsal part of the base of the heart. Its auricle lies caudal and to the left of the pulmonary trunk and conus and its apex covers the proximal part of the ventral interventricular groove (Plate 70, fig. 1). Four openings of pulmonary veins are seen in the wall of the left atrium. The atrium opens to the left ventricle by the atrioventricular orifice which has a valve described below. Its walls are smooth and the interatrial septum has a thin area, the fossa ovalis, with a fold of the septal wall ventrad. This represents a valve of the foramen ovale. The pectinate muscles are confined to the auricle.

The **ventricles** together form the bulk of the heart and together they form a cone-shaped body, the apex of which is the apex of the left ventricle (Plate 70, fig. 1). The right ventricle is arciform in shape as it makes a half turn about the left ventricle from the right lateral side at the apex to the ventral side at the base. The left ventricle is conical in shape with its apex forming the apex of the heart.

The **right ventricle** communicates with the right atrium craniodorsad through the right atrioventricular orifice. The part of the ventricle between this orifice and the pulmonary trunk is the **conus arteriosus.** Blood coming into the ventricle from the right atrium is discharged through the valve of the pulmonary trunk.

The inner walls of the ventricles are marked by myocardial ridges, more common on the outer wall of the ventricle than on the interventricular septum. They tend to converge from the base of the ventricle to the apex of the heart. They are the **trabeculae carneae.** Depressions of various length and depth lie between the trabeculae.

Special conical-shaped musclar projections of the ventricular walls, the **papillary muscles,** give rise to fibromuscular cords at their apices called **chordae tendineae.** The chordae tendineae attach in turn to the **cusps** of the right and left atrioventricular valves (*tricuspid and bicuspid*). When the ventricles contract, the cusps close the atrioventricular ostia and are kept from everting into the atria by the tension placed on them by the papillary muscle—chordae tendineae mechanism. Blood therefore can leave the ventricles only through their arteries, the pulmonary trunk and the aorta.

There are three prominent papillary muscles in the right ventricle though there are frequent variations. Chordae tendineae arise from trabeculae or from small papillae in the ventricular wall. Two large papillary muscles are found in the left ventricle and additional chordae tendineae arise from the ventricular wall.

The aorta and pulmonary trunk have semilunar valves which allow free flow of blood from the ventricles but prevent it from returning. They are called the **aortic** and **pulmonary valves.** Their under-surfaces are shown in figures 1 and 2, while figure 3 shows the base of the aorta cut and spread open to reveal clearly the three little half cups or **valvulae** which together constitute this type of valve. When blood moves into the valvulae, their edges come into contact and close the orifice.

Opposite each valvula, the wall of the vessel is slightly expanded forming the **sinuses of the aorta.** The **nodule** at the center of the free border of each semilunar cusp serves to close the space which would occur in the center of the vessel when the three curved surfaces come together.

Right and left heart valves are essentially alike as indicated above. However, since the work done by the "left heart" exceeds that done by the right, all of its structures described above are more strongly constructed.

lumen
tunica intima
tunica media
external elastic membrane
tunica externa
vasa vasorum

fig. 1

fig. 2

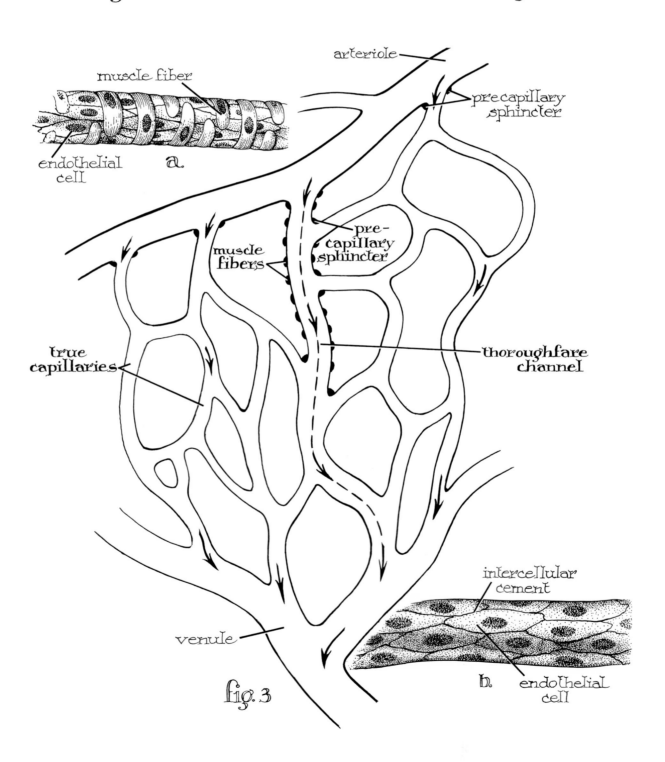

muscle fiber

endothelial cell

a.

arteriole

pre-capillary sphincter

pre-capillary sphincter

muscle fibers

thoroughfare channel

true capillaries

venule

fig. 3

intercellular cement

b. endothelial cell

(Crouch: Functional Human Anatomy, Lea & Febiger)

This plate illustrates an artery and vein surrounded by loose connective tissue with many adipose cells (figs. 1 and 2). Figure 3 is a schematic presentation of a capillary bed with an entering arteriole and a venule leaving the area. Inserts a and b show in more detail and more realistically the histology of a thoroughfare channel and a true capillary.

Figure 1 is a photomicrograph of a muscular artery. It shows very clearly the three tunics characteristic of both arteries and veins and some of the differences within the tunics. These tunics are tunica intima (*interna*) next to the lumen of the vessel; tunica media the thick middle layer and tunica externa (*adventitia*), the outer layer. The **tunica intima** is thin and has a surface layer of smooth endothelium, which is a simple epithelium of flat or squamous cells. Beneath the endothelium is a layer of connective tissue and an internal elastic membrane (*not labeled*). The **tunica media** is a thick muscular layer in which the muscle is of the smooth type and is circularly arranged. Some connective tissue, mostly of the elastic type, is found among the muscle tissue. The elastic tissue continues outward to form an **external elastic membrane** at the boundary of the muscular layer. The outer layer in the wall of the artery is the **tunica externa** (*adventitia*), composed of loose connective tissue with its collagenous and elastic fibers which assume a tangential or longitudinal direction. In the artery in figure 1 the tunica externa is thinner than the media. In some muscular arteries it may be thicker. The externa passes over into the surrounding connective tissue without clear boundaries. Nerve fibers and blood vessels may be found in the externa, the latter being called the **vasa vasorum** (*blood vessels of blood vessels*). The externa, with its loose connective tissue, allows an artery to move within certain limits and accommodates to the constantly changing size of the lumen.

Of very great importance from the standpoint of circulation is the variation that we find in the tunica media of arteries as we go from the larger proximal vessels to the smaller distal ones. The larger vessels, like the aorta, pulmonary artery, subclavian and common carotid, have a very thick tunica media with some smooth muscle but with so much elastic tissue that the muscle fibers are obscured. There are also prominent internal and external elastic membranes. When these arteries, often called **elastic arteries,** expand under the influence of the contraction of the left ventricle, they essentially storeup energy from the heart beat so that when the ventricle is relaxing and refilling, the stretched elastic tissue of the vessel wall recoils and presses upon the contained blood thus maintaining blood pressure and blood flow. They act like *shock absorbers* in smoothing out the flow of blood which would otherwise be intermittent, as indeed it may become when through aging or other processes, the elastic walls become hardened, a condition called arteriosclerosis. **Muscular arteries,** like those of the limbs, play an active role in blood flow and blood pressure by contraction and relaxation of the tunica media.

Figure 2, a photomicrograph of a vein, shows the same layers as the artery as stated above. They differ, however, in thickness and in composition and the lumen of a vein is larger than that of a corresponding artery. Veins when empty collapse. Veins are also provided with valves as seen in figure 16. The tunica intima may contain only the endothelium or there may be a small amount of connective tissue. The tunica media contains smooth muscle but is very thin. The tunica externa, usually the thickest layer in a vein, is composed of connective tissue and usually some smooth muscle. Elastic tissue is at a minimum. Blood flow in veins is influenced largely by action of other muscles, as those of the limbs, which in contracting press on the vessels and move the blood. Since the veins have valves, so arranged as to allow blood to move only toward the heart, venous return is accomplished. The aspirating effect of the low pressures around the heart in the thorax also aids flow.

Figure 3 shows schematically the kinds of structures which lie between the arteries and the veins in the tissues. They serve as the means by which blood makes available to the cells the nutrients that they need and the means by which waste materials can be returned to the blood for transportation to the tissues which remove them (*excretion*). **Arterioles,** which are arteries of very small caliber, carry blood either into the **true capillaries** through the **precapillary sphincters** or into a **thoroughfare channel** which has smooth muscle fibers and which carries blood through a set of capillaries to a venule without its passing through a true capillary. True capillaries (fig. 3b) are tiny vessels 8 to 10 microns in diameter—about the diameter of a red corpuscle. Their walls consist of endothelial cells forming a layer one cell thick. The blood passes over the inner surface of the endothelium; tissue fluid over the outer surface. It is here that the exchanges of nutrients, waste products, water and hormones takes place. Though the capillaries are small they form extensive networks in most of the tissues of the body. If they should all be dilated at one time, they would hold all of the blood in the body or if they could all be put into one mass, they would be the largest organ in the body, twice the size of the liver. These conditions, plus a slow flow through the capillaries, allows for an efficient exchange between the blood and the tissue fluid and cells.

203

Note too that there are true capillaries leading off of the thoroughfare channels each guarded by a precapillary sphincter. The thoroughfare channel itself is made up of endothelial cells and numerous muscle fibers. These muscle fibers and the precapillary sphincters may account for the "movement of capillaries" which we frequently hear mentioned. These structures—the true capillaries, thoroughfare channels and sphincters—make up a capillary bed.

Arterioles, structured like arteries, lose their tunics gradually as they approach and contribute to a capillary bed. Conversely venules, draining capillary beds take on the character of tiny veins as they unite to form the venous side of the cardiovascular system.

Arteries and Veins of the Head and Neck
of the Cat—Ventral View
PLATE 73

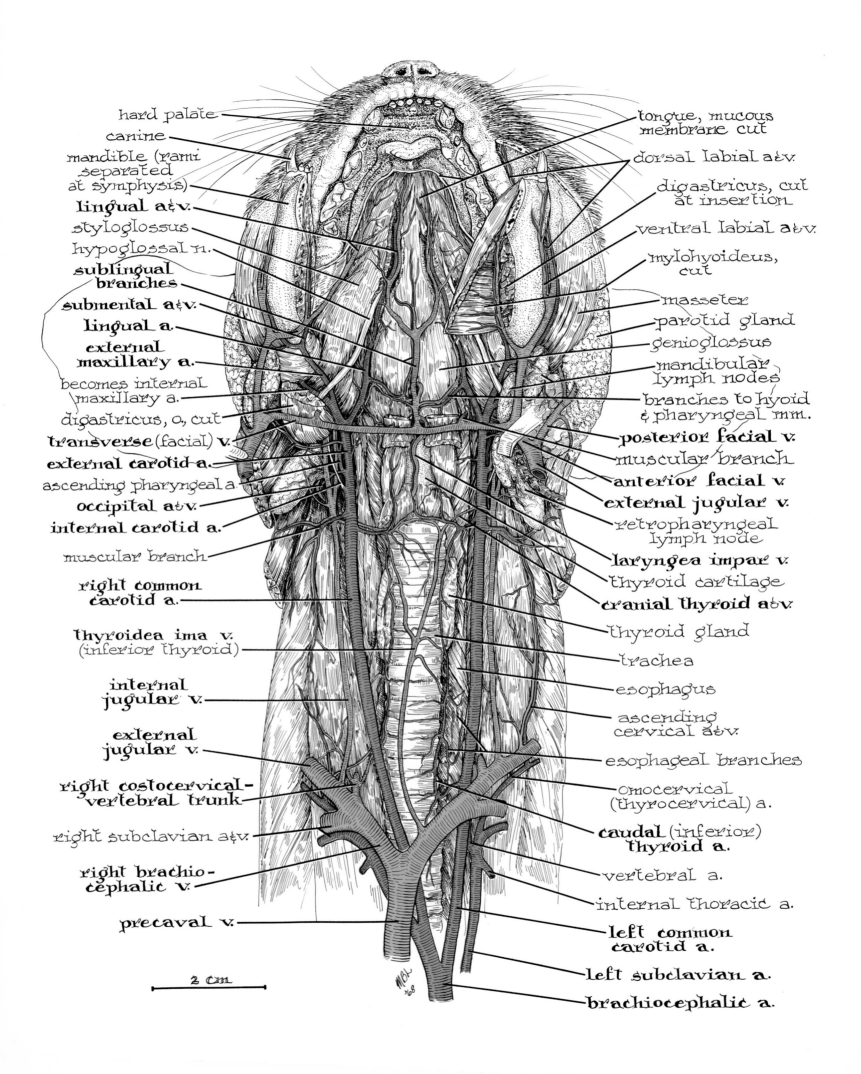

hard palate

canine

mandible (rami separated at symphysis)

lingual a&v.

styloglossus

hypoglossal n.

sublingual branches

submental a&v.

lingual a.

external maxillary a.

becomes internal maxillary a.

digastricus, o, cut

transverse (facial) **v.**

external carotid a.

ascending pharyngeal a.

occipital a&v.

internal carotid a.

muscular branch

right common carotid a.

thyroidea ima v. (inferior thyroid)

internal jugular v.

external jugular v.

right costocervical-vertebral trunk

right subclavian a&v.

right brachio-cephalic v.

precaval v.

tongue, mucous membrane cut

dorsal labial a&v.

digastricus, cut at insertion

ventral labial a&v.

mylohyoideus, cut

masseter

parotid gland

genioglossus

mandibular lymph nodes

branches to hyoid & pharyngeal mm.

posterior facial v.

muscular branch

anterior facial v.

external jugular v.

retropharyngeal lymph node

laryngea impar v.

thyroid cartilage

cranial thyroid a&v.

thyroid gland

trachea

esophagus

ascending cervical a&v.

esophageal branches

omocervical (thyrocervical) a.

caudal (inferior) **thyroid a.**

vertebral a.

internal thoracic a.

left common carotid a.

left subclavian a.

brachiocephalic a.

2 cm

Arteries and Veins of the Head and Neck
of the Cat—Ventral View
PLATE 73

This illustration shows the relationship of some of the important and more superficial arteries and veins of the head, neck and cranial part of the thorax. The mylohyoideus muscle has been cut, the mandible separated at its symphysis and these structures pulled laterally to reveal the tongue. The mucosa on the ventral side of the tongue has been removed to uncover the lingual and sublingual vessels. A section of the external jugular veins has been removed and the cranial ends pulled laterally placing the transverse (facial) vein in a stretched condition.

The arteries which branch off of the aortic arch are shown—the **brachiocephalic** and **left subclavian arteries.** The brachiocephalic gives rise to the left common carotid artery and farther craniad to the right common carotid and right subclavian arteries. The four branches of the **subclavian arteries** are the internal thoracic (*mammary*), omocervical (*thyrocervical*), vertebral, and costocervical. These vessels will be studied in other plates but it is advisable to view them here. Note the ascending cervical artery, a branch of the omocervical which passes craniad, the main vessel passing into the shoulder region (Plate 84). The vertebral arteries, discussed on pages 213 and 231, and the common carotid arteries are the primary sources of blood to the head.

Follow the **common carotid arteries** forward and note that they give off first a small **caudal** (*inferior*) **thyroid** artery which passes mediad and then craniad giving off branches to the esophagus, trachea, and thyroid glands. The caudal thyroid then continues craniad where it anastomoses with the cranial thyroid artery. The caudal thyroid artery is sometimes called the thyroidea ima. The **cranial thyroid artery** arises in this instance at about the level of the cricoid cartilage of the larynx. It passes mediad and then caudad to supply the thyroid gland and some musclar branches to the sternohyoideus and sternothyroideus. It also sends a branch to the external muscles of the larynx, the cranial laryngeal. Opposite the cranial thyroid artery the common carotid gives off a muscular branch or branches to the muscles on the dorsal side of the neck.

Continuing forward the common carotid artery gives rise to a very tiny **internal carotid artery** and then, slightly craniad, to **occipital** and **ascending pharyngeal** arteries. These vessels and their distribution are shown in Plates 76 and 77. The main vessel is now called the external carotid artery.

The **external carotid artery** now passes craniad giving off the **lingual artery** to the tongue and a few small muscular branches. Its next large branch is the **external maxillary artery** which in turn gives rise to the submental artery which runs forward between the digastricus and mylohyoideus which it serves. The external maxillary also gives off the dorsal and ventral labial arteries which go to upper and lower lips respectively. Though not shown in this plate the external carotid artery next gives rise to the posterior auricular and then the superficial temporal artery after which it is called the internal maxillary artery. The internal maxillary is thus the terminal branch of the external carotid. These last named branches of the external carotid artery are shown in Plates 76, figure 1 and 77, figure 1 and are described there with emphasis on the internal maxillary artery.

Most of the veins shown in this plate will be illustrated and described in Plates 74 and 75. Again note their relationship to the arteries. The **internal jugular vein** is shown better in this plate than elsewhere. It drains blood from the brain and cranium and deep veins of the head. Follow it caudad to its point of juncture with the external jugular. Notice its cranial thyroid and muscular branches accompanying the arteries of the same name. The thyroidea ima vein is also well shown here draining blood from the trachea, thyroid, and laryngeal area. It empties into the **left brachiocephalic vein.** The two brachiocephalic veins come together to form the **precava** which empties into the right atrium of the heart. Notice also the **costocervical-vertebral trunk** bringing blood from the costocervical and vertebral veins into the brachiocephalics and the subclavians coming in from the forelimbs.

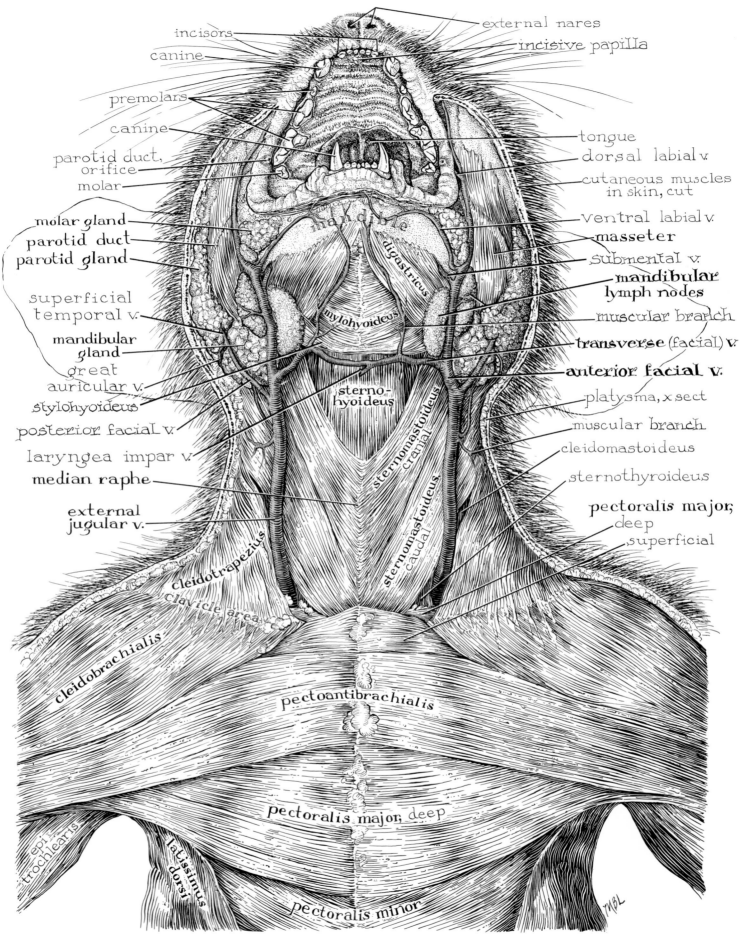

incisors

canine

premolars

canine

parotid duct,
orifice

molar

molar gland

parotid duct

parotid gland

superficial
temporal v.

mandibular
gland

great
auricular v.

stylohyoideus

posterior facial v.

laryngea impar v.

median raphe

external
jugular v.

cleidotrapezius

clavicle area

cleidobrachialis

epi-
trochlearis

latissimus
dorsi

external nares

incisive papilla

tongue

dorsal labial v.

cutaneous muscles
in skin, cut

ventral labial v.

masseter

submental v.

mandibular
lymph nodes

muscular branch

transverse (facial) v.

anterior facial v.

platysma, x sect.

muscular branch

cleidomastoideus

sternothyroideus

pectoralis major,
deep
superficial

mandible

digastricus

mylohyoideus

sterno-
hyoideus

sternomastoideus
cranial

sternomastoideus
caudal

pectoantibrachialis

pectoralis major, deep

pectoralis minor

MBL

2 cm

Superficial Veins of the Head and Neck of the Cat—Ventral View
PLATE 74

This plate illustrates the superficial veins of the head and neck in their normal relationship to muscles, lymph nodes, salivary glands and other structures. Only the superficial fascia has been dissected away following the removal of the skin. It provides one the opportunity to review the muscles which are shown here with a minimal amount of displacement.

This plate emphasizes the tributaries of the anterior and transverse facial veins. In the next plate, figure 1 the posterior facial vein and its tributaries are shown to better advantage.

The **external jugular vein** is formed by the union of the anterior facial and posterior facial veins which unlike the condition in man is the main vein from the head. It crosses the sternomastoideus muscle, giving off a muscular branch, and disappears into a triangular interval between the sternomastoideus and cleidotrapezius muscles. Other tributaries of the external jugular vein can be seen in Plate 75.

A part of the drainage area of the **anterior facial vein** is shown here. It receives blood from the face, tongue and other parts. Some of those not shown here are seen in Plates 73 and 75. Here we see the dorsal and ventral labial veins from the lip areas and the submental vein from the area ventral and caudal to the mandible. Just before the anterior and posterior facial veins join to form the external jugular the large **transverse** (*facial*) **vein** enters the anterior facial. It receives a large vessel from the larynx the laryngeal impar, a pair of muscular branches and though not shown in this plate (see Plate 73) a branch of varying size from beneath the tongue, the sublingual. The transverse facial connects the two anterior facial veins.

The **posterior facial vein** is shown only as it joins the external jugular having crossed over the mandibular salivary gland. Its tributaries are shown more completely in the next plate.

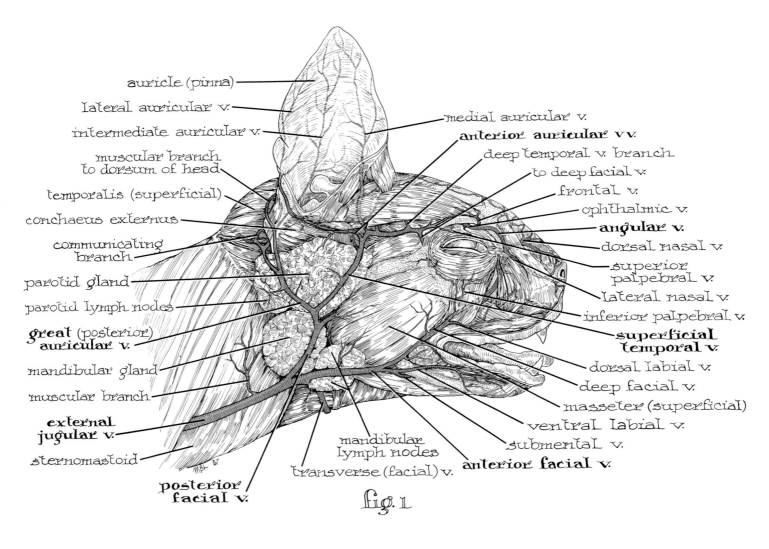

auricle (pinna)

lateral auricular v.

intermediate auricular v.

muscular branch
to dorsum of head

temporalis (superficial)

conchaeus externus

communicating
branch

parotid gland

parotid lymph nodes

**great (posterior)
auricular v.**

mandibular gland

muscular branch

**external
jugular v.**

sternomastoid

medial auricular v.

anterior auricular vv.

deep temporal v. branch

to deep facial v.

frontal v.

ophthalmic v.

angular v.

dorsal nasal v.

superior
palpebral v.

lateral nasal v.

inferior palpebral v.

**superficial
temporal v.**

dorsal labial v.

deep facial v.

masseter (superficial)

ventral labial v.

submental v.

anterior facial v.

mandibular
lymph nodes

transverse (facial) v.

**posterior
facial v.**

fig. 1

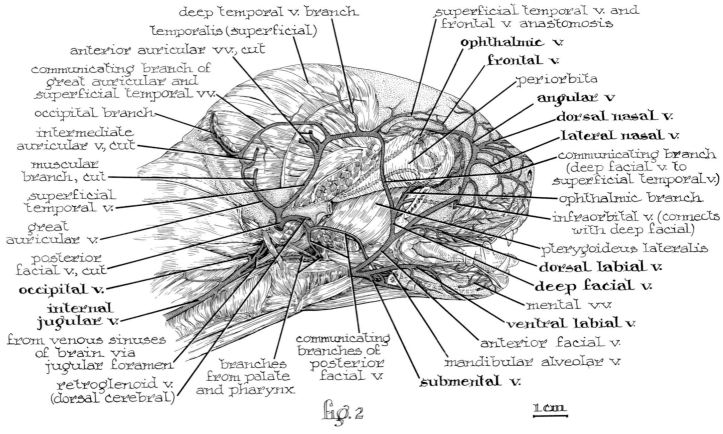

deep temporal v. branch

temporalis (superficial)

anterior auricular vv, cut

communicating branch of
great auricular and
superficial temporal vv.

occipital branch

intermediate
auricular v, cut

muscular
branch, cut

superficial
temporal v.

great
auricular v.

posterior
facial v, cut

occipital v.

**internal
jugular v.**

from venous sinuses
of brain via
jugular foramen

retroglenoid v.
(dorsal cerebral)

branches
from palate
and pharynx

communicating
branches of
posterior
facial v.

superficial temporal v. and
frontal v. anastomosis

ophthalmic v.

frontal v.

periorbita

angular v

dorsal nasal v.

lateral nasal v.

communicating branch
(deep facial v. to
superficial temporal v.)

ophthalmic branch

infraorbital v. (connects
with deep facial)

pterygoideus lateralis

dorsal labial v.

deep facial v.

mental vv.

ventral labial v.

anterior facial v.

mandibular alveolar v.

submental v.

fig. 2

1cm

This plate shows in lateral view most of the important tributaries of the anterior and posterior facial veins which in turn are shown entering the **external jugular vein.** The superficial part of the parotid salivary gland has been cut away in figure 1 to show the passage of the tributaries of the posterior facial through that gland. Some of the tributaries of the anterior and posterior facial veins have been shown by dashed lines as they pass under the more superficial muscles. In figure 2 the dissection is deeper and shows very clearly the deeper veins which enter the anterior and posterior facial veins or their tributaries. Note particularly the number of anastomoses which tie the tributaries of these two important veins together as essentially one system.

In figures 1 and 2 trace the following tributary veins into the anterior facial vein starting with the small frontal vein above the orbit. The **frontal vein** anastomoses with the superficial temporal, receives superior palpebral veins from the eyelid and a branch from the orbit. It continues forward around the cranial angle of the eye as the angular vein. The **angular vein** receives **dorsal and lateral nasal veins** from the external surface of the nose, an infraorbital vein from the infraorbital foramen and an inferior palpebral vein from the lower eyelid. The infraorbital vein connects with the deep facial medial to the zygomatic arch.

The angular vein now becomes the **anterior facial vein** and receives a **dorsal labial vein** from the upper lip and a large deep facial branch from the orbital area. The **deep facial** (*reflex*) **vein** receives an ophthalmic branch from the inside of the periorbita which receives blood from the orbital plexus. The deep facial also collects blood from the soft and hard palates. It has a communicating branch to the superficial temporal vein.

The anterior facial vein next receives a ventral labial vein from the lower lip and branching from the ventral labial or arising independently from the anterior facial vein is the submental vein. The **submental vein** is also shown in Plate 73 and should be studied there as well as in figures 1 and 2 of this plate. It receives a lingual vein from the tongue, and from between the mandible and

pterygoid muscles a branch which connects with the posterior facial vein. This latter branch (*communicating*) receives the mandibular alveolar vein from the mandibular canal. Finally, just before joining the posterior facial vein to form the external jugular vein, the anterior facial receives the transverse (*facial*) vein described in connection with the previous plate.

Follow now the tributaries of the **posterior facial vein** which drain blood from the pterygoid, masseter, and temporal muscles of the back of the head and external ear. Starting with the **superficial temporal vein** which we said above anastomoses with the frontal vein, and moving caudad we see a communicating branch between the superficial temporal vein and the deep facial vein. Farther caudad a branch is received from the temporal muscle, the deep temporal vein. The superficial temporal vein next receives a communicating branch which connects it with the great auricular vein (fig. 2). Its next two tributaries are the **anterior auricular veins,** one draining blood from the cranial part of the auricle or pinna by way of the medial auricular vein, the other passing ventral to the ear opening and by way of the lateral auricular vein draining blood from the caudal part of the auricle. The two anterior auricular veins may come into a common vein before joining the superficial temporal vein. The superficial temporal vein now passes ventrad through the parotid gland and near its ventral border is joined by the great (*posterior*) auricular vein with which it forms the posterior facial vein.

The **great auricular vein** starts out from the intermediate auricular vein of the auricle, the communicating branch to the superficial temporal, the occipital branch and a muscular branch (figs. 1 and 2).

In figure 2 note the retroglenoid (*dorsal cerebral*) vein, a tributary of the posterior facial vein, which sends not only a communicating branch to the submental vein, but also a branch which passes ventrad and then caudad to join the internal jugular vein. Note that the **internal jugular vein** receives blood from the palate and pharynx; from the venous sinuses of the brain (*inferior cerebrals*); and from an **occipital vein** and muscular branches.

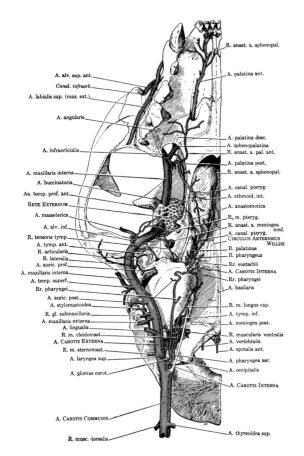

R. anast. a. sphenopal.

A. alv. sup. ant.
Canal. infraorb.
A. labialis sup. (max. ext.)

A. angularis

A. palatina ant.

A. infraorbitalis

A. palatina desc.
A. sphenopalatina
R. anast. a. pal. ant.
A. palatina post.
R. anast. a. sphenopal.

A. maxillaris interna
A. buccinatoria
Aa. temp. prof. ant.
RETE EXTERNUM
A. masseterica
A. alv. inf.
R. tensoris tymp.
A. tymp. ant.
R. articularis
R. lateralis
A. auric. prof.
A. maxillaris interna
A. temp. superf.
Rr. pharyngei
A. auric. post.
A. stylomastoidea
R. gl. submaxillaris
A. maxillaris externa
A. lingualis
R. m. cleidomast.
A. CAROTIS EXTERNA
R. m. sternomast.
A. laryngea sup.
A. glomus carot.

A. canal. pteryg.
A. ethmoid. int.
A. anastomotica
R. m. pteryg.
R. anast. a. meningea med.
A. canal. pteryg.
CIRCULUS ARTERIOSUS WILLISI
R. palatinus
R. pharyngeus
Rr. eustachii
A. CAROTIS INTERNA
Rr. pharyngei
A. basilaris

R. m. longus cap.
A. tymp. inf.
A. meningea post.
R. muscularis ventralis
A. vertebralis
A. spinalis ant.
A. pharyngea asc.
A. occipitalis
A. CAROTIS INTERNA

A. CAROTIS COMMUNIS

R. musc. dorsalis

A. thyreoidea sup.

FIG. 1

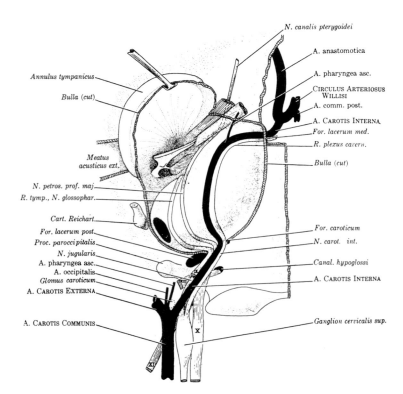

N. canalis pterygoidei

Annulus tympanicus

Bulla (cut)

A. anastomotica
A. pharyngea asc.
CIRCULUS ARTERIOSUS WILLISI
A. comm. post.
A. CAROTIS INTERNA
For. lacerum med.
R. plexus cavern.

Meatus acusticus ext.

Bulla (cut)

N. petros. prof. maj.
R. tymp., N. glossophar.

Cart. Reichart.
For. lacerum post.
Proc. paroccipitalis.
N. jugularis
A. pharyngea asc.
A. occipitalis
Glomus caroticum
A. CAROTIS EXTERNA

For. caroticum
N. carot. int.

Canal. hypoglossi

A. CAROTIS INTERNA

A. CAROTIS COMMUNIS

Ganglion cervicalis sup.

FIG. 2

Arteries of the Head Region of the Cat No. 1
PLATE 76

(Reprinted by permission from Carotid Circulation in the Domestic Cat by Davis and Story, Field Museum of Natural History, Zoological Sciences, Vol. 28) (R = ramus or branch)

This plate shows in figure 1 the arteries which carry blood to the head region and those within the head itself. Blood is carried to the head by the common carotid and the vertebral arteries. The **common carotid arteries** are bound, along with the internal jugular veins and the vagosympathetic trunks, into common sheaths one on each side of the trachea. At the back of the head, the common carotids give off a small degenerate **internal carotid artery,** an **ascending pharyngeal artery** and an **occipital artery.** The main vessel continues craniad as the external carotid artery. The ascending pharyngeal and occipital arteries may derive from the common carotid individually or as a single vessel, as shown here. At this juncture is found the **carotid sinus** which is an important receptor for the regulation of blood pressure. On the wall of the carotid sinus is a small nodule or **glomus** the carotid body, which is a chemoreceptor probably sensitive to the changing levels of oxygen in the blood and from which respiration is, in part, reflexly regulated.

Since the internal carotid artery is degenerate and part of it completely occluded in the adult cat, blood is carried to the brain largely through the ascending pharyngeal, the vertebral and the external carotid by way of the **internal maxillary artery** and the **rete externum** (*carotid plexus*). The internal carotid artery, it should be noted by reference to figure 2, is a functional artery in the fetal cat. Notice how it passes through the bulla and joins the circle of Willis craniad. It is functional also in the dog, seal, tiger and lion, but not functional in rodents such as the squirrel and guinea pig and among some of the artiodactyla.

The vertebral arteries, you should recall, are derived from the subclavian arteries. They enter the transverse foramina of the cervical vertebrae and pass craniad and at the atlas traverse the atlantal foramina and enter the vertebral (*neural*) canal where the right and left vessels join on the ventral side of the spinal cord to form the basilar artery. The **basilar artery** (fig. 1) passes craniad on the ventral midline of the spinal cord and brainstem giving off some branches and finally joining the circle of Willis to which it contributes. The **circle of Willis** is an important arterial circle or interchange on the ventral side of the brain in the region of the hypophysis. Note in figure 1 that the anterior spinal artery joins the vertebrals.

Notice each of the following arteries in figure 1. The ascending pharyngeal artery sends branches to the muscles of the pharynx and the auditory (*Eustachian*) tube and at its distal end utilizes the end of the internal carotid artery to join the circle of Willis. The occipital artery supplies the deep muscles of the neck and the sternomastoid.

Some branches of the **external carotid artery** shown here are the sternomastoid artery to the muscle of the same name and to muscles of the cranial part of the larynx. The **lingual artery** is a fairly large branch which follows the hypoglossal nerve (Plate 73) along the border of the masseter muscle to supply the tongue. The **external maxillary artery,** with a branch to the submaxillary gland, arises near the angle of the jaw and gives off submental and labial arteries to the symphysis of the lower jaw and to the upper and lower lips. A **postauricular artery** supplies the area caudad of the ear and by a cranial auricular branch the area craniad of the ear. Finally, the **superficial temporal artery** to the temporal region behind and above the eye and to the dorsal surface of the nose is the last branch of the external carotid artery which now becomes the internal maxillary artery.

The **internal maxillary artery** forms an S-shaped curve and continues craniad and enters the pterygoid fossa. At this point many small branches form a plexus around it, the rete externum or carotid plexus. Notice its connections to the arterial circle of Willis through the anastomotica artery.

Study the branches of the internal maxillary artery and the rete externum as seen in figure 1. Many of the names are descriptive of the structures which the vessels supply. Note especially the following. The **inferior alveolar artery** after supplying blood to the pterygoid muscle enters the mandibular foramen and supplies blood to the lower teeth and emerges craniad through the mental foramina. It then spreads over the lower lips and gums and anastomoses with its opposite number. The **masseteric artery** supplies the masseter and part of the temporalis muscles. The **anastomotic artery** is made up of several branches from the external rete which after passing through the orbital fissure into the cranial cavity join together. The large anastomotic artery joins the terminal portion of the internal carotid (*ascending pharyngeal*) to enter the **circle of Willis** (fig. 2). The **middle meningeal artery** enters the foramen ovale to supply the dura mater. The impressions of this artery are the ones to be seen on the inner surface of the skull. Other arteries to note are the buccinator, deep temporal and superior alveolar.

Craniad the internal maxillary artery has several terminal branches. The **infraorbital artery** is lateral and accompanies the nerve of the same name through the infraorbital foramen and supplies the lateral side of the nose and eye structures. The **sphenopalatine artery** is medial and with a nerve of the same name passes through the sphenopalatine foramen and supplies the median nasal septum and nasal cavities. The **descending palatine artery** branches off of the sphenopalatine just before the latter enters the sphenopalatine foramen. It enters the posterior palatine canal along with the anterior palatine nerve and then emerges on the ventral surface of the hard palate through the posterior palatine foramen. It is now called the **anterior palatine artery.** Upon reaching the palate the anterior palatine gives off an anastomotic branch which runs caudad to anastomose with the posterior palatine artery. The anterior palatine artery continues forward on the hard palate along with the anterior palatine nerves. It enters the anterior palatine (*incisive*) foramen and anastomoses with the sphenopalatine artery. The **posterior palatine artery** arises near the end of the internal maxillary artery. Along with the posterior palatine nerve, it goes to the internal pterygoid muscle then to the soft palate. It anastomoses with the ascending palatine artery and with the palatine branch of the ascending pharyngeal.

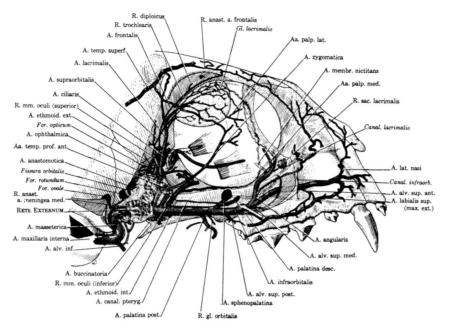

Fig. 1

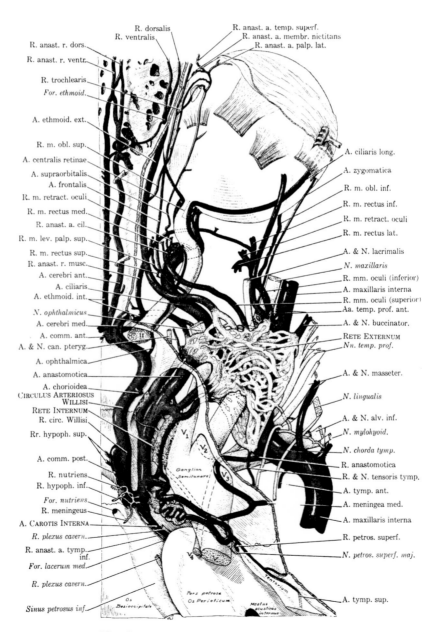

Fig. 2

*(Reprinted by permission from Carotid Circulation in the Domestic Cat
by Davis and Story, Field Museum of Natural History, Zoological Sciences, Vol. 28)*

Figure 1 illustrates the right internal maxillary artery and its many branches as seen in lateral view. It reviews many of the vessels described in connection with the previous plate but shows them in a different perspective with a view to improving the readers understanding of relationships. Additional vessels and anastomoses on the dorsal side are shown which should enable one to realize more fully the extensive ramifications of the internal maxillary artery and its branches.

Notice again the S-shaped curve in the internal maxillary artery and the external rete which it gives rise to and through which it passes. This complex network of fine arteries is interlaced with the sinus-like veins of the pterygoid plexus (*not shown*). A number of vessels described in the previous plate have their origins from this rete complex, for example the anastomotic and middle meningeal. The vessels of an "orbital artery complex" also derive from the internal maxillary artery and the anterior part of the external rete. These vessels include the ciliary, ethmoidal, deep anterior temporal, lacrimal, frontal, zygomatic and supraorbital arteries and their branches. The structures supplied by these are clearly suggested in their names as well as from what is shown in figures 1 and 2 of this plate. The relationship of this whole complex of arteries to the optic (II) and trigeminal (V) nerves should be noted as well as the fact that there is a small **internal rete** close to the semilunar ganglion of the fifth nerve and its ophthalmic branch (fig. 2). Figure 1 shows some of the anastomoses which occur among these vessels. The **lacrimal artery** not only sends a branch to the lacrimal gland but also one which joins the **superficial temporal**

artery. The **zygomatic artery** has anastomotic branches to the superior temporal and lateral palpebral and through the latter to the lacrimal artery. The **frontal artery** anastomoses with the superficial temporal, lateral palpebral, angular and supraorbital arteries. Follow the **internal ethmoidal artery** and note its anastomotic muscular branch to the branch of the inferior oculi artery which supplies the inferior rectus muscles and craniad the multiple anastomoses between the **anterior cerebral artery** and the internal ethmoidal. Right and left internal ethmoidal arteries join into a common vessel within the cranial cavity which terminally anastomoses with the **external ethmoids** from either side.

Craniad of the external rete note the posterior palatine artery described before and entering the sphenopalatine foramen and the posterior palatine canal the sphenopalatine and descending palatine arteries respectively. These were described in reference to the last plate. Follow the **infraorbital artery** craniad and notice that before passing through the infraorbital foramen, it gives off an **angular artery** which in turn gives rise to nictitating membrane and medial palpebral arteries and a branch which anastomoses with the lateral nasal arteries and proceeding craniad and dorsad with the frontal and superficial temporal arteries dorsad of the orbit. The infraorbital artery also gives off posterior, middle and anterior superior alveolar arteries and at its terminus gives rise to the **lateral nasal arteries** and anastomoses with the **superior labial artery** which is the cranial end of the **external maxillary artery.**

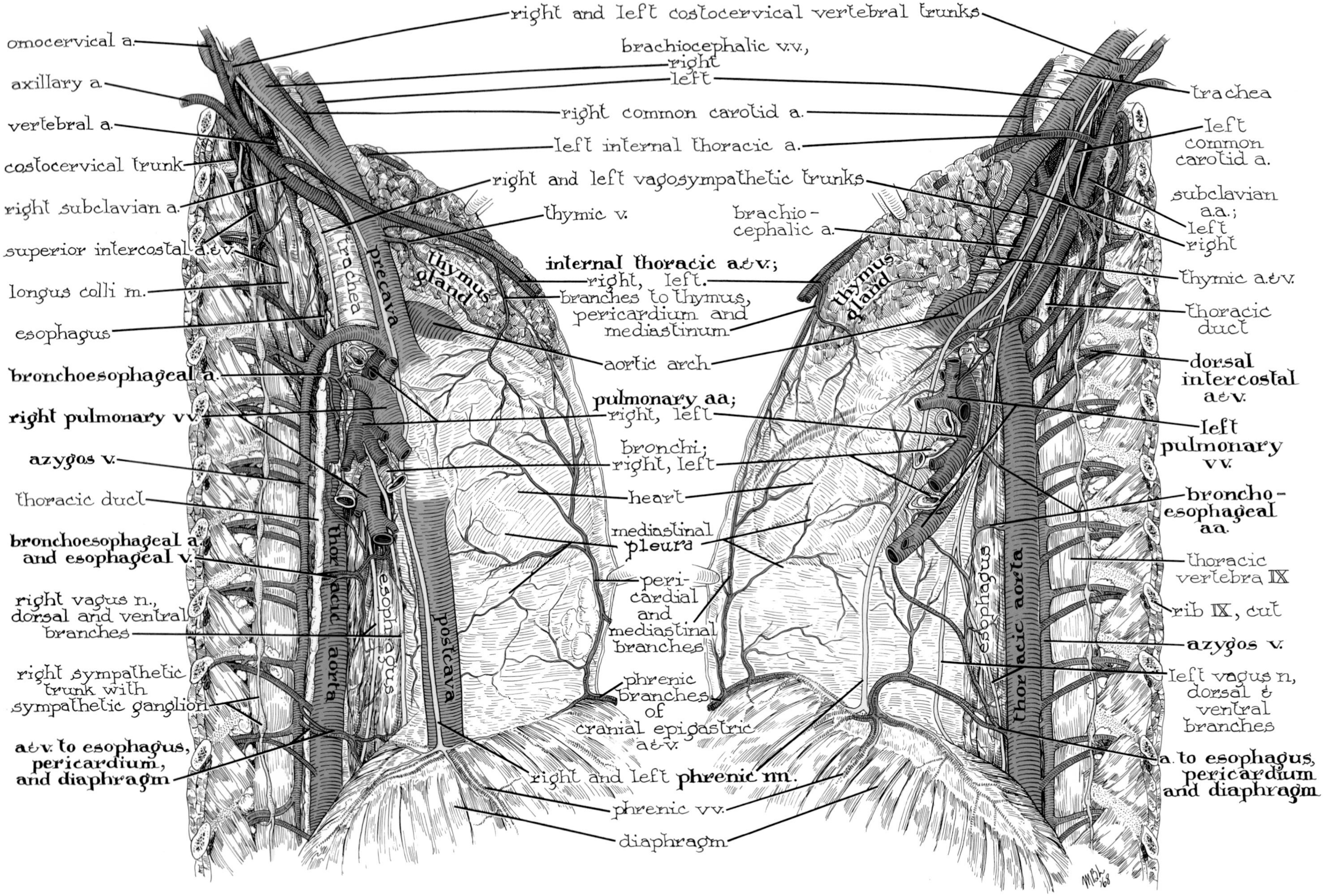

right and left costocervical vertebral trunks

trachea

left
common
carotid a.

subclavian
aa.;
left
right

thymic a.&v.

thoracic
duct

dorsal
intercostal
a.&v.

left
pulmonary
vv.

broncho-
esophageal
aa.

thoracic
vertebra IX

rib IX, cut

azygos v.

left vagus n,
dorsal &
ventral
branches

a. to esophagus,
pericardium
and diaphragm

Thoracic aorta

esophagus

thymus
gland

brachiocephalic v.v.,
right
left

right common carotid a.

left internal thoracic a.

right and left vagosympathetic trunks

thymic v

brachio-
cephalic a.

internal thoracic a.&v.;
right, left.
branches to thymus,
pericardium and
mediastinum

aortic arch

pulmonary aa.;
right, left

bronchi;
right, left

heart

mediastinal
pleura

peri-
cardial
and
mediastinal
branches

phrenic
branches
of
cranial epigastric
a.&v.

right and left phrenic m.

phrenic vv.

diaphragm

fig. 2

1 cm

fig. 1

thymus
gland

precava

trachea

thoracic aorta

postcava

esophagus

omocervical a.

axillary a

vertebral a.

costocervical trunk

right subclavian a.

superior intercostal a.&v.

longus colli m.

esophagus

bronchoesophageal a.

right pulmonary vv.

azygos v.

thoracic duct

bronchoesophageal a.
and esophageal v.

right vagus n,
dorsal and ventral
branches

right sympathetic
trunk with
sympathetic ganglion

a.&v. to esophagus,
pericardium,
and diaphragm

Right and Left Lateral Views of the Thoracic Cavity and Its Contents Exclusive of the Lungs
PLATE 78

In this plate showing lateral views of the thorax and its contents, only the lungs, pulmonary artery and veins have been dissected away. The heart remains in the middle mediastinum and the aortic arch and thoracic aorta are intact as are the precava and postcava.

The **internal thoracic artery** and **vein** and some of their branches are shown cranially and then they are cut. They are shown, however, in the next plate figure 1.

In figure 1 the **azygos vein** and its relationship anterior to the heart should be noted particularly as it crosses craniad of the root of the right lung, to the right of the trachea and runs dorsad then caudad to reach its position to the right of the thoracic aorta which it parallels through the thorax. The branches of these two vessels are the **bronchoesophageal arteries** at two points, the **esophageal vein,** and caudally an artery and vein which reach by their branches, the **pericardium, diaphragm** and **esophagus.** The **right and left dorsal intercostal arteries and veins** are also shown. These can be seen on both figures 1 and 2.

In the caudal part of both figures the **phrenic nerves** are shown, the right one traveling along the postcava, the left one in the **mediastinal pleura.** Also in this region phrenic veins and arteries are shown as branches of the **cranial epigastric artery** and also phrenic veins from other sources. Pericardial and mediastinal branches of the internal thoracic arteries and veins are seen in this area.

The vagosympathetic trunk is shown cranially in figures 1 and 2 and caudally the right and left sympathetic trunks with ganglia are seen—labeled only on the right. Further, the right and left dorsal and ventral branches of the **vagus nerves** are seen in relationship to the **esophagus.**

The **thymus gland** is a prominent feature in the cranial mediastinum as are the **trachea, precava,** esophagus, the cranial portions of the internal thoracic arteries and veins, the superior intercostal arteries and veins, and the subclavian arteries.

The branching of the subclavian arteries anteriorly is well shown in these illustrations especially in figure 1. The branching of the precava into right and left brachiocephalic veins and in turn their branching is shown here but is better seen in Plate 79, figure 2.

It is very important to study this plate in conjunction with Plate 79. It shows many of the same structures from different points of view in one of the most complex areas of the body. Plate 79, figure 1 will help too in understanding the relationships of the various epigastric arteries and veins. There is no better help than to have a well-dissected cat to compare with the illustrations. If you find discrepancies between the illustrations and the dissected cat, it can mean some error in the illustration or it can merely emphasize a reality that cats do vary in the details of their anatomy.

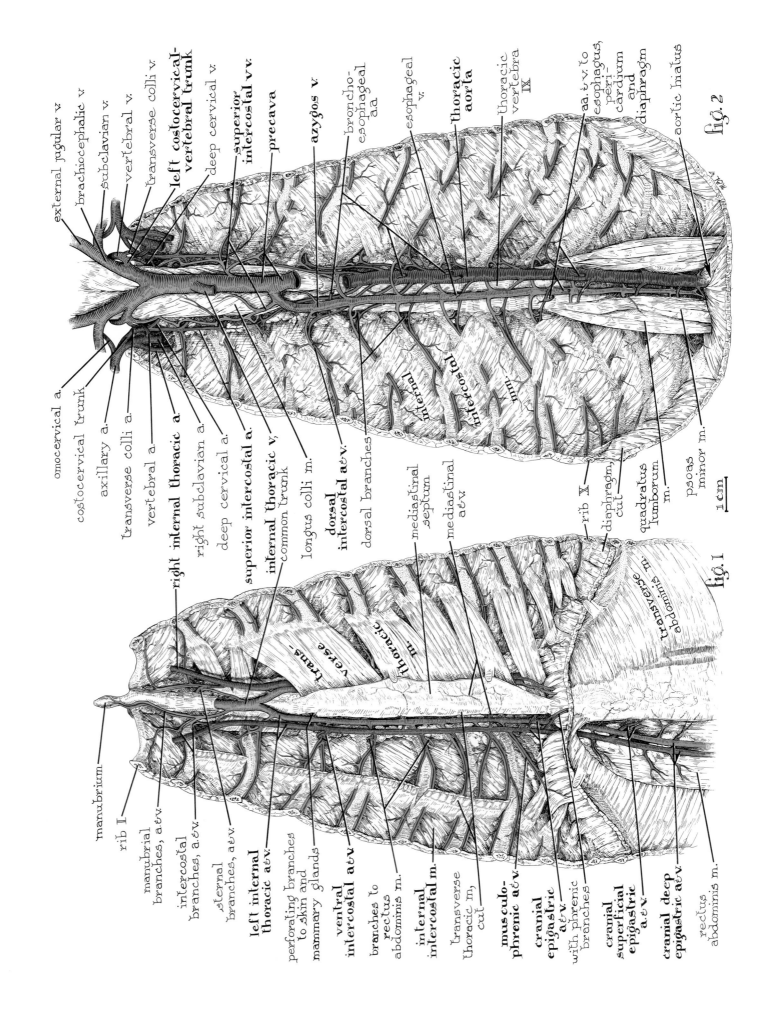

external jugular v.
brachiocephalic v.
subclavian v.
vertebral v.
transverse colli v.
left costocervical-vertebral trunk
deep cervical v.
superior intercostal vv.
precava
azygos v.
broncho-esophageal a.a.
esophageal v.
thoracic aorta
thoracic vertebra IX
aa. & v. to esophagus, pericardium and diaphragm
aortic hiatus

internal
intercostal
mm.

fig. 2

omocervical a.
costocervical trunk
axillary a.
transverse colli a.
vertebral a.
right internal thoracic a.
right subclavian a.
deep cervical a.
superior intercostal a.
internal thoracic v,
common trunk
longus colli m.
dorsal intercostal a. & v.
dorsal branches

mediastinal septum
mediastinal a. & v.

rib X
diaphragm, cut
quadratus lumborum m.
psoas minor m.

1 cm

manubrium
rib I
manubrial branches, a. & v.
intercostal branches, a. & v.
sternal branches, a. & v.
left internal thoracic a. & v.
perforating branches to skin and mammary glands
ventral intercostal a. & v.
branches to rectus abdominis m.
internal intercostal m.
transverse thoracic m, cut
musculo-phrenic a. & v.
cranial epigastric a. & v.
with phrenic branches
cranial superficial epigastric a. & v.
cranial deep epigastric a. & v.
rectus abdominis m.

trans-verse

thoracic m.

transverse m.
abdominis

fig. 1

Figure 1 shows the dorsal surface of the ventral wall of the thorax, a cut section of the diaphragm and the transverse abdominal muscle below the diaphragm. A "window" has been made in the transverse abdominal muscle and aponeurosis on the left side to show the **cranial deep epigastric artery and vein**. Also the **cranial superficial epigastric artery and vein** are shown penetrating the ventral wall. A portion of the mediastinal septum (*pleura*) is shown and on the right side the **transverse thoracic muscle** is featured. It consists of five or six flat bands of musclar tissue lying on the inner surfaces of the sternum and the costal cartilages. Its origin is on the lateral borders of the dorsal surface of the sternum, its insertion into the cartilages of the ribs near their attachment to the bony ribs and on to the fascia covering the intervening **internal intercostal muscles.** On the left side the transverse thoracic muscle is removed to show to better advantage the extent of the **internal intercostal muscle** and the internal thoracic artery and vein and their branches. At the cranial end the manubrium of the sternum is shown and caudad of the manubrium the first rib marking the inlet to the thorax.

The **internal thoracic artery** arises from the caudo-ventral side of the subclavian artery at the level of the first rib (fig. 2). It passes ventrocaudad in the cranial mediastinum and reaches the sternum opposite the third intercostal space (fig. 1). It passes caudad running parallel to the sternum and dorsal to the sternal ends of the costal cartilages and the interchondral spaces. It lies ventral to the transverse thoracic muscle which has been removed on the left side. At the level of the eighth costal cartilage it gives off a **musculophrenic artery** which passes caudo-dorsolaterad and runs in the angle between the diaphragm and lateral body wall. It then penetrates the diaphragm and comes to lie under the parietal peritoneum (See Plate 57, fig. 3). It gives rise to numerous small branches including small phrenics and ventral intercostal arteries eight, nine and ten. Its terminal branch anastomoses with the eleventh dorsal intercostal caudal to the diaphragm.

Beyond the musculophrenic a. the internal thoracic becomes the **cranial epigastric artery.** It penetrates the diaphragm, giving off phrenic branches, and then divides into two branches. One is the **cranial superficial epigastric artery** which penetrates the abdominal wall and comes to lie in the subcutaneous tissue where it supplies blood primarily to the cranial abdominal mamma (See Plate 66, fig. 2). It anastomoses with the caudal superficial epigastric artery. The other branch is the **cranial deep epigastric artery** which runs in the rectus abdominis and anastomoses with the caudal deep epigastric artery.

Returning now to the internal thoracic artery notice that it sends a number of small branches craniad near its origin which go to the manubrium, sternum, thymus and to the first and second intercostal spaces. A regular series of **ventral intercostal arteries** arise from the internal thoracic artery as far as the eighth costal cartilage. The remaining ventral intercostal arteries come from the musculophrenic artery. Branches to the mediastinum and perforating branches to the skin are present. The latter form ventral cutaneous branches and also branches to adjacent muscles and in a lactating cat branches to the mammae (See Plate 66, fig. 2).

The **internal thoracic vein** arises as a single vessel from the precava and then divides, its branches lying adjacent to the internal thoracic arteries and following them as satellites in all of their peripheral distribution.

Figure 2 is a ventral view of the dorsal wall of the thorax. Internal intercostal muscles are evident between the ribs, longus colli muscles extend into the thorax at the inlet and psoas minor and quadratus lumborum muscles lie against the dorsal wall caudally.

Cranially the main branches of the **precava** and **brachiocephalic** veins are shown. Note that the costocervical arterial and venous trunks give off transverse colli vessels to the longus colli muscles. The first large vein to enter the precava is the **azygos** at the level of the right third intercostal space. It has its origin in the abdominal region from the confluence of right and left lumbar veins of the dorsal wall. It enters the thorax between the crura of the diaphragm, inclines slightly to the right and lies in the angle formed by the bodies of the vertebrae and the **thoracic aorta.** The azygos receives the **dorsal intercostal veins** separately from the caudal part of the thorax. Cranially on the right side, the first three dorsal intercostals join a common trunk which may be called the superior intercostal vein which enters the azygos. On the right side there is a similar arrangement except that the superior intercostal vein enters the precava directly.

The **thoracic aorta** is cut at about the level of the sixth rib and the cranial portion is removed. Dorsal intercostal arteries are given off from the dorsal side of the thoracic aorta, each passing into an intercostal space, and divides into three branches: one follows the rib, another enters the spinal cord through the intervertebral foramen, the third goes to the deep muscles of the back. The first two or three dorsal intercostal arteries arise from the subclavian or the costocervical axis.

Bronchoesophageal, esophageal, and pericardial branches of the thoracic aorta supply those areas. These are better illustrated in Plate 78, figures 1 and 2.

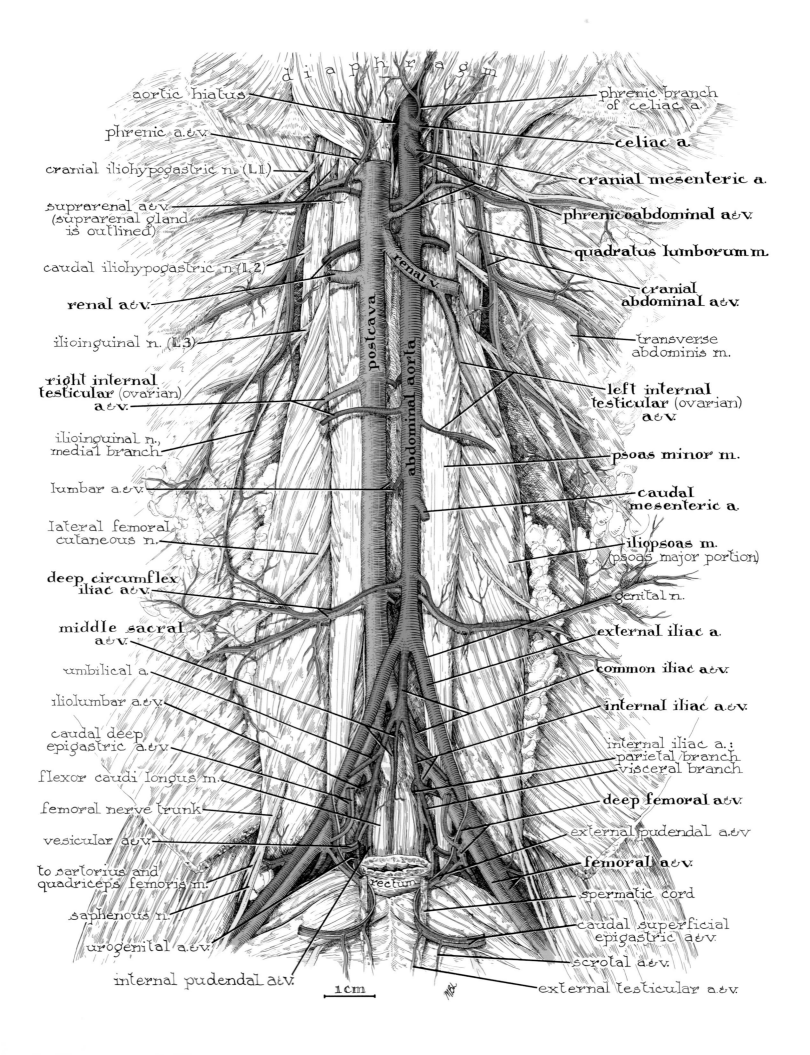

diaphragm

aortic hiatus

phrenic a.&v.

cranial iliohypogastric n. (L1)

suprarenal a.&v.
(suprarenal gland
is outlined)

caudal iliohypogastric n. (L2)

renal a.&v.

ilioinguinal n. (L3)

right internal
testicular (ovarian)
a.&v.

ilioinguinal n.,
medial branch

lumbar a.&v.

lateral femoral
cutaneous n.

deep circumflex
iliac a.&v.

middle sacral
a.&v.

umbilical a.

iliolumbar a.&v.

caudal deep
epigastric a.&v.

flexor caudi longus m.

femoral nerve trunk

vesicular a.&v.

to sartorius and
quadriceps femoris m.

saphenous n.

urogenital a.&v.

internal pudendal a.&v.

phrenic branch
of celiac a.

celiac a.

cranial mesenteric a.

phrenicoabdominal a.&v.

quadratus lumborum m.

cranial
abdominal a.&v.

transverse
abdominis m.

left internal
testicular (ovarian)
a.&v.

psoas minor m.

caudal
mesenteric a.

iliopsoas m.
(psoas major portion)

genital n.

external iliac a.

common iliac a.&v.

internal iliac a.&v.

internal iliac a.:
parietal branch
visceral branch

deep femoral a.&v.

external pudendal a.&v.

femoral a.&v.

spermatic cord

caudal superficial
epigastric a.&v.

scrotal a.&v.

external testicular a.&v.

postcava

renal v.

abdominal aorta

rectum

1 cm

Ventral View of the Postcava and the
Abdominal Aorta and their Branches
PLATE 80

The purpose of this illustration is to show the general arrangement of the blood vessels which originate from the **abdominal aorta** and those which enter the **postcava**. The principal viscera have therefore been removed as have the parietal peritoneum and fascia from the dorsal and lateral walls of the peritoneal cavity. Enough of the "background" organs are included to show meaningful relationships. The vessels originating from the abdominal aorta fall into two groups, paired parietal and visceral vessels and single median visceral vessels.

The paired parietal arteries of the abdominal aorta are the **phrenicoabdominal**, the **deep circumflex iliac** (*ileo-lumbar*) and the **external iliac**. The paired visceral branches are the **renal** and **internal testicular** (*ovarian*) **arteries.** The single median visceral arteries are the **celiac** and the **cranial** and **caudal mesenterics.**

The veins entering the postcava are paired parietal and visceral. Paired parietal veins are the **phrenicoabdominal, deep circumflex iliac** and the **common iliac.** Paired visceral veins are the **renals.** It should be noticed that the **internal testicular** (*ovarian*) **veins** are asymmetrical, the right one entering the postcava, the left one entering the left renal vein. The postcava receives no single medial visceral veins. One might ask, what happens to the blood which is delivered to the viscera by the single median visceral arteries? It, of course, is picked up by veins which lead into the portal vein to the liver (Plates 54 and 81).

This plate shows in some detail the blood vessels of the pelvic region and should be referred to when these vessels are emphasized in other illustrations and discussions. Notice that the **common iliac artery** is an extension of the abdominal aorta and the **middle sacral artery** in turn an extension of the common iliac. The common iliac artery gives rise to the **internal iliac arteries** which supply most of the parietal and visceral structures in the pelvic region.

Also emphasized in this plate are the muscles of the dorsal abdominal wall, the psoas minor, iliopsoas, and the quadratus lumborum.

The **psoas minor m.** lies just outside the parietal peritoneum and fascia and extends from the thoracic vertebrae to the ilium. It originates from the caudal border of the bodies (*centra*) of the last two thoracic and the first three or four lumbar vertebrae. It is a broad muscle cranially but gradually tapers to a slender tendon which inserts into the iliopectineal (*arcurate*) line just craniad of the acetabulum on the ilium. Its function is to flex the back in the lumbar region.

The **iliopsoas m.** has a complicated origin since it is comparable to two muscles, the iliacus and psoas, as seen in man. The psoas portion arises by ten heads, the first five coming from the five cranial tendons of the psoas minor, the sixth from the transverse process of the fifth lumbar vertebra, the remaining four by fleshy fibers from the ventral surfaces of the centra of the last four lumbar vertebrae. The iliacus portion arises from the ventral border of the ilium from opposite the auricular impression to the iliopectineal eminence. All parts of the muscle converge to a conicle tendon and fleshy fibers which insert into the apex of the lesser trochanter of the femur. It rotates the thigh and flexes it.

The **quadratus lumborum m.** lies on the transverse processes of the lumbar vertebrae and has its origin craniad on the ventral surface of the last two thoracic vertebrae and by a few fibers from the last rib. It passes caudad attaching to each of the transverse processes of the lumbar vertebrae. It inserts into the anterior inferior spine of the ilium. It abducts the vertebral column or in some situations fixes the lumbar spine.

Finally, some of the **lumbar spinal nerves** are shown as they emerge through or between the ventral vertebral muscles and branch into the body wall or in the case of the femoral nerve trunk go into the thigh. These will be seen to greater advantage in Plate 99.

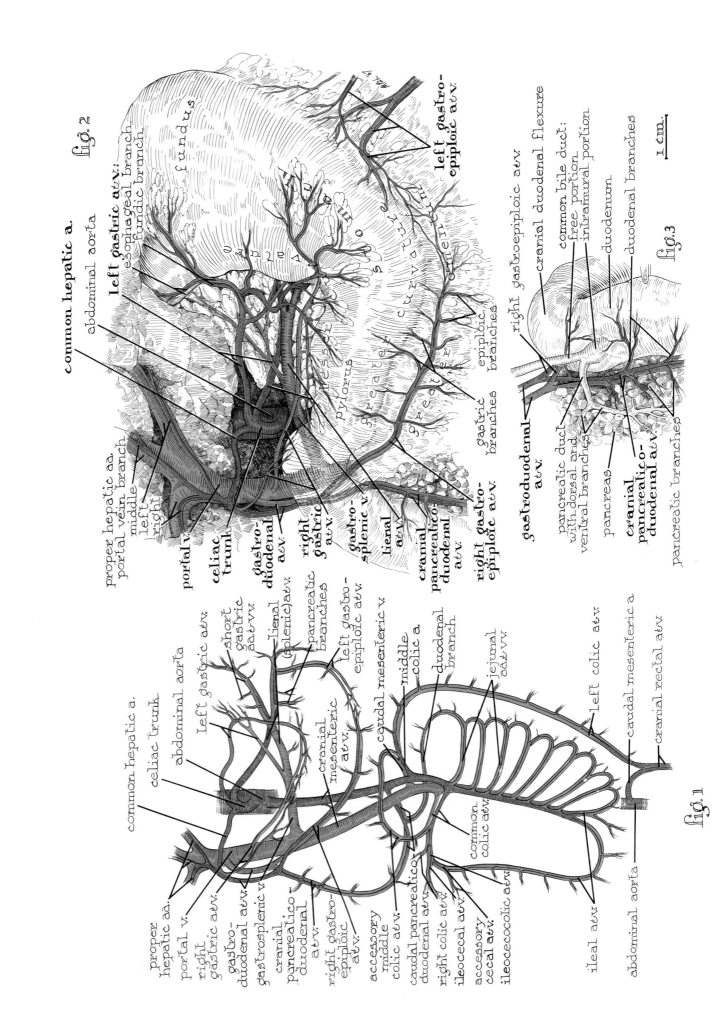

common hepatic a.   fig. 2

abdominal aorta

Left gastric a&v:
esophageal branch
fundic branch

fundus

left gastro-epiploic a&v.

right gastroepiploic a&v.

cranial duodenal flexure

common bile duct:
free portion
intramural portion

duodenum

duodenal branches

fig. 3

proper hepatic aa.
portal vein branch
middle
left
right

portal v.

celiac trunk

gastro-duodenal a&v.

right gastric a&v.

gastro-splenic v.

lienal a&v.

cranial pancreatico-duodenal a&v.

gastric branches

epiploic branches

right gastro-epiploic a&v.

gastroduodenal a&v.

pancreatic duct
with dorsal and
ventral branches

pancreas

cranial pancreatico-duodenal a&v.

pancreatic branches

1 cm.

proper hepatic aa.

portal v.

right gastric a&v.

gastro-duodenal a&v.

gastrosplenic v.

cranial pancreatico-duodenal a&v.

right gastro-epiploic a&v.

accessory middle colic a&v.

caudal pancreatico-duodenal a&v.

right colic a&v.

ileocecal a&v.

accessory cecal a&v.

ileocecocolic a&v.

common hepatic a.

celiac trunk

abdominal aorta

Left gastric a&v.

short gastric a&vv.

lienal (splenic) a&v.

pancreatic branches

left gastro-epiploic a&v.

cranial mesenteric a&v.

middle colic a&v.

duodenal branch

caudal mesenteric v.

common colic a&v.

jejunal a&vv.

left colic a&v.

caudal mesenterica a.

cranial rectal a&v.

ileal a&v.

abdominal aorta

fig. 1

This and the following two plates will show the distribution of the three single median visceral arteries of the abdominal aorta, the celiac and cranial and caudal mesenteric. Also the veins of the portal system which carry blood from these visceral organs to the portal vein and hence to the liver are illustrated.

Figure 1 is a schematic representation of these vessels and their interrelationships. It should be studied carefully before going on to the more detailed drawings. Note that the cranial pancreaticoduodenal artery, a part of the celiac circuit, becomes continuous with the caudal pancreaticoduodenal artery coming from the cranial mesenteric artery. Also the left colic branch of the caudal mesenteric artery connects through the middle colic artery to the cranial mesenteric artery. Veins which lead directly or indirectly into the portal vein accompany most of the arteries.

Figure 2, with lesser omentum cut away, shows in detail the celiac artery (trunk) and its branches and most of the organs which are thus supplied. The **celiac artery** or trunk is very short dividing almost immediately into three branches; the common hepatic artery, the most cranial; the middle one, the left gastric artery; the largest and most caudal, the lienal (splenic) artery. Small phrenic, pancreatic or gastric arteries may arise from the celiac.

The **common hepatic artery** runs cranioventrad and to the right. It lies along with the portal vein and the common bile duct in the ventral boundary of the epiploic foramen (Plate 54, fig. 3). At the porta of the liver it gives off middle, left and right proper hepatic arteries which go along with similar branches of the portal vein to the liver lobes. The left proper hepatic gives off a cystic branch (not shown) to the gallbladder. These branches furnish nutrition and oxygen to the liver. The common hepatic artery finally gives off a small right gastric and a large gastroduodenal artery.

The **right gastric** (*pyloric*) **artery** supplies both surfaces of the pylorus, the lesser omentum and lesser curvature of the stomach and anastomoses with the much larger left gastric artery on the pylorus (see next plate).

The **gastroduodenal artery** gives off a few small branches to the pylorus and pancreas and then divides giving rise to the right gastroepiploic and the cranial pancreaticoduodenal arteries. These branches are shown more clearly in a dorsal view of the area (fig. 3). Pancreatic and common bile duct relationships are also more clearly shown.

The **right gastroepiploic artery** follows a course in the ascending (*parietal*) limb of the greater omentum close to the greater curvature of the stomach. It sends numerous gastric branches to the stomach which at the greater curvature send branches to the parietal and visceral surfaces of the stomach. These anastomose with gastric branches from the right and left gastric arteries which lie on the lesser curvature of the stomach. The right gastroepiploic artery also sends epiploic branches to the greater omentum and anastomoses with the left gastroepiploic artery which is derived from the lienal (*splenic*) artery. Recall that the epiploic arteries are mostly covered with columns of fat (Plate 48).

The **cranial pancreaticoduodenal artery** enters the ventral (right) limb of the pancreas providing it with branches and also sending branches to the duodenum. Its continuation can best be studied in the next illustration, Plate 82.

The **lienal** (*splenic*) **artery** is best described in relationship to the next illustration, Plate 82.

The **left gastric artery,** the middle branch of the celiac trunk, runs craniad in the lesser omentum to the fundus of the stomach and the cardia. It sends branches to both surfaces of the stomach and anastomotic branches to the right gastric artery along the lesser curvature of the stomach. Small epiploic branches go to the lesser omentum and one or more esophageal branches reach the abdominal portion of the esophagus and pass through the esophageal hiatus to the thoracic esophagus where they anastomose with esophageal branches from the thoracic aorta.

Satellite veins accompany most of the arteries named above which are roots of the portal vein. Also lymphatic vessels, not visible here, accompany most of these branches of the celiac trunk and the radices of the portal vein.

Celiac ganglia and nerve plexuses which normally occur about the celiac and cranial mesenteric arteries have all been removed to show more clearly the relationships described above.

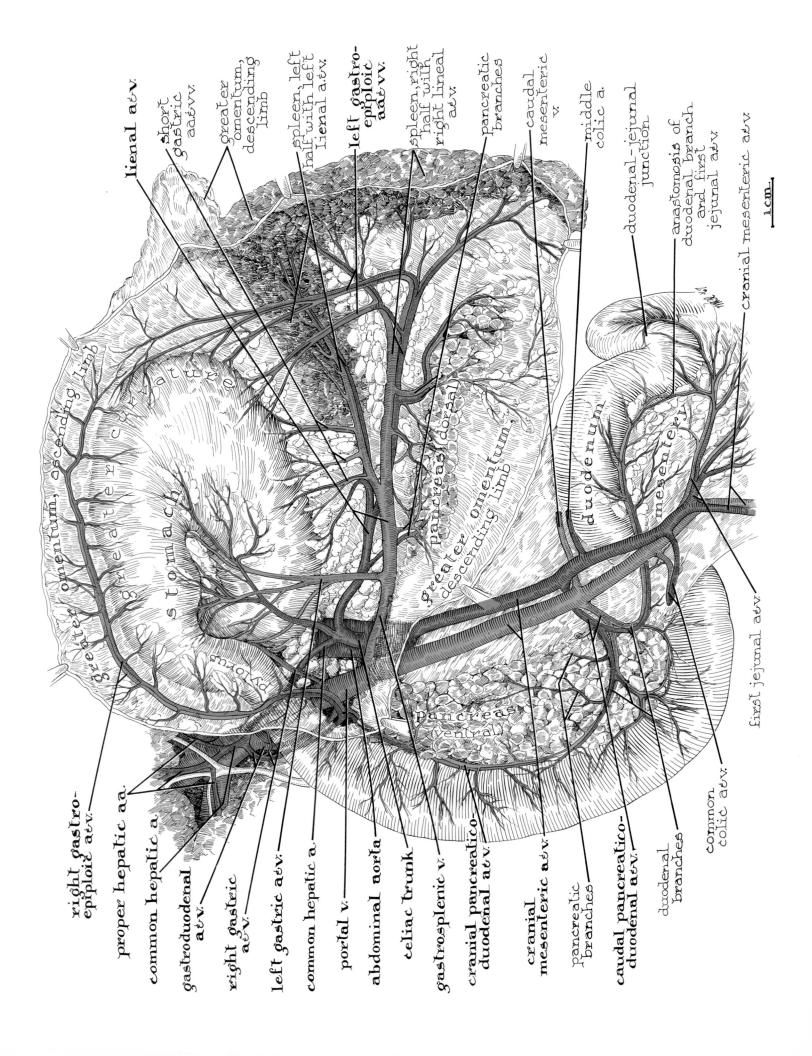

lienal a.&v.

short gastric aa.&vv.

greater omentum, descending limb

spleen, left half with left lienal a.&v.

left gastro-epiploic aa.&vv.

spleen, right half with right lienal a.&v.

pancreatic branches

caudal mesenteric v.

middle colic a.

duodenal-jejunal junction

anastomosis of duodenal branch and first jejunal a.&v.

cranial mesenteric a.&v.

1 cm.

greater omentum, ascending limb

greater curvature

stomach

pancreas, dorsal

greater omentum, descending limb

duodenum

mesentery

pylorus

pancreas (ventral)

first jejunal a.&v.

right gastro-epiploic a.&v.

proper hepatic a.

common hepatic a.

gastroduodenal a.&v.

right gastric a.&v.

left gastric a.&v.

common hepatic a.

portal v.

abdominal aorta

celiac trunk

gastrosplenic v.

cranial pancreatico-duodenal a.&v.

cranial mesenteric a.&v.

pancreatic branches

caudal pancreatico-duodenal a.&v.

duodenal branches

common colic a.&v.

The greater omentum has been cut away except for a few centimeters which remain attached at the greater curvature of the stomach. A small part of the ascending (*parietal*) limb is shown at the top of the illustration; considerably more of the descending (*visceral*) limb is shown below the stomach. The stomach has been lifted ventrad and cephalad to expose important branches of the celiac and the cranial mesenteric artery and its branches.

In the previous plate we had left the **cranial pancreaticoduodenal artery** in the ventral pancreas where it gives off branches to both the pancreas and the duodenum. In the caudal half of the ventral pancreas it has duodenal branches which anastomose with branches from the caudal pancreaticoduodenal artery and a little beyond this point the two main vessels anastomose. From one of the duodenal vessels a small branch is derived which runs to the right in the mesentery of the duodenum to the region of the duodenal flexure and the duodenojejunal junction where it anastomoses with the first jejunal artery.

Returning now to the **lienal artery** from the celiac trunk it can be traced to the left lying in a groove in the dorsal (*left*) limb of the pancreas. Its first branch is a pancreatic branch which anastomoses with a branch which entered the pancreas from the cranial pancreaticoduodenal artery. Other pancreatic branches are usually present some extending into the descending (*visceral*) limb of the greater omentum.

The lienal artery next forms two branches, the right and left lienals which pass toward the hilus of the spleen. The left lienal gives off numerous branches to the greater curvature of the stomach, the short gastric arteries, and then sends splenic branches into the left side of the hilus of the spleen. The short gastric arteries anastomose in the stomach wall with branches from the left gastric artery.

The right lienal artery sends branches to the pancreas and descending limb of the greater omentum and numerous branches to the right half of the spleen. It gives rise finally to the left gastroepiploic arteries which go to the greater curvature of the stomach in the gastrosplenic ligament giving off gastric branches which anastomose with branches of the left gastric artery and epiploic branches to the greater omentum. It anastomoses with the right gastroepiploic artery near the pylorus of the stomach.

The **cranial mesenteric artery** is a large artery arising from the ventral wall of the abdominal aorta a centimeter or less caudad of the celiac trunk. Like the celiac trunk it is surrounded by a complex net of autonomic fibers forming the cranial mesenteric plexus. It has been dissected away in the illustration to better show the blood vessels. The cranial mesenteric extends ventrocaudad in the mesentery. During development it acts as an axis around which the whole small and large intestine rotate. It supplies them and the dorsal pancreas with blood. Toward the bottom of this plate it gives off a middle colic artery, a **caudal pancreaticoduodenal artery,** a common colic artery and the first jejunal branch which anastomoses, as stated above, with a duodenal branch of the caudal pancreaticoduodenal artery.

The remainder of the cranial mesenteric artery is shown on the following illustration, Plate 83.

Note again that most of the arteries described are accompanied by satellite veins—the roots of the portal vein.

227

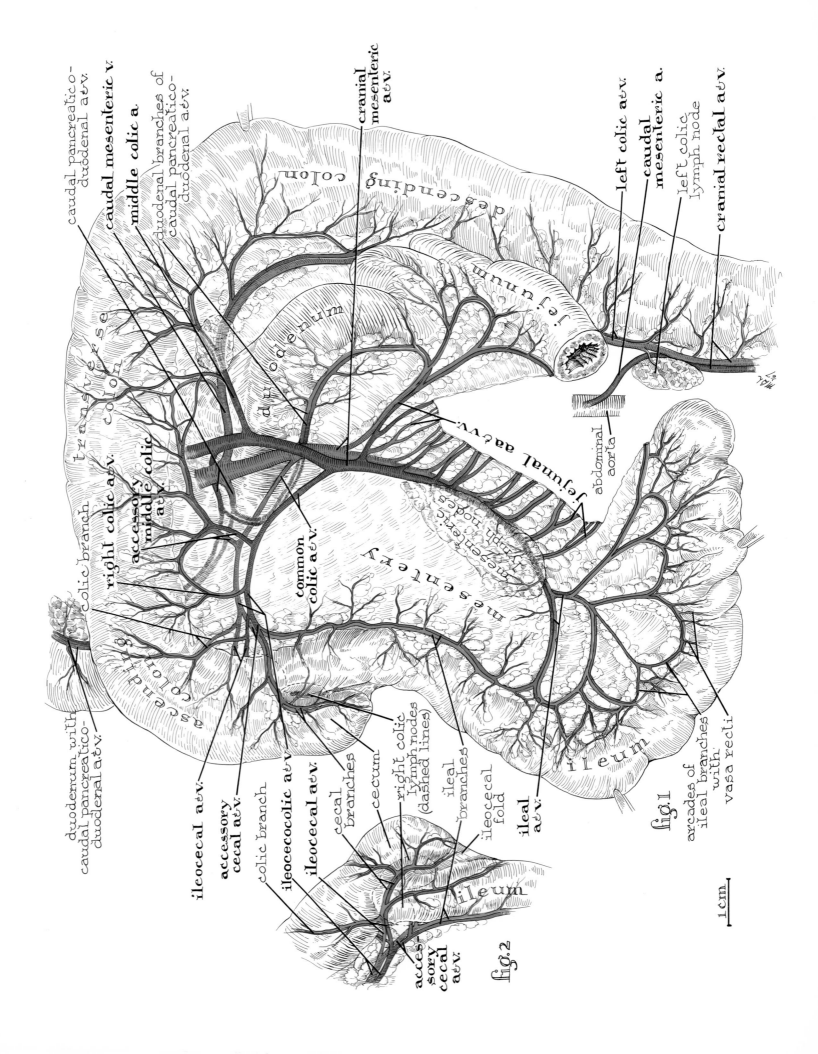

caudal pancreatico-duodenal a&v.

caudal mesenteric v.

middle colic a.

duodenal branches of caudal pancreatico-duodenal a&v.

cranial mesenteric a&v.

left colic a&v.

caudal mesenteric a.

left colic lymph node

cranial rectal a&v.

descending colon

jejunum

transverse colon

duodenum

abdominal aorta

jejunal a&vv.

accessory middle colic a&v.

right colic a&v.

colic branch

mesenteric lymph nodes

mesentery

common colic a&v.

ascending colon

ileum

ileocecal a&v.

accessory cecal a&v.

colic branch

ileocecocolic a&v.

ileocecal a&v.

cecal branches

cecum

right colic lymph nodes (dashed lines)

ileal branches

ileocecal fold

ileal a&v.

fig.1

arcades of ileal branches with vasa recti

duodenum with caudal pancreatico-duodenal a&v.

ileum

accessory cecal a&v.

fig.2

1 cm

This plate, in its upper portion, duplicates what was seen in the lower part of the last plate which followed the cranial mesenteric artery down to the point where it had given off the common colic, and first jejunal arteries. The remainder of the branches of the cranial mesenteric artery, the sixteen jejunals and the two or three ileal branches and their arcades terminate the vessel.

The colon is shown here in a more typical position than in Plates 46 and 47. The distribution of the middle colic to the transverse and descending colon, the accessory middle colic to the transverse colon and the right colic branch to the transverse and ascending colon and their anastomoses are all clearly shown.

Note that the **accessory middle colic** and the **right colic branches** are both branches of the **common colic artery.** The common colic next becomes the **ileocecocolic artery** which divides into colic branches to the ascending colon, an **ileocecal branch** to the dorsal side of the cecum as shown in figure 2, an **accessory cecal artery** to the ventral side of the cecum and finally the **ileal branches** which run down the mesenteric side of the ileum to join the ileal

arcades. The ileal arcades give off vasa recti to the ileum.

The **caudal mesenteric artery** shown in the lower right side of the illustration runs in the left mesocolon to the mesenteric border of the left or descending colon where it divides into left colic and cranial rectal arteries. The **left colic artery** runs proximally along the mesenteric border of the descending colon and anastomoses with the descending **middle colic artery** at about the midway region of the colon. The **cranial rectal** (*hemorrhoidal*) **artery** descends along the mesenteric border of the colon and rectum supplying many branches to the rectum which may anastomose with branches of the middle and caudal rectal arteries.

While the cranial mesenteric artery is accompanied by the cranial mesenteric plexus of nerves, the intestinal lymph trunk and the cranial mesenteric vein, the caudal mesenteric artery has only a caudal mesenteric nerve plexus.

Finally, notice the mesenteric, left colic and right colic lymph nodes, the only parts of the lymphatic system illustrated in this plate.

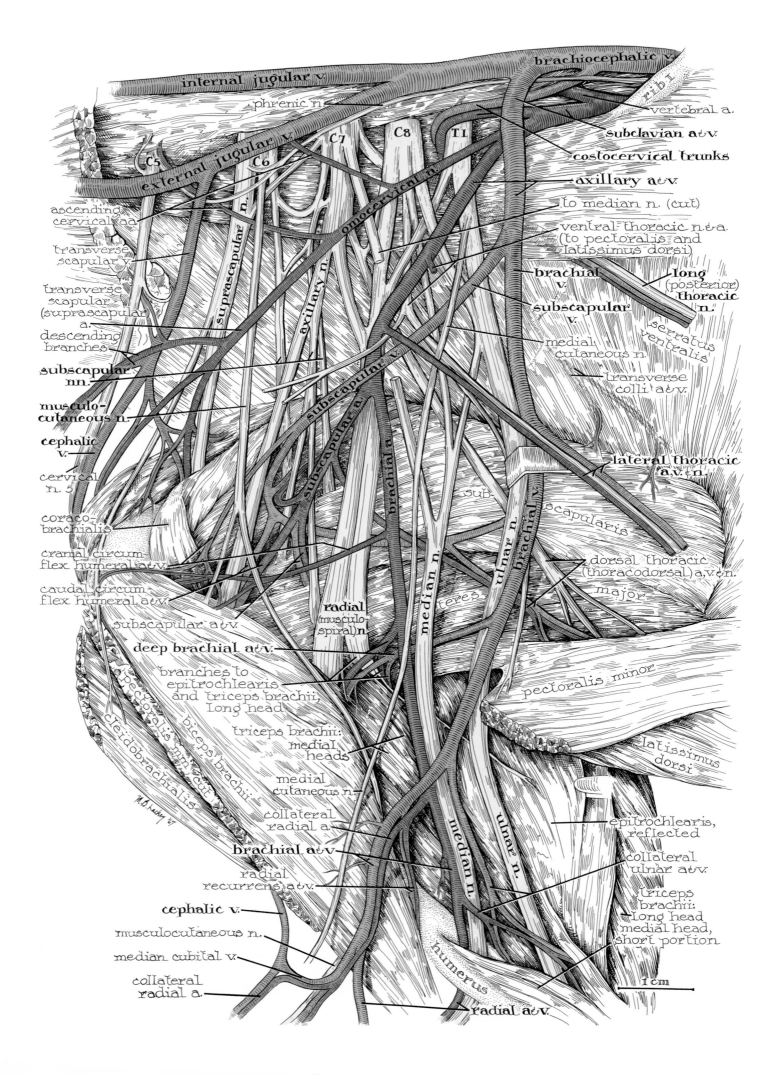

internal jugular v.

brachiocephalic v.

phrenic n.

rib I

vertebral a.

C5 C6 C7 C8 T1

subclavian a&v.

external jugular v.

costocervical trunks

axillary a&v.

omocervical n.

to median n. (cut)

ascending cervical aa.

ventral thoracic n.&a. (to pectoralis and latissimus dorsi)

transverse scapular v.

suprascapular n.

axillary n.

brachial v.

long (posterior) thoracic n.

transverse scapular (suprascapular) a. descending branches

subscapular v.

serratus ventralis

subscapular nn.

subscapular a.

medial cutaneous n.

transverse colli a&v.

musculo-cutaneous n.

brachial a.

cephalic v.

lateral thoracic a.&v.

cervical n. 5

sub-

scapularis

coraco-brachialis

ulnar n.

brachial v.

dorsal thoracic (thoracodorsal) a,v.&n.

cranial circum-flex humeral a&v.

major

caudal circum-flex humeral a&v.

radial (musculo-spiral) n.

median n.

teres

subscapular a&v.

deep brachial a&v.

pectoralis minor

branches to epitrochlearis and triceps brachii long head

triceps brachii: medial heads

pectoralis minor cut

latissimus dorsi

cleidobrachialis

biceps brachii

medial cutaneous n.

collateral radial a.

epitrochlearis, reflected

brachial a&v.

ulnar n.

collateral ulnar a&v.

radial recurrens a&v.

median n.

triceps brachii: long head medial head, short portion

cephalic v.

musculocutaneous n.

median cubital v.

collateral radial a.

humerus

1 cm

radial a&v.

The Brachial Plexus and the Arteries and Veins
of the Axilla and Arm
PLATE 84

The brachial plexus is illustrated in this plate along with the nerves which are derived from it. Note particularly the musculocutaneous, radial, median, and ulnar nerves which pass into the arm and forearm. These will be carried farther distad in the next plate. Also shown are the main arteries and veins of the area starting with the subclavian artery and vein and moving distad to the vessels of the arm and shoulder and in the next plate the continuation of these vessels into the forearm and hand. The nerves and vessels are shown against a background of muscles which they cross or supply. The relationships of structures are somewhat modified due to the necessity of cutting the pectoralis muscles and of pulling the arm and shoulder into an abnormal position to reveal the vessels and nerves in the axilla. Much fat and fascia have been removed. The illustration will be most useful to those who do the dissection and arrange the parts of their cats in a similar position. You may find variations from this illustration in your animal.

The **brachial plexus** is derived from the ventral rami of the fifth, sixth, seventh and eighth cervical and the first thoracic spinal nerves. Only a small part of the fifth nerve enters into the plexus, namely that part which, with a contribution from the sixth nerve, forms the phrenic nerve. Reference to Plate 78 will show the rest of the phrenic nerve as it passes caudad to the diaphragm. Though there is variation in the connecting branches or ansae of the brachial plexus it usually follows closely the pattern shown in this illustration. No effort will be made to describe it in detail. Note that all of the nerves contributing to the brachial plexus emerge craniad of the first rib.

The **fifth cervical nerve** after supplying the branch to form the phrenic nerve continues into the neck to supply the muscles and integument of that area and a main branch follows the cephalic vein to supply the integument over the shoulder region.

The **sixth cervical nerve** contributes to the phrenic nerve, the **suprascapular**, the cranial one of the **subscapular nerves**, and the **musculocutaneous**. By a branch to the seventh nerve it contributes to the **axillary nerve**. Note the branch of the suprascapular nerve going to the cleidobrachialis muscle (*and to the integument*).

The **seventh cervical nerve**, the largest entering into the plexus, contributes to the **axillary**, the **subscapulars** (*3 branches*), the **long** (*posterior*) **thoracic, radial** and **median nerves** and in some specimens the **musculocutaneous nerve.**

The **eighth cervical nerve** furnishes large branches to the **radial, median, ulnar nerves,** and smaller branches to the ventral and dorsal thoracic nerves.

The **first thoracic nerve** contributes to the **radial** and **median nerves** and especially to the **ulnar nerve.** It also forms a medial cutaneous nerve which passes to the medial side of the biceps brachii muscle and to the integument of the distal end of the arm and the ulnar side of the forearm to the wrist.

The arteries presented in this illustration are the **subclavian** and the proximal parts of its branches, the vertebral and **costocervical arteries** and a third branch the omocervical artery whose branches in turn are shown here in some detail. A fourth branch of the subclavian artery the internal thoracic (*mammary*) is not shown here but can be studied in Plates 78 and 79. The subclavian artery after passing through the body wall craniad of the first rib and reaching the axillary region becomes the axillary artery.

The vertebral artery after arising from the subclavian artery passes craniad and dorsad and enters the transverse foramen of the sixth cervical vertebra. Passing craniad in the transverse canal and giving off branches through the intervertebral foramina to muscles of the back of the neck it reaches the atlas, enters the atlantal foramen and goes to the midventral side of the spinal cord where at the foramen magnum it joins the other vertebral artery to form the basilar artery. Turn to Plate 76 and follow this vessel forward to the arterial circle on the ventral side of the brain.

The **omocervical** (*thyrocervical*) **artery** arises from the subclavian artery. Passing dorsad and craniolaterad over the mediocranial side of the brachial plexus it soon gives off ascending cervical branches which supply the sternohyoid, sternomastoid, the cervical portion of the scalenus and sometimes other cervical muscles. Reaching the cranial border of the scapula the artery takes the name of transverse scapular (*suprascapular*) artery. It soon divides into three branches—one is the descending branch, a second going craniad of the shoulder, pierces the supraspinatus and a third passes between the supraspinatus and subscapularis muscles which it supplies. Acromiotrapezius, cleidotrapezius, levator scapulae ventralis, splenius, occipitoscapularis and rhomboideus muscles are all supplied by these vessels.

The **costocervical trunk** arises from the subclavian artery. It gives off a superior intercostal artery (see Plate 79) which supplies the first and second intercostal spaces and then penetrates to the deep muscles of the neck. A second branch passes dorsad between the heads of the first and second ribs and enters the deep muscles of the neck to supply the complexus and others as far forward as the atlas. A third branch, though it may have a common origin with the second, is shown in part in this illustration. It is the transverse colli which sends two branches, cranial and caudal, into the levator scapulae and serratus ventralis muscles. The cranial branch is labeled, the caudal accompanies the long thoracic nerve.

The **axillary artery,** the continuation of the subclavian outside of the thorax, runs laterad, and in the living subject, caudad and parallel to the brachial plexus. In the illustration the artery has been pulled craniad. The first branch of the axillary is the ventral thoracic artery which supplies the medial ends of the pectoralis and the latissimus dorsi (*in part*) muscles. The next branch of the axillary is the **lateral** (*long*) **thoracic artery** which passes caudad to the middle portions of the pectoralis muscles and to the inner surface of the latissimus dorsi. The axillary artery now divides into two main vessels, the more cranial subscapular and the brachial. The latter is a continuation of the axillary into the arm.

The **subscapular artery,** passing laterad, gives off the dorsal thoracic (*thoracodorsal*) artery which passes over the teres major muscle to the latissimus dorsi and the epitrochlearis. It supplies branches to all of these muscles. The caudal circumflex humeral artery branches from the subscapular artery and passes between the subscapularis and teres major muscles close to the biceps brachii. It supplies the spinodeltoideus, acromiodeltoideus, and the lateral and long heads of the triceps brachii. The subscapular artery now continues between the subscapularis and teres major muscles and sends branches to the long head of the triceps, latissimus dorsi, and subscapular. The artery turns craniad at the border of the scapula crosses over the lateral surface of the infraspinatus muscle and across the spine to the supraspinatus fossa where it sends branches in both directions parallel to the spine. All of these muscles receive branches from it as do the acromiotrapezius and spinotrapezius. These branches anastomose with the transverse scapular artery. Though the cranial circumflex humeral and deep brachial arteries are sometimes reported as branches of the subscapular they are more commonly, as in this illustration, branches of the brachial artery. (See Plates 85 and 86.)

The **brachial artery,** the continuation of the axillary artery into the arm as stated above, passes along the side of the biceps. It is accompanied in the living subject by the median nerve and the brachial vein and passes with them through the bicipital arch. It also passes with the median nerve through the supracondyloid foramen. The brachial vein does not pass through this foramen. The first branch of the brachial artery is usually the cranial circumflex humeral artery which goes to the origin of the biceps muscle and sends a branch to the head of the humerus. The **deep brachial artery** is the next branch which passes with the radial nerve on to the dorsal side of the humerus where it supplies the triceps, epitrochlearis and latissimus dorsi. Note the collateral radial artery which passes over the distal surface of the biceps along with the brachial and median cubital vein. It gives off branches to the pectoantibrachialis, clavobrachialis and extensor muscles of the forearm. On the ventroradial border of the forearm it runs parallel to the **cephalic vein** to the wrist giving off integumentery branches (see Plate 86). Finally, note the collateral ulnar artery which branches from the brachial artery just proximad of the supracondylar foramen. It supplies the structures around the olecranon such as the fat and the capsule of the elbow joint. It sends branches distally (Plate 85). The radial artery and vein can be seen at the bottom of the illustration.

The veins of this region follow closely the pattern described for the arteries except that the **cephalic vein** has no arterial counterpart. It connects (*anastomoses*) with the brachial vein through the median cubital and usually has connections with the subscapular and becomes the transverse vein which enters the **external jugular vein.** Tributaries of the transverse scapular travel with the branches of the omocervical artery.

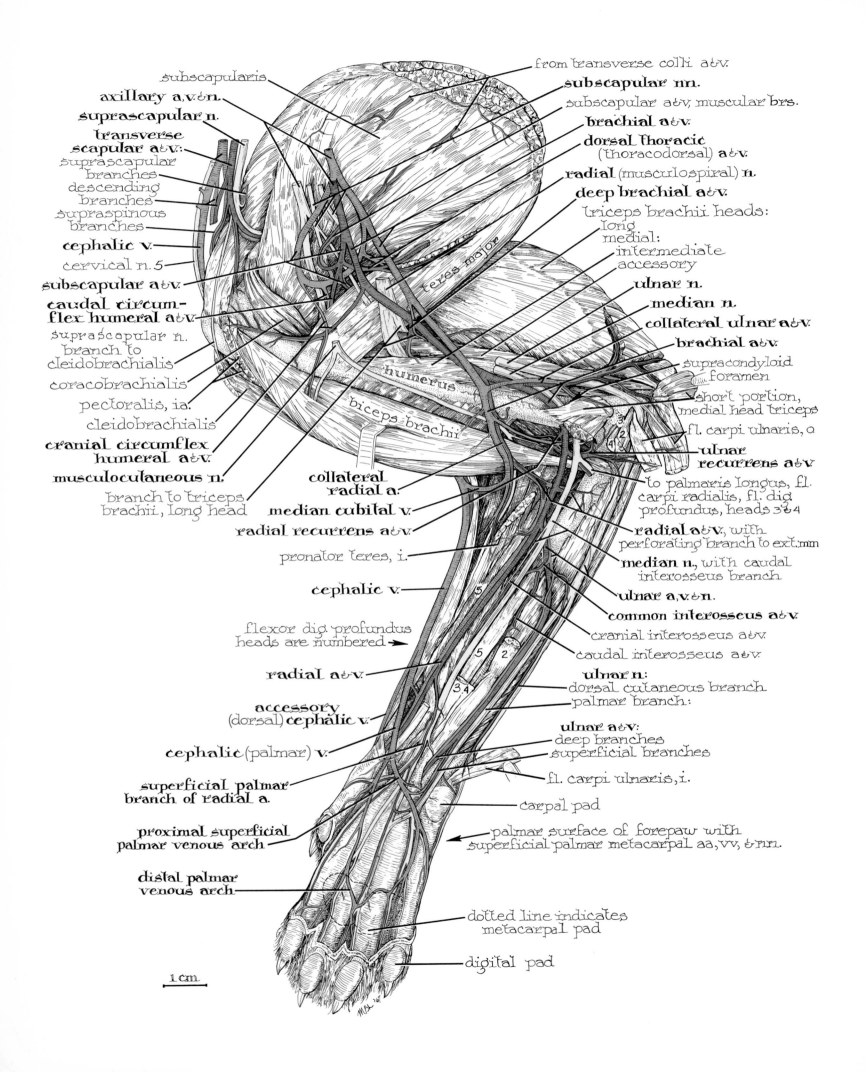

subscapularis

axillary a., v. & n.

suprascapular n.

transverse
scapular a. & v.:
suprascapular
branches
descending
branches
supraspinous
branches

cephalic v.

cervical n. 5

subscapular a. & v.

caudal circum-
flex humeral a. & v.

suprascapular n.
branch to
cleidobrachialis

coracobrachialis

pectoralis, ia.

cleidobrachialis

cranial circumflex
humeral a. & v.

musculocutaneous n.

branch to triceps
brachii, long head

median cubital v.

radial recurrens a. & v.

pronator teres, i.

cephalic v.

flexor dig. profundus
heads are numbered

radial a. & v.

accessory
(dorsal) cephalic v.

cephalic (palmar) v.

superficial palmar
branch of radial a.

proximal superficial
palmar venous arch

distal palmar
venous arch

1 cm

collateral
radial a.

from transverse colli a. & v.

subscapular nn.

subscapular a. & v., muscular brs.

brachial a. & v.

dorsal thoracic
(thoracodorsal) a. & v.

radial (musculospiral) n.

deep brachial a. & v.

triceps brachii heads:
long
medial:
intermediate
accessory

ulnar n.

median n.

collateral ulnar a. & v.

brachial a. & v.

supracondyloid
foramen

short portion,
medial head triceps

fl. carpi ulnaris, o

ulnar
recurrens a. & v.

to palmaris longus, fl.
carpi radialis, fl. dig.
profundus, heads 3 & 4

radial a. & v., with
perforating branch to ext. mm

median n., with caudal
interosseus branch

ulnar a., v. & n.

common interosseus a. & v.

cranial interosseus a. & v.

caudal interosseus a. & v.

ulnar n:
dorsal cutaneous branch
palmar branch:

ulnar a. & v:
deep branches
superficial branches

fl. carpi ulnaris, i.

carpal pad

palmar surface of forepaw with
superficial palmar metacarpal aa, vv, & nn.

dotted line indicates
metacarpal pad

digital pad

teres major

humerus

biceps brachii

Before getting into a critical study of this plate and the next one, one should compare them to the preceding plate to determine at precisely which points the nerves, arteries and veins connect. These plates together enable one to trace the important nerves and arteries from their origins proximad to their distal terminations and the veins from their distal origins to their proximal connections with the major veins of the shoulder and neck.

Follow the nerves first on this plate. A small section of the fifth cervical nerve is shown coming around the cranial aspect of the shoulder accompanying the cephalic vein. It supplies the integument over the shoulder. The **suprascapular nerve** gives off a small branch which, passing around the shoulder, penetrates the cleidobrachialis muscle and is distributed to the skin on the cranial surface of the upper arm. The main nerve, accompanied by the **transverse** (*suprascapular*) **scapular artery and vein,** penetrates the intermuscular septum between the subscapularis and supraspinatus muscles and provides motor branches to them.

There are usually three **subscapular nerves** which can be seen here penetrating the subscapularis and teres major muscles.

The **axillary nerve** is shown giving off subscapular branches and then continuing distad it passes into the intermuscular septum between the subscapularis and teres major muscles along with the caudal circumflex humeral vessels. It supplies a branch to these muscles and one to the teres minor (*not shown*) and to the cleidobrachialis (*deltoideus*) muscles and the capsule of the shoulder joint (Plate 84).

Turn next to the **musculocutaneous** (*external cutaneous*) **nerve.** It sends branches to the coracobrachialis and the biceps brachii muscles. The main nerve continues distad along the caudal side of the biceps brachii and near the elbow joint gives off a branch to the brachialis muscle. Here also, in the caudal angle of the elbow, it passes lateral to the distal end of the biceps brachii and reaches the integument. It supplies the skin on the craniomedial side of the forearm as far as the wrist.

The **radial** (*musculospiral*) **nerve** is a large structure seen here as it passes over the medial surface of the teres major muscle. It sends branches to the long and medial heads of the triceps brachii and to the epitrochlearis muscles. It then penetrates between the intermediate and accessory portions of the medial head of the triceps following the deep brachial artery and vein and emerges on the lateral side of the humerus. The remainder of this important nerve will be considered in reference to the next plate which is a lateral view of the forelimb.

Follow now the **median nerve,** the cut end of which is seen medial to the medial head of the triceps muscle. Follow it distad and note that it passes, along with the brachial artery, through the supracondyloid foramen of the humerus. It gives off branches to the flexor and pronator muscles of the elbow region (*except the flexor carpi ulnaris*) and continues into the antebrachium (*forearm*). In the proximal third of the antebrachium a caudal interosseous branch is given off by the median nerve which supplies the pronator quadratus and the flexor digitorum superficialis. The main nerve then passes distad, under the transverse ligament and in the foot gives off three main branches. These, while shown here, are easier to study in reference to Plate 87.

The **ulnar nerve,** shown lying next to the median nerve, passes over the medial epicondyle of the humerus and reaches the caudal border of the forearm. It supplies branches to the remaining flexors (*not supplied by median n.*), the flexor carpi ulnaris and the ulnar head of flexor digitorum profundus. It then, near the middle of the antebrachium, divides into dorsal cutaneous and palmar branches. The dorsal cutaneous branch follows around the ulnar side of the wrist to the dorsum of the forefoot and divides into twigs going to fourth and fifth digits. The palmar branch sends twigs to the flexor surface of the lower forearm and then divides into deep and superficial palmar branches. These distal cutaneous and muscular branches of the ulnar nerve are best studied in Plate 87 and its description.

Observe now the main arteries of the forelimb as shown in this plate. Follow the **axillary artery** distad to note its main branches. It divides almost immediately into subscapular and brachial arteries. A glance at Plate 84 will remind you of the origin of this vessel.

The **subscapular artery** passes laterad (see Plate 84) and sends muscular branches to the subscapularis muscle; a **dorsal thoracic** (*thoracodorsal*) **artery** to the teres major, latissimus dorsi and epitrochlearis muscles; and a **caudal circumflex humeral artery** to the lateral and long head of the triceps, acromiodeltoideus and spinodeltoideus. On the **lateral side** of the scapula (Plate 86) the subscapular artery gives off a circumflex scapular artery to the infraspinatus and other muscles. The subscapular artery next turns craniad over the lateral surface of the infraspinatus which it supplies and then crossing the spine of the scapula sends branches in both directions parallel to the spine which supply the supraspinatus, spinodeltoideus, and acromiodeltoideus and then anastomoses with branches of the **transverse scapular artery.**

The **brachial artery** continues distad giving off the **cranial circumflex humeral,** the **deep brachial, collateral radial** and finally the **collateral ulnar arteries** before it passes with the median nerve through the supracondyloid foramen. It now becomes the **radial artery** and gives off immediately the **radial recurrens artery,** then the **ulnar recurrens** and muscular branches. At about the middle of the forearm it gives off a large ulnar artery and then continues distad and near the wrist gives off a **superficial palmar branch** which may send a branch to the ulnar artery or continue independently onto the palmar surface of the forefoot. The main vessel (*radial artery*) now turns gradually dorsad under the tendon of the extensor brevis pollicis and onto the dorsum of the forefoot where it anastomoses with the ulnar artery to form the **palmar arch.**

The **ulnar artery** passes distad giving off a **common interosseus artery** which divides into cranial and caudal interosseus branches. It divides near the wrist into deep and superficial branches. The terminal branches of these vessels and their interconnections, while shown here, are better and more fully shown and labeled on Plate 87.

Most of the veins which drain the forelimb accompany the arteries already named. An important exception is the superficial **cephalic vein.** It is formed by a number of vessels which come from the palmar surface of the forefoot from **proximal and distal palmar venous arches** and from similar vessels on the dorsal surface which come together into an **accessory** (*dorsal*) **cephalic vein.** This vein joins the main cephalic vessel on the radial side of the forearm. The cephalic vein passes proximad along with the radial nerve and collateral radial artery to the elbow where it is joined by the **median cubital vein** which connects it with the **brachial vein.** The cephalic vein now continues proximad onto the arm following along the cleido-brachialis muscle and at the shoulder joint sends a branch to join the **caudal circumflex humeral vein** (see Plate 84). The main vessel continues over the superficial surface of the shoulder to empty into the **transverse scapular vein** which in turn joins the **external jugular vein** (Plate 84).

Note the carpal, metacarpal and digital pads on the palmar surface of the forefoot. Review also the many muscles which are shown on this plate where they are shown in relationship to vessels and nerves. Remember, however, that there is some distortion because of the dissection.

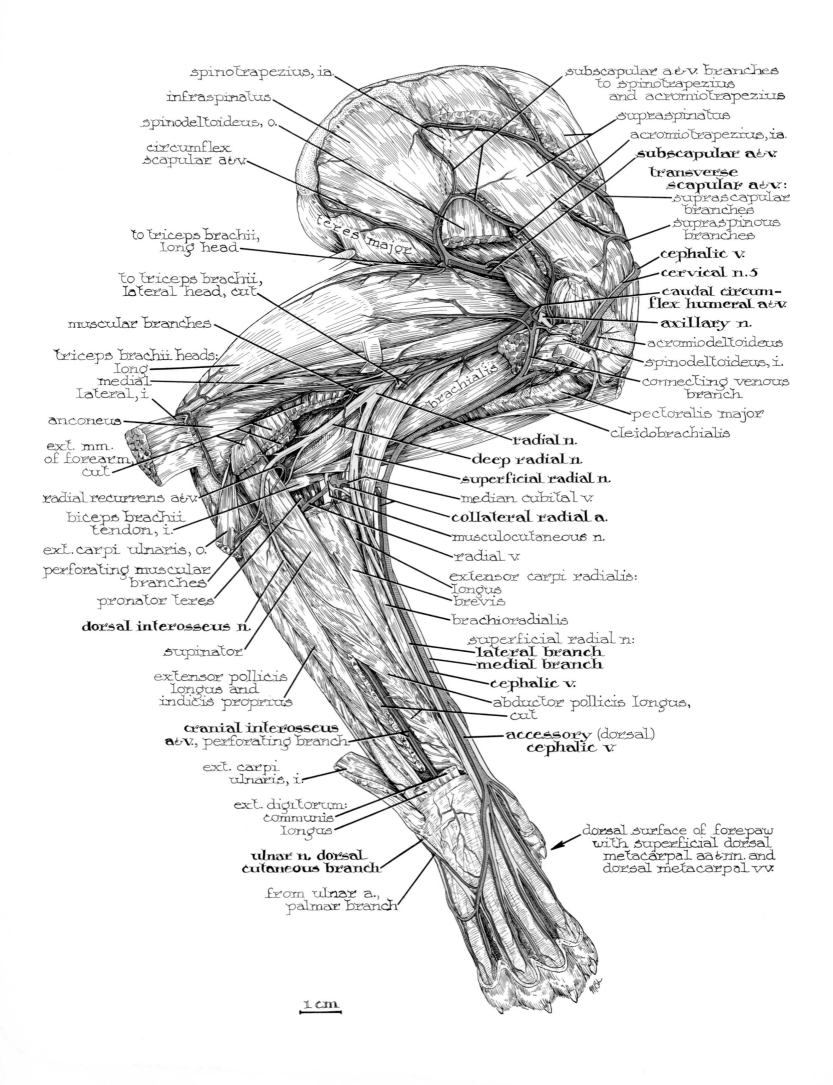

spinotrapezius, ia.

infraspinatus

spinodeltoideus, o.

circumflex
scapular a&v.

to triceps brachii,
long head

to triceps brachii,
lateral head, cut

muscular branches

triceps brachii heads:
long
medial
lateral, i

anconeus

ext. mm.
of forearm,
cut

radial recurrens a&v.

biceps brachii
tendon, i.

ext. carpi ulnaris, o.

perforating muscular
branches

pronator teres

dorsal interosseus n.

supinator

extensor pollicis
longus and
indicis proprius

**cranial interosseus
a&v., perforating branch**

ext. carpi
ulnaris, i.

ext. digitorum:
communis
longus

**ulnar n. dorsal
cutaneous branch**

from ulnar a.,
palmar branch

teres major

brachialis

subscapular a&v. branches
to spinotrapezius
and acromiotrapezius

supraspinatus

acromiotrapezius, ia.

subscapular a&v.

**transverse
scapular a&v.:**
suprascapular
branches
supraspinous
branches

cephalic v.

cervical n.5

**caudal circum-
flex humeral a&v.**

axillary n.

acromiodeltoideus

spinodeltoideus, i.

connecting venous
branch

pectoralis major

cleidobrachialis

radial n.

deep radial n.

superficial radial n.

median cubital v.

collateral radial a.

musculocutaneous n.

radial v.

extensor carpi radialis:
longus
brevis

brachioradialis

superficial radial n:
lateral branch
medial branch

cephalic v.

abductor pollicis longus,
cut

**accessory (dorsal)
cephalic v.**

dorsal surface of forepaw
with superficial dorsal
metacarpal aa&nn. and
dorsal metacarpal vv.

1 cm

This plate illustrates the main nerves and blood vessels of the forelimb as seen from a lateral view and with a minimal amount of dissection of the muscles. In comparing this plate with the previous one, one realizes that there are far fewer vessels and nerves to be seen here. Again note the muscles and their relationship to vessels and nerves. The vessels and nerves of the forefoot (*hand*) are not shown in as much detail as they are in Plate 87, nor are they fully labeled. Efforts have been made to give continuity to vessels and nerves throughout the forelimb.

The major nerve to concern us in this illustration is the **radial**, the origin of which was shown in Plate 84 and its more distal part in Plate 87. Here it is seen as it emerges from between the intermediate and accessory portions of the medial head of the triceps brachii. It divides almost immediately into two branches; a deep radial nerve which is motor in function and a superficial radial nerve which is sensory. The **deep radial nerve** travels along the brachialis muscle and between it and the extensor muscles of the forearm. It innervates these extensors and the supinator. The deep radial nerve continues into the antebrachium as the **dorsal interosseus nerve** innervating the abductor pollicis longus, extensor pollicis longus and indicis proprius.

The **superficial radial nerve** travels across the lower third of the brachialis muscle and emerges from between it and the lateral head of the triceps. It follows along the brachioradialis muscle and then divides into **lateral** and **medial** branches which become superficial and pass caudad, one on each side of the cephalic vein and with the dorsal tributary of this vein, pass onto the dorsum of the hand (*forepaw*). It supplies the integument of the distal end of the arm, the forearm, and dorsum of the hand having divided at the wrist into seven terminal branches which are more completely shown and labeled in Plate 87. Note also small sections of the axillary, musculocutaneous and ulnar nerves which show here.

The arteries shown here on the lateral surface of the scapula, the subscapular, transverse scapular and circumflex scapular have been described in reference to the preceding plate.

The **caudal circumflex humeral artery** can be seen in this illustration as it passes along the medial head of the triceps, the lateral head of the triceps having been cut away to reveal the artery. It provides branches proximally to the spinodeltoideus and acromiodeltoideus and more distally to the long and lateral heads of the triceps.

The radial recurrens artery had its origin, as seen in Plate 85, from the radial artery. It supplies structures on the medial side of the concavity of the elbow and then passes to the lateral (*radial*) side to supply the brachialis, extensor digitorum communis, and extensor carpi radialis.

Note the **collateral radial artery** which initially accompanies the median cubital vein to the cephalic vein and then passes distad along that vein and the superficial radial nerve to the wrist sending off branches to the integument. At the wrist the collateral radial artery passes onto the dorsum of the hand and crosses to the ulnar side giving off a branch (*dorsal metacarpal artery*) into each of the spaces between metacarpal bones. These branches become the dorsal digital arteries which anastomose with branches coming from the palm of the hand (Plate 87). Notice the palmar branch of the ulnar artery coming around the ulnar side of the hand.

Again, study the veins which accompany most of the arteries and the **cephalic vein** especially which can be followed for most of its length here. The **accessory cephalic vein** is formed by a number of dorsal metacarpal veins which travel with the superficial dorsal metacarpal arteries and nerves and drain blood from the digits by way of the dorsal digital veins.

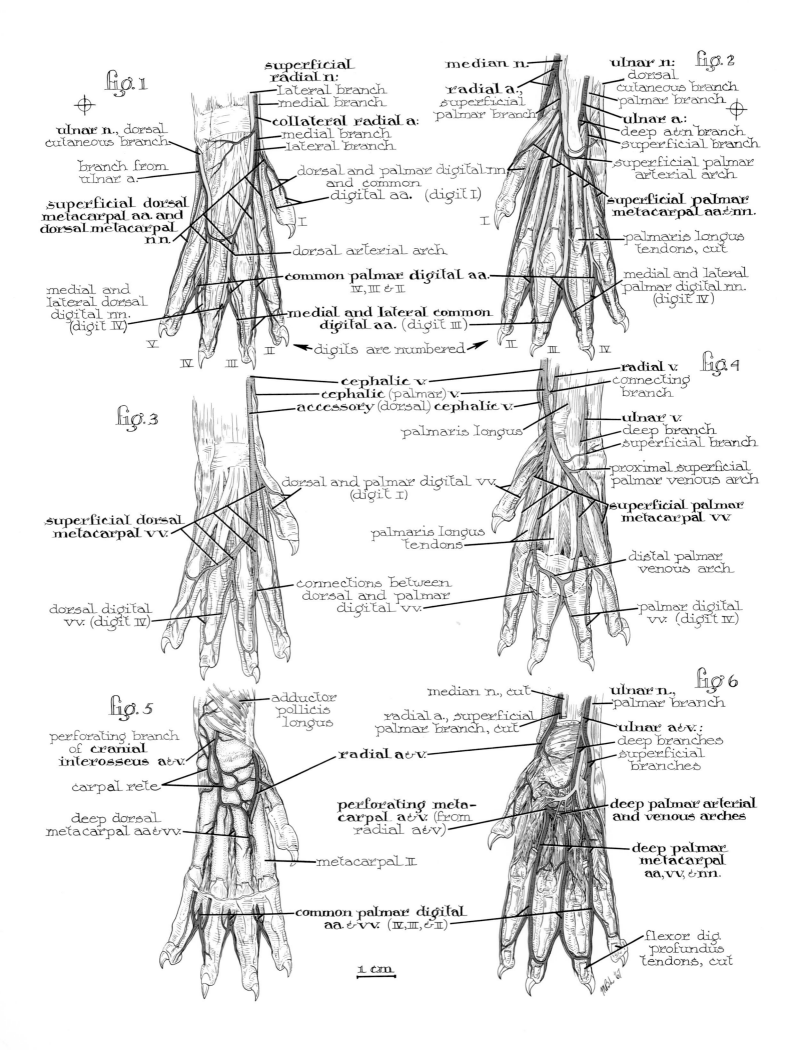

fig. 1

superficial
radial n.:
— lateral branch
— medial branch
collateral radial a.:
— medial branch
— lateral branch

ulnar n., dorsal
cutaneous branch

branch from
ulnar a.

**superficial dorsal
metacarpal aa. and
dorsal metacarpal
nn.**

dorsal and palmar digital nn.
and common
digital aa. (digit I)

dorsal arterial arch

common palmar digital aa.
IV, III & II

medial and
lateral dorsal
digital nn.
(digit IV)

**medial and lateral common
digital aa. (digit III)**

I

V   IV   III   II

← digits are numbered →

fig. 2

median n.   ulnar n.:
— dorsal
cutaneous branch
— palmar branch
radial a.,
superficial
palmar branch
ulnar a.:
— deep a. & n. branch
— superficial branch

superficial palmar
arterial arch

**superficial palmar
metacarpal aa. & nn.**

palmaris longus
tendons, cut

medial and lateral
palmar digital nn.
(digit IV)

I

II   III   IV

fig. 3

cephalic v.
cephalic (palmar) v.
accessory (dorsal) cephalic v.

palmaris longus

dorsal and palmar digital vv.
(digit I)

**superficial dorsal
metacarpal vv.**

dorsal digital
vv. (digit IV)

connections between
dorsal and palmar
digital vv.

fig. 4

radial v.
connecting
branch

ulnar v.
deep branch
superficial branch

proximal superficial
palmar venous arch

**superficial palmar
metacarpal vv.**

palmaris longus
tendons

distal palmar
venous arch

palmar digital
vv. (digit IV)

fig. 5

adductor
pollicis
longus

perforating branch
of **cranial
interosseus** a. & v.

carpal rete

deep dorsal
metacarpal aa. & vv.

radial a. & v.

perforating meta-
carpal a. & v. (from
radial a. & v.)

metacarpal II

common palmar digital
aa. & vv. (IV, III, & II)

1 cm

fig. 6

median n., cut   ulnar n.,
palmar branch
radial a., superficial
palmar branch, cut
ulnar a. & v.:
deep branches
superficial
branches

**deep palmar arterial
and venous arches**

**deep palmar
metacarpal
aa., vv., & nn.**

flexor dig.
profundus
tendons, cut

The six figures on this plate illustrate the nerves and blood vessels on the forefoot (*hand*) or paw. Figures 1 and 2 show the more superficial nerves and arteries on dorsal and palmar surfaces respectively. Figures 3 and 4 do the same thing for the superficial veins. Figures 5 and 6 show the deep arteries, veins and nerves. Reference should be made to the two preceding plates on the medial and lateral views of the nerves and vessels of the arms and to Plate 92 which shows the corresponding vessels on the hindfoot. There are marked similarities between the nerves and vessels on the hindfoot and those on the forefoot. By means of these nerves and vessels the complicated musculature, foot pads and skin areas are served.

Figure 1, a dorsal view of the forepaw (*hand*) shows the more superficial nerves and arteries. The dorsal cutaneous branch of the **ulnar nerve** is seen coming from the palmar side of the forearm onto the dorsum of the hand on the lateral side. Here the nerve divides into two **dorsal metacarpal nerves,** one to the ulnar side of the fifth digit and one which divides to send digital nerves to the contiguous sides of digits IV and V. Lateral and medial branches of the **superficial radial nerve** come onto the dorsum of the hand on the medial side. The medial branch divides into two **dorsal metacarpal nerves,** one of which goes to the first digit (*thumb*) and divides into two dorsal digital nerves, one to the integument on each side of the thumb, the other passing down the medial side of metacarpal II to form its dorsal digital nerve. The lateral branch of the superficial radial nerve forms two **dorsal metacarpal nerves** which divide to form branches to the contiguous sides of digits II and III, and III and IV. The superficial radial nerve thus ends in seven dorsal digital nerves.

The superficial arteries on the dorsum of the hand in the cat used for this drawing followed the same pattern as the nerves just described except for a connecting branch forming a dorsal arterial arch. A branch from the ulnar artery comes onto the dorsum of the hand along with the dorsal cutaneous branch of the ulnar nerve and follows its distribution. The **collateral radial artery** with medial and lateral branches follows the branches of the superficial radial nerve. Reighard and Jennings indicate that all of these superficial dorsal arteries come from the collateral radial artery, mentioning no branch from the ulnar artery. The terminal branches of these arteries are the **medial and lateral common digital arteries.**

Figure 2 shows the superficial nerves and blood vessels on the palmar side of the hand. The median nerve and the superficial palmar branch of the ulnar nerve provide the innervation in this illustration. The **median nerve** is seen coming onto the palmar surface of the wrist on the

medial side. It divides into three main branches, the **superficial palmar metacarpal nerves.** The first of these divides into two branches, one which again divides going to each side of the integument of the thumb, the other to the medial side of the second digit. The second superficial palmar metacarpal divides to send branches to the skin of the contiguous sides of digits II and III; the third divides sending branches to the skin of the contiguous sides of digits III and IV. These terminal nerve branches are called medial and lateral palmar digital nerves. The metacarpal nerves also give off branches to the trilobed palmar pad, and probably to the three radial lumbricales muscles. The superficial palmar branch of the ulnar nerve coming onto the lateral side of the palm forms the last two **superficial palmar metacarpal arteries,** one going to the lateral side of digit V, the other supplying the contiguous sides of the IV and V digits with digital nerves to the integument. Note also another more medial branch of the ulnar nerve, the deep palmar branch which passes into the palm and can be seen in figure 6 as it breaks into a number of branches supplying the short, deep muscles in the palm.

Figures 3 and 4 show the dorsal and palmar superficial veins of the hand. Note that on the dorsal side they drain into the **accessory** (*dorsal*) **cephalic vein** and on the palmar side into the **cephalic** (*palmar*) **vein.** Above the wrist on the radial side the accessory cephalic joins the cephalic which we traced proximad in the description of Plate 85.

In figure 3 note the distribution of the dorsal digital veins and the connections between them and the palmar digital veins forming the digital venous arcs. The digital veins lead into the **superficial dorsal metacarpal veins** which come to focus in the accessory cephalic vein as stated above.

In figure 4 note the palmar digital veins, their connections to the dorsal digital veins and the distal palmar venous arch from which they are derived. The arch also gives off twigs to the metacarpal footpad and its blood is then passed into the **superficial palmar metacarpal veins** which in turn come to focus in the **cephalic** (*palmar*) **vein.** Note also that proximad a branch from the metacarpal vein joins the superficial branch of the ulnar vein to form a proximal superficial palmar venous arch. The deep branch of the ulnar vein disappears into the palm but can be seen distally in figure 6 and will be described there. A connection between the radial and cephalic vein should also be noticed in this figure.

Figures 5 and 6 emphasize the deep arteries, veins and nerves of the hand. Figure 5 is the dorsal side and one should notice the radial artery and vein which appear on the medial side but then disappear as they perforate to the palmar side between the second and third metacarpals (see fig. 6). Toward the lateral side the perforating branches of the **cranial interosseous artery** and vein are seen entering and leaving the hand. They form a network of vessels on the wrist constituting the carpal rete. From the carpal rete small deep dorsal metacarpal arteries and veins pass to and from the interosseous muscles.

Figure 6 shows the deep palmar nerves and vessels. The palmar branch of the ulnar nerve is seen on the lateral side of the wrist. Its superficial branch enters the palm of the hand and passes to the integument of the ulnar side of the fifth digit. The deep palmar branch of the **ulnar nerve** swings abruptly toward the radial side of the wrist and gives off a number of deep palmar metacarpal nerves which supply the short muscles in the palm. The cut end of the median nerve is shown here in the lower arm. Its distribution is shown in figure 2.

Turn now to the ulnar artery as shown in figure 6. It and its accompanying vein form deep and superficial branches, the latter shown in figure 2. The deep branch of the ulnar artery gives off a branch to the wrist and at the proximal end of the metacarpals forms, with the **perforating metacarpal artery** (*radial artery*), the **deep palmar arterial arch.** From this arch three **deep palmar metacarpal arteries** (*palmar interosseas*) appear which at the metacarpophalangeal region become the **common palmar digital arteries.** Another branch from the radial end of the deep palmar arch goes to the ulnar side of the thumb and the radial side of the second metacarpal and digit. This branch is small and is sometimes called the princeps pollicis et indicis artery. There are other small branches from the arch, some distad, some proximad, which with the main vessel supply the deep muscles of the hand.

All of the arteries described in the above paragraph are accompanied by veins of the same names.

Nerves and Blood Vessels of the Groin and Thigh—
Ventral View—Right Side
PLATE 88

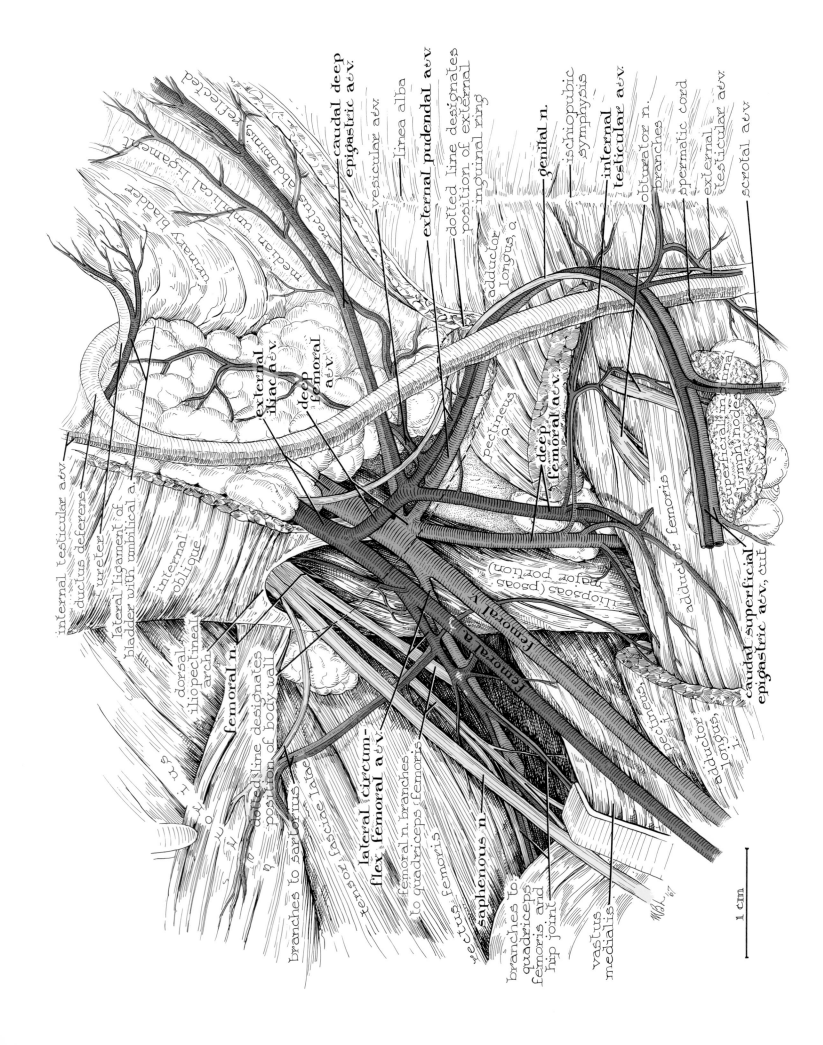

internal testicular a&v.

ductus deferens

ureter

lateral ligament of
bladder with umbilical a.

internal
oblique

dorsal

iliopectineal
arch

sinus

femoral n.

dotted line designates
position of body wall

branches to sartorius

tensor fasciae latae

lateral circum-
flex femoral a&v.

femoral n. branches
to quadriceps femoris

rectus femoris

saphenous n.

branches to
quadriceps
femoris and
hip joint

vastus
medialis

adductor
longus, n.

pectineus, n.

femoral a.

femoral n.

iliopsoas (psoas
major portion)

deep
femoral a&v.

pectineus,
a.

caudal superficial
epigastric a&v., cut

adductor femoris

median umbilical ligament

rectus abdominis reflected

urinary bladder

caudal deep
epigastric a&v.

vesicular a&v.

linea alba

external pudendal a&v.

dotted line designates
position of external
inguinal ring

adductor
longus, a.

genital n.

ischiopubic
symphysis

internal
testicular a&v.

obturator n.
branches

spermatic cord

external
testicular a&v.

scrotal a&v.

superficial inguinal
lymph nodes

external
iliac a&v.

deep
femoral
a&v.

Mbl '67

1 cm

This illustration shows some of the ventral peripheral nerves which derive from the lumbosacral plexus and the blood vessels which are branches and tributaries of the **external iliac and femoral arteries and veins.** Reference should be made to Plates 80 and 99 which show the proximal connections of these nerves and vessels and to Plate 89 for the peripheral distribution of some of them. The dotted line running across the illustration designates the position of the body wall and the dotted circle the position of the external inguinal ring. Note the **external pudendal artery and vein,** the **genital nerve** (see also Plate 99), the **internal testicular artery and vein** and the ductus deferens all of which pass through the external inguinal ring. Recall also that the ductus deferens and the internal testicular vessels and associated tunics make up the spermatic cord. The ductus deferens can be seen looping around the ureter in the upper part of the illustration. At the right side of the plate the linea alba and the position of the ischiopubic symphysis are shown to aid in your orientation. The right thigh has been abducted to show the more superficial muscles and the nerves and vessels which supply them or relate to them. The important ones are labeled.

The large **femoral nerve** is shown emerging through the dorsal iliopectineal arch and lying on the ventral surface of the iliopsoas muscle. It was formed, as shown in Plate 99, by contributions from lumbar nerves 5 and 6. It divides into three or four branches, in this case four. One (*sometimes two*) goes to the sartorius muscle; two go between the rectus femoris and vastus medialis muscles. The latter breaks into a number of branches supplying the above muscles and the vastus intermedius. The remaining branch is the **saphenous or long saphenous nerve** which passes distad along with the femoral artery and vein. It gives off a few twigs to the integument and can be followed in Plate 89 into the leg along with the long saphenous artery.

Note the branches of the obturator nerve as they come from the dorsal side of the adductor longus and across and into the adductor femoris muscle on its ventral surface. Reference to Plate 99 will show the origin of this nerve from lumbar 6 and 7.

The **genital nerve** is shown here as it leaves the body cavity through the external inguinal ring and travels along with the external pudendal artery and vein and the caudal superficial epigastric artery and vein. Its origin is from lumbar 4 seen in Plate 99. It sends branches to lymph nodes, spermatic cord and related structures in the inguinal region and ends in branches to the skin on the medial surface of the thigh.

Follow the **external iliac artery** distad and note that just inside of the body wall it gives off the deep femoral artery and then continues through the body wall to become the femoral artery. The **deep femoral artery** passes caudad and after about 1 centimeter gives off a number of branches the pattern of which varies in different specimens. Here a branch, the **caudal deep epigastric** passes directly to the dorsal surface of the rectus abdominis muscle. It passes craniad sending branches to the abdominal muscles and anastomosing with the cranial deep epigastric artery, a branch of the internal thoracic (Plate 79).

Another branch of the deep femoral, the **external pudendal artery,** leaves the abdominal cavity through the external inguinal ring. It travels with the genital nerve and gives off external testicular and scrotal branches and gives rise to the **caudal superficial epigastric artery.** Cranially the caudal and cranial superficial epigastrics anastomose and both serve the mammary glands. In this instance the vesicular artery to the urinary bladder is a branch of the external pudendal artery. The remaining branch of the deep femoral artery, sometimes called the medial circumflex femoral artery gives off branches to the adductor muscles and to the semimembranosus.

The femoral artery continues into the thigh on the medial surface and lies in a triangular depression called the femoral triangle or the iliopectineal fossa (*Scarpa's triangle*). The fossa contains also the femoral vein and the saphenous nerve. The femoral artery as it continues distad goes between the semimembranosus and vastus medialis and finally into the popliteal space where it becomes the popliteal artery (Plate 91). One large branch of the femoral artery is seen on this plate, the lateral circumflex femoral. Other branches are shown in Plate 89.

The **lateral circumflex femoral artery** (*cranial femoral*) arises from the femoral artery about 1 centimeter from its emergence from the abdominal cavity. Passing laterad and craniad it sends a large branch to the inner surface of the sartorius and to the tensor fasciae latae muscles of the thigh. The main branch of the lateral circumflex femoral passes between the rectus femoris and vastus medialis muscles to which it gives off branches. It also sends an ascending branch to structures around the hip joint.

The veins of this area can be seen following the arteries as described above. They have the same names as the accompanying arteries and drain blood from the limb into the **external iliac vein.**

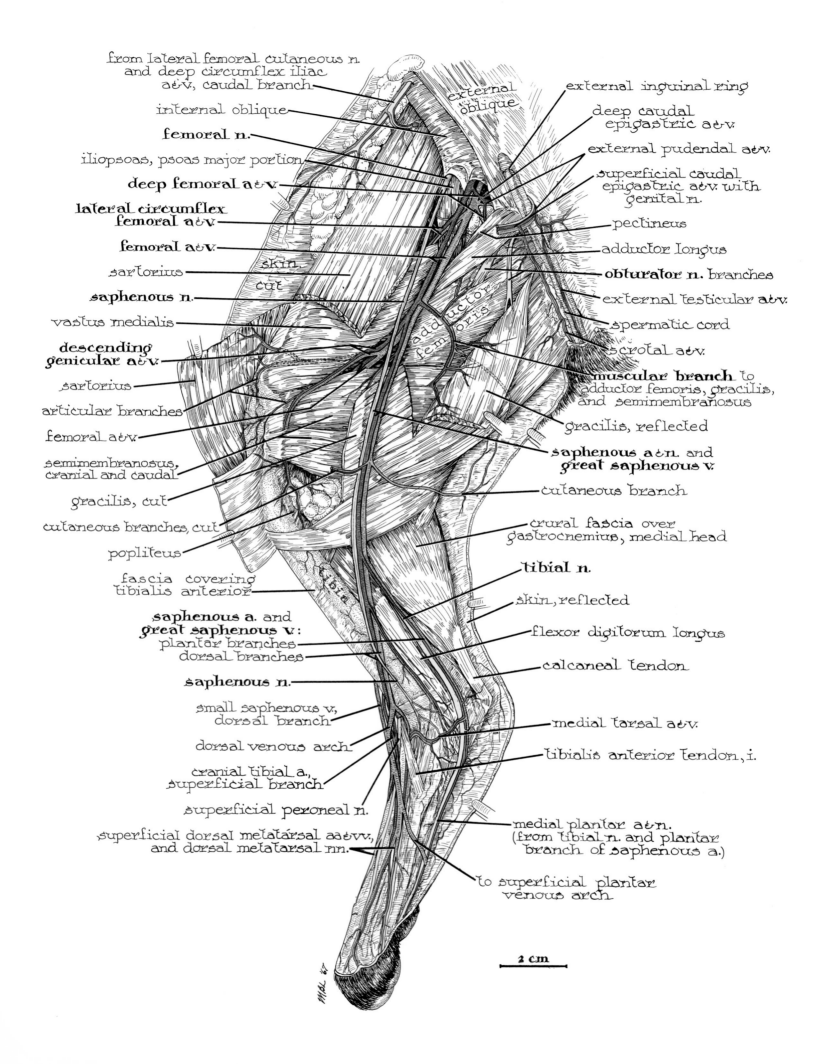

from lateral femoral cutaneous n.
and deep circumflex iliac
a&v. caudal branch

internal oblique

femoral n.

iliopsoas, psoas major portion

deep femoral a&v.

lateral circumflex
femoral a&v.

femoral a&v.

sartorius

saphenous n.

vastus medialis

descending
genicular a&v.

sartorius

articular branches

femoral a&v.

semimembranosus,
cranial and caudal

gracilis, cut

cutaneous branches, cut

popliteus

fascia covering
tibialis anterior

saphenous a. and
great saphenous v.:
plantar branches
dorsal branches

saphenous n.

small saphenous v.,
dorsal branch

dorsal venous arch

cranial tibial a.,
superficial branch

superficial peroneal n.

superficial dorsal metatarsal aa&vv.,
and dorsal metatarsal nn.

external
oblique

skin,
cut

adductor
femoris

tibia

external inguinal ring

deep caudal
epigastric a&v.

external pudendal a&v.

superficial caudal
epigastric a&v. with
genital n.

pectineus

adductor longus

obturator n. branches

external testicular a&v.

spermatic cord

scrotal a&v.

muscular branch to
adductor femoris, gracilis,
and semimembranosus

gracilis, reflected

saphenous a&n. and
great saphenous v.

cutaneous branch

crural fascia over
gastrocnemius, medial head

tibial n.

skin, reflected

flexor digitorum longus

calcaneal tendon

medial tarsal a&v.

tibialis anterior tendon, i.

medial plantar a&n.
(from tibial n. and plantar
branch of saphenous a.)

to superficial plantar
venous arch

2 cm

This illustration should be related directly to Plate 88 which shows in detail the vessels and nerves of the groin, the superficial ones of which are now, in this plate, carried through the limb to the hindfoot. The superficial muscles such as the sartorius and gracilis are cut and reflected.

First examine the three "openings" in the abdominal wall through which various structures from the trunk region gain access to the hindlimb. The most lateral of these is sometimes called the muscular lacuna and is set off by the iliopectineal arch from the middle opening, the vascular lacuna. The third opening, the external inguinal ring, is set off from the vascula lacuna by the aponeurosis of the external oblique muscle. Through the muscular lacuna pass the femoral nerve and the iliopsoas muscle; through the vascular lacuna the femoral artery and vein and lymphatics; and through the external inguinal ring the spermatic cord. Through the lacunae the femoral artery and vein and the saphenous nerve come into a triangular depressed area on the proximal part of the thigh, the **femoral triangle** (*iliopectineal fossa; Scarpa's triangle*). This fossa is bounded by the sartorius muscle cranially and by the pectineus muscle caudally. The iliopsoas muscle forms its floor proximally, the vastus medialis distally. In the living subject the triangle is covered by the medial femoral fascia and thin skin. Since the femoral vein and artery are so superficial here, it is a good place to take the animal's pulse or to draw blood.

The main nerve to be considered here is the **saphenous** nerve which is a branch of the femoral. It passes with the femoral artery and vein through the femoral triangle giving off a few twigs to the skin. At the point where the saphenous artery branches from the femoral artery, the saphenous nerve follows that artery along the medial side of the thigh and leg. At the knee the saphenous nerve gives off cutaneous branches and just distal of the middle of the leg it divides into two main branches (only one branch shown here). These form numerous branches in the region of the concavity of the ankle joint and some extend onto the dorsum of the foot and may even extend to the digits. The more proximal branches of the femoral nerve were shown on the previous plate and described.

Note also the **obturator nerve** on the adductor femoris muscle and refer to the last plate for more information. The **tibial nerve** can be seen as it emerges from the musculature of the leg and passes distad on the surface of the flexor hallucis longus. At the ankle it passes between the heel and the medial malleolus to the plantar surface of the foot. It sends small branches to the plantar surface of the heel and divides into medial and lateral plantar branches. The medial plantar branch is shown here. The distribution of both the lateral and medial plantar nerves can be seen on Plate 92.

The arteries and veins of the hindlimb should now be studied. The **deep femoral artery** is derived from the external iliac artery just inside of the body wall. It and its branches were studied in relationship to the previous plate. The **femoral artery,** a continuation of the external iliac artery outside of the body wall, passes distad into the femoral triangle. It gives off the **lateral circumflex femoral artery,** a **muscular branch** to the adductor femoris, gracilis, and semimembranosus, a saphenous artery and a descending genicular artery to the medial side of the knee. The femoral artery now passes deep between the vastus medialis and semimembranosus muscles to the politeal space where it becomes the popliteal artery. See the next two plates for a study of this artery.

The **saphenous artery** passes distad across the gracilis muscle along with the saphenous nerve and great saphenous vein. It gives off twigs to the gracilis and semimembranosus muscles and at the knee sends cutaneous branches craniad and distad to supply the skin in those areas. The cranial cutaneous vessel is sometimes called the medial genicular artery. Farther distad on the medial side of the leg the saphenous artery divides into a small dorsal branch and a large plantar branch.

The **dorsal branch** of the saphenous artery is accompanied by the saphenous nerve and great saphenous vein and passes around the medial border of the anterior (*cranial*) tibial muscle, and passes distad in the fascia of this muscle to the flexor (*dorsal*) surface of the tarsus. It gives off branches to the medial side of the ankle. The dorsal branch of the saphenous artery anastomoses in the tarsal region with the superficial branch of the cranial tibial artery forming a dorsal arterial arch. This can be seen in Plate 92 where one can also trace the terminal branches of this vessel.

The **plantar branch of the saphenous artery** passes distad craniad of the medial head of the gastrocnemius and on the surface of the flexor hallucis longus. It is accompanied by the tibial nerve. It passes around the medial side of the ankle joint giving off superficial and deep branches to structures around the ankle, one of these is the medial tarsal artery. On the plantar surface of the foot it gives rise to a lateral plantar artery which, with the terminal branches of the medial plantar artery, are best seen in Plate 92, figure 2.

In general the veins follow the arteries which have been described. It should be noticed, however, that the dorsal branch of the small saphenous vein does join the dorsal branch of the **great saphenous vein** to form the dorsal venous arch of the foot. The vessels of the hindfoot shown here in general pattern are shown in greater detail on Plate 92. An effort should be made to relate these vessels to the more complete drawings.

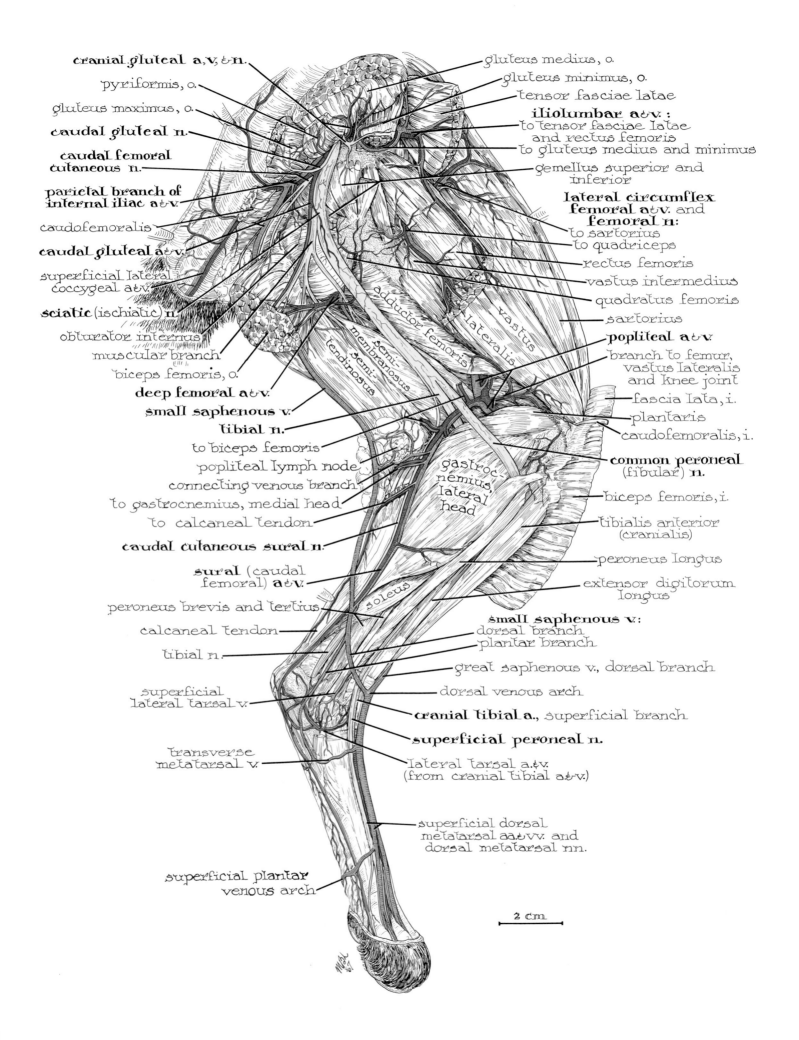

cranial gluteal a., v., & n.

pyriformis, o.

gluteus maximus, o.

caudal gluteal n.

caudal femoral
cutaneous n.

parietal branch of
internal iliac a. & v.

caudofemoralis

caudal gluteal a. & v.

superficial lateral
coccygeal a. & v.

sciatic (ischiatic) n.

obturator internus

muscular branch

biceps femoris, o.

deep femoral a. & v.

small saphenous v.

tibial n.

to biceps femoris

popliteal lymph node

connecting venous branch

to gastrocnemius, medial head

to calcaneal tendon

caudal cutaneous sural n.

sural (caudal
femoral) a. & v.

peroneus brevis and tertius

calcaneal tendon

tibial n.

superficial
lateral tarsal v.

transverse
metatarsal v.

superficial plantar
venous arch

gluteus medius, o.

gluteus minimus, o.

tensor fasciae latae

iliolumbar a. & v. :
to tensor fasciae latae
and rectus femoris

to gluteus medius and minimus

gemellus superior and
inferior

lateral circumflex
femoral a. & v. and
femoral n:
to sartorius
to quadriceps

rectus femoris

vastus intermedius

quadratus femoris

sartorius

popliteal a. & v.

branch to femur,
vastus lateralis
and knee joint

fascia lata, i.

plantaris

caudofemoralis, i.

common peroneal
(fibular) n.

biceps femoris, i.

tibialis anterior
(cranialis)

peroneus longus

extensor digitorum
longus

small saphenous v:
dorsal branch
plantar branch

great saphenous v., dorsal branch

dorsal venous arch

cranial tibial a., superficial branch

superficial peroneal n.

lateral tarsal a. & v.
(from cranial tibial a. & v.)

superficial dorsal
metatarsal aa. & vv. and
dorsal metatarsal nn.

adductor femoris

vastus lateralis

semimembranosus

semitendinosus

gastrocnemius, lateral head

soleus

2 cm

Before attempting to follow the nerves and vessels in this view of the hindlimb it might be well to review the muscles, especially those of the gluteal region, which are small and complex in their interrelationships. To know the muscles will help to understand the arrangement and the nomenclature of the nerves and vessels. It might also be helpful to refer to Plate 99 to review the origin of the sciatic and other nerves which, with their branches, provide innervation for many of the muscles on this aspect of the limb.

The **sciatic** (*ischiatic*) **nerve** is the largest nerve in the body and arises from the lumbosacral cord, the first sacral and by a branch from the second sacral spinal nerves. By passing dorsocaudad it leaves the pelvic cavity by crossing through the great sciatic notch. It makes its appearance in this illustration as it emerges from between the pyriformis and gemellus superior muscles. Some consider only that part of the nerve outside of the pelvic cavity as the true sciatic nerve. It passes caudad over the tendon of the obturator internus giving off a large **muscular branch** to the biceps femoris, semimembranosus and semitendinosus. The caudal gluteal artery and vein pass along caudad of this part of the nerve and its muscular branch. Continuing caudad and medial to the biceps femoris and lateral to the semimembranosus muscles, the sciatic nerve gives off near its terminal branches a slender **caudal cutaneous sural nerve** which passes lateral to the semimembranosus muscle and onto the plantar surface and then the lateral surface of the lateral head of the gastrocnemius. Distally the caudal cutaneous sural divides into two branches, one passes over the tendon of Achilles and ramifies in the region of the calcaneus, the other goes to the lateral surface of the foot and supplies the skin over the tarsus and metatarsus. The sciatic nerve now divides into terminal, tibial and common peroneal (*fibular*) branches. The tibial nerve disappears behind the lateral head of the gastrocnemius muscle and is followed caudad in Plate 91, figure 1 and Plate 92. The **common peroneal** (*fibular*) **nerve** passes along the medial surface of the biceps femoris to the lateral surface of the lateral head of the gastrocnemius and then passes under the slip of the gastrocnemius which goes to the fascia of the shank. It next passes from view behind the peroneus longus muscle but can be followed caudad by turning to Plate 91, figure 2. Its superficial branch can be seen on this plate as it emerges in the lower part of the leg and passes over the dorsum of the foot to form the dorsal metatarsal nerves. These can be seen to better advantage on Plate 92, figure 1.

The **cranial gluteal nerve** is derived from the lumbosacral cord and from the first sacral nerve. It is shown here as it emerges from beneath the pyriformis muscle. It passes between the gemellus superior and the gluteus minimus sending branches to these muscles and to the gluteus medius and the tensor fasciae latae.

The **caudal gluteal nerve** is also from the lumbosacral cord and the first sacral nerve. It comes from the pelvis by way of the great sciatic notch and travels along the dorsal side of the sciatic nerve. It is shown here as it, with the sciatic nerve, comes from under the pyriformis muscle. It sends branches to the gluteus maximus (*superficialis*) and caudofemoralis muscles.

The **caudal femoral cutaneous nerve** arises from the second and third sacral nerves and runs in close relationship to the pudendal nerve and then to the caudal gluteal artery and vein. It then divides sending one branch (*perineal nerves*) into the fat near the legs and the other onto the biceps femoris. It supplies the skin on the caudal proximal part of the thigh and the adjacent medial and lateral surfaces. It may send a branch as far distad as the popliteal region.

Note that branches of the femoral nerve on the medial side of the thigh do come through to the lateral surface along with branches of the lateral circumflex femoral artery and vein. They supply sartorius and quadriceps femoris muscles.

A number of important arteries with their satellite veins are seen on this aspect of the hindlimb. Most of them have already been discussed in conjunction with the preceding plate. At the top of the plate note the **cranial gluteal artery** which appears from under the pyriformis muscle, turns craniad and dorsad and divides into two branches, one supplying the gluteus medius and anastomosing with branches from the iliolumbar artery, the other turns dorsad and caudad sending branches to the gluteus maximus and pyriformis.

The **iliolumbar artery** arises from the internal iliac artery. Passing out of the pelvis it sends a branch caudad around the gluteus minimus (*profundus*) muscle and on to the gluteus medius which it supplies and where it anastomoses with the cranial gluteal artery. The other branch sends branches to the tensor fasciae latae, sartorius and quadriceps femoris muscles.

The **parietal branch** or continuation of the internal iliac artery passes caudad after leaving the pelvis and has two terminal branches. One is the superficial lateral coccygeal which runs lateral to the tail giving off cutaneous branches to the skin dorsally and ventrally. It runs to the end of the tail. The other branch essentially a continuation of the parietal branch is the **caudal gluteal artery.** It provides branches to muscles of the gluteal region, the main part supplying the biceps femoris and semitendinosus where it anastomoses with branches from the deep femoral (seen here) and with branches from the sural artery.

Turn now to the popliteal region and note the small section of the **popliteal muscle** which is exposed and from which branches arise going to femur, vastus lateralis, biceps femoris, and the knee joint. The largest artery to arise here from the popliteal artery is the **sural** (*caudal femoral*) **artery** and one should follow it distad as it travels along with the **caudal cutaneous sural nerve** and gives off branches to the hamstring muscles, vastus lateralis and to both heads of the gastrocnemius and to the calcaneal tendon. The sural artery becomes cutaneous. The artery has anastomosed with the caudal gluteal, with the deep femoral and with branches of the femoral artery which reach the hamstrings.

A section of the **cranial tibial artery** can be seen on the cranial side of the lower part of the leg. It passes onto the dorsum of the foot and gives rise with the dorsal branch of the saphenous artery to superficial dorsal metatarsal arteries better studied on Plate 92, figure 1.

Again veins follow most of the arteries described but special attention should be given to the **small** (*lateral*) **saphenous vein** clearly shown here through much of its length. Its dorsal and plantar branches are shown distally as is the joining of its dorsal branch with the **great saphenous vein** in the ankle region. The common vessel on the dorsum of the foot receives the superficial dorsal metatarsal vein connecting the plantar branch of the small saphenous vein with the common vessel on the dorsum of the foot. Finally, notice the complex of tarsal veins and distally the superficial plantar venous arch. Refer to the next plate for details of vessels and nerves on the hindfoot.

Deep Nerves and Vessels of the Popliteal
and Knee Regions and the Leg
PLATE 91

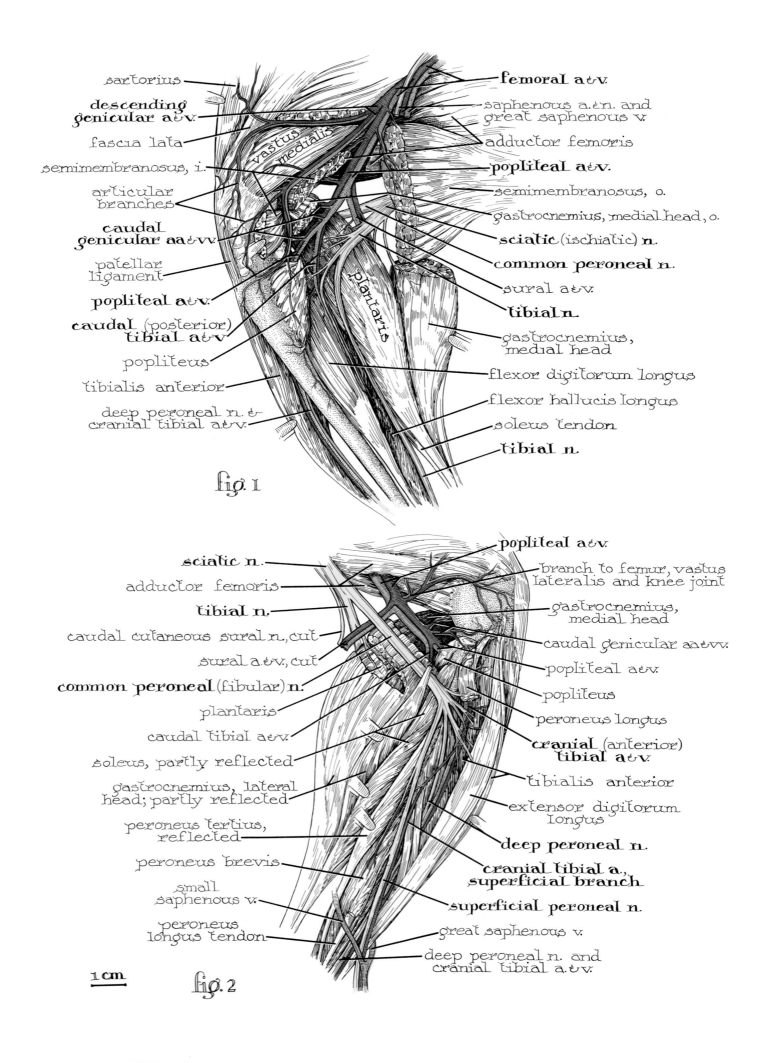

sartorius

descending genicular a & v.

fascia lata

semimembranosus, i.

articular branches

caudal genicular aa & vv.

patellar ligament

popliteal a & v.

caudal (posterior) tibial a & v.

popliteus

tibialis anterior

deep peroneal n. & cranial tibial a & v.

vastus medialis

plantaris

femoral a & v.

saphenous a. & n. and great saphenous v.

adductor femoris

popliteal a & v.

semimembranosus, o.

gastrocnemius, medial head, o.

sciatic (ischiatic) n.

common peroneal n.

sural a & v.

tibial n.

gastrocnemius, medial head

flexor digitorum longus

flexor hallucis longus

soleus tendon

tibial n.

fig. 1

sciatic n.

adductor femoris

tibial n.

caudal cutaneous sural n., cut

sural a & v., cut

common peroneal (fibular) n.

plantaris

caudal tibial a & v.

soleus, partly reflected

gastrocnemius, lateral head; partly reflected

peroneus tertius, reflected

peroneus brevis

small saphenous v.

peroneus longus tendon

popliteal a & v.

branch to femur, vastus lateralis and knee joint

gastrocnemius, medial head

caudal genicular aa & vv.

popliteal a & v.

popliteus

peroneus longus

cranial (anterior) tibial a & v.

tibialis anterior

extensor digitorum longus

deep peroneal n.

cranial tibial a., superficial branch

superficial peroneal n.

great saphenous v.

deep peroneal n. and cranial tibial a. & v.

1 cm

fig. 2

Deep Nerves and Vessels of the Popliteal
and Knee Regions and the Leg
PLATE 91

This illustration shows the deep vessels and nerves in the regions indicated which were not clearly seen in the previous two plates. Deep dissection has been performed in both the popliteal area and the leg.

Figure 1 is the medial aspect of the right hindlimb showing the distal part of the thigh and the lower leg. The saphenous nerve, a continuation of the femoral nerve, is shown at the top of the illustration. It is cut off, but in the living subject is distributed to the integument of the medial surface of the leg and the dorsum of the foot. The **sciatic** (*ischiatic*) **nerve** is shown as it divides into the lateral common peroneal nerve whose course is shown better in figure 2, and the medial tibial nerve. The **tibial nerve** passes between the heads of the gastrocnemius muscle and then extends distad between the plantaris and the medial head of the gastrocnemius and to the space between the flexor hallucis longus and plantaris. It provides branches to the popliteus, gastrocnemius, plantaris, soleus, flexor digitorum longus, flexor hallucis longus and tibialis posterior—all muscles of the caudal side of the leg. It sends branches also to the knee (*stifle*), tarsal and digital joints. Terminally it passes around the medial side of the ankle and to the plantar surface of the foot dividing into lateral and medial plantar nerves as shown in Plate 92.

Note also the blood vessels shown in figure 1. The femoral artery is seen at the top of the illustration in the distal part of the thigh. It gives off a saphenous artery which, though cut off here, in life passes over the medial surface of the gracilis muscle and after giving off branches to the knee and neighboring muscles it divides in the lower leg to form dorsal and plantar branches to the foot (Plate 92). The **femoral artery,** after giving off the saphenous, gives rise to the **descending genicular** (*superior articular*) **artery.** Sometimes these two arteries have a common branch from the femoral. The descending genicular passes between the vastus medialis and semimembranosus muscles into which it sends branches and continues to the knee region. It sends articular branches into the stifle (*knee*) joint and its capsule.

The **femoral artery** now passes between the semimembranosus and the vastus medialis muscles and through the distal portion of the adductor femoris to reach the popliteal space. The artery is now called the popliteal and gives off a number of branches before it disappears under the popliteus muscle and finally continues as the cranial tibial artery as seen here and to better advantage in figure 2.

The main branches of the **popliteal artery** are the sural, caudal genicular, muscular, and caudal tibial arteries. The sural artery (cut here) passes distad from the caudal side of the popliteal artery sending branches to the fat in the popliteal space and to the biceps femoris. It passes onto the caudal border of the lateral head of the gastrocnemius and sends branches to it and the medial head and passes to the lateral side of the tendon of Achilles and to the foot (Plate 92). **Caudal genicular arteries** can be seen passing to the knee joint and muscular branches (*not labeled*) to muscles about the popliteal space. The large **caudal tibial artery** branches from the popliteal just before the latter passes under the popliteus muscle. The caudal tibial passes distad across the popliteus muscle to the surface of the flexor hallucis longus into which it ramifies also sending branches to the soleus and gastrocnemius. It does not pass onto the foot.

The veins can be seen to be satellites of the arteries and most of them have the same names. The vein accompanying the saphenous artery is called the great saphenous vein to distinguish it from the small saphenous vein shown in the lower part of figure 2.

Figure 2, a lateral view of the distal end of the thigh and the lower leg, shows many of the same nerves and vessels as seen, in part, in figure 1. Again we see the **sciatic nerve** as it divides in the popliteal space into the **tibial** and common peroneal (*fibular*) nerves.

The **common peroneal nerve** passes medial to the distal end of the biceps femoris and over the lateral surface of the proximal end of the lateral head of the gastrocnemius. At a point just distal to the head of the fibula it passes under a slip of the gastrocnemius and between the soleus and peroneus longus. Here it provides a number of branches to the muscles craniad of the tibia and fibula and to the peroneus longus. The common peroneal then divides into superficial and deep peroneal nerves.

The **superficial peroneal nerve** travels distad between the peroneus longus and tertius muscles which it supplies. Near the ankle region it comes close to the surface, crosses the transverse ligament, sends small branches to the ankle and divides into four parts which supply the toes as shown in Plate 92.

The **deep peroneal nerve** travels distad between the tibialis anterior and the extensor digitorum longus muscles to which it furnishes branches. It then passes distad accompanying the cranial tibial artery and vein and lying on the tibialis anterior muscle. It passes onto the dorsum of the foot and is distributed to the extensor digitorum brevis and the contiguous sides of digits four and five (Plate 92). Note the caudal cutaneous sural nerve branching from the posterior side of the tibial nerve. It goes to the integument of the lower leg.

The **popliteal artery** is seen here from the lateral side. A branch is shown which divides to supply the femur, knee joint and vastus lateralis. Opposite this branch is the sural artery described above and distad of the sural the caudal genicular artery branches from the popliteal artery to go to the knee. The **caudal tibial artery** is again shown as it branches from the popliteal just before the latter goes behind the popliteus muscle. The **cranial tibial artery** is a continuation of the popliteal. It is illustrated here as it passes distad between the tibialis anterior and the extensor digitorum longus. It passes, with the tendons of these muscles, under the transverse ligament to the dorsum of the foot. Its distribution to the foot is shown in Plate 92. A **superficial branch** of the cranial tibial artery joins the superficial peroneal nerve and on the dorsum of the foot is connected to the saphenous artery.

The veins follow closely as satellites of the arteries. Note especially the union of the great and small saphenous veins at the distal end of the leg.

Nerves and Blood Vessels of the Right Hindfoot—
Dorsal and Ventral Aspects
PLATE 92

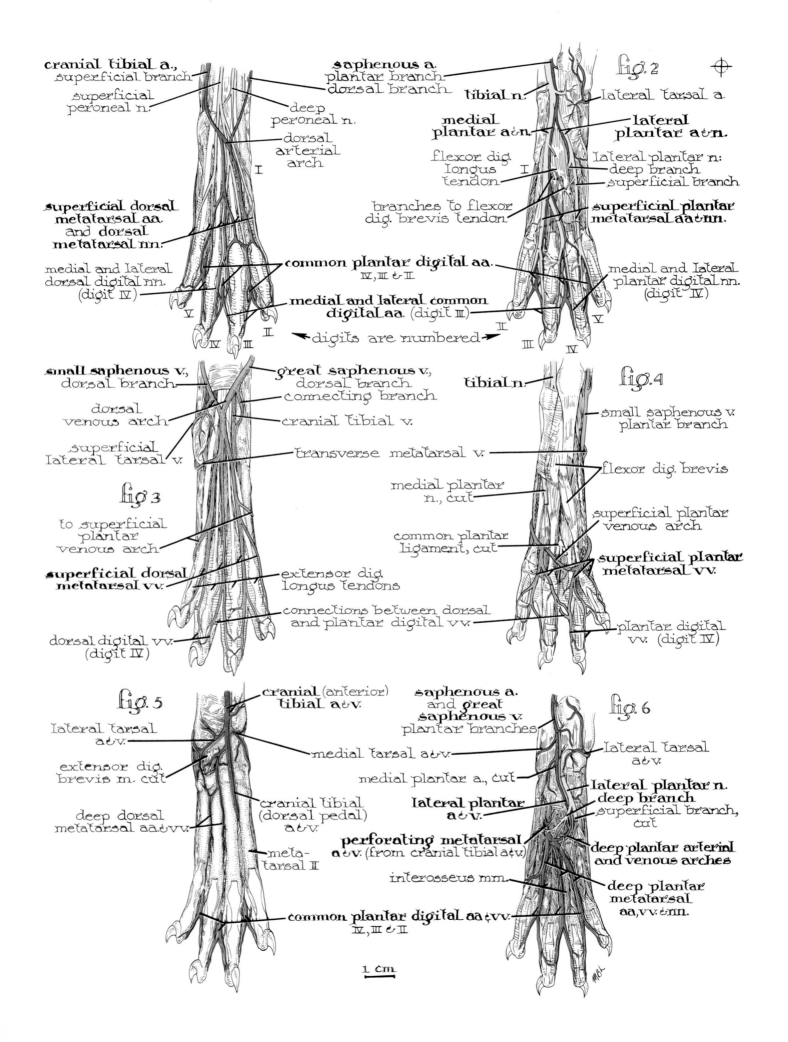

cranial tibial a.,
superficial branch
superficial peroneal n.
deep peroneal n.
dorsal arterial arch
I
superficial dorsal metatarsal aa. and dorsal metatarsal nn.
medial and lateral dorsal digital nn. (digit IV)
V
II
IV III

saphenous a.
plantar branch
dorsal branch
tibial n.
medial plantar a.&n.
flexor dig. longus tendon
I
branches to flexor dig. brevis tendon
common plantar digital aa. IV, III & II
medial and lateral common digital aa. (digit III)
← digits are numbered →
II
III IV

fig. 2
lateral tarsal a.
lateral plantar a.&n.
lateral plantar n.: deep branch
superficial branch
superficial plantar metatarsal aa.& nn.
medial and lateral plantar digital nn. (digit IV)
V

small saphenous v., dorsal branch
dorsal venous arch
superficial lateral tarsal v.

fig 3

to superficial plantar venous arch
superficial dorsal metatarsal vv.
extensor dig. longus tendons
dorsal digital vv. (digit IV)
connections between dorsal and plantar digital vv.

great saphenous v., dorsal branch
connecting branch
cranial tibial v.
transverse metatarsal v.
medial plantar n., cut
common plantar ligament, cut

tibial n.
fig. 4
small saphenous v. plantar branch
flexor dig. brevis
superficial plantar venous arch
superficial plantar metatarsal vv.
plantar digital vv. (digit IV)

fig. 5
lateral tarsal a.&v.
extensor dig. brevis m. cut
deep dorsal metatarsal aa.& vv.
cranial (anterior) tibial a.&v.
medial tarsal a.&v.
cranial tibial (dorsal pedal) a.&v.
meta- tarsal II
common plantar digital aa.& vv. IV, III & II

saphenous a. and great saphenous v. plantar branches
medial tarsal a.&v.
medial plantar a., cut
lateral plantar a.&v.
perforating metatarsal a.&v. (from cranial tibial a.&v.)
interosseus mm.

fig. 6
lateral tarsal a.&v.
lateral plantar n. deep branch
superficial branch, cut
deep plantar arterial and venous arches
deep plantar metatarsal aa., vv.& nn.

1 cm

This plate illustrates by six drawings the distribution of nerves and vessels on the hindfoot or paw. Figures 1 and 2 emphasize the more superficial nerves and arteries, figure 1 a dorsal view and figure 2 a plantar view. Figures 3 and 4 emphasize the superficial veins on dorsal and plantar surfaces of the foot. Figures 5 and 6 emphasize the deep arteries, veins and nerves. To gain a good understanding of these nerves and vessels constant reference should be made to Plates 89 and 90 of the medial and lateral views of the hindlimb. One should realize also that these structures are very difficult to dissect in the cat and are subject to considerable variation. While it is unlikely that most students would be expected to learn all of these nerves and vessels or to dissect them, there is much to be gained in the way of general understanding and appreciation just to see them. One should note also the similarities and differences in nerves and blood vessels on the forefoot and hindfoot (Plate 87).

Figure 1 shows the superficial peroneal nerve coming onto the dorsum of the foot and dividing into two branches to each side of each digit. These nerves in the metatarsal region are called the (*superficial*) dorsal metatarsal nerves and those on the toes the median and lateral dorsal digital nerves. The deep peroneal nerve is also shown on the dorsum of the foot. It supplies the tibialis anterior (ventralis) and some of the dorsal foot muscles.

Also in figure 1 follow the superficial branch of the **cranial tibial artery** and the dorsal branch of the **saphenous artery** distad to where they join to form the dorsal arterial arch. From this arch, though usually considered from the saphenous artery, superficial dorsal metatarsal arteries pass distad. They in turn send a digital artery along the medial side of the medial digit and digital arteries to the contiguous sides of digits II-V. These are called the **medial and lateral common digital arteries** because they connect with branches of the **common plantar digital arteries IV, III, and II** which will be considered particularly in reference to figure 6 of this plate.

Figure 2 shows the **tibial nerve** as it comes onto the plantar surface of the foot and divides into medial plantar and lateral plantar nerves. The **lateral plantar nerve** divides into deep and superficial branches, the latter forming a superficial plantar metatarsal nerve which divides distally to form a medial digital nerve for the fifth digit and a lateral digital nerve for the fourth digit. The deep branch of the lateral plantar will be considered in conjunction with figure 6. The **medial plantar nerve** divides into two superficial plantar metatarsal nerves each of which divides providing the remaining medial and lateral plantar digital nerves. Note that branches to the flexor digitorum brevis and flexor digitorum longus muscle

tendons are given off from the lateral and medial plantar nerves.

Follow now the **plantar branch of the saphenous artery** in figure 2 as it comes onto the plantar surface of the foot accompanying the tibial nerve. It divides into medial and lateral plantar arteries. The **lateral plantar artery** follows the deep branch of the lateral plantar nerve and will be seen again in figure 6. The **medial plantar artery** divides into **superficial plantar metatarsal arteries.** These in turn divide into **medial and lateral common digital arteries** connecting by small branches to the **common plantar digital arteries.**

Figures 3 and 4 emphasize the superficial veins of the hindfoot, the main ones being the dorsal branches of the **great and small saphenous veins.** The cranial tibial vein should be noted with its connecting branch to the dorsal branch of the great saphenous vein (fig. 3). The role of the cranial tibial vein is best studied in connection with figures 5 and 6. The great and small saphenous veins come together to form the dorsal venous arch (fig. 3). Their branches in the foot follow the general pattern of the arteries described in figures 1 and 2 since they form **superficial dorsal metatarsal veins** and they in turn dorsal digital veins. The dorsal digital veins have connecting veins to the plantar digital veins. Note the superficial lateral tarsal vein in figure 3, a branch of the small saphenous to the tarsal region and the branches from the medial and lateral metatarsals in figure 3 which connect on the plantar surface in figure 4 to form the **superficial plantar venous arch.** From the arch **superficial plantar metatarsal veins** provide plantar digital veins to the digits. Note the cut median plantar nerve in figure 4.

Figures 5 and 6 emphasize the deeper arteries and veins of the hindfoot. It requires a very careful dissection in order to uncover these deep vessels. The deep vessels which have been mentioned in our discussions of figures 1 to 4 above are the sources of these deep vessels.

Figure 5 shows the **cranial tibial artery** and vein coming in on the dorsum of the foot. This is the deep branch of the cranial tibial artery in contrast to the superficial branch seen in figure 1. Medial and lateral tarsal arteries and veins leave and enter the cranial tibial artery and vein. They supply structures in the tarsal or ankle region—or drain them. Note that deep dorsal metatarsal arteries and veins are present between the third and fourth and fourth and fifth metatarsals and that they are derived from a branch of the dorsal pedal (*cranial tibial*) artery and vein—a branch often called the transverse metatarsal artery and vein. The (*deep*) cranial tibial artery and vein become the **dorsal pedal artery and vein** which dip between the second and third metatarsal on the dorsum of the foot to become the **perforating metatarsal artery** and vein as seen on the plantar surface of the foot in figure 6. Here these vessels form **deep plantar arterial and venous arches** with the deep branches of the **lateral plantar artery and vein** which in turn came from the **saphenous artery and great saphenous vein** (see fig. 2). From the deep plantar arterial and venous arches the **deep plantar metatarsal arteries and veins** are derived and from these in turn the **common plantar digital arteries and veins IV, III, II.** We have seen the latter in figures 1, 2 and 5.

Finally in figure 6 the **deep branch of the lateral plantar nerve** (*tibial n.*), seen in part in figure 2, is shown as it forms **deep plantar metatarsal nerves.**

These nerves and vessels provide the complex musculature, skin, footpads and other structures of the foot with sensory and motor nerve fibers, with arteries for nutrition, oxygenation and general supply and with veins to carry away the products of metabolism.

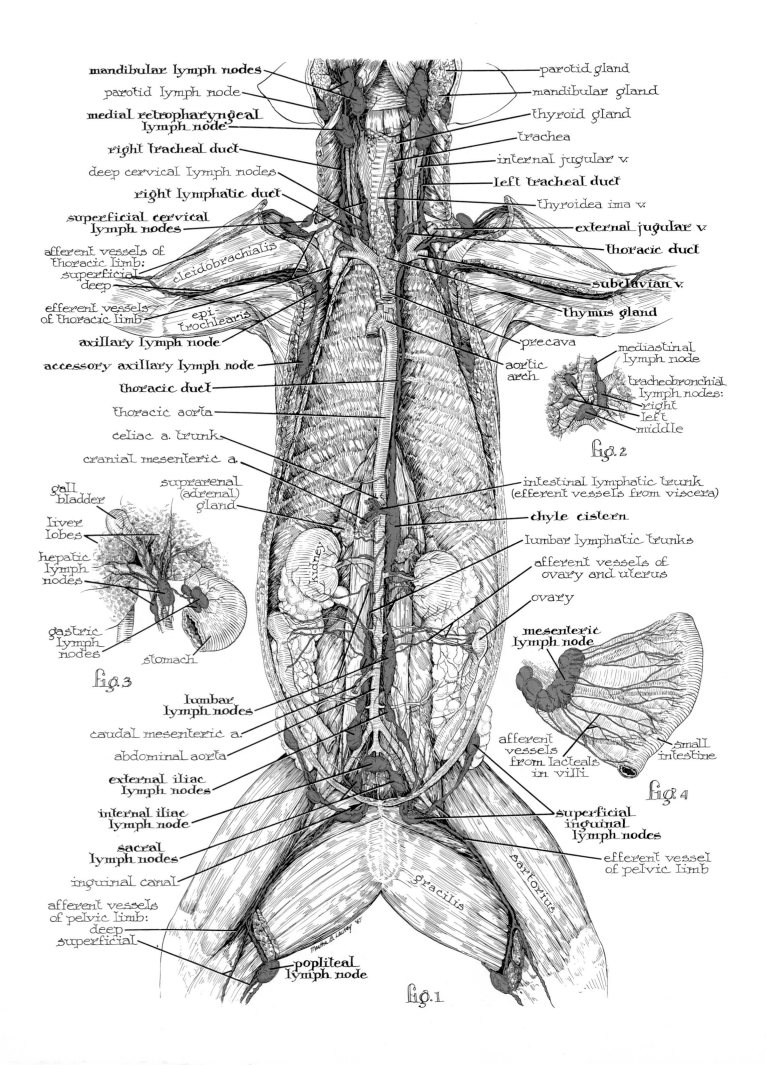

mandibular lymph nodes
parotid lymph node
medial retropharyngeal lymph node
right tracheal duct
deep cervical lymph nodes
right lymphatic duct
superficial cervical lymph nodes
afferent vessels of thoracic limb: superficial deep
efferent vessels of thoracic limb
axillary lymph node
accessory axillary lymph node
thoracic duct
thoracic aorta
celiac a. trunk
cranial mesenteric a.
suprarenal (adrenal) gland

cleidobrachialis
epitrochlearis

parotid gland
mandibular gland
thyroid gland
trachea
internal jugular v.
left tracheal duct
thyroidea ima v.
external jugular v.
thoracic duct
subclavian v.
thymus gland
precava
aortic arch

mediastinal lymph node
tracheobronchial lymph nodes: right left middle

fig. 2

intestinal lymphatic trunk (efferent vessels from viscera)
chyle cistern
lumbar lymphatic trunks
afferent vessels of ovary and uterus
ovary

gall bladder
liver lobes
hepatic lymph nodes
gastric lymph nodes
stomach
kidney

fig. 3

mesenteric lymph node
afferent vessels from lacteals in villi
small intestine

fig. 4

lumbar lymph nodes
caudal mesenteric a.
abdominal aorta
external iliac lymph nodes
internal iliac lymph node
sacral lymph nodes
inguinal canal
afferent vessels of pelvic limb: deep superficial

superficial inguinal lymph nodes
efferent vessel of pelvic limb

gracilis
sartorius

popliteal lymph node

fig. 1

As stated in the introductory comments on the circulatory system the lymphatic system may be considered as a subdivision of that system or as a separate system. However one considers it, it serves as does the venous system as a means of returning tissue fluid to the heart. It does this by returning it to the large veins craniad of the heart. Besides this function it serves, by means of the lymph nodes, a protective function and as a place for the manufacture of lymphocytes. Through its lymph capillaries, the lacteals in the intestinal villi, it aids in the absorption of fatty acids and glycerine.

This plate shows some of the main lymphatic vessels and lymph glands of the cat. One must remember that since lymphatic vessels are extremely thin walled, they do not show well in the ordinary embalmed animal. Even when efforts are made to inject these vessels, the results are not as satisfactory as obtained in injecting arteries and veins.

While it may be said that in general lymphatic vessels accompany veins, it is not necessarily always the case. There are no lymphatics in the brain and spinal cord or in the marrow of bones. There are none in skeletal muscle although they are found in the fascial planes between muscles. They are numerous in mucous membranes and in the skin.

The lymphatic vessels come to focus at two points in the venous system as seen clearly in this plate—at the junction of right and left subclavian and external jugular veins. While these points of entry may vary in detail and in the number of connections made, they are most always in the general area. Likewise the lymph nodes are found most frequently in places where they interfere least with bodily movement and where they receive maximum protection. They are found in deposits of fat at the flexor angles of joints, in the mesenteries and mediastinum and in the angles formed by the origin of many of the larger blood vessels. View this plate initially with these generalizations in mind.

Starting at the cranial end note the parotid and **mandibular lymph nodes** of the head. The superficial afferent lymphatic vessels of the head drain into these nodes. There are efferent vessels from the parotid to the mandibular nodes and the mandibular nodes themselves have interconnecting vessels. The efferent vessels of the mandibular nodes go to the large **medial retropharyngeal node** on the same side and may have branches to the opposite side.

The **medial retropharyngeal node** is the largest in the head and neck. Its afferent vessels come from the deep structures of the head and it also receives the efferent ducts from parotid and mandibular nodes. The **right and left tracheal ducts** arise at the caudal pole of the medial retropharyngeal lymph node of the same side and serve as efferent vessels. The tracheal ducts pass caudad along the internal jugular veins and the common carotids. The left tracheal duct joins the thoracic duct; the right tracheal duct reaches the right lymphatic duct.

Two **superficial cervical lymph nodes** are seen craniad of the shoulder and between it and the neck and under the cleidobrachialis and cleidotrapezius muscles. Their numbers may vary from one to three, two being most common. The afferent vessels drain most of the forelimb, the skin of the back of the head, pharyngeal region and the lateral surface of the neck. Their efferent vessels connect the nodes themselves and send one or more branches ventrad to join the tracheal ducts or to enter directly the right lymphatic duct or the thoracic duct.

A number of deep cervical lymph nodes can be observed along the trachea on each side. They are small in size and variable in number. Their afferent vessels come from the trachea, larynx, thyroid gland and esophagus. Some may come from even deeper structures. The most cranial of the deep cervical nodes receives efferent vessels from the mandibular and retropharyngeal nodes and each deep cervical node's efferent vessels become the afferent vessels of the next node caudad. The most caudal deep cervical node sends its efferent vessels into the right lymphatic duct or on the left into the thoracic duct.

The **axillary and accessory axillary lymph nodes** as indicated lie in relationship to the pectoralis and shoulder muscles. The axillary node is craniad of the accessory axillary node and the two are connected by lymph vessels. The axillary nodes collect from the superficial and deep structures of the forelimb and from the thoracic wall including the thoracic and cranial abdominal mammary glands. The efferent vessels of the axillary nodes pass craniad around the first rib and empty into the right lymphatic duct or on the left, the thoracic duct.

A number of additional nodes are present in the thoracic region. In general they classify as parietal, those associated with the body wall, or visceral, associated with the internal organs. Those shown here are in figure 2 which is a view of the dorsal aspect of the tracheobronchial region. One mediastinal lymph node is shown in addition to right, left and middle tracheopulmonary nodes. The afferent vessels of the mediastinal nodes come from many structures both in the thoracic wall and the thoracic viscera. Their efferent vessels pass craniad to empty into the right lymphatic duct or the thoracic duct. The tracheobronchial lymph nodes receive afferent vessels primarily from the lungs and bronchi, but also from the thoracic portions of the esophagus, trachea and aorta and probably from the heart mediastinum and diaphragm. On the efferent side they drain craniad, very likely through the mediastinal nodes and to the thoracic duct and right lymphatic duct.

A number of lymph nodes are present in the abdominal cavity some associated with the viscera, their afferent vessels draining lymph from those organs and their efferent vessels passing into the **chyle cistern** which is an enlargement at the caudal end of the thoracic duct. Follow the thoracic duct forward from between the crura of the diaphragm to the point, already described, where it enters the venous system craniad of the heart. Note that in its course it divides and rejoins one or more times. Notice also its bead-like nature which indicates the presence of a whole series of internal valves. This can also be detected in other large lymphatic vessels like the right lymphatic and tracheal ducts. Some of these visceral nodes are the gastric and hepatic nodes shown in figure 3 and the large **mesenteric lymph node,** the largest in the body, shown in figure 4, which receives its afferent vessels from the small intestine. Other lymph nodes, not shown here, occur along the large and small intestines in their mesenteries. Some can be seen in Plates 82 and 83.

Other abdominal and also pelvic lymph nodes collect from the body wall and to some degree from the pelvic and abdominal viscera. Among these are the **lumbar, external iliac, internal iliac** and **sacral lymph nodes.** Their efferent vessels drain craniad each more caudal node connected by efferent vessels with the node craniad of it; all emptying ultimately through lumbar lymphatic trunks into the chyle cistern.

A number of nodes are also shown which drain lymph from the hindlimb. Coming from the feet afferent vessels lead into the **popliteal lymph node** in back of the knee joint; other deeper vessels join the efferent vessels of the popliteal node and lead into **superficial inguinal nodes.** From these inguinal nodes efferent lymph vessels pass through the inguinal canal and in the pelvic region join vessels and nodes already described. It is clear then that the thoracic duct receives lymph not only from the left side of the head and thorax and the forelimb, but from the entire body caudad of the diaphragm.

Finally, take note of the **thymus gland** shown here as it appears in an adult cat. It is relatively large at birth and is best developed and reaches its maximum size just before sexual maturity. After this it gradually decreases in size but some still remains in most old cats. It is an elongated and laterally flattened organ which lies in the precardial mediastinal septum between the two lungs and may extend craniad beyond the first rib into the neck region. Its caudal end is usually bifurcated and fits over the cranial end of the pericardium. This gives an indistinct two-lobed appearance to the gland. Dorsally it lies against the trachea; ventrally against the sternum.

The tissue of the thymus is largely lymphoid and for this reason it is considered here as a part of the lymphatic system. Its lymphatic vessels probably empty into mediastinal and sternal lymph nodes.

The thymus is a source of lymphocytes early in life. These cells travel to the blood, spleen and lymph nodes and from them come the cells which are responsible for the body's development of immunity to disease. They produce antibodies. Also there is evidence that the thymus may produce a hormone-like factor which, reaching the spleen and lymph nodes, stimulates these organs to produce lymphocytes.

Other lymphoid tissues, namely the tonsils are seen in Plate 53.

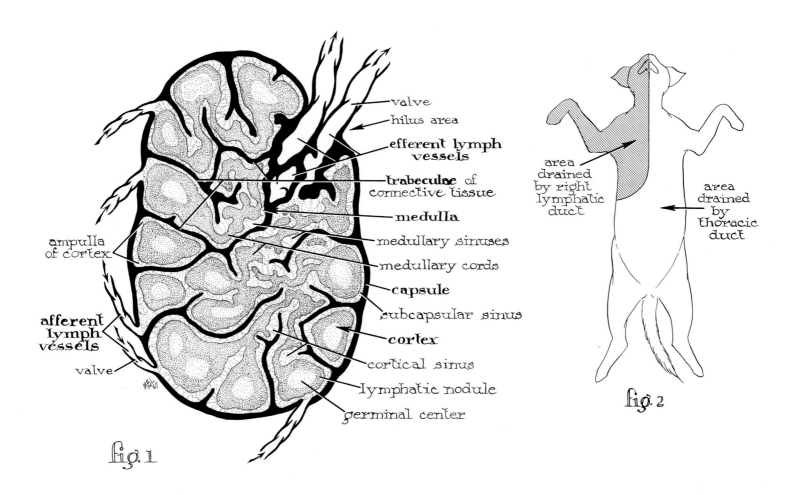

valve
hilus area
efferent lymph vessels
trabeculae of connective tissue
medulla
medullary sinuses
medullary cords
capsule
subcapsular sinus
cortex
cortical sinus
lymphatic nodule
germinal center

ampulla of cortex

afferent lymph vessels

valve

fig. 1

area drained by right lymphatic duct

area drained by thoracic duct

fig. 2

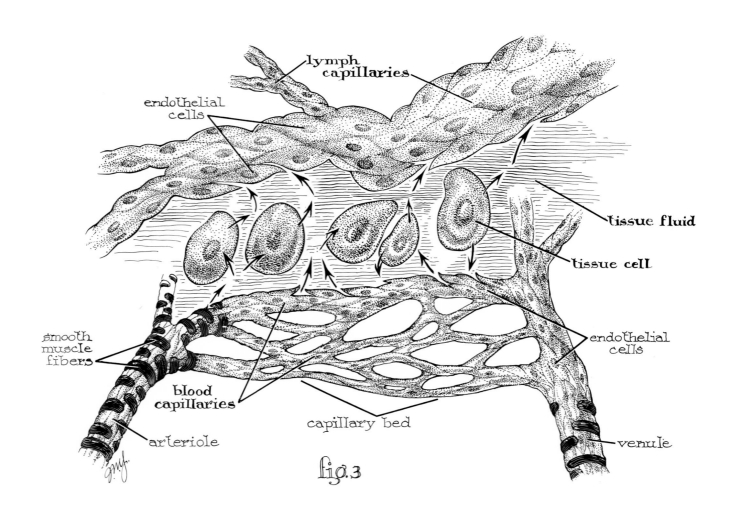

lymph capillaries

endothelial cells

tissue fluid

tissue cell

endothelial cells

smooth muscle fibers

blood capillaries

capillary bed

arteriole

venule

fig. 3

(In part from Crouch: Functional Human Anatomy, Lea & Febiger)

Figure 1 is a section through a lymph node. Recall the numbers of these nodes that you have seen in your dissection and that they are placed in the pathways of lymphatic vessels which with the veins make up the system by means of which tissue fluid is returned to the heart. Here you see the **afferent lymph vessels** entering a node and the **efferent lymph vessels** leaving at the hilus. Note the valves in these vessels and the arrows indicating direction of flow of the lymph.

The node consists of a **capsule** of connective tissue from which **trabeculae** push into the node to divide the outer part of the node or **cortex** into lymphatic nodules. In the center of the node, the **medulla,** the trabeculae anastomose around the medullary cords and sinuses.

The lymphatic nodules of the cortex are made up of lymph cells, mostly small lymphocytes, but in their lighter germinal centers there are larger lymphoblasts. The lymph cells in the medulla of the node form medullary cords. These formations of medullary cords and lymphatic nodules are surrounded by medullary, cortical and subcapular sinuses which contain relatively few lymph cells.

The hilus of the node is not only where the efferent lymp(hatic) vessels leave the node, but also the point where arteries and veins pierce the capsule of the node.

Note that there are more afferent lymph vessels than there are efferent and that they are placed around the convex periphery of the node. The lymph passes from these vessels into the subcapular sinus and then through the other loose spaces and sinuses of the medulla and cortex. The walls of the sinuses of the nodes are not continuous nor are they composed only of endothelium. There are reticular cells and fixed macrophages supported by reticular fibers. As the lymph stream moves slowly through the node lymphocytes manufactured in the germinal centers move out into the lymph by ameboid action and also the reticular cells act as phagocytes to pick up foreign particles which may be in the circulating lymph. The lymph finally makes its way to the hilus and leaves the node by the efferent lymph vessels. Two functions have been served by the node: (1) the manufacture of lymphocytes and (2) the removal of foreign particles and pathogenic organisms. In addition dead red cells are removed by the reticular cells and like other lymphoid tissues antibodies may be formed.

Figure 2 indicates in schematic fashion the areas drained by the right lymphatic duct and the large thoracic duct. There may be variations from this general pattern.

Figure 3 shows schematically the relationships of blood and lymph capillaries to the tissue fluid and the direction of flow of the blood, tissue fluid and lymph. The blood brings oxygen, nutrients, hormones and other substances into the blood capillaries. These materials move out into the tissue fluid on the arterial side and are made available to tissue cells. The tissue cells in turn contribute their waste or other products to the tissue fluid and it may return to the blood on the venous side of the capillary bed or it may go instead into a lymphatic vessel and reach the venous system ultimately through either the thoracic or right lymphatic duct as indicated in figure 2 or better in Plate 93.

Functions. Our study of the cat thus far has dealt with those systems having to do with protection, support, movement, transportation and the maintenance of the individual and the species. Emphasis has been given to interrelationships and interdependence of systems. The endocrine and nervous systems are the ones, however, by which control, coordination and integration are truly achieved within the organism. The endocrine system works through the medium of chemical agents, the hormones, which, by entering the blood, are distributed through the body and to their target organ or tissue where they exercise their functions. The nervous system is more rapid in its action and more critical in finding its targets than is the endocrine system. It also adds two other dimensions to an organism's capacity to cope with its environment. Through its receptors, its organs of general and special sense, it achieves a greater variety of responses to environmental situations; a better orientation; a greater capacity to live in various ecological situations. Also inherent within the nervous system as a whole, but centered in the cerebral cortex of the brain, is the capacity for learning and for information storage. A cat's behavior is modified by experience. The extent of its capacity for learning is difficult to judge.

General Plan. It is only for purposes of discussion that we subdivide the nervous system as we do. Such subdivision is arbitrary because actually there is functional continuity throughout the nervous system.

It is subdivided as follows:

Central nervous system
  Brain
  Spinal cord
Peripheral nervous system
  Cranial nerves
  Spinal nerves
Autonomic nervous system
  Visceral afferent neurons
  Sympathetic division
  Parasympathetic division

The term **voluntary** nervous system is sometimes used, when thinking in functional terms, for the central and peripheral systems. It is through the somatic afferent neurons that the animal is kept in touch with the external environment, through the central system that information is integrated, and through the somatic efferent neurons to the skeletal muscles that willful and purposeful action is taken. The autonomic nervous system, however, dealing with visceral activity is called **involuntary.**

The **central nervous system** is composed of the brain which is housed in the cranial cavity of the skull. It is continuous at the foramen magnum with the spinal cord which is enclosed within the vertebral canal of the backbone.

In the **peripheral system** the **cranial nerves** emerge from the ventral side of the brain and make their way through the various foramina of the skull to supply the structures of the head region or to bring information to the brain from the receptors of that region. One of these nerves, the vagus or vagrant, supplies structures as far as the left colic flexure of the colon and carries also afferent neurons. This, however, is primarily an autonomic function.

The **spinal nerves** emerge segmentally from the spinal cord and pass through the intervertebral foramina to be distributed to various parts of the body below the head region.

The **autonomic nervous system** consists of **visceral afferent neurons** which have their cell bodies in the cerebrospinal ganglia and serve to convey impulses from the viscera. The **sympathetic** and **parasympathetic** divisions are made up of **visceral efferent neurons** which consist of preganglionic neurons, ganglia and postganglionic neurons which carry impulses to smooth and cardiac muscle and glands throughout the body.

Nervous tissue. Nervous tissue is made up of nerve cells which are called neurons and of supporting tissue called neuroglia or glial cells.

Neurons are the basic structural units of the nervous system (Plate 95). They are specialized in the physiological properties of **irritability** and **conductivity.** By irritability we refer to their capacity to respond to stimulation by a change in structure and activity of the cell at the point which is stimulated. The spread of the activity is called conduction; the activity, a self-propagating physiochemical disturbance, constitutes the **nerve impulse.**

Nerves may be defined as bundles of nerve fibers coursing together outside of the central nervous system (Plate 97). They are held together by well-organized connective tissue sheaths which give protection and strength to what would otherwise be very weak and therefore vulnerable structures. It is because of the connective tissue encasements that tiny axons (*fibers*) can travel great distances in the body to innervate muscles and other structures. A fine sheath of reticular fibers in loose connective tissue cover the individual nerve fibers. These constitute the **endoneurium.** Many of these nerve fibers are collected together into **fasciculi** (*bundles*) under an outer wrapping of areolar connective tissue called the **perineurium.** A number of bundles, together constituting a nerve, have an outside sheath called the **epineurium.** The connective tissue elements of these various sheaths

267

are continuous one into another. In the sheaths will be found blood vessels and lymphatics appropriate to the size of the nerves.

Nerves vary greatly in size and in the nature of their contained fibers. Spinal and cranial nerves of the peripheral system are large and contain large numbers of myelinated fibers as well as some unmyelinated fibers. Many nerves of the autonomic system have small fibers which are unmyelinated and in sections of these nerves it is sometimes difficult to differentiate between the fibers and the sheath structures.

Collections of nerve fibers in the central system are called **tracts.** They, too, are highly organized and have supporting neuroglial cells.

Collections of nerve cell bodies occurring outside of the central nervous system are called **ganglia** (Plates 95 to 98). Some ganglia are large, others very small and in organs of the digestive tract single cells occur. The misconception is common that individual nerve cell bodies occur at intervals along the nerves but they are found only in the gray matter of the central system and in ganglia. Two kinds of ganglia are found in the body—the sensory and autonomic.

Sensory ganglia are found as spinal ganglia on the dorsal roots of spinal nerves and on the sensory roots of some of the cranial nerves (Plate 98). They are invested with a tough capsule of dense connective tissue which is continuous with the epineurium of the nerve. From the capsule numerous septa penetrate the ganglion carrying blood vessels and lymphatics. The nerve cells are often grouped around the periphery with some centers of deep lying cells. Fasciculi of fibers run through the ganglion. The nerve cells in spinal ganglia are pear-shaped with a single process which divides near the cell and sends a central branch to the spinal cord and a peripheral branch to the skin or muscle. Capsules present around the ganglion cells are continuous with the neurilemma of their fibers. Both myelinated and unmyelinated neurons are represented in the spinal ganglia.

Autonomic (sympathetic) ganglia occur as swellings along the sympathetic nerve trunks which extend through the abdomen, thorax, and neck, one trunk to each side of the vertebral column. They are found also in association with splanchnic nerves in the abdomen (Plate 93). Autonomic ganglion cells are multipolar, being the cell bodies of unmyelinated postganglionic neurons. Myelinated preganglionic neurons synapse with the ganglion cells or pass through the ganglion. Sheath cells form capsules around the ganglion cells.

The term **nucleus** is commonly used to designate groups of cell bodies within the central system. They are circumscribed in gray areas. A nerve center is a functional designation for a nucleus. We speak of respiratory and cardiac centers in the medulla which function in control of breathing and heart action respectively.

The endings of the peripheral nerve fibers range from simple naked fibers to those with highly complex accessory structures such as the eye. They may be divided into two functional groups, the receptor (*sensory*) and effector (*motor*) endings. The **receptors** are considered in the next section.

Effector nerves are supplied to muscle tissues and glands. Those to **involuntary muscle** (*smooth and cardiac*) and to **glands** are derived from the autonomic system and are composed mostly of unmyelinated fibers. While relatively little is known about the nerve endings to these involuntary structures, it appears that near their terminations, these nerves form numerous branches which communicate and form intimate plexuses near or around the muscles and glands. These plexuses give off tiny branches which in turn break up into fibrillae which run between the involuntary muscles and gland cells to terminate in their surfaces.

Myelinated efferent axons (*fibers*) **to skeletal or voluntary muscles** have their cell bodies in the gray matter of the spinal cord and brain and travel outward in the spinal and cranial nerves. Each axon may supply just one or two fibers in the more critical and precise acting muscles as those of the fingers or those moving the eye, whereas in the supporting and postural muscles one axon may supply a hundred or more muscle fibers. An axon and the skeletal muscle fibers it innervates constitute a **motor unit.**

As a nerve fiber enters a muscle fasciculus (*bundle*), it branches repeatedly and each terminal branch makes a functional connection with a muscle fiber to form a **myoneural junction** or **motor end plate.** Here the nerve impulse activates the muscle fiber (Plate 95, fig. 1).

THE NERVE IMPULSE

The nature of the nerve impulse is not fully understood, so that any attempt to define it must be considered tentative. The most widely held idea of the nature of the nerve impulse is called the **membrane theory.** According to this, the membrane (*cell membrane*) of the neuron is semipermeable and electrically polarized. Specifically, there is a layer of positively charged ions on the outside and a layer of negatively charged ions on the inside of the cell membrane. Presumably, the membrane is impermeable to these ions so polarization is maintained. Polarization is, conversely, supposed to be in some way important in maintaining the semipermeability of the membrane. A breakdown in one would involve a breakdown in the other. A stimulus applied to the nerve fiber is believed to bring about this breakdown by causing the membrane at the point of stimulation to become permeable. The positive and negative ions move through the permeable gap in the membrane and neutralize one another. This point in the nerve fiber is now electrically negative to adjacent areas and they depolarize, and a wave of depolarization thus moves along the entire fiber. This is the nerve impulse, and may be defined as a physicochemical change in the nerve fibers which, once initiated, is self-propagating. Both direct and indirect evidence support this theory, but it is beyond our scope to consider it. However, as you learn facts about the nerve impulse, see whether you can explain them on the basis of this theory.

Nerve impulses are all alike in kind. The sensation which results, pain, cold, etc., depends on the part of the brain receiving the impulses. The same impulses, if ending in muscle, would cause it to contract, or if ending in a gland, would make it secrete.

The strength of contraction or the intensity of sensation is not dependent upon the magnitude of the nerve impulses, but upon the frequency with which they follow one after the other. The more intense the stimulus, within limits, the greater the frequency of the impulses.

The rate at which impulses travel along nerve fibers depends on the diameter of the fiber; the larger the fiber, the greater the rate of conduction, other factors remaining the same. The largest myelinated fibers conduct at a rate of about 120 meters per second, 270 miles per hour; the smallest ones at about 6 meters per second or $13\frac{1}{2}$ miles per hour. Myelinated fibers conduct more rapidly than unmyelinated and with less expenditure of energy.

Nerve fibers also behave according to the all-or-none phenomenon; *i.e.*, if they respond at all to a stimulus, the response is maximal for the condition of the fiber at that time.

There is nothing within a nerve fiber or neuron to determine the direction of passage of the impulse. A fiber stimulated at midpoint would conduct equally well in either direction. Yet, we know that in the nervous system, fibers conduct in one direction only. Where is direction determined?

THE SYNAPSE

The terminal branches of neurons either come into functional contact with effectors or with other neurons. The area of functional continuity between neurons is the **synapse.** The synapse, among other functions, determines the direction of passage of nerve impulses in the nervous system. What is its nature (Plate 109)?

As we understand it, each tip of the many terminal branches of an axon forms a terminal bulb or **bouton terminale** in contact with the dendrites and cell body of the next neuron or neurons in the nerve pathway. This contact is not a point of structural continuity between the neurons, nor does the impulse pass directly from one to another. Rather, the arrival of an impulse at a terminal bulb causes some change or condition which serves to stimulate the next neuron, make it more, or in some cases less, susceptible to stimulation by the terminal bulbs of other neurons present at the synaptic area. It is likely that the resulting excitation or inhibition is mediated by the release of chemical agents though other means are not excluded. Acetylcholine and adrenalin-like substances have been collected in synaptic areas in the autonomic system as well as at junctions between certain efferent nerves and their effectors. Recent evidence indicates that these substances are being formed all the time at nerve muscle junctions, and that a nerve impulse does not initiate this secretion, but changes the rate of secretion. Assuming that the dendrites do not have this capacity to secrete substances into the synapse, whereas axons do, we can understand how synapses serve to establish direction of conduction in the intact nervous system. We know, too, that conduction is slowed at the synapse and that the synapse is responsible for other important phenomena. These facts and suggestions should help to encourage your continued interest and study.

THE SPINAL CORD

The **spinal cord** occupies the vertebral canal from the foramen magnum, where it is continuous with the brain, to the caudal region where it ends in a slender fibrous cord, the **filum terminale.** In cross section it is oval in form.

The cord has two enlargements, the cervical and lumbar. The **cervical enlargement** is in the area where spinal nerves originate which go to the forelimb. It is bounded by the fourth cervical vertebra craniad and by the first thoracic caudad. Caudad of the cervical enlargement the cord remains nearly uniform in diameter until it reaches the **lumbar enlargement,** where spinal nerves going to the hindlimb originate and which extends from the third to the seventh lumbar vertebrae. The cord begins to diminish in size in the lower lumbar and sacral regions. This tapered portion is the **conus medullaris** which ends in the filum terminale mentioned above.

The surface of the cord is marked by a number of longitudinal sulci and fissures. For a discussion and illustrations of these refer to Plate 98.

The autonomic nervous system is considered by many authorities to consist only of visceral pathways and the ganglia in which they synapse by which the heart, blood and lymphatic vessels, and the viscera are innervated. They do not consider the visceral afferent pathways a part of this system. In this book the two are considered as one system.

This system is sometimes called the involuntary, vegetative or visceral nervous system because it controls functions which the individual animal cannot ordinarily control or manipulate at will. Very specifically it innervates the smooth and cardiac muscle and the glands of the body. And since these tissues are found in the structure of every system including the integument, the functional ramifications of the autonomic system are complex. In contrast the neurons of the somatic efferent system innervate the skeletal muscles, a function over which one can exercise conscious control as an author does in the use of the muscles of his hand in the process of writing.

The autonomic nervous system as considered here is composed of the

> (1) visceral afferent neurons
> (2) visceral efferent neurons
>> *a.* sympathetic
>> *b.* parasympathetic

Visceral afferent neurons, like somatic afferent neurons, have their cell bodies in the sensory ganglia of the spinal and cranial nerves. Their difference lies in the peripheral and central distribution of their dendrites and axons. They travel not only in the regular autonomic channels, but also in the spinal and cranial nerves. They carry impulses which may initiate either autonomic or somatic reflex action, or may travel upward to the hypothalamus of the brain or even to the cortex of the frontal lobe. They do not provide the clear localization of feeling such as pain which is provided in the somatic sensory system. Referred pain—"pain" initiated in the heart, for example—may be felt in the left axilla or down the ulnar side of the left arm rather than in the heart itself.

Visceral efferent neurons can be divided into two kinds on sound anatomical and physiological bases. Because of this we speak of the sympathetic and parasympathetic divisions of the autonomic nervous system. Both divisions are alike in that their efferent pathways are always composed of two neurons. The first one known as the **preganglionic neuron** has its cell body in the central nervous system, the spinal cord or brain. It has a **myelinated axon** which synapses peripherally, usually in a ganglion, with the second or **postganglionic neuron** whose **nonmyelinated axon** ends in smooth or cardiac muscle or in glands.

The ganglia in which synapses occur between preganglionic and postganglionic neurons are classified as sympathetic trunk (*vertebral*) ganglia, collateral (*prevertebral*) ganglia and terminal ganglia. **Sympathetic trunk ganglia** occur in a pair of more or less segmental chains, the **sympathetic trunks,** along the vertebral column. The **collateral ganglia** are more peripheral and are related to the large branches of the aorta and are so named, the **celiac, cranial** and **caudal mesenteric ganglia.** The **terminal ganglia** are small and are located in or near body organs and may in some cases carry special names. Sympathetic trunk and collateral ganglia are generally considered to belong to the sympathetic division and terminal ganglia to the parasympathetic.

Both divisions of the autonomic nervous system are considered to be under the control of the hypothalamus of the brain. This has been demonstrated in the cat by Beattie *et al.* (1930). They were able to trace efferent fibers from the hypothalamus to the second lumbar segment of the spinal cord. The rostral part of the hypothalamus controls parasympathetic activity, the caudal part sympathetic activity. The hypothalamus is in turn under the control of certain areas of the cerebrum. Other parts of the brain also influence autonomic activity by sending impulses to preganglionic neruons, much like the brain influences the lower motor neurons of the somatic system.

The **parasympathetic division** is sometimes referred to as the craniosacral division because it has preganglionic axons which leave the brain through cranial nerves III, VII, IX and X; and because parasympathetic preganglionic axons also leave the sacral cord through the ventral roots of sacral nerves two and three and become, through the pelvic nerve, a part of the pelvic plexus.

The parasympathetic components of cranial nerves III, VII and IX are illustrated on Plate 109, figure 1 and discussed on page 315. Nerve X, the vagus, is shown in part on Plate 108 and discussed in connection with that plate on page 313. The same plate shows the sacral parasympathetic outflow.

The **sympathetic division** of the autonomic system is sometimes called the thoracolumbar division because its primary tie with the rest of the nervous system is through the thoracic and first four lumbar spinal nerves and the related segments of the spinal cord. These connections are made through gray and white communicating rami. Gray rami alone connect with the cervical, the last lumbar and the sacral spinal nerves.

The basic plan of this part of the autonomic system is illustrated in Plate 109, figure 2 and described in conjunction with that plate. Study also Figure 17 which shows diagrammatically the general arrangements of the whole

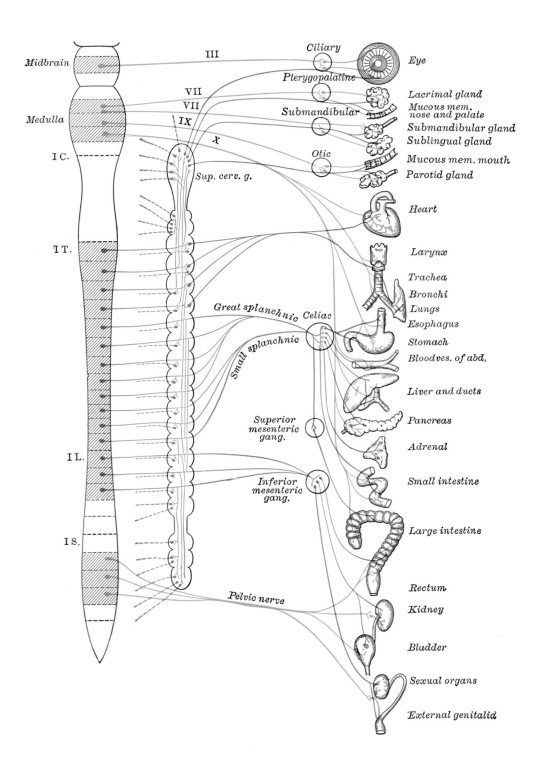

Fig. 17. *Diagram of efferent autonomic nervous system. Blue, cranial and sacral outflow, parasympathetic. Red, thoracolumbar outflow, sympathetic. ----------, postganglionic fibers to spinal and cranial nerves to supply vasomotors to head, trunk and limbs, motor fibers to smooth muscles of skin and fibers to sweat glands. (Modified after Meyer and Gottlieb.) This is only a diagram and does not accurately portray all the details of distribution. (Gray's Anatomy, Lea & Febiger.)*

efferent autonomic nervous system. It represents the situation in man which, however, is not basically different than we find in the cat. The superior cervical ganglion of man is called the cranial cervical ganglion in the cat. In the cat the first three thoracic sympathetic trunk ganglia combine to form the stellate ganglion. Note particularly in Figure 17 how through the cranial (*superior*) cervical ganglion postganglionic sympathetic fibers reach the effectors of the autonomic system in the head—the smooth muscles of the iris, and ciliary muscles of the lens of the eye, the smooth muscle of the blood vessels, and the glands.

Beyond the illustrations and descriptions presented, the autonomic system in its finest detail is far more complex. There are many small ganglia in the course of autonomic pathways, some microscopic, which we have not discussed. It is said that the distribution of autonomic nerves is as extensive as the vessels of the circulatory system. These complexities, however, are not deviations of the general plan of the system as presented.

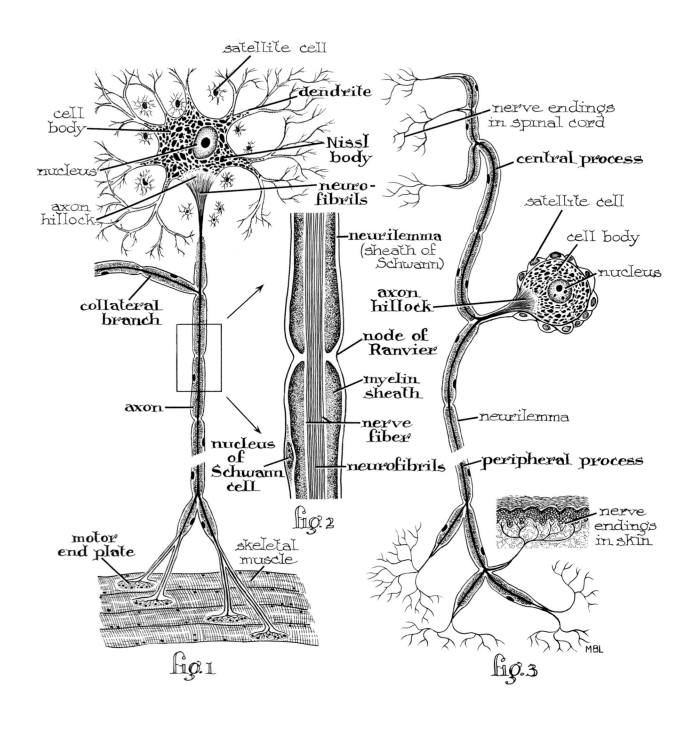

satellite cell

dendrite

cell body

Nissl body

nucleus

neuro-fibrils

axon hillock

nerve endings in spinal cord

central process

neurilemma (sheath of Schwann)

satellite cell

cell body

nucleus

collateral branch

axon hillock

node of Ranvier

myelin sheath

axon

nerve fiber

nucleus of Schwann cell

neurofibrils

neurilemma

peripheral process

fig. 2

motor end plate

skeletal muscle

nerve endings in skin

fig. 1

fig. 3

MBL

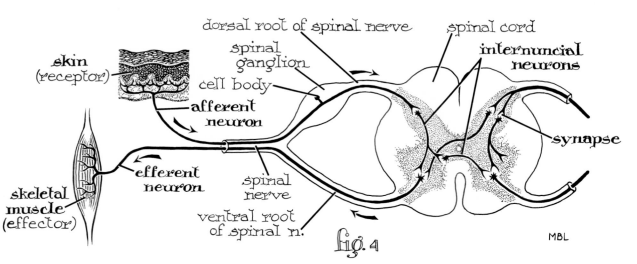

dorsal root of spinal nerve

spinal cord

spinal ganglion

internuncial neurons

skin (receptor)

cell body

afferent neuron

synapse

efferent neuron

spinal nerve

skeletal muscle (effector)

ventral root of spinal n.

fig. 4

MBL

(*Crouch: Functional Human Anatomy, Lea & Febiger*)

Nervous tissue is made up of nerve cells which are called neurons and of supporting tissue, the neuroglia. **Neurons** are the basic structural units of the nervous system. Neurons synapse one with another to form chains for the transmission of nerve impulses. These synapses, defined as areas of functional continuity between neurons, are located in the central gray matter of the central nervous system or in ganglia which lie in the peripheral nervous system. Neurons are most often related in such a way as to form **reflex arcs** which may be called the functional units of the nervous system.

This illustration shows the structure of two kinds of neurons and a reflex arc. Figure 1 is an **efferent** or **motor neuron** or more specifically a lower efferent or motor neuron terminating by **motor end plates** in skeletal or voluntary muscle. One is seen in its proper relationship in figure 4 on this plate. Figure 3 is an **afferent** or **sensory neuron** originating in the receptors of the skin, carried in spinal or cranial nerves and with its cell body in one of the cerebrospinal ganglia. One such neuron is seen in its normal relationship in a spinal nerve and reflex arc in figure 4 of this plate.

A neuron is composed of a cell body (*perikaryon*) and its processes. The **cell body** contains a large, vesicular nucleus with a prominent nucleolus. Within the cytoplasm (*neuroplasm*) are found the mitochondria, Golgi bodies, fat droplets and pigments. In addition, there are the specialized structures of nerve cell bodies: the chromophil substance in the form of **Nissl bodies** and fine, longitudinally arranged fibrils, the **neurofibrils.**

The **processes** of cell bodies are thread-like cytoplasmic extensions of two kinds, dendrites and axons. **Dendrites** (*dendrons*) receive and conduct impulses toward the cell body. There may be one or more dendrites and they vary as to size, shape, and the extent of their branching. Figure 1 is an example of a neuron with numerous branching dendrites. In figure 3 the structure called the **peripheral process** answers our definition of a dendrite.

Axons (*axis cylinders*) conduct impulses away from the cell body and are commonly called nerve fibers. Each neuron has only one axon, but it may give off a number of branches called **collaterals.** The **central process** in figure 3 corresponds to the axon. Axons are of uniform diameter and contain neurofibrils and mitochondria but no Nissl bodies or granules. At the point of origin of the axon from the cell body there is an elevated area, the **axon hillock,** which is usually devoid of Nissl bodies. At their free ends, axons branch out and either end in effector organs such as muscle tissue (fig. 1), glands, or contact the dendrites and cell body of another neuron or neurons to form a **synapse** (fig. 4). Since many axons have their cell bodies

in the gray matter of the spinal cord or brain and extend to peripheral muscles as far away as the feet or hands, they must be very long. Others, of course, are shorter.

Axons may be covered with one or two sheaths: an inner, relatively thick myelin or medullary sheath and an outer, thin, cellular sheath, the neurilemma (fig. 2). Note that the **myelin sheath** is not a continuous covering, but is broken at intervals into separate segments by constrictions, the **nodes of Ranvier.** The segments between nodes are called **internodes.** The axons of some neurons are without myelin. We, therefore, speak of fibers as **myelinated** (*medullated*) or **unmyelinated** (*nonmedullated*). Myelinated fibers are found in the central nervous system where they give to the areas in which they are concentrated a whitish or yellowish color, hence the term white matter. The myelin sheaths of the central nervous system are more continuous than shown in figures 1 and 3. They tend to have nodes only where they form collaterals (*branches*). Figures 1 and 3 show myelinated fibers from peripheral nerves where nodes are spaced at regular intervals and the internodes are relatively short. Unmyelinated fibers are found in the autonomic nervous system, mostly in the postganglionic neurons and in the gray areas of the central nervous system.

The **neurilemma** (*sheath of Schwann*) is found on fibers of the peripheral and autonomic nervous system and probably on those of the central nervous system (figs. 1 to 3). It is composed of a single layer of flattened cells, one cell per internode in myelinated fibers and it is continuous in the nodes of Ranvier. In unmyelinated fibers the neurilemma is difficult to distinguish from the axon on which it rests, except by the presence of the nuclei. A silver stain blackens the nerve fibers and then the neurilemma stands out in contrast.

Plate 96 shows some of the different types of neurons and some neuroglial cells. Plate 97 gives a modern view of the relationship of the myelin and neurilemma and also how nerve fibers (*axons and dendrites*) relate to nerves.

Figure 4 is a diagram of a simple **reflex arc**—the functional unit of the nervous system. It provides the opportunity to see how neurons of different types relate to each other, to the peripheral and central nervous system, and to receptor and effector structures.

The parts of the reflex arc as shown in figure 4 are as follows.

(1) **receptor**—in this case nerve endings in the skin.
(2) **afferent neuron**—or sensory neuron conveying impulses to the spinal cord. Its peripheral process travels in a spinal nerve; its cell body is in the spinal ganglion on the dorsal root of a spinal nerve; and

275

its central process (*axon*) enters the dorsal column of gray matter and synapses with an

(3) **internuncial neuron** or association neuron which makes various connections with other neurons in the cord and finally with an

(4) **efferent neuron** in the ventral column of the gray matter. This neuron leaves the cord through the ventral root of the spinal nerve and through the spinal nerve reaches an

(5) **effector** which responds to the impulses.

Figure 4 shows clearly that many connections can be made by an afferent nerve and by internuncial neurons coming into the central gray matter of the spinal cord—or the brain stem. This diagram can show them, however, at only one level of the cord. Not only do the internuncial neurons cross to the other side making a bilateral response possible, but they may extend to different levels of the cord or brain. An afferent neuron may also have ascending and descending branches which synapse at various levels.

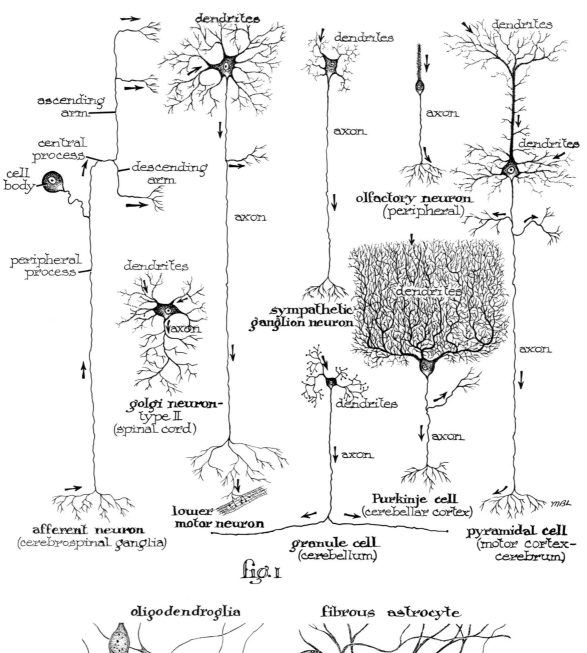

dendrites

ascending
arm

central
process

cell
body

descending
arm

peripheral
process

dendrites

axon

golgi neuron-
type II
(spinal cord)

axon

afferent neuron
(cerebrospinal ganglia)

lower
motor neuron

dendrites

axon

sympathetic
ganglion neuron

dendrites

axon

granule cell
(cerebellum)

dendrites

olfactory neuron
(peripheral)

dendrites

axon

Purkinje cell
(cerebellar cortex)

dendrites

dendrites

axon

MBL

pyramidal cell
(motor cortex-
cerebrum)

fig. 1

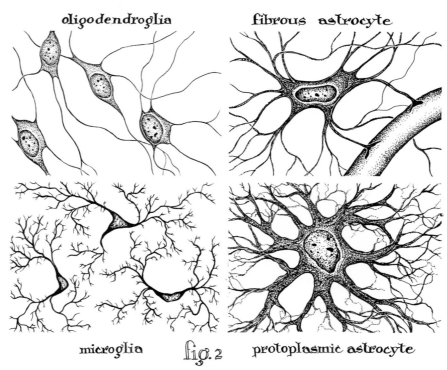

oligodendroglia

fibrous astrocyte

microglia   fig. 2   protoplasmic astrocyte

(Fig. 1. Modified from Bailey—fig. 2. Crouch, Functional Human Anatomy, Lea & Febiger)

This plate illustrates a number of different types of neurons and neuroglial cells. The neurons in figure 1 are shown schematically. No effort has been made to show myelin and neurilemma but rather to indicate differences in form of the cells and in the number and type of branching of cell processes.

Neurons of three general types are shown in figure 1; unipolar, bipolar, and multipolar. **Unipolar neurons** are exemplified by the afferent neuron in figure 1 which has a single process on the cell body from which branch the peripheral and central processes. These are considered by some authors as bipolar on the basis of embryological development, the single process forming by the joining of two processes. They are found in the spinal nerves with their cell bodies in the spinal ganglion and in certain sensory cranial nerves. A **bipolar neuron,** with two separate processes is seen in the **olfactory** neuron in figure 1 and they are also found in the ganglia of the vestibulo-cochlear (*auditory, statoacoustic*) nerve and in the retina of the eye. **Mutipolar neurons** are represented by the remainder of the illustrations in figure 1. They are by far the most numerous of the three types. The position in the nervous system of those illustrated is apparent either by their names or by the words in parenthesis under each diagram.

Drawings of four types of cells found in the supporting or **neuroglia tissue** of the nervous system are shown in figure 2. Neuroglia, a word taken from the Greek, means nerve glue. The **oligodendroglia** (*oligodendrocytes*) are small cells with relatively few processes. They are found in close association with small blood vessels and with large nerve cells and in the white matter of the central system they lie between bundles of fibers. Their processes frequently clasp the nerve fibers. Oligodendrocytes are the "satellite cells" (see Plate 95, figs. 1 and 2) of nerve cells and may serve in the formation and preservation of the myelin sheaths of fibers of the central nervous system, thus assuming the role of the neurilemma of peripheral nerves. **Microglia** (*microgliocytes*) are found throughout the gray and white matter of the central system. They are small cells whose two or more processes are finely branched and feathery. They are phagocytic and thus serve to remove dead or dying parts of tissues and foreign materials. The microgliocytes are of mesodermal origin in contrast to the other glial cells which are ectodermal and often called collectively macroglia. Astrocytes are of two types, fibrous, and protoplasmic. They are relatively large cells with many radiating processes whose terminal portions may have expansions which attach to the pia mater or to blood vessels. **Fibrous astrocytes** are found mostly in the white matter of the central nervous system. They have fibers which run through the cytoplasm of their cell bodies—hence their name. The **protoplasmic astrocytes** occur mainly in the gray matter. The astrocytes contribute to the repair processes in the central nervous system and help to support nervous structures.

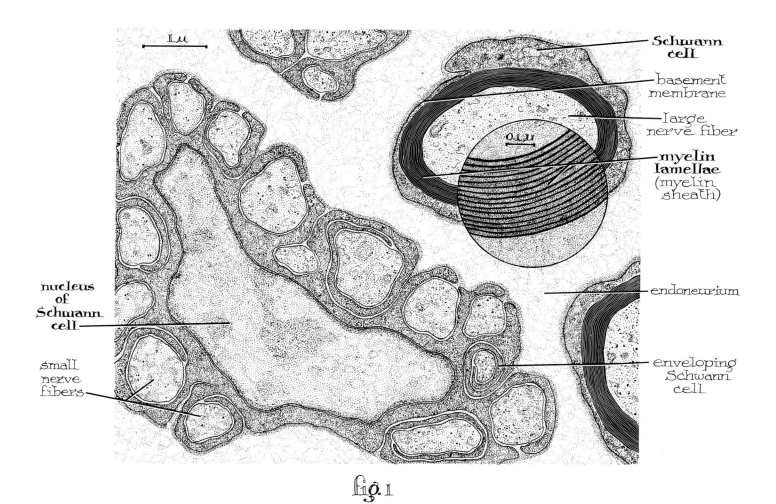

Schwann cell

basement membrane

large nerve fiber

myelin lamellae (myelin sheath)

1 μ

0.1 μ

endoneurium

nucleus of Schwann cell

small nerve fibers

enveloping Schwann cell

fig. 1

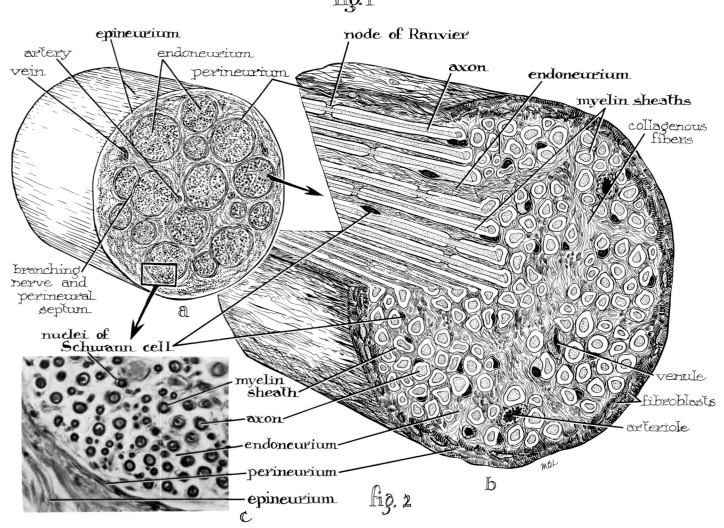

epineurium

artery

vein

endoneurium

perineurium

node of Ranvier

axon

endoneurium

myelin sheaths

collagenous fibers

branching nerve and perineural septum

a

nuclei of Schwann cell

myelin sheath

axon

endoneurium

perineurium

epineurium

c

venule

fibroblasts

arteriole

b

fig. 2

Electron Micrograph of Nerve Fibers and Schwann Cells; Study of Microscopic Structure of a Nerve
PLATE 97

(Figure 1—Modified from Porter and Bonneville)

Figure 1 is a drawing made from an electron micrograph of peripheral nerve fibers (*axons*) and their sheaths. The overall magnification is 13,750×, that of the insert in the circle at the upper right is 77,750×. The structure in the lower left part of the illustration consists of many small nerve fibers disposed around the large nucleus of a Schwann cell, each one enveloped by cytoplasmic extensions of the Schwann or sheath cell. Careful examination of the plate will reveal places where the plasma membrane of the Schwann cell can be traced without a break as it folds in from the surface and embraces the fiber. If the fiber is deeply embedded in the Schwann cell, the opposition of the plasma membranes, as they fold in around the fiber, leave a channel, the mesaxon (*see top center of fig. 1*).

At the upper right in figure 1 is a large nerve fiber with a thick myelin sheath, the lamella of which represent successive layers of the plasma (*cell*) membrane of a Schwann cell. This interpretation of myelin and the neurilemma or sheath of Schwann as one structure is different than that in the description of Plate 95 where we talked of two sheaths. This more recent view has come about as a result of studies of nerve fibers with the electron microscope. Yet the older nomenclature still prevails and no doubt will, until there is further clarification of this problem.

Figure 2 is a study of the microscopic structure of a nerve. A nerve may be defined as bundles of nerve fibers coursing together outside of the central nervous system.

The fibers are held together by well organized connective tissue sheaths which give protection and strength to what would otherwise be very weak and therefore vulnerable structures. It is because of the connective tissue encasements that tiny axons (*fibers*) can travel great distances in the body to innervate muscles and other structures.

Figure 2a is a piece of a nerve cut in cross section. It shows the outer fibrous covering of the nerve, the **epineurium.** Scattered through the nerve are a number of **fasciculi** (*bundles*) of nerve fibers; these bundles are covered with an areolar connective tissue, the **perineurium.** The fibers constituting the fasciculi are, in turn, covered with a fine sheath of reticular fibers in areolar connective tissue, the **endoneurium.** The connective tissue elements of these various sheaths are continuous one with another. In the sheaths are found blood vessels and lymphatics appropriate to the size of the nerves.

Figure 2b is one nerve fascicle (*bundle*) shown in longitudinal and cross section. It is greatly magnified to show not only the **perineurium** and **endoneurium** as described above but individual **axons** (*fibers*) with their **myelin sheaths, Schwann cells with nuclei,** and **nodes of Ranvier.** Arterioles, venules, collagenous fibers and fibroblasts are also present.

Figure 2c is a photomicrograph of a small section of one fascicle of the same nerve. Osmic acid stain makes the myelin black. Notice again the same structures as mentioned above.

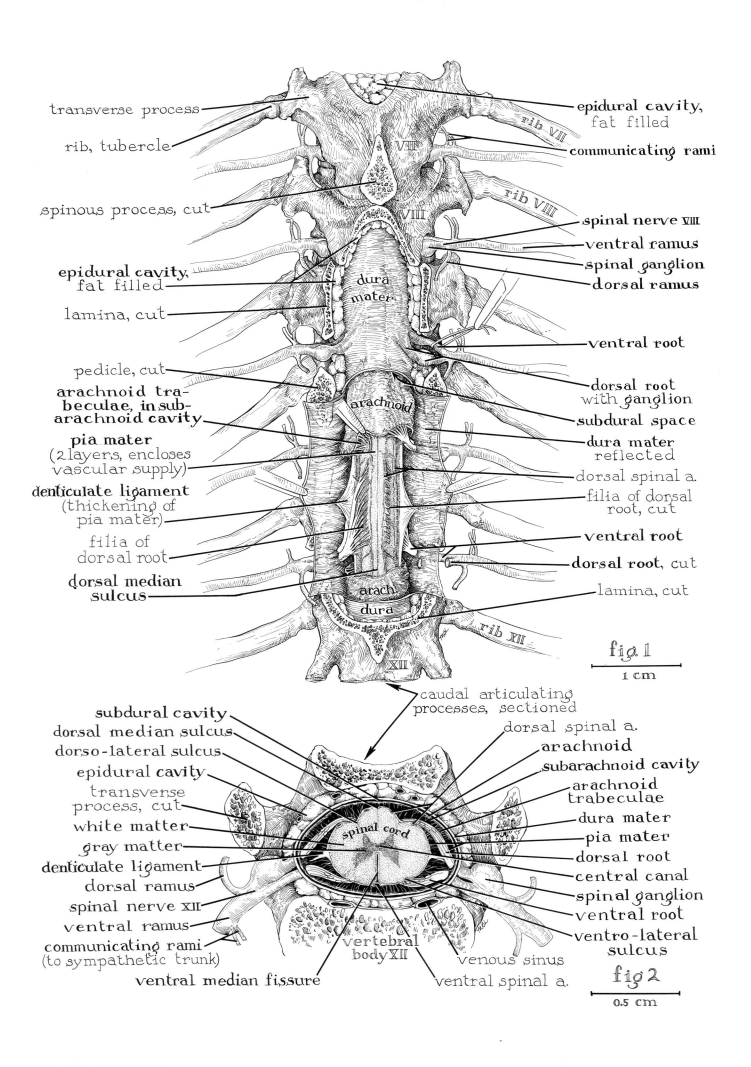

transverse process

rib, tubercle

spinous process, cut

epidural cavity, fat filled

lamina, cut

pedicle, cut

arachnoid trabeculae, in sub-arachnoid cavity

pia mater (2 layers, encloses vascular supply)

denticulate ligament (thickening of pia mater)

filia of dorsal root

dorsal median sulcus

epidural cavity, fat filled

communicating rami

rib VII

rib VIII

spinal nerve VIII

ventral ramus

spinal ganglion

dorsal ramus

ventral root

dorsal root with ganglion

subdural space

dura mater reflected

dorsal spinal a.

filia of dorsal root, cut

ventral root

dorsal root, cut

lamina, cut

VII

VIII

dura mater

arachnoid

arach.

dura

XII

rib XII

fig. 1

1 cm

caudal articulating processes, sectioned

subdural cavity

dorsal median sulcus

dorso-lateral sulcus

epidural cavity

transverse process, cut

white matter

gray matter

denticulate ligament

dorsal ramus

spinal nerve XII

ventral ramus

communicating rami (to sympathetic trunk)

ventral median fissure

dorsal spinal a.

arachnoid

subarachnoid cavity

arachnoid trabeculae

dura mater

pia mater

dorsal root

central canal

spinal ganglion

ventral root

ventro-lateral sulcus

spinal cord

vertebral body XII

venous sinus

ventral spinal a.

fig 2

0.5 cm

Figure 1, a lengthwise view, and figure 2, a cross section of the vertebral column, spinal cord and its meninges and spinal nerves, serve to show the relationships of these important structures.

Figure 1 deals with the area between the seventh and twelfth vertebrae. The vertebral arches and spines have been removed from the eighth to the forepart of the twelfth vertebra to reveal the vertebral canal and its contents—the meninges and spinal cord. The meninges have been removed in successive layers to show all of the meninges and their intermediate spaces. You may wish to review the structure of the vertebral column by referring to Plates 12 and 13.

Figures 1 and 2 will be described together as they relate so closely. Note first that the **spinal cord** with its **meninges** passes through the vertebral canal and sends its spinal nerves, at segmental intervals, out through the intervertebral foramina. The **spinal ganglia** on the dorsal roots of the spinal nerves lie barely to the outside of the intervertebral foramina and the **dorsal** and **ventral roots** join just beyond the ganglion to form the **spinal nerve**. Most of the spinal nerves divide almost immediately into **dorsal, ventral** and **communicating rami** as seen well in figure 2. The spinal nerve roots, dorsal and ventral, each arise by a series of strands or filia as seen best, at least for the dorsal root, in figure 1.

The spinal meninges consist of three membranes, separated by spaces, which enclose and protect the spinal cord. Cranial meninges, which cover the brain, are continuous with them at the foramen magnum of the skull (Plate 104).

The outer meninx, the **dura mater,** is of a tough, fibrous nature. It extends out on to the roots of the spinal nerves where it contributes to the nerve covering, the epineurium. Between the dura mater and the bony wall of the vertebral canal is a space, the **epidural space,** filled with fat. It also contains venous sinuses. Beneath the dura mater is a narrow *subdural space* containing a tiny bit of fluid.

The next spinal meninx is the **arachnoid,** a very thin translucent membrane which follows the pattern of the dura mater. Under the arachnoid is a **subarachnoid space** which is filled with cerebrospinal fluid and is connected to the third and final meninx, the pia mater, by connective tissue, **arachnoid trabeculae,** which divide the cavity into numerous spaces. This fluid-filled space is like a cushion around the spinal cord and reduces danger to the cord.

The **pia mater** meninx is a thin, transparent but tough membrane which adheres closely to the spinal cord and extends into the sulci and fissures of the cord. It contains all the blood vessels adjacent to the spinal cord.

The **denticulate ligament** is a lateral thickening of the pia mater whose edges are free except between the spinal nerves where it has tooth-like serrations that connect to the arachnoid and dura mater. It serves to hold the spinal cord in the center of the subarachnoid cavity.

The spinal cord has a number of surface markings. Along its mid-dorsal line is a **dorsal median sulcus,** a slight indentation dividing the cord into right and left halves. At the bottom of this sulcus, a dorsal median septum composed of nervous supporting tissue, called neuroglia, goes ventrally into the cord almost to its **central canal.** This septum is shown in figure 2 beneath the median sulcus but is not labeled. Lateral to the dorsal median sulcus on each side is a **dorsolateral sulcus.** As seen in figure 1, the filia of the dorsal root of the spinal nerve attach along this sulcus. In the cervical and part of the thoracic region, a dorsointermediate sulcus lies between the other two mentioned above (See Plate 101, fig. 1). It marks the position between important nerve tracts in the cord, the median fasciculus gracilis and the lateral fasciculus cuneatus.

A **ventral median fissure** divides the ventral part of the cord into two halves and the ventral spinal artery can be seen at the peripheral portion of this fissure (fig. 2). A **ventrolateral sulcus** can be seen where the filia of the ventral root of the spinal nerve attach.

The cut section of the spinal cord in figure 2 shows clearly a **central canal** and an H-shape in the center which is **gray matter.** The gray matter is composed largely of the cell bodies of neurons and unmyelinated nerve fibers.

The **white matter,** which surrounds the gray, is made up of myelinated and some unmyelinated nerve fibers and supporting neuroglia cells. It is organized into a complex of nervous pathways connecting different levels of the nervous system. They would divide roughly into ascending and descending tracts.

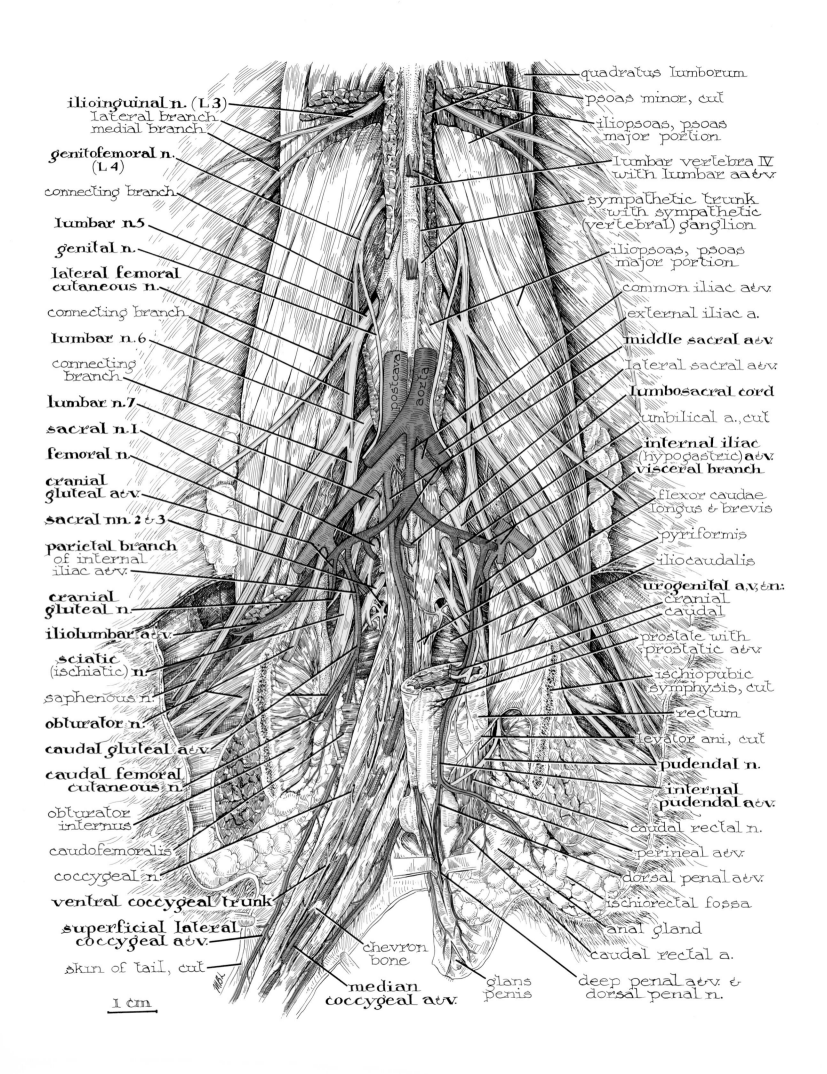

ilioinguinal n. (L3)
lateral branch
medial branch

genitofemoral n.
(L4)

connecting branch

lumbar n.5

genital n.

lateral femoral
cutaneous n.

connecting branch

lumbar n.6

connecting
branch

lumbar n.7

sacral n.1

femoral n.

cranial
gluteal a&v.

sacral nn.2&3

parietal branch
of internal
iliac a&v.

cranial
gluteal n.

iliolumbar a&v.

sciatic
(ischiatic) n.

saphenous n.

obturator n.

caudal gluteal a&v.

caudal femoral
cutaneous n.

obturator
internus

caudofemoralis

coccygeal n.

ventral coccygeal trunk

superficial lateral
coccygeal a&v.

skin of tail, cut

1 cm

postcava

aorta

chevron
bone

median
coccygeal a&v.

glans
penis

quadratus lumborum

psoas minor, cut

iliopsoas, psoas
major portion

lumbar vertebra IV
with lumbar aa&v.

sympathetic trunk
with sympathetic
(vertebral) ganglion

iliopsoas, psoas
major portion

common iliac a&v.

external iliac a.

middle sacral a&v.

lateral sacral a&v.

lumbosacral cord

umbilical a., cut

internal iliac
(hypogastric) a&v.
visceral branch

flexor caudae
longus & brevis

pyriformis

iliocaudalis

urogenital a,v,&n.:
cranial
caudal

prostate with
prostatic a&v.

ischiopubic
symphysis, cut

rectum

levator ani, cut

pudendal n.

internal
pudendal a&v.

caudal rectal n.

perineal a&v.

dorsal penal a&v.

ischiorectal fossa

anal gland

caudal rectal a.

deep penal a&v. &
dorsal penal n.

This plate shows the dorsal and part of the lateral walls of the peritoneal cavity and the proximal portions of the thighs and tail. The peritoneum and fascia have been removed to reveal the ventral vertebral muscles (*psoas major, psoas minor (cut) and quadratus lumborum*). Parts of the postcava and abdominal aorta and adjacent blood vessels are left in position to show relationships of vessels and spinal nerves. A part of the sympathetic trunk with sympathetic trunk (*vertebral*) ganglia are shown. The **lumbosacral plexus** is the main point of emphasis in this plate.

The **third lumbar** or **ilioinguinal nerve,** seen at the top of the plate, is not involved in the lumbosacral plexus. It divides into lateral and medial branches. The lumbosacral plexus is made up of the ventral rami and interconnecting branches of the fourth through the seventh lumbar and the first through the third sacral. The lumbar plexus, considered separately, is made up of lumbar nerves four through seven. The sacral plexus involves lumbar nerves six and seven and sacrals one through three.

Lumbar 4 forms the **genitofemoral nerve** which gives off a **genital nerve** and a branch which joins **lumbar 5** with which it forms a **lateral femoral cutaneous nerve.** **Lumbar 5,** besides contributing to the lateral cutaneous nerve, also contributes to the **femoral nerve** and sends a branch to lumbar 6. **Lumbar 6** is a large nerve which forms the major part of the femoral and the **obturator nerves** and sends a large branch to **lumbar 7.** The latter by its connection with lumbar 6 contributes to the obturator nerve and then passes caudad with a part of lumbar 6 to connect with the sacral nerves. This constitutes the **lumbosacral cord** which joins the ventral ramus of the first sacral nerve to form the **sciatic** (*ischiatic*) **nerve** and the **cranial** and caudal (not shown) **gluteals.** It also sends a branch to **sacral 2.** Sacrals **2 and 3** are small forming a network with the branch from sacral 1 from which arise the **pudendal, caudal femoral cutaneous,** caudal rectal and a small branch to the sciatic nerve.

The ventral roots of the sacral nerves forming the sacral plexus described above leave the sacral vertebral canal through the two ventral sacral foramina and through the foramen between the sacrum and the first caudal vertebra. The dorsal rami of the sacral spinal nerves leave the vertebral canal through corresponding foramina on the dorsal side.

The coccygeal nerves emerge from the intervertebral foramina of the first seven or eight caudal vertebrae. The dorsal rami go to the skin and dorsal muscles of the tail. The ventral rami are joined to each other and to the last sacral nerve to form the **ventral coccygeal trunk** from which coccygeal branches innervate the skin and muscles of the ventral side of the tail.

The femoral, obturator, pudendal and other nerves are traced peripherally in Plates 89 and 90. The **genital nerve** cut in this illustration is picked up in Plate 88 and carried out to more distal structures.

The blood vessels of the pelvis are shown better here than in Plates 66 and 80. Note the **middle sacral artery and vein,** the former a continuation of the common iliac artery which in turn is a continuation of the aorta. The latter vein is a tributary of the right common iliac vein though it more frequently joins the left common iliac. These two vessels pass together on to the ventral side of the sacrum and tail, passing through the hemal arches and then between the chevron bones. They are called, in the tail region, the **median coccygeal artery and vein.** The middle sacrals give off or recieve one branch the lateral sacral artery and vein which enters the ventral sacral foramen and is distributed to the sacral canal. They send branches out of the dorsal sacral foramen to the dorsal muscles.

Note the **internal iliac arteries and veins** and their branches which supply the pelvic organs and adjacent areas. The first branch of the internal iliac artery is the umbilical artery to the urinary bladder. In the fetus the umbilical carries blood to the placenta and is the main artery of the pelvis. In the adult the umbilical provides branches to the urinary bladder and the distal part of the vessel occludes and becomes the lateral umbilical ligament.

The internal iliac artery and vein give rise to cranial gluteal and iliolumbar vessels. The former reach the gluteal muscles; the latter, the muscles of the rump and thigh and the gluteal muscles. Cranial gluteal and iliolumbar vessels anastomose caudally. The iliolumbar anastomoses with the superficial circumflex iliac cranially. The internal iliacs terminate in parietal and visceral branches. The **parietal branch** gives rise to the **caudal gluteal** to the gluteal and neighboring muscles and to the **superficial lateral coccygeal,** supplying the tail or in the case of the vein draining it. The **visceral branch** gives rise to cranial and caudal **urogenital vessels** and small branches to the prostate. It continues caudad as the **internal pudendal** which sends branches to the recutm, perineum and penis.

Nomenclature varies in regard to the branches of the internal iliac arteries and veins. The iliolumbar is often considered a branch of the cranial gluteal. The visceral branch is the same as the middle hemorrhoidal (*rectal*) of Reighard and Jennings.

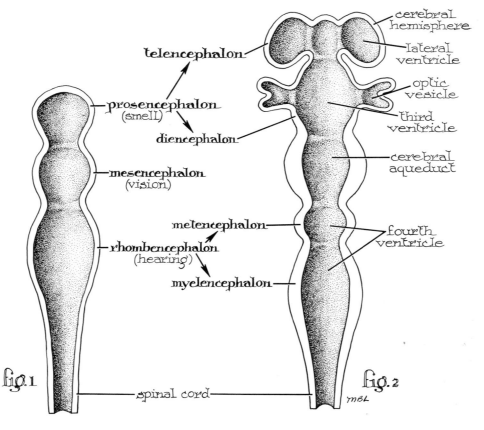

telencephalon

prosencephalon
(smell)

diencephalon

mesencephalon
(vision)

metencephalon

rhombencephalon
(hearing)

myelencephalon

cerebral
hemisphere

lateral
ventricle

optic
vesicle

third
ventricle

cerebral
aqueduct

fourth
ventricle

spinal cord

fig. 1

fig. 2

MBL

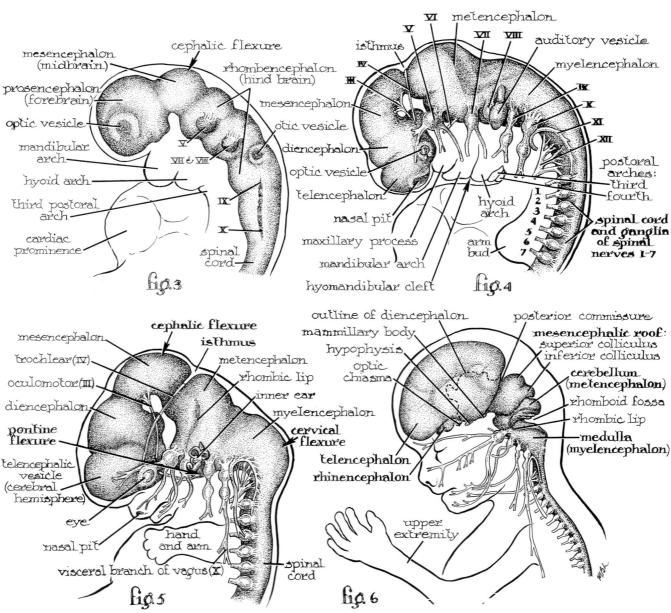

mesencephalon
(midbrain)

cephalic flexure

rhombencephalon
(hind brain)

prosencephalon
(forebrain)

optic vesicle

mesencephalon

mandibular
arch

otic vesicle

hyoid arch

diencephalon

third postoral
arch

optic vesicle

cardiac
prominence

telencephalon

V

VII & VIII

IX

X

spinal
cord

fig. 3

VI   metencephalon

V   VII VIII   auditory vesicle

isthmus

IV   myelencephalon

III

IX

X

XI

XII

postoral
arches:
third
fourth

mesencephalon

diencephalon

optic vesicle

telencephalon

nasal pit

maxillary process

mandibular arch

hyomandibular cleft

hyoid
arch

arm
bud

1
2
3
4
5
6
7

spinal cord
and ganglia
of spinal
nerves 1-7

fig. 4

mesencephalon

cephalic flexure

isthmus

trochlear (IV)

metencephalon

oculomotor (III)

rhombic lip

diencephalon

inner ear

pontine
flexure

myelencephalon

telencephalic
vesicle
(cerebral
hemisphere)

cervical
flexure

eye

nasal pit

hand
and arm

visceral branch of vagus (X)

spinal
cord

fig. 5

outline of diencephalon

posterior commissure

mammillary body

mesencephalic roof:
superior colliculus
inferior colliculus

hypophysis

optic
chiasma

cerebellum
(metencephalon)

rhomboid fossa

rhombic lip

telencephalon

medulla
(myelencephalon)

rhinencephalon

upper
extremity

MBL

fig. 6

The central nervous system, both brain and spinal cord, develops from a hollow neural tube derived from a thickened neural plate on the dorsal side of the ectoderm of the embryo. This plate depresses centrally to form a groove; the groove deepens as the edges of the plate elevate; the edges close over to form the tube; the ectoderm closes over the tube externally. The brain, with which we are concerned in this plate, differentiates from the cranial portion of the neural tube by rapid growth, thickening, flexing and evagination.

This illustration shows six stages in the development of the vertebrate brain and cranial nerves, the last four as seen in **man.** The process of development does not differ significantly from that in the cat.

Figures 1 and 2 are schematic presentations of very early stages in the development of the brain. Figure 1 shows a stage where the neural tube has, by expansion, formed three enlargements or vesicles: the **prosencephalon** (*forebrain*), the **mesencephalon** (*midbrain*), and the **rhombencephalon** (*hindbrain*). The functions of smell, vision and hearing respectively may be assigned to these parts. Figure 2 shows further growth and differentiation by thickening, evagination and constriction. The prosencephalon has now formed two structures: the **telencephalon** with lateral outgrowths, the cerebral hemispheres and lateral ventricles within; the **diencephalon** with lateral outgrowths, the optic vesicles and inside, the third ventricle.

The mesencephalon does not divide but the cavity within is the cerebral aqueduct.

The rhombencephalon forms two parts, a cranial **metencephalon** and a caudal **myelencephalon.** The cavity within these parts is the fourth ventricle. The structures so far are best seen in the brain of lower vertebrates although man, as seen in the remaining figures, is not very different except in degree of development. These five parts constitute the brain stem with the outgrowths, the cerebral hemispheres, marking the beginning of the superstructure of the brain, sometimes called the brain proper. The brain stem is continuous caudad as the spinal cord with its central canal. The system of ventricles shown in figure 2 persists with some modification and can be demonstrated in the brain of the adult (See Plate 102).

Figure 3 is a drawing of the brain of a 3.5-millimeter, 3 to 4 week-old-human embryo. This and figures 4, 5 and 6 are all drawn in body outlines of constant size and the brain and nerves are drawn in proportion to the outlines. In the 3.5-millimeter human brain (fig. 3) the three primary divisions of the brain are shown as described above. The

rhombencephalon has started to divide and the prosencephalon is bent forward at the **cephalic flexure.** The whole brain is starting to bend forward at its point of continuation with the spinal cord, the **cervical flexure.** These flexures are labeled in boldface type on figure 5 along with a **pontine flexure** near the cranial end of the rhombencephalon—a curvature in the opposite direction from the other two. The developing optic vesicle is apparent on the prosencephalon; the fifth cranial nerve is appearing at the cranial end of the rhombencephalon; the seventh and eighth nerves are seen developing just caudad of the fifth; an otic vesicle lies in the midpart of the rhombencephalon and cranial nerves nine and ten are developing from the caudal portion of the rhombencephalon. The spinal cord is to be noted at the bottom of the figure. By examining the outline of the embryo one can become aware of some of the relationships of structures in this cranial end of the animal. The developing mandibular, hyoid and third postoral arches are shown and caudad and ventrad of them the prominence which marks the position of the heart.

In figure 4, an embryo 7 millimeters long and 5 weeks old shows considerable change beyond the condition of the 3 to 4-week embryo. The cervical and cephalic flexures (not labeled) are more prominent. The prosencephalon has differentiated into telencephalon and diencephalon; a nasal pit has formed; the remaining cranial nerves have appeared; and a maxillary arch has developed craniad of the mandibular arch. An arm bud is shown at the bottom of the figure and a series of 7 spinal nerves and ganglia have developed from the spinal cord.

Figure 5, a 17-millimeter 7-week embryo, shows all three flexures, the **cephalic, pontine** and **cervical,** the cephalic and cervical being very pronounced. The eye shows clearly; the ear structure is well developed; the cranial nerves have elongated and the hand is developing on the arm.

The brain in figure 6 (*60-millimeter 11-week embryo*) is in a position more similar to that of the adult, the pontine flexure having deepened to lift the forebrain dorsad and craniad. The telencephalon has enlarged considerably and an olfactory area, the **rhinencephalon,** is seen on the ventral side along with mammillary bodies, hypophysis and optic chiasma. The **roof of the mesencephalon** shows two pairs of elevations, the superior and inferior colliculi, known collectively as the corpora quadrigemina which have to do with auditory and visual functions. From the metencephalon, a **cerebellum,** and from the myelencephalon, a **medulla** have differentiated.

Derivations of primary and secondary divisions of the embryonic brain—a summary.

Prosencephalon (forebrain)	Telencephalon . . .	Cerebral hemispheres; olfactory bulbs; basal nuclei; lateral ventricles	Mesencephalon . . . . (midbrain)		Corpora quadrigemina; cerebral peduncles; cerebral aquaduct
	Diencephalon	Epithalamus; thalamus; hypothalamus; infundibulum; pineal body; third ventricle	Rhombencephalon (hindbrain)	Metencephalon . . .	Pons; cerebellum; part of fourth ventricle
				Myelencephalon . .	Medulla oblongata; part of fourth ventricle.

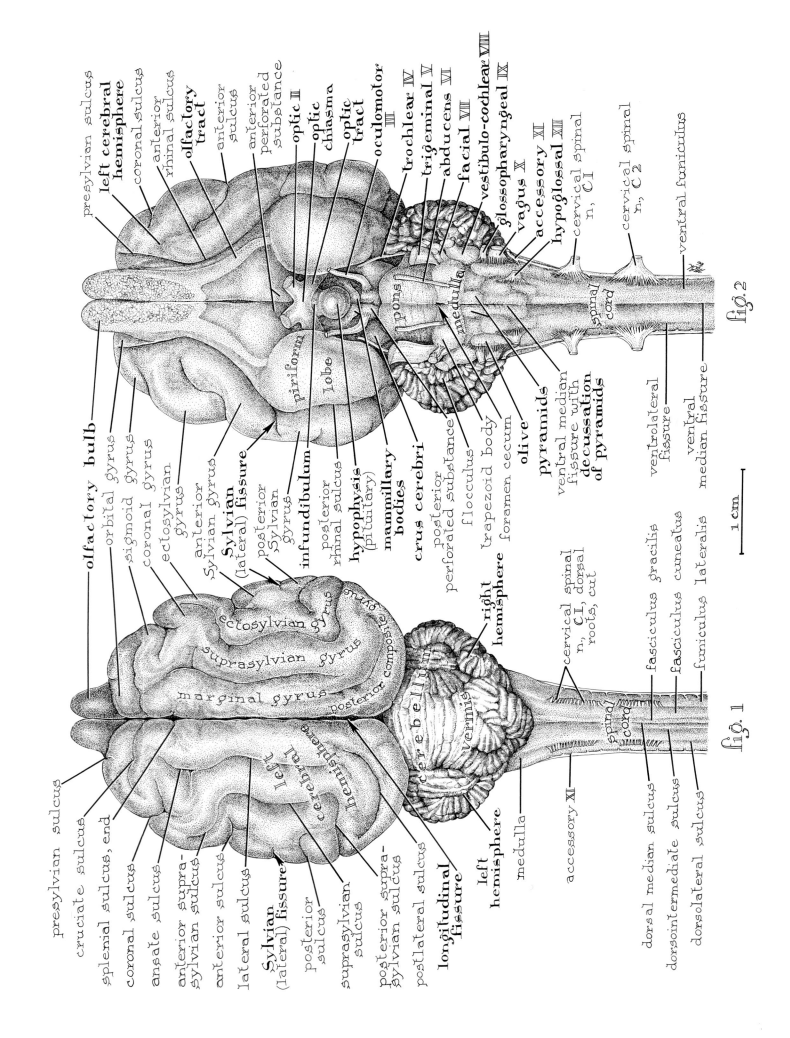

presylvian sulcus

cruciate sulcus

splenial sulcus, end

coronal sulcus

ansate sulcus

anterior supra-sylvian sulcus

anterior sulcus

lateral sulcus

Sylvian (lateral) fissure

posterior sulcus

suprasylvian sulcus

posterior supra-sylvian sulcus

postlateral sulcus

longitudinal fissure

left hemisphere

medulla

accessory XI

dorsal median sulcus

dorsointermediate sulcus

dorsolateral sulcus

ectosylvian gyrus

suprasylvian gyrus

marginal gyrus

posterior composite gyrus

left cerebral hemisphere

cerebellum

vermis

spinal cord

fig. 1

presylvian sulcus

left cerebral hemisphere

coronal sulcus

anterior rhinal sulcus

olfactory tract

anterior sulcus

anterior perforated substance

optic II

optic chiasma

optic tract

oculomotor III

trochlear IV

trigeminal V

abducens VI

facial VII

vestibulo-cochlear VIII

glossopharyngeal IX

vagus X

accessory XI

hypoglossal XII

cervical spinal n., CI

cervical spinal n., C2

ventral funiculus

olfactory bulb

orbital gyrus

sigmoid gyrus

coronal gyrus

ectosylvian gyrus

anterior Sylvian gyrus

Sylvian (lateral) fissure

posterior Sylvian gyrus

posterior rhinal sulcus

piriform lobe

infundibulum

hypophysis (pituitary)

mammillary bodies

crus cerebri

posterior perforated substance

flocculus

trapezoid body

foramen cecum

olive

pyramids

ventral median fissure with **decussation of pyramids**

ventrolateral fissure

ventral median fissure

funiculus lateralis

fasciculus cuneatus

fasciculus gracilis

cervical spinal n., CI, dorsal roots, cut

right hemisphere

pons

medulla

spinal cord

1 cm

fig. 2

The brain is shown here in its natural relationship to the spinal cord—as a continuation of it. Like the spinal cord it is differentiated from the neural tube of the embryo. Its growth was more rapid, its differentiation and specialization greater. The first pair of cervical spinal nerves mark as well as any feature the end of the spinal cord cranially. The medulla is the most caudal part of the brain. It belongs to the brain stem or "old brain" as some call it. The remainder of the brain stem in figure 1 is concealed by the large cerebellum and the even larger cerebrum.

The cerebellum has many folds, called folia, and grooves, called fissures, as seen here in the dorsal view. They are the reflection of a very great superficial growth which had to fold in order to accommodate to the available space. This superficial part is the cortex. It is divided into three parts, a central vermis and right and left cerebellar hemispheres.

The **vermis** is so called because of its segmented appearance which reminds one of a segmented worm. It extends forward under the cerebral hemispheres and is also the most caudal part of the cerebellum. The **lateral hemispheres**, right and left, spread well beyond the sides of the medulla oblongata. Between the cerebrum and the cerebellum is a deep transverse fissure which goes all the way to the brain stem.

The **cerebrum** is the largest single part of the brain and because of a very extensive cortex it is folded into ridges called gyri and grooves called **sulci** or fissures. While most of these are named on the drawings, they will not be described except for a few rather conspicuous landmarks. The cerebrum is divided into right and left cerebral hemispheres by the **longitudinal fissure.** This fissure cuts very deep as can be seen in Plate 102, figure 2.

The **Sylvian** (*lateral*) **fissure** is an important reference point if one wishes to study and learn the gyri and sulci. The cerebral hemispheres are often divided into lobes and in so doing the Sylvian fissure is the dividing line between an anterior frontal lobe and a lateral temporal lobe. The posterior part of the hemisphere is the occipital lobe or by some considered a parietal lobe roughly lying under the parietal bone and the occipital lobe in back of that. Such lobes are not labeled here.

A further extension of the cerebrum anteriorly is the **olfactory bulb** whose ventral portion lies against the cribriform plate of the cerebrum. It is partly divided into right and left lobes.

Figure 2 is a ventral view of the brain and shows the brain stem to much greater advantage. The lateral hemispheres of the cerebellum are seen at the sides of the medulla oblongata and pons, their most ventral portion

the flocculus lying right next to the superficial origin of the **trigeminal nerve.**

The **piriform lobes,** the most ventral lobes of the cerebrum, lie medial to the posterior rhinal sulcus. This sulcus, which is continuous with the anterior rhinal sulcus, sets off the olfactory or older parts of the cerebrum from the so-called more recent new or neopallium of the cortex. The rhinal sulcus is one of the most constant grooves of the cerebral cortex.

The **Sylvian** (*lateral*) **fissure** is again seen laterally. Anteriorly the ventral aspect of the olfactory bulbs and the **olfactory tracts** leading caudad into the olfactory cortex are clearly shown.

Returning now to the brain stem, the medulla oblongata is the most caudal part. It is quite flat ventrally and widens gradually as it passes from the cord to terminate at the caudal edge of the pons. It has a ventral median fissure which is continuous with that of the spinal cord and terminates at the pons in a deep foramen cecum. The fissure is bounded by the **pyramids** (*fiber tracts*). Some of the pyramidal fibers can be seen crossing in the median fissure—the **decussation of the pyramids.** The **olive** and the trapezoid body are other prominent features of the ventral surface of the medulla.

The last six pairs of the twelve pairs of cranial nerves have their origin on the medulla; the **abducens (VI)**, medial to the trapezoid body; the **facial (VII)**, lateral to the trapezoid body; the **vestibulocochlear (VIII)** (*statoacoustic*), caudal to the facial near the trapezoid body; the **glossopharyngeal (IX)** lateral to the olive; the **vagus (X)**, behind the glossopharyngeal and from the olive; the **accessory (XI)**, from the lateral surface of the medulla to the level of the sixth or seventh cervical spinal nerves; the **hypoglossal (XII)**, from the ventral surface of the medulla.

The **pons** (*bridge*) is a mass of transverse fibers associated with the cerebellum. It also contains other fiber masses and nuclei among them the nuclei of origin of the **trigeminal nerve (V)** which comes out at the caudoventrolateral aspect of the pons. The **trochlear (IV) nerve,** the smallest of the cranial nerves, arises from the dorsal surface of the brain stem and appears ventrally lateral and anterior to the pons.

Anterior to the pons, forming a part of the floor of the midbrain, are the **crus cerebri** consisting of large groups of descending cortical nerve fibers. They constitute the ventral part of the cerebral peduncles. The tegmentum is the dorsal portion.

Between the crus cerebri (*cerebral peduncles*) and caudal to the mammillary bodies is a small triangular area marked by a median longitudinal sulcus. In this area known as the posterior perforated substance numerous blood vessels pass into the brain.

291

The **oculomotor (III) nerve** leaves the brain stem at the medial side of the crus cerebri.

Examine now the floor of the brain lying between the crus cerebri and the anterior **optic tracts, chiasma** and **nerves** (**II**). Anterior to the crus cerebri are the paired **mammillary bodies** and in front of them a rounded elevation the tuber cinereum (not labeled) bearing on its ventral surface the **infundibulum** with the **hypophysis.** The infundibulum is a hollow extension of the floor of the tuber cinereum, while the hypophysis (*pituitary*) is an endocrine gland of great importance. It is lodged in the hypophyseal fossa of the basisphenoid bone.

Anteriorly the **olfactory bulbs** with the **olfactory** (**Nerve I) tracts** are seen.

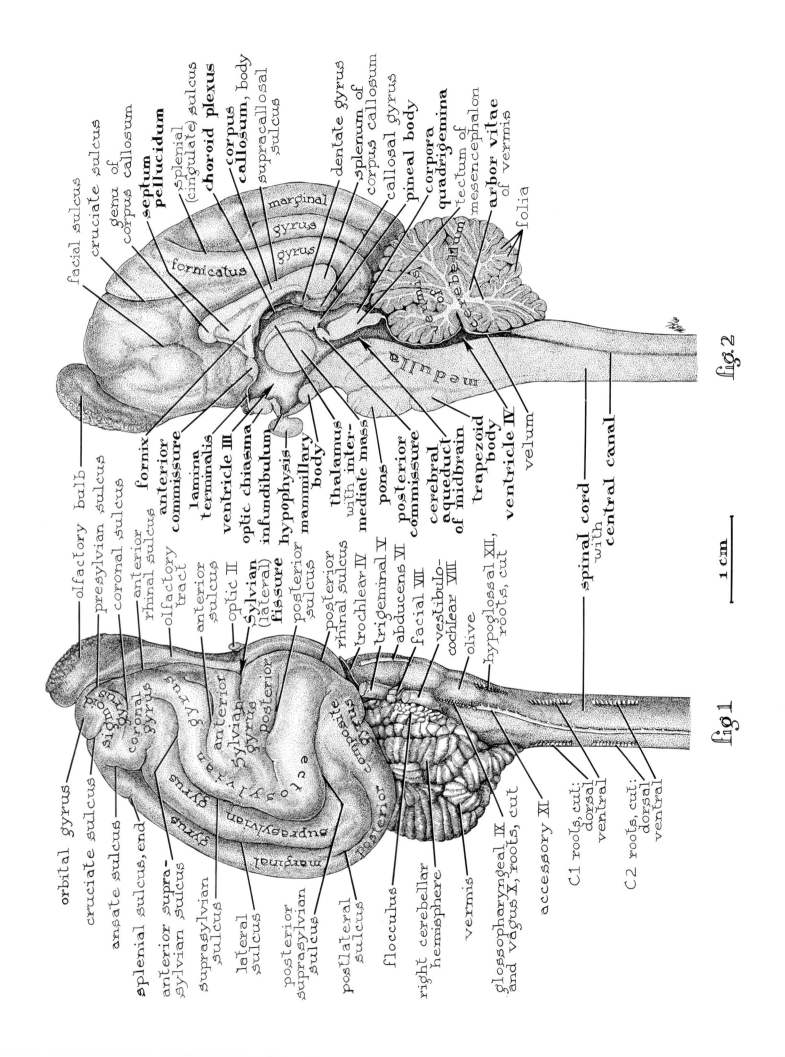

facial sulcus
cruciate sulcus
genu of corpus callosum
septum pellucidum
splenial (cingulate) sulcus
choroid plexus
corpus callosum, body
splenium of corpus callosum
supracallosal sulcus
callosal gyrus
dentate gyrus
pineal body
corpora **quadrigemina**
tectum of mesencephalon
arbor vitae of vermis
folia

marginal
gyrus
gyrus
fornicatus

vermis of cerebellum

medulla

fig. 2

orbital gyrus
cruciate sulcus
ansate sulcus
anterior supra-sylvian sulcus
splenial sulcus, end
suprasylvian sulcus
lateral sulcus
posterior suprasylvian sulcus
postlateral sulcus
flocculus
right cerebellar hemisphere
vermis
glossopharyngeal IX and vagus X, roots, cut
accessory XI
C1 roots, cut: dorsal ventral
C2 roots, cut: dorsal ventral

olfactory bulb
presylvian sulcus
coronal sulcus
anterior rhinal sulcus
fornix
anterior commissure
lamina terminalis
ventricle III
optic chiasma
infundibulum
hypophysis
mammillary body
thalamus with inter-mediate mass
pons
posterior commissure
cerebral **aqueduct of midbrain**
trapezoid body
ventricle IV
velum

olfactory tract
anterior sulcus
optic II
Sylvian (lateral) fissure
posterior sulcus
posterior rhinal sulcus
trochlear IV
trigeminal V
abducens VI
facial VII
vestibulo-cochlear VIII
olive
hypoglossal XII, roots, cut

spinal cord with **central canal**

proreus gyrus
sigmoid gyrus
coronal gyrus
anterior Sylvian gyrus
posterior Sylvian gyrus
composite gyrus
ectosylvian
suprasylvian
marginal

1 cm

fig 1

Figure 1, a lateral view of the brain, shows again the relationships of spinal cord to brain stem and of the cerebellum and cerebrum to each other and to the brain stem. The extent of the origin of the accessory nerve (XI) is more clearly indicated as is the positioning of other cranial nerves which originate from the lateral border of the medulla oblongata.

Figure 2 is a median section of the brain and cranial end of the spinal cord. The medial face of the cerebral cortex is here shown with its gyri and sulci. This is not a cut surface for, as you should recall, a longitudinal fissure separates the two cerebral hemispheres. The cut surface begins just below this at the **corpus callosum.** Anteriorly is shown the olfactory bulb and caudad of the cerebral hemisphere the cut surface of the cerebellum, the cut being through the **vermis.** The tree-like arrangement of the gray and white matter shown here is called **arbor vitae,** the tree of life.

The structures on the ventral side of the brain stem as seen here were described in connection with Plate 101, figure 2, and should be reviewed from this different point of view. Of great importance now is to note the **central canal** of the **spinal cord** and the **ventricles** of the brain with which it is continuous. Both the ventricles and the central canal have their origin in the neural tube which appears early in embryological development as a rolled up portion of the mid-dorsal ectoderm. The neural tube in becoming a central canal remained relatively undifferentiated whereas the ventricles represent a vast change in size, form and relationships. Since the central canal leads into the fourth ventricle, we will consider it first. The **fourth ventricle** represents a broadening out of the central canal in the medulla oblongata. Dorsoventrally it is very shallow, its roof being the cerebellum and its supporting velar membranes. In the roof of the ventricle there is a mass of tortuous blood vessels mostly of capillary size forming a choroid plexus which secrete cerebrospinal fluid into the ventricle. These are not shown here but may be seen in Plate 104. Laterally in the wall of the ventricle on each side is a lateral aperture (not shown—see Plate 104) which carries the fluid into the subarachnoid space. When these openings are not open, the cerebrospinal fluid accumulates within the ventricles of the brain causing hydrocephalus or water on the brain which may be fatal or at least damaging to brain tissue.

The fourth ventricle leads anteriorly into a narrow **cerebral aqueduct** which runs through the **midbrain** and leads into the third ventricle. The **third ventricle** is very deep dorsoventrally, but very narrow and lies in the diencephalon part of the brain. In its roof is a **choroid plexus** which also forms cerebrospinal fluid. It opens on either side anteriorly through interventricular foramina into the lateral ventricles which lie in the cerebral hemispheres. They also have choroid plexuses which are continuous with the one in the third ventricle. These also are shown in Plate 104. The third ventricle terminates anteriorly at the **lamina terminalis** and **anterior commissure** which mark the end of the brain stem. Only the cerebral hemispheres and olfactory bulbs extend farther forward than this.

This section shows the roof of the midbrain or mesencephalon which is called the tectum. In it are two pairs of elevations known collectively as the **corpora quadrigemina** which contain reflex centers mediating functions related to vision, hearing and tactile sensations.

Just anterior to the corpora quadrigemina and in the back part of the brain called the diencephalon is a structure called the **posterior commissure** and above it a projection from the brain roof, the **pineal body.** Anterior to it is the third ventricle already described. Its lateral walls are very thick and constitute the thalamus, an important relay station in the sensory pathways to the cerebral cortex. In the center, the walls of the third ventricles come together forming the **intermediate mass.** In the roof of the third ventricle above the choroid plexus is a curved ridge of white matter, the **fornix,** and above it a larger mass of white matter, the **corpus callosum,** consisting of fibers which connect the right and left cerebral hemispheres. At their anterior ends the fornix and corpus callosum diverge and between them is a very thin membrane, the **septum pellucidum** (*lucidum*) which separates the right and left ventricles of the cerebrum.

You might review the structures in the floor of the brain stem described in Plate 101, figure 2—the **optic chiasma, infundibulum, hypophysis, mammillary bodies, pons** and **trapezoid body.**

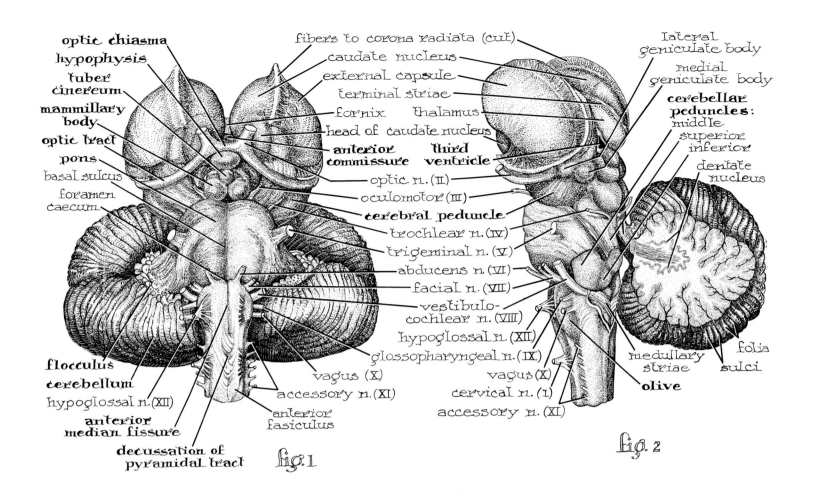

optic chiasma

hypophysis

tuber cinereum

mammillary body

optic tract

pons

basal sulcus

foramen caecum

flocculus

cerebellum

hypoglossal n. (XII)

anterior median fissure

decussation of pyramidal tract

fibers to corona radiata (cut)

caudate nucleus

external capsule

terminal striae

fornix   thalamus

head of caudate nucleus

anterior commissure   **third ventricle**

optic n. (II)

oculomotor (III)

cerebral peduncle

trochlear n. (IV)

trigeminal n. (V)

abducens n (VI)

facial n. (VII)

vestibulo-cochlear n. (VIII)

hypoglossal n. (XII)

glossopharyngeal n. (IX)

vagus (X)

accessory n. (XI)

anterior fasiculus

fig. 1

lateral geniculate body

medial geniculate body

cerebellar peduncles:
middle
superior
inferior

dentate nucleus

medullary striae

folia
sulci

vagus (X)

cervical n. (1)

accessory n. (XI)

olive

fig. 2

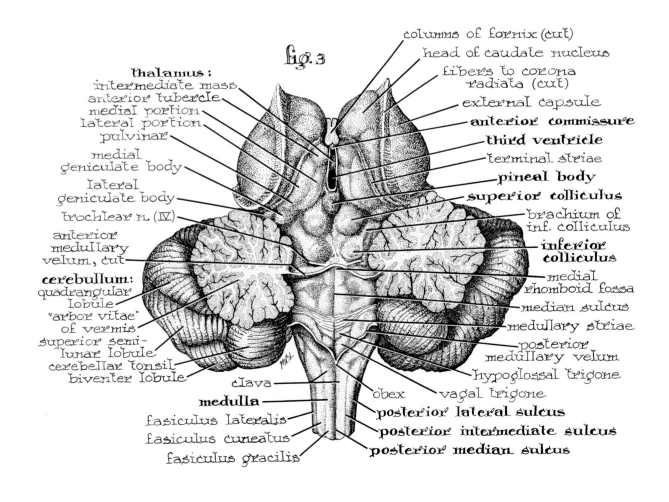

fig. 3

thalamus:
intermediate mass
anterior tubercle
medial portion
lateral portion
pulvinar

medial geniculate body

lateral geniculate body

trochlear n. (IV)

anterior medullary velum, cut

cerebellum:
quadrangular lobule
"arbor vitae" of vermis
superior semi-lunar lobule
cerebellar tonsil
biventer lobule

clava

medulla

fasiculus lateralis

fasiculus cuneatus

fasiculus gracilis

columns of fornix (cut)

head of caudate nucleus

fibers to corona radiata (cut)

external capsule

anterior commissure

third ventricle

terminal striae

pineal body

superior colliculus

brachium of inf. colliculus

inferior colliculus

medial

rhomboid fossa

median sulcus

medullary striae

posterior medullary velum

hypoglossal trigone

vagal trigone

obex

posterior lateral sulcus

posterior intermediate sulcus

posterior median sulcus

(Crouch: Functional Human Anatomy, Lea & Febiger)

This series of drawings of the brain stem and cerebellum of man serves to reinforce and to clarify some of the structures studied in Plates 101 and 102 of the cat's brain. It reminds one also of the essential similarity of the structures of these two or almost any two mammals.

Figure 1 is a ventral view of the brain stem and the overlying cerebellum. The attachments of cranial nerves II to XII, with the exception of IV, which attaches dorsally, are shown. Cranial nerve IV shows on figures 2 and 3, arising just below the inferior colliculi. Note carefully the points of attachment of all of the cranial nerves and compare them to those of the cat on Plate 101.

The lower part of the brain stem is the medulla oblongata, shown here but not labeled as such. Note the cranial nerves attached to it. It is divided into halves by the **anterior median fissure** and shows clearly the **decussation of the pyramidal tracts.** These are the descending lateral corticospinal fibers, called often pyramidal because they arise in the pyramidal cells of the cerebral cortex. They end around the anterior horn cells of the gray matter of the spinal cord. Some of these fibers which do not cross in the medulla cross farther down in the cord. Near the upper end of the medulla and laterally there is a swelling which because of its shape is called the olive, shown on figure 1 but labeled only on figure 2. The olive is a nucleus (*group of nerve cell bodies*) which sends a large bundle of fibers across the medulla where it enters the cerebellum through the inferior cerebellar peduncles. The olive also receives fibers from the frontal cerebral cortex, the globus pallidus, and the central gray substance of the midbrain. It thus links all of these structures to the cerebellum.

The dorsal side of the **medulla,** seen in figure 3, shows clearly the **posterior median sulcus** dividing the medulla into halves; the **posterior intermediate sulcus** lateral to the former and the **posterior lateral sulcus.** These divide the medulla from the medial line outward into the fasciculus gracilis, fasciculus cuneatus and fasciculus lateralis. The clava marks the position, beneath the surface, of the nucleus gracilis. The obex marks the medial caudal angle of the roof of the fourth ventricle. The fourth ventricle is here shown opened up by cutting in the median plane through the vermis of the cerebellum and spreading the halves laterad and by cutting the anterior medullary velum and the posterior medullary velum. In this way the floor of the fourth ventricle is fully exposed. It is called the rhomboidal fossa. In this floor can be seen transverse folds, the medullary striae, which represent the position of transverse nerve fibers; two triangular elevated areas the vagal and hypoglossal trigones which mark the position internally of the dorsal motor nucleus of the vagus nerve and the nucleus of the hypoglossal nerve. The rhomboid fossa is divided by a prominent medial sulcus.

The **pons** is a "bridge" of large bundles of transverse axons which arise from masses of cell bodies, the pontine nuclei, and pass into each half of the cerebellum. They form the main mass of the **middle cerebellar peduncles** which are labeled in figure 2. They relay impulses from the cerebral cortex to the cerebellum. To the ventrolateral aspect of the pons are attached the motor and sensory roots of the trigeminal (V) nerve and from the transverse groove between the pons and medulla the abducens (VI), facial (VII), and vestibulocochlear (VIII) cranial nerves emerge. The dorsal part of the pons forms the upper half of the floor of the fourth ventricle. In this area a number of longitudinal fiber tracts pass through, and here the nuclei of the fifth through the eighth cranial nerves have their nuclei. Two fiber tracts pass through the substance of the pons, the corticospinal and corticobulbar; one ends in the pontine nuclei, the corticopontine tract.

The next part of the brain stem is the midbrain, lying between the pons and the diencephalon. Within it is a narrow canal connecting the third and fourth ventricle, the cerebral aqueduct (*aqueduct of Sylvius*). The **cerebral peduncles** are the only parts of the midbrain visible on the ventral side. Between them the oculomotor nerves arise. Laterally between the cerebral peduncles and the pons the trochlear (IV) cranial nerve appears. The cerebral peduncles are made up mostly of motor fibers which descend from the cerebrum to the lower parts of the nervous system. Bundles of ascending (*sensory*) fibers traveling upward to the thalamus of the diencephalon lie deep in relationship to the cerebral peduncles. Nuclei of the third, fourth and anterior part of the fifth nerves are located in the midbrain. The red nucleus contains the cells of origin of the rubrospinal tract and is a relay station for impulses from the cerebellum and higher brain centers. The fibers to the red nucleus from the cerebellum constitute the **superior cerebellar peduncle** as seen in figure 2.

On the dorsal surface of the midbrain are the corpora quadrigemina consisting of **superior colliculi** and **inferior colliculi** as seen in figures 3 and 4 and labeled in the latter. The superior colliculi connect laterad by their brachia to the lateral geniculate bodies; the inferior colliculi connect superolaterally by their brachia to the medial geniculate bodies. The superior colliculi are reflex centers for visual, auditory and tactile impulses, while the inferior colliculi are reflex centers for auditory functions.

The **diencephalon** lies between the midbrain and the telencephalon (*cerebrum*). Its forward limit is at the interventricular foramina (*foramen of Monro*) which connects its cavity, the third ventricle, with the lateral ventricles of the cerebral hemispheres (Plate 104). The **third ventricle** is shown only in figures 2 and 3 here but should be reviewed in Plate 104 where it is opened. The parts of the diencephalon are the epithalamus or roof marked in figures 2 and 3 by the **pineal body;** the thalamus, seen best in figure 3 and marking the lateral walls of the diencephalon and third ventricle; and the hypothalamus or floor of the diencephalon.

The thalamus is an important receiving station for sensory information from receptors throughout the body and except for primary stimuli is a relay station to the cerebral cortex. It also distributes impulses to the rest of the brain and is likely an integrating center. Pain and perhaps other "sensations" are likely felt at the thalamic level. The pulvinar and other nuclei (basal nuclei) closely related to the thalamus cannot be discussed here in detail. The caudate nucleus, external capsule, and fibers of the corona radiate are shown, however, in figures 1 to 3 and through these structures the brain stem and cerebrum are related structurally and functionally.

The structures of the hypothalamus were described briefly in relationship to Plate 102.

The Meninges of the Brain and Spinal Cord and their Relationship to
the Ventricles and to the Circulation of Cerebrospinal Fluid
PLATE 104

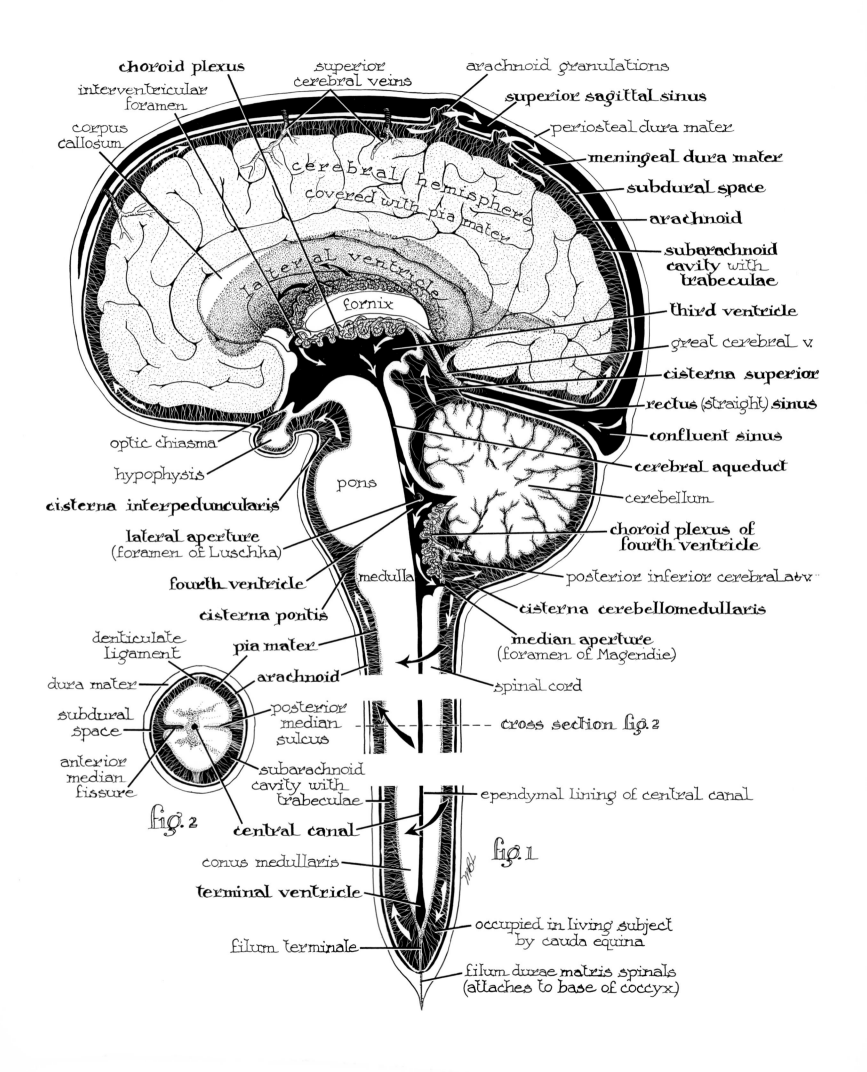

choroid plexus

superior cerebral veins

arachnoid granulations

interventricular foramen

superior sagittal sinus

corpus callosum

periosteal dura mater

cerebral hemisphere covered with pia mater

meningeal dura mater

subdural space

arachnoid

subarachnoid cavity with trabeculae

lateral ventricle

fornix

third ventricle

great cerebral v.

cisterna superior

rectus (straight) sinus

confluent sinus

optic chiasma

hypophysis

cisterna interpeduncularis

pons

cerebral aqueduct

cerebellum

lateral aperture (foramen of Luschka)

choroid plexus of fourth ventricle

fourth ventricle

medulla

posterior inferior cerebral a.v.

cisterna pontis

cisterna cerebellomedullaris

denticulate ligament

pia mater

median aperture (foramen of Magendie)

dura mater

arachnoid

spinal cord

subdural space

posterior median sulcus

anterior median fissure

subarachnoid cavity with trabeculae

cross section fig. 2

fig. 2

ependymal lining of central canal

central canal

fig. 1

conus medullaris

terminal ventricle

occupied in living subject by cauda equina

filum terminale

filum durae matris spinals (attaches to base of coccyx)

The Meninges of the Brain and Spinal Cord and their Relationship to the Ventricles and to the Circulation of Cerebrospinal Fluid
PLATE 104

(Crouch: Functional Human Anatomy, Lea & Febiger)

This plate illustrates, in figure 1, a median section of the brain and spinal cord of man. Sections of the spinal cord are cut out at two points to reduce its length. The central canal of the spinal cord, the ventricles of the brain, the meninges of the spinal cord and brain and their related spaces, and a few of the venous sinuses are also shown. Figure 2 is a cross section of the spinal cord and its meninges. In both figures the pia mater is shown in light stipple; ventricles and other cavities are in black except the lateral ventricle which is heavily stippled. The arrows in figure 1 indicate the direction of flow of the cerebrospinal fluid and at the top of the drawing show how cerebrospinal fluid, through the arachnoid granulations, seeps into the superior sagittal (*venous*) sinus.

The spinal cord with its central canal, its meninges and related spaces have already been studied in Plate 98 and the accompanying description. Review these structures as seen here in figure 1 which relates them to similar structures in and around the brain. Note also the inferior (*caudal in cat*) end of the spinal cord as it tapers off as the conus medullaris and finally into the filum terminale and filum durae matris which attaches at the base of the coccyx in man. Notice also the **terminal ventricle** at the end of the **central canal** of the spinal cord. The rather large subarachnoid cavity at and beyond the end of the cord is occupied in the living subject by the lower spinal nerves which constitute the cauda equina.

The ventricles of the brain, which are modifications of the neural canal of the embryo, are shown clearly in figure 1. Some of them have been described in relation to the cat's brain, Plate 102, figure 2. The central canal extends into the medulla oblongata and then widens out into the diamond-shaped, shallow **fourth ventricle.** Its floor is the rhomboid fossa of the brain stem; its roof is the cerebellum and anterior and posterior medullary vella (Plate 103). It leads into the cerebral aqueduct at its superior (*cranial*) end. From the roof of the fourth ventricle, the **choroid plexus** projects into and beyond the lateral recesses of the ventricle. The choroid plexus consists of an elongated mass of tortuous blood vessels, mostly of capillary size. The capillaries belong to the pia mater, but they are covered with the epithelial ependymal lining of the ventricle. They secrete cerebrospinal fluid into the ventricle from which it may, as indicated by the arrows, pass out through a **median aperture** (*foramen of Magendie*) or through **lateral apertures** (*foramen of Luschka*) into the **cisterna cerebellomedullaris** and subarachnoid cavity or space. Closure of these apertures causes the cerebrospinal fluid to accumulate in excess resulting in hydrocephalus or water on the brain. This condition, sometimes seen in children,

may be fatal or at least damaging to the brain due to the high pressure which develops. The cerebrospinal fluid, leaving through the above apertures, circulates in the subarachnoid space upward around the cerebellum, downward on the posterior side of the spinal cord and upward on the anterior side of the cord and brain stem and then around the cerebrum.

The **cerebral aqueduct** (*aqueduct of Sylvius*) of the midbrain leads into the third ventricle of the diencephalon. Review the structures of the midbrain by reference to Plate 103.

The **third ventricle** is narrow but deep and ends anteriorly at the lamina terminalis and the anterior commissure (Plate 102). In the roof of the third ventricle is a **choroid plexus** which secretes cerebrospinal fluid. Posteriorly one sees the pineal body and the posterior commissure. Near its anterior termination the third ventricle opens into the lateral ventricles of the cerebrum by the **interventricular foramina** (*foramina of Monro*).

The **lateral ventricles,** shown here in heavy stipple, are ventricles one and two though that terminology is not commonly used. These ventricles are medially placed within the cerebral hemisphere being separated from each other by the very thin vertical partition, the septum pellucidum (Plate 102). They are lined with ependyma and each has a central part and three extensions: the anterior, posterior, and inferior horns (*not labeled*). There is a **choroid plexus** in the floor of the central part of each lateral ventricle which is continuous through the interventricular foramina with the choroid plexus of the third ventricle. It continues also into the inferior horn.

The meninges of the brain are the same as those around the spinal cord which, as stated above, have been illustrated and described (Plate 98). There are some variations, however, which should be noted on this plate. The **dura mater** is a single meninx around the spinal cord. It becomes double at the foramen magnum forming a periosteal dura mater and a meningeal dura mater (*labeled at top of fig. 1*), the former being closely related to the inner periosteum of the cranial bones. The **meningeal dura mater** becomes folded on itself and passes down between the right and left cerebral hemispheres to form the falx cerebri and between the cerebellum and cerebral hemispheres to form the transverse tentorium cerebelli. The tentorium cerebelli is a bony structure in the cat, a part of the parietal bone (Plate 6). It is membranous in man. The falx cerebri is not shown in this plate.

The **cranial arachnoid** is a delicate, avascular, transparent membrane which loosely surrounds the brain. It is connected to the pia mater, through the **subarachnoid** cavity by many thin connective tissue **trabeculae.** A narrow **subdural space** or cavity separates the meningeal dura mater from the arachnoid. Follow the subarachnoid space as seen around the brain in figure 1 and note that it widens out in certain areas to form cisterns. The **cisterna cerebellomedullaris** is the largest of these located in the angle between the cerebellum and the medulla oblongata. Other cisterns shown in figure 1 are the **cisterna interpeduncularis** below the midbrain and the **cisterna superior** above the midbrain.

The cranial subarachnoid space communicates with the spinal subarachnoid space at the foramen magnum, with the ventricular system of the brain through the **lateral apertures** (*foramina of Luschka*) and in man through a **median aperture** also—the foramen of Magendie. By means of the **arachnoid granulations** (*pacchionian granulations*), which are enlargements of arachnoid villi, the meningeal dura mater is penetrated and communication is established with the superior sagittal and other venous sinuses. In figure 1 the **superior sagittal, confluent** and **rectus** (*straight*) **sinuses** are shown. In this manner then, circulation is possible not only between subarachnoid spaces, cisterns and ventricles but a means is provided for the draining of cerebrospinal fluid into the venous system.

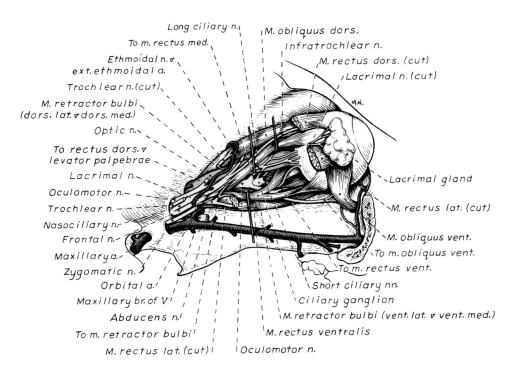

Long ciliary n.
To m. rectus med.
Ethmoidal n. v
ext. ethmoidal a.
Trochlear n.(cut)
M. retractor bulbi
(dors. lat. v dors. med.)
Optic n.
To rectus dors. v
levator palpebrae
Lacrimal n.
Oculomotor n.
Trochlear n.
Nasociliary n.
Frontal n.
Maxillary a.
Zygomatic n.
Orbital a.
Maxillary br. of V
Abducens n.
To m. retractor bulbi
M. rectus lat. (cut)

M. obliquus dors.
Infratrochlear n.
M. rectus dors. (cut)
Lacrimal n. (cut)
M.N.
Lacrimal gland
M. rectus lat. (cut)
M. obliquus vent.
To m. obliquus vent.
To m. rectus vent.
Short ciliary nn.
Ciliary ganglion
M. retractor bulbi (vent. lat. v vent. med.)
M. rectus ventralis
Oculomotor n.

FIG. 1

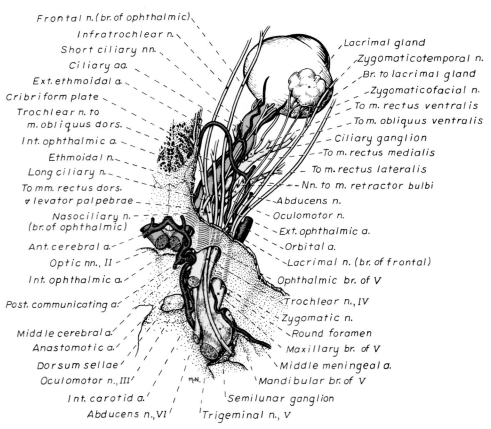

Frontal n. (br. of ophthalmic)
Infratrochlear n.
Short ciliary nn.
Ciliary aa.
Ext. ethmoidal a.
Cribriform plate
Trochlear n. to
m. obliquus dors.
Int. ophthalmic a.
Ethmoidal n.
Long ciliary n.
To mm. rectus dors.
v levator palpebrae
Nasociliary n.
(br. of ophthalmic)
Ant. cerebral a.
Optic nn., II
Int. ophthalmic a.
Post. communicating a.
Middle cerebral a.
Anastomotic a.
Dorsum sellae
Oculomotor n., III
Int. carotid a.
Abducens n., VI

Lacrimal gland
Zygomaticotemporal n.
Br. to lacrimal gland
Zygomaticofacial n.
To m. rectus ventralis
To m. obliquus ventralis
Ciliary ganglion
To m. rectus medialis
To m. rectus lateralis
Nn. to m. retractor bulbi
Abducens n.
Oculomotor n.
Ext. ophthalmic a.
Orbital a.
Lacrimal n. (br. of frontal)
Ophthalmic br. of V
Trochlear n., IV
Zygomatic n.
Round foramen
Maxillary br. of V
Middle meningeal a.
Mandibular br. of V
M.N.
Semilunar ganglion
Trigeminal n., V

FIG. 2

Nerves of the Eye and the Orbit of the Dog
PLATE 105

(Miller, Christensen and Evans: Anatomy of the Dog, W. B. Saunders Co.)

This plate illustrates the orbit containing the eye, its extrinsic muscles, the nerves supplying the eye structures and some of the major arteries. The terminology for the muscles used here is slightly different than that used for the human eye on Plate 114, figures 1 and 2. It is the terminology used by students of veterinary anatomy. The superior oblique and superior rectus of human anatomy are called dorsal oblique and dorsal rectus. The inferior oblique and inferior rectus are called ventral oblique and ventral rectus. Note also that there is in the dog and also in the cat a retractor oculi muscle.

While this plate deals with the dog, it should be understood that the dog's eye and its extrinsic muscles, vessels, and nerves vary in no major way from that of the cat. Indeed the eyes of all vertebrates are much the same—varying only in details.

Figure 1 is a lateral view of the eye in which the rectus lateralis and rectus dorsalis (*superior*) have been partly cut away and the dorsal medial part of the retractor bulbi has been pulled dorsad. This enables one to see the other eye muscles and the nerves to better advantage.

The seven extrinsic muscles of the eyeball can be seen in figure 1. They are the four recti muscles, two obliques, and the retractor bulbi. A muscle closely associated with these but not attached to the eyeball is the levator palpebrae dorsalis (*superioris*) which can be seen in figure 1 of Plate 114. It inserts into the upper eyelid and is used to lift the lid, *i.e.* to open the eye. The muscles of the eyeball all insert into the fibrous sclerotic coat. The recti muscles insert forward of the equator of the eyeball and the retractor bulbi a little farther caudad. The oblique muscles insert between these other two groups. The origins of the muscles of the eyeball, except for the ventral (inferior) oblique come from the margin area around the optic foramen and orbital fissure. The **ventral oblique** originates in a fossa at the junction of palatine, lacrimal and maxillary bones. The muscle passes beneath the eyeball, below the tendon of the ventral rectus and then divides, one part going over and the other under the lateral rectus to insert into the sclera. Its action is to rotate the eyeball around its polar axis pulling the ventral part mediad and dorsad. The **dorsal oblique,** originating on the medial side of the optic canal, passes over the periorbita to a cartilaginous pulley, the trochlea, which is located on the medial wall of the orbit near the medial canthus of the eye. It passes through the trochlea and then turns sharply dorsolaterad, passes under the tendon of the rectus dorsalis and inserts into the sclera. Its contraction rotates the eyeball around an axis through its poles which brings the dorsal part of the eye mediad and ventrad.

The arrangement and naming of the **recti muscles** is such that their actions can be readily analyzed. The lateral and medial recti rotate the eye around a vertical axis through the equator, while the dorsal and ventral recti rotate the eye around a horizontal axis through the equator of the eyeball.

The **retractor bulbi** muscle originates at the apex of the orbit deep to the recti muscles at the orbital fissure and lateral to the optic nerve. The muscle breaks into four fasciculi, two dorsal and two ventral, which diverge to insert on the equator of the eyeball caudad and deep to the recti muscles. The optic nerve passes between the dorsal and ventral portions. As the name suggests, the retractor bulbi muscle retracts or draws back the eyeball or bulb and also may bring about oblique eye movements.

Figures 1 and 2 both show the principal nerves to the eye structures and a few arteries. The distribution of the nerves is indicated in figure 1 and their foramina of exit from the cranial cavity in figure 2. Reference to Plate 101 will show you the points of origin of these cranial nerves from the brain stem. The **oculomotor** (III) is the main nerve to the muscles of the orbit. It leaves the cranial cavity through the orbital fissure (Table 2, p. 16) and divides immediately into dorsal and ventral rami or branches. The smaller dorsal ramus sends branches to the dorsal rectus and to the levator palpebrae. The ventral branch passes rostrally and supplies the medial rectus, ventral rectus and ventral oblique muscles. All of the above nerve branches carry general somatic efferent (*motor*) fibers. A final branch which is derived from the ventral ramus is general visceral efferent and goes to the **ciliary ganglion.** It carries preganglionic parasympathetic nerve fibers and is further described in connection with Plate 109.

The **trochlear nerve** (IV) is the smallest of the twelve cranial nerves and the only one to arise from the dorsal side of the brain stem. It can be seen in figure 2 as it transverses the semilunar ganglion and passes with the oculomotor, ophthalmic branch of the trigeminal (V) and abducens (VI) nerves through the orbital fissure to supply the dorsal oblique muscle.

The **abducens nerve** (VI) leaves the cranial cavity through the orbital fissure and supplies the lateral rectus and retractor bulbi muscles with somatic motor fibers.

Note the **trigeminal nerve** (V), the largest cranial nerve, and its **semilunar** (*trigeminal*) **ganglion** as shown in figure 2. Its three main branches or divisions are well shown: the **ophthalmic** passing through the orbital fissure into the orbit and dividing into three main branches, the lacrimal, frontal and nasociliary, all seen in figures 1 and 2; the **maxillary** passing from the semilunar ganglion through the round foramen and sending branches to the skin of the

305

cheek, side of nose and muzzle and mucous membranes of nasopharynx, soft and hard palates and the teeth and gums of the upper jaw; and the **mandibular,** from the lateral side of the semilunar ganglion whose distribution is suggested by the names of its main branches, the pterygoid, buccal, masseteric, deep temporal, auriculotemporal, mylohyoid, mandibular alveolar, and lingual nerves. More detailed distribution of the components of the trigeminal nerve are discussed in reference to the next plate.

Note the optic nerve (II) entering the optic foramen.

This plate shows some of the branches of the (internal) maxillary artery which supply structures of the orbit and eye. Both figures show the **orbital branch** of the maxillary artery. Figure 2 shows the internal carotid artery as it forms a part of the circle of Willis and the internal **ophthalmic artery** as it enters the orbit.

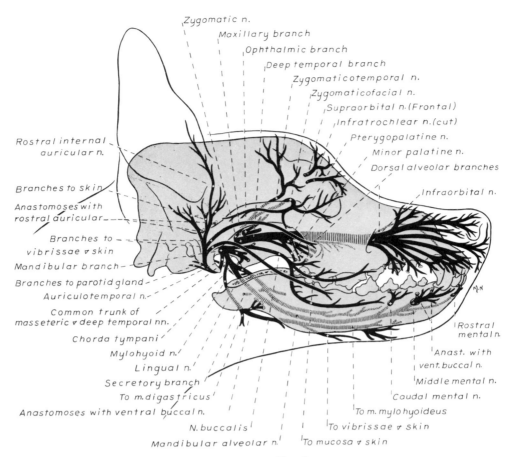

Zygomatic n.
Moxillary branch
Ophthalmic branch
Deep temporal branch
Zygomaticotemporal n.
Zygomaticofacial n.
Supraorbital n. (Frontal)
Infratrochlear n. (cut)
Pterygopalatine n.
Minor palatine n.
Dorsal alveolar branches
Infraorbital n.

Rostral internal auricular n.

Branches to skin
Anastomoses with rostral auricular
Branches to vibrissae & skin
Mandibular branch
Branches to parotid gland
Auriculotemporal n.
Common trunk of masseteric & deep temporal nn.
Chorda tympani
Mylohyoid n.
Lingual n.
Secretory branch
To m. digastricus
Anastomoses with ventral buccal n.
N. buccalis
Mandibular alveolar n.

Rostral mental n.
Anast. with vent. buccal n.
Middle mental n.
Caudal mental n.
To m. mylohyoideus
To vibrissae & skin
To mucosa & skin

FIG. 1

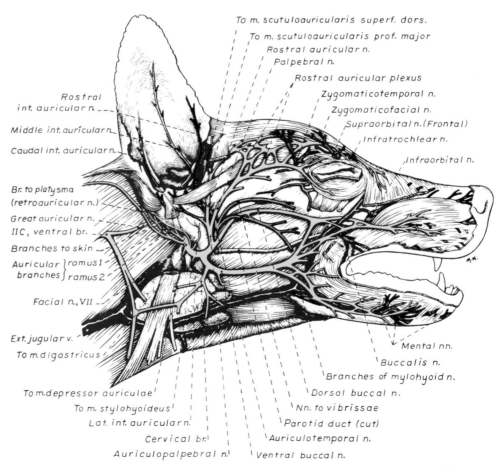

To m. scutuloauricularis superf. dors.
To m. scutuloauricularis prof. major
Rostral auricular n.
Palpebral n.
Rostral auricular plexus
Zygomaticotemporal n.
Zygomaticofacial n.
Supraorbital n. (Frontal)
Infratrochlear n.
Infraorbital n.

Rostral int. auricular n.
Middle int. auricular n.
Caudal int. auricular n.

Br. to platysma (retroauricular n.)
Great auricular n.
IIC, ventral br.
Branches to skin
Auricular branches { ramus 1 ramus 2
Facial n., VII

Ext. jugular v.
To m. digastricus

To m. depressor auriculae
To m. stylohyoideus
Lat. int. auricular n.
Cervical br.
Auriculopalpebral n.

Mental nn.
Buccalis n.
Branches of mylohyoid n.
Dorsal buccal n.
Nn. to vibrissae
Parotid duct (cut)
Auriculotemporal n.
Ventral buccal n.

FIG. 2

(Miller, Christensen and Evans: Anatomy of the Dog, W. B. Saunders Co.)

Figure 1 is a schematic presentation of the **trigeminal nerve (V)** showing its three divisions and the areas supplied by them through their many branches. Refer back to the description of the last plate for a consideration of the **semilunar or trigeminal ganglion** and its relationship to the three divisions of the trigeminal nerve. The ophthalmic branch is better shown in Plate 105, figure 2. The maxillary and mandibular branches should be carefully studied here. Most of the names of the nerves indicate the structures served by the nerve. In general the maxillary nerve deals with structures in and above the level of the upper jaw or maxillary bone. The **mandibular nerve** services structures in and around the mandible and also the ear region.

Figure 2 emphasizes the distribution superficially of the facial nerve (VII). A few branches of the trigeminal (stippled) are present, such as the auriculotemporal and its branches, the supraorbital or frontal, and the zygomaticotemporal and zygomaticofacial.

The **facial nerve (VII)**, since it contains special visceral efferent and afferent fibers and visceral efferent and afferent fibers, is a mixed nerve. The **facial nucleus,** lying in the rostroventral part of the medulla oblongata, contains the cell bodies of the special visceral efferent fibers. These fibers make up the bulk of the facial nerve. The remaining parts of the nerve, the special visceral afferent and the visceral efferent and afferent, make up what is often called the **nervous intermedius.** The special visceral efferent fibers are distributed to the musculature of the ear and the face and are the ones which make up the nerves shown here in figure 2. Locate the facial nerve on the illustration. It is seen here just as it emerges from the stylomastoid foramen along with some of its auricular branches. Examine these auricular branches and note that ramus 1 gives off a branch to the platysma muscle—the retroauricular nerve. Other branches go to others of the retroauricular muscles. The lateral and middle internal auricular nerves can be seen arising from the facial nerve, passing into the cartilage of the ear and emerging to supply the skin of the ear canal and the auricular cartilage. A very short branch is given off of the facial to the caudal belly of the digastricus muscle and another to the stylohyoideus. The main facial nerve now divides into three terminal branches, the auriculopalpebral, dorsal buccal, and ventral buccal nerves.

The **auriculopalpebral nerve** as the name suggests goes to the rostral auricular muscles and to the muscles of the eyelids. The main nerve divides near its origin into the rostral auricular nerve and the palpebral nerve. The **rostral auricular nerve** has a number of anastomoses with the auriculotemporal branch of the trigeminal nerve. It supplies facial muscles near the ear. The palpebral nerve passes dorsorostrally and contributes to the formation of a rostral auricular plexus. It supplies the corrugator supercilii and orbicularis oculi muscles and its terminal branch goes to the maxillonasolabialis and the levator nasolabialis muscles. It anastomoses with a branch of the rostral auricular nerve.

The **dorsal buccal nerve** also anastomoses with branches of the auriculotemporal nerve and other branches of the trigeminal. It arches upward and turns rostrally following the zygomatic arch and then anastomoses with the ventral buccal nerve caudad and dorsad to the commissure of the mouth. Note that it passes through the orbicularis oris muscle where it anastomoses with the buccal and infraorbital nerves of the trigeminal. It supplies the maxillonasolabialis muscle.

The **ventral buccal nerve** passes ventrad and rostrally from the facial nerve and gives off ventrocaudally the cervical branch which supplies the sphincter colli primitivus and the depressor auriculae. This branch has several branches which anastomose with branches of the second cervical spinal nerve. The ventral buccal nerve then passes rostrally over the lateral surface of the masseter muscle and ventral to the parotid duct. It divides just before reaching the commissure of the mouth and sends several branches into the ventral portion of the orbicularis oris muscle. It receives anastomosing branches of the mylohyoid nerve of the trigeminal and terminally anastomoses with the mental nerves which are the terminal branches of the mandibular alveolar nerve—also from the trigeminal.

The close relationship of the facial nerve and its branches with the trigeminal is quite apparent in this illustration. Plate 109, figure 1 should be studied at this point because it shows some of the **parasympathetic** components of the facial nerve.

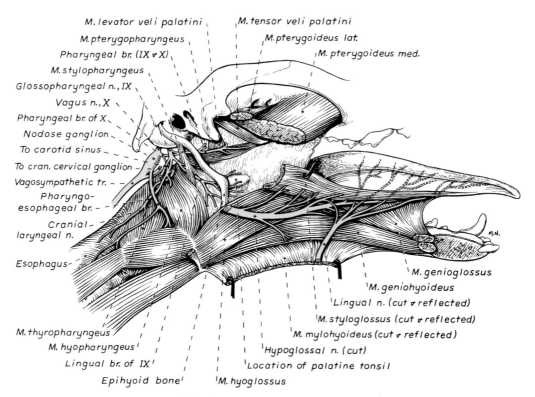

M. levator veli palatini
M. pterygopharyngeus
Pharyngeal br. (IX ᵥ X)
M. stylopharyngeus
Glossopharyngeal n., IX
Vagus n., X
Pharyngeal br. of X
Nodose ganglion
To carotid sinus
To cran. cervical ganglion
Vagosympathetic tr.
Pharyngo-esophageal br.
Cranial-laryngeal n.
Esophagus
M. thyropharyngeus
M. hyopharyngeus
Lingual br. of IX
Epihyoid bone

M. tensor veli palatini
M. pterygoideus lat.
M. pterygoideus med.

M. genioglossus
M. geniohyoideus
Lingual n. (cut ᵥ reflected)
M. styloglossus (cut ᵥ reflected)
M. mylohyoideus (cut ᵥ reflected)
Hypoglossal n. (cut)
Location of palatine tonsil
M. hyoglossus

Fig. 1

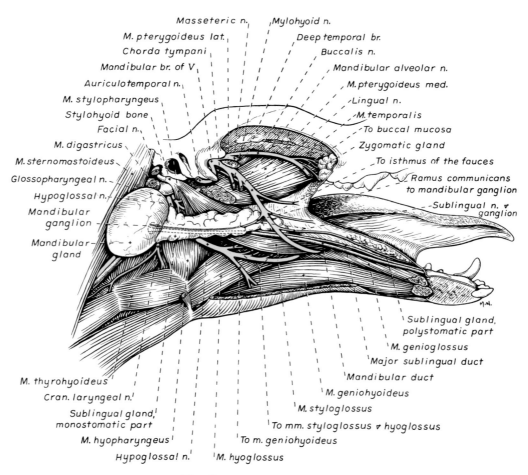

Masseteric n.
M. pterygoideus lat.
Chorda tympani
Mandibular br. of V
Auriculotemporal n.
M. stylopharyngeus
Stylohyoid bone
Facial n.
M. digastricus
M. sternomastoideus
Glossopharyngeal n.
Hypoglossal n.
Mandibular ganglion
Mandibular gland

Mylohyoid n.
Deep temporal br.
Buccalis n.
Mandibular alveolar n.
M. pterygoideus med.
Lingual n.
M. temporalis
To buccal mucosa
Zygomatic gland
To isthmus of the fauces
Ramus communicans to mandibular ganglion
Sublingual n. ᵥ ganglion

Sublingual gland, polystomatic part
M. genioglossus
Major sublingual duct
Mandibular duct
M. geniohyoideus
M. styloglossus
To mm. styloglossus ᵥ hyoglossus
To m. geniohyoideus
M. hyoglossus

M. thyrohyoideus
Cran. laryngeal n.
Sublingual gland, monostomatic part
M. hyopharyngeus
Hypoglossal n.

Fig. 2

Nerve Distribution in the Pharyngeal Region and the Tongue and the Medial Side of the Mandible of the Dog
PLATE 107

(Miller, Christensen and Evans: Anatomy of the Dog, W. B. Saunders Co.)

While we are mainly concerned with nerves in our study of this plate, one should take advantage of an opportunity to review the muscles of the tongue and those of the pharyngeal region. In so doing you come to realize how much alike the musculature is in the cat and dog.

The principal nerves seen here are the glossopharyngeal (IX) and the hypoglossal (XII) cranial nerves. Small parts of three others are shown, the facial (VII), the vagus (X) and the spinal accessory (XI). The lingual nerve is a branch of the mandibular division of the trigeminal nerve (V) and is seen here. Also in figure 1 the vagosympathetic trunk is shown caudally. Again, be aware of the interrelationships of the cranial nerves as presented here.

Figure 1 is an illustration of a deep dissection of the pharyngeal region and of the tongue. In the caudodorsal part of the illustration note the jugular process, the tympanic bulla and the stylohyoid bone pointing into the stylomastoid foramen from which, in figure 2, the facial nerve is seen emerging. Note that the vagus (X) and glossopharyngeal (IX) nerves appear in this area having emerged from the jugular foramen which lies medial to the tympanic bulla.

The **nodose ganglion** of the **vagus nerve** (X) is shown here. From its proximal end a pharyngeal branch arises which is joined by a branch from the glossopharyngeal (IX). The combined nerve forms many smaller branches which form the **pharyngeal plexus.** From the plexus the pharyngoesophageal nerve arises which supplies the caudal pharyngeal muscles and the forepart of the esophagus. The cranial cervical (*sympathetic*) ganglion (*not shown*) supplies branches to the plexus and its trunk joins that of the vagus to form the **vagosympathetic trunk.** Arising more distally from the nodose ganglion of the vagus is the cranial laryngeal nerve which soon divides into external and internal branches. The external branch in turn sends another branch deep to supply the cricothyroideus muscle; the main branch continues caudad to the region of the thyroid gland. The internal branch of the cranial laryngeal nerve passes between the thyropharyngeus (*pharyngeal constrictor*) and hyopharyngeus (*pharyngeal constrictor*) muscles, gives off a branch which anastomoses with the caudal laryngeal nerve and then enters the larynx where it is distributed to the mucosa. Note also the lingual branch of the glossopharyngeal nerve. It gives off branches to the stylopharyngeal muscle as it passes through and terminates in branches in the lateral pharyngeal mucosa and the palatine tonsil.

Follow the **hypoglossal (XII) nerve,** which has been cut proximally. In figure 2 its more proximal parts are shown, while in figure 1 its terminal ramifications are seen. It passes forward on the lateral surfaces of the hyopharyngeus and hyoglossus muscles and provides branches to the styloglossus, hyoglossus, genioglossus and geniohyoideus muscles. Branches are also sent to the intrinsic muscles of the tongue. A few anastomoses occur between the lingual and hypoglossal nerves.

Finally, notice the **lingual nerve** which has been cut and reflected in this illustration (fig. 1); only its distal portions are shown. Its fibers are distributed to the dorsal mucosa of the tongue, rostral to the circumvallate papillae, and they conduct somatic afferent impulses to the brain stem by way of the mandibular division of the trigeminal (V) nerve. Special visceral afferent fibers from the taste buds leave the lingual nerve to join the chorda tympani nerve, a part of the facial.

Follow the lingual nerves proximal portion now as it is seen in figure 2. Notice the sublingual nerve and its ganglion as it branches from the lingual nerve to supply the sublingual gland and also the communicating branch to the mandibular ganglion near the mandibular salivary gland. Branches are also seen which go to the isthmus of the fauces. Proximad, on the surfaces of the medial pterygoid muscle, it gives off mylohyoid and mandibular alveolar nerves and receives the chorda tympani branch from the facial (VII) nerve. It joins the mandibular division of V at about the point where that division gives rise to auriculotemporal, masseteric and buccal nerves. The remaining nerves in figure 2 have already been observed in figure 1 and described. For further information on the innervation of the salivary glands shown here refer to Plate 109, figure 1 and the accompanying description.

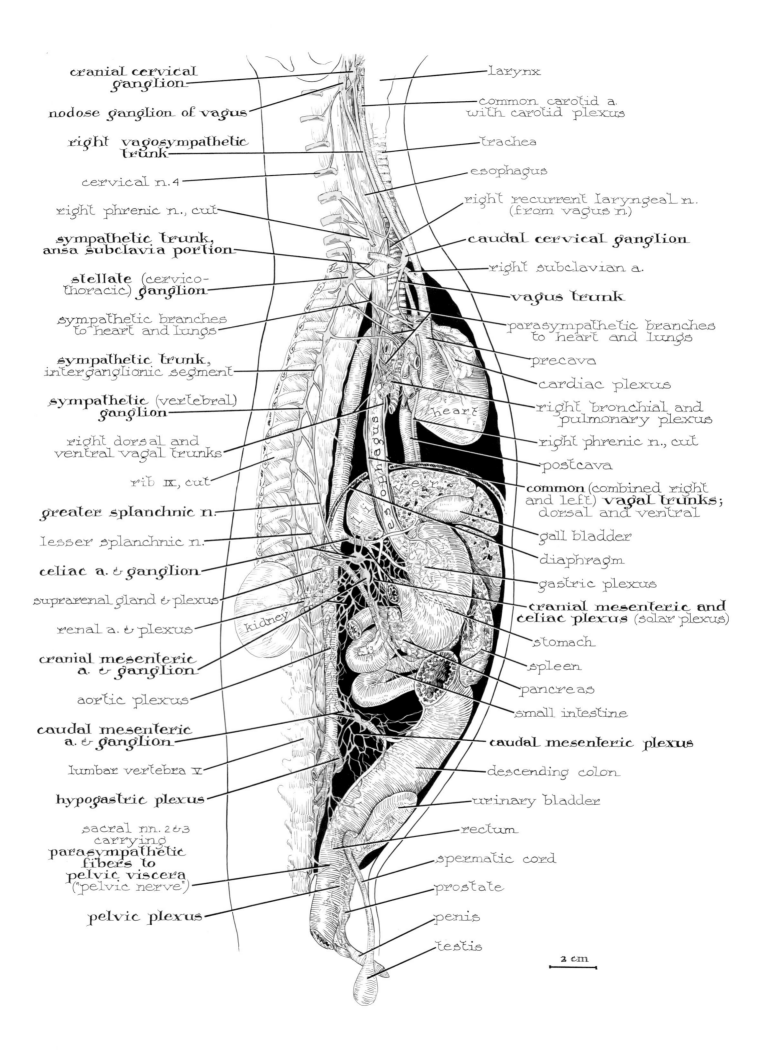

cranial cervical ganglion

nodose ganglion of vagus

right vagosympathetic trunk

cervical n. 4

right phrenic n., cut

sympathetic trunk, ansa subclavia portion

stellate (cervico-thoracic) ganglion

sympathetic branches to heart and lungs

sympathetic trunk, interganglionic segment

sympathetic (vertebral) ganglion

right dorsal and ventral vagal trunks

rib IX, cut

greater splanchnic n.

lesser splanchnic n.

celiac a. & ganglion

suprarenal gland & plexus

renal a. & plexus

cranial mesenteric a. & ganglion

aortic plexus

caudal mesenteric a. & ganglion

lumbar vertebra V

hypogastric plexus

sacral nn. 2&3 carrying parasympathetic fibers to pelvic viscera ("pelvic nerve")

pelvic plexus

larynx

common carotid a. with carotid plexus

trachea

esophagus

right recurrent laryngeal n. (from vagus n.)

caudal cervical ganglion

right subclavian a.

vagus trunk

parasympathetic branches to heart and lungs

precava

cardiac plexus

right bronchial and pulmonary plexus

right phrenic n., cut

postcava

common (combined right and left) vagal trunks; dorsal and ventral

gall bladder

diaphragm

gastric plexus

cranial mesenteric and celiac plexus (solar plexus)

stomach

spleen

pancreas

small intestine

caudal mesenteric plexus

descending colon

urinary bladder

rectum

spermatic cord

prostate

penis

testis

heart

liver

esophagus

kidney

2 cm

This plate considers both the sympathetic and parasympathetic parts of the autonomic nervous system but the emphasis is on the **sympathetic.** The parasympathetic parts of the **vagus nerve** receive some attention as do the **sacral** parasympathetics.

In the upper part of the plate are seen the **nodose** (*inferior*) **ganglion of the vagus** and the **cranial** (*superior*) **cervical ganglion** lying medial to it. Generally each has its own epineurium but sometimes there may be fusion. Each can be clearly separated out as shown here and the same careful dissection will reveal that each has its own fibers running caudad from it. The vagus and sympathetic fibers soon join, however, into a common sheath as the **vagosympathetic trunk** and bound in with them are the common carotid artery and the internal jugular vein.

The right recurrent laryngeal nerve leaves the vagus at the line of the **caudal** (*middle*) **cervical ganglion** of the sympathetic system and runs craniad along the dorsal side of the trachea. The vagus and sympathetic trunks also separate at about this level. The vagus trunk now sends preganglionic fibers to the heart and to the trachea, lungs and esophagus. These enter into complicated cardiac, bronchial and pulmonary plexuses into which postganglionic sympathetic fibers enter and become involved. Afferent vagus fibers are also included. The vagus preganglionic fibers enter the heart and synapse (*terminal ganglia*) within the heart wall with postganglionic neurons. Terminal ganglia are also found in the lungs. The vagus in the general region of the caudal cervical ganglion also has connections with that ganglion and with the **ansa subclavia** where the sympathetic trunk divides around the subclavian between the caudal cervical and the **stellate** (*cervicothoracic*) **ganglion.**

Caudad of the root of the lung the right vagus divides into right dorsal and ventral vagal trunks in relationship to the esophagus. These fuse farther caudad with their counterparts of the left vagus nerve to form **common vagal trunks,** one dorsal, the other ventral. These give off branches to the esophagus and then pass through the esophageal hiatus of the diaphragm. The ventral branch becomes plexiform upon entering the abdominal cavity at the lesser curvature of the stomach and branches from this plexus to supply mainly the liver and the ventral part of the stomach. The dorsal branch goes initially to the cardiac stomach and then to its dorsal surface and greater curvature where it forms a plexus and supplies nerves to the pyloric region of the stomach and fibers which anastomose with the celiac plexus. No effort will be made to trace the vagus nerves farther, but evidence indicates that they supply the viscera as far caudad as the left colic flexure or the proximal descending colon.

Returning now to look at the sympathetic division of the autonomic system, we see that below the **stellate ganglion** the **sympathetic trunk** is a series of segmentally arranged **sympathetic trunk ganglia.** The stellate ganglion incorporates the first three thoracic sympathetic trunk ganglia.

The first thoracic ganglion (*part of the stellate*) sends a cardiac nerve (*postganglionic*) to join the vagus and thereby to reach the heart. Also fibers from the first four thoracic ganglia (three a part of the stellate) send postganglionic fibers to the root of the lungs where with vagus fibers they enter into the bronchopulmonary plexus. The preganglionic fibers of the thoracic nerves, five through nine, pass through their respective sympathetic trunk ganglia to form the **greater splanchnic nerve** which synapses in a collateral ganglion, the celiac, with postganglionic fibers to supply most of the abdominal visceral above the colon. The remaining thoracic nerves, ten through thirteen, send preganglionic neurons through their respective sympathetic trunk ganglia to form the **lesser splanchnic nerve** which passes to the **celiac** and **cranial mesenteric ganglia** where it synapses with postganglionic neurons which supply the colon and other abdominal structures.

The sympathetic trunk passes into the abdominal cavity beside the crus of the diaphragm. The right and left trunks lie closer together as we follow them deeper into the abdominal cavity and they get smaller in the pelvis and are lost as they approach the caudal region. The number of ganglia here correspond to the number and position of the lumbar and sacral spinal nerves. They send gray rami to all of the spinal nerves but receive white preganglionic fibers from only the first three or four lumbar spinal nerves. These go to the **caudal mesenteric ganglion** for synapse with postganglionic neurons which supply the colon, rectum and urogenital organs.

The postganglionic neurons which have been described above also form complicated plexuses throughout the abdominal and pelvic regions in relationship to the large arteries supplying the organs and are named for the vessels with which they are involved. Hence we have the **celiac** and **cranial mesenteric plexus,** the **caudal mesenteric plexus,** gastric plexus, renal plexus, suprarenal plexus and others. These plexuses involve neurons not only of the sympathetic system, but of the parasympathetic as well.

Finally, notice the **parasympathetic** contributions from second and third sacral nerves which give rise to the pelvic nerve. This nerve in turn gives rise to plexuses such as the **pelvic plexus** and a number of ganglia have been described in the area from which postganglionic neurons arise to supply pelvic organs as high at least as the left colic flexure.

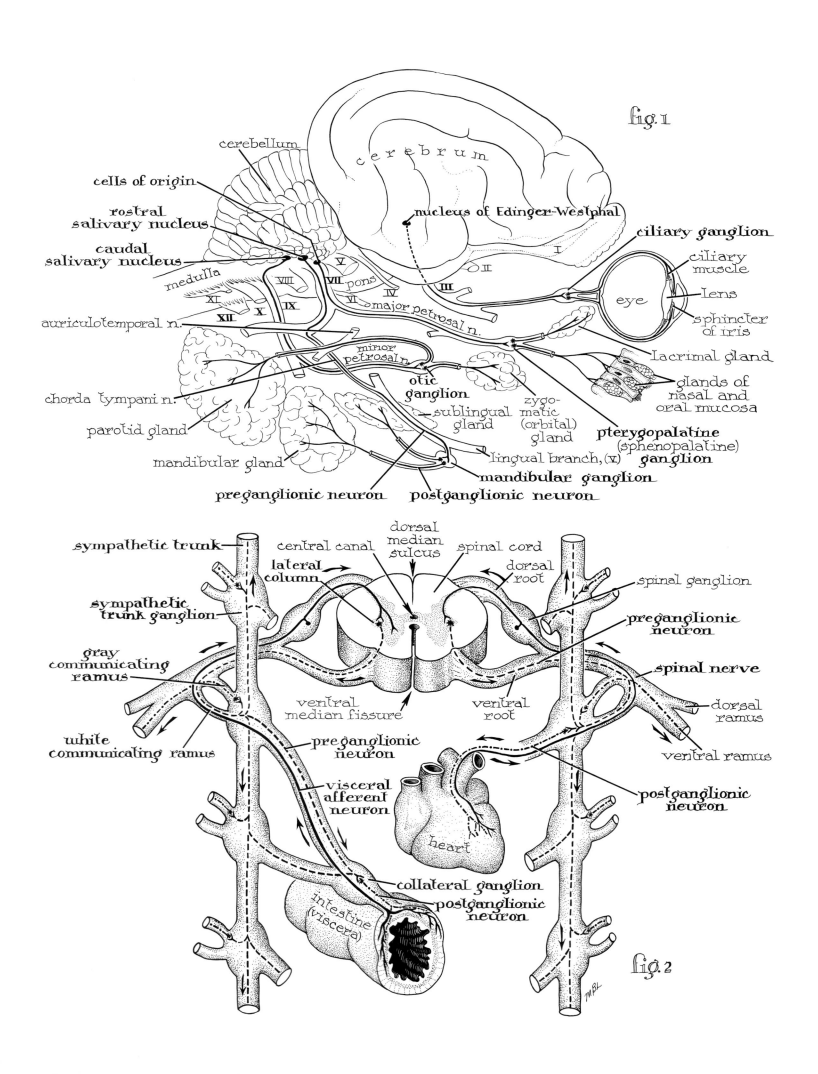

fig. 1

cerebellum

c e r e b r u m

cells of origin

rostral salivary nucleus

nucleus of Edinger-Westphal

ciliary ganglion

caudal salivary nucleus

ciliary muscle

medulla

V

VIII

VII

pons

I

II

III

IV

VI

major petrosal n.

eye

Lens

sphincter of iris

XI

IX

X

XII

auriculotemporal n.

minor petrosal n.

otic ganglion

zygo-matic (orbital) gland

Lacrimal gland

glands of nasal and oral mucosa

chorda tympani n.

parotid gland

sublingual gland

pterygopalatine (sphenopalatine) ganglion

mandibular gland

lingual branch, (v)

mandibular ganglion

preganglionic neuron

postganglionic neuron

sympathetic trunk

dorsal median sulcus

central canal

spinal cord

dorsal root

spinal ganglion

lateral column

sympathetic trunk ganglion

preganglionic neuron

gray communicating ramus

spinal nerve

dorsal ramus

white communicating ramus

ventral median fissure

ventral root

preganglionic neuron

ventral ramus

visceral afferent neuron

postganglionic neuron

heart

collateral ganglion

intestine (viscera)

postganglionic neuron

fig. 2

M.Bl.

Figure 1 is a schematic drawing to show the distribution of parasympathetic fibers in the head region. The fibers are carried in the oculomotor III, facial VII, and glossopharyngeal IX cranial nerves. The preganglionic cell bodies for nerve III are found in the **nucleus of Edinger-Westphal** in the mesencephalon. They go to the **ciliary ganglion** near the eye where synapse is made with the postganglionic fibers some of which terminate in the ciliary muscle which is used in changing the shape of the lens of the eye and thereby focusing; others go to the smooth muscle of the iris, the sphincter, which regulates pupil size and therefore determines the amount of light reaching the retina.

The nucleus for the cells of origin for the parasympathetic pathway to the lacrimal gland and the mucosa of the nose and mouth is in the medulla. The preganglionic fibers pass through nerve VII, to the major petrosal nerve and the **pterygopalatine ganglion** where synapse with the postganglionic neurons takes place.

The **rostral salivary nucleus** is the site of the preganglionic cell bodies for the pathway to the sublingual and mandibular salivary glands. The **preganglionic neurons** pass through the chorda tympani branch of the **facial (VII) nerve** to the **mandibular ganglion** where they synapse with the **postganglionic neurons.**

The parotid and zygomatic (*orbital*) salivary glands are activated by a parasympathetic pathway whose preganglionic cell bodies are in the **caudal salivary nucleus** of the **glossopharyngeal (IX) nerve.** The preganglionic fibers travel through the minor petrosal branch of nerve IX to the **otic ganglion** where synapse is made with the postganglionic neurons.

Figure 2 is a schematic drawing showing basic connections in the sympathetic part of the autonomic nervous system. Notice that afferent visceral neurons are considered here as a part of the autonomic system.

The spinal cord with its central gray and peripheral white areas and central canal are shown in cross section. Note that the **lateral column** of the gray matter is the area from which the **preganglionic neurons** take origin. They pass through the ventral root of the **spinal nerve** and on reaching the spinal spinal nerve enter the **white communicating ramus** to join one of the **sympathetic trunk ganglia.** At this point the preganglionic neuron may synapse with a **postganglionic neuron** as in figure 2 the one going to the heart or it may pass through the sympathetic trunk ganglia without synapse and go to a **collateral ganglion** and there synapse with a **postganglionic neuron** going to the intestine or some other visceral organ. A third possibility is that the preganglionic neuron will synapse in a sympathetic trunk ganglion with a postganglionic neuron which returns to the spinal nerve through the **gray communicating ramus** and through the spinal nerve reach the body wall to innervate the smooth muscle of a blood vessel, an arrector pili muscle or perhaps a skin gland. Since the sympathetic trunk ganglia are all connected by the **sympathetic trunk,** neurons can pass up or down the trunk and thereby nervous communication in this system can be made from one level of the body to another.

Finally, notice that two **visceral afferent neurons** are shown; one from the heart (not labeled) and one from the intestine. They pass through the white communicating ramus, have their cell bodies in the **spinal ganglion** and synapse in the central gray matter with a preganglionic neuron or may make other connections. A reflex mechanism is thus established as is also the basis for feeling what happens in the viscera.

Adjustment to environment is essential to survival. A critical sensitivity to the environment is prerequisite to adjustment. The receptor "organs" provide this sensitivity and each functions to **generate** nervous impulses in response to selected stimuli. The sensations which arise are the functions of the brain, and the responses that the animal makes are the result of the activity of the nervous and endocrine system. The sense "organs" are the "scouts" for these integrating systems providing them with vital information.

Receptors or sense "organs" are composed of specialized sensory cells which are highly sensitive to some particular stimulus, the **adequate stimulus,** but are much less sensitive to other stimuli. The adequate stimulus of the retina of the eye is light waves; of the spiral organ of Corti, sound waves; of the taste buds, dissolved substances. The sense "organ" may be only a naked free nerve ending as in those for pain, or it may be as complicated as the eye or ear. The receptors may have accessory structures which protect the more delicate sense cells; which intensify the force of the stimulus, as in the ear; concentrate or bring the stimulus to focus upon the sensory cells, as in the eyes; or transform the character of the stimulus to the end that it may act more effectively upon the sense organ proper, as in the ear.

It should be emphasized that nerve impulses do not necessarily occur in receptor or sense cells. In the photoreceptor cells, the rods and cones, of the vertebrate eye, for example, no one has ever demonstrated a nerve impulse. Yet, when stimulated by light they set up the physicochemical conditions which initiate impulses in the nerve cells associated with them.

As sensitive as our receptors are they operate within definite limitations. These limitations differ among the vertebrate animals and also vary to some extent among individual men. Our eyes are not sensitive to the long infrared or the short ultraviolet rays of light; the human ear detects frequencies between 16 cycles and 20,000 cycles per second, but is insensitive to the high frequency sounds that bats or dogs can detect; our chemical senses, olfaction and taste, are similarly limited. We are, in terms of our sense organs alone, insensitive to vast worlds of experience. As human animals we can compensate for some of this by the development of instruments which can detect and record data from these worlds which are to us otherwise "extrasensory."

We should appreciate also that these limitations are not necessarily unfortunate. Many are indeed a physical necessity. The fact that our ears are least sensitive at low frequencies protects us from hearing our own body sounds. Stick a finger in each ear and by thus closing out air-borne sounds, you will hear the sounds produced by the contracting muscles of your arms and fingers. To have ears more sensitive in the lower frequencies would mean that these and other body sounds would always be present to annoy us.

CLASSIFICATION OF RECEPTORS

Treating the sense "organs" as a system raises the same problems that we experience with the endocrine "system." They are widely distributed over the body; they are diverse in structure, and there is no physical continuity among them. Yet they do serve a common function in the collecting of data which the nervous and endocrine systems require in order to perform their orienting and integrating roles. It is more logical perhaps to think of them as parts of the nervous system since with it they are structurally continuous.

Similar problems face us in attempting to classify the receptors. It is common to speak, as the title of this chapter suggests, of the general and special sense organs. Those are called **general** which have a wide distribution throughout the body such as the receptors for heat, cold, pain, touch and muscle or tendon sense (*proprioception*). The **special** sense organs are found in the head region only and are more advanced and specialized. They are the eye, ear, taste "buds" and olfactory organs, and by many authors are treated separately from the general sense structures.

Perhaps the best way to classify receptors is on the basis of the part of the environment they sample or sense, as follows:

1. **Exteroceptors**
   a. receive stimuli from external environment
   b. located in skin and its apertures
   c. include organs of special senses except taste; also those for pain, temperature, touch and pressure.

2. **Interoceptors**
   a. receive stimuli from "internal environment"
   b. lie within the walls of the digestive tube or its derivatives and in walls of other internal organs
   c. includes sense organs of taste, pain, pressure, etc.

3. **Proprioceptors**
   a. receive stimuli from the true internal environment
   b. lie within the muscles and tendons and around joints
   c. include neurotendinous and neuromuscular spindles.

Classified on the basis of location, receptors fall into four groups: 1. **general somatic sensory**—the widely dis-

317

tributed sense organs of skin and skeletal muscles. 2. **special somatic sensory**—those of limited distribution at body surfaces—the eyes, ears, olfactory organs. 3. **general visceral sensory**—widely distributed in organs of the digestive tract and other viscera. 4. **special visceral sensory**—sense organs of limited distribution in digestive tract—as the taste buds.

ORGANS OF GENERAL SENSE

The structure of these receptors can best be learned and understood by reference to the diagrams. They range from simple free nerve endings to organs of quite complex design (Fig. 18).

Free Nerve Endings (Fig. 18A). These consist of dendrites which have lost their myelin and neurilemma and whose fibrillae anastomose and end in knobs or discs between epithelial cells. All the **pain** receptors and some of those for touch, temperature, and muscle sense (*proprioceptive*) are of this type.

End Bulbs of Krause (Fig. 18B). They occur quite widely over the body and vary in shape from cylindrical to oval. They consist of a capsule from the connective tissue sheath of a myelinated nerve fiber. Inside they contain a soft material in which the nerve fiber ends in either a bulb or a coiled mass. It has been suggested that they are receptors for **cold.**

Brushes of Ruffini. These were described by Ruffini from the subcutaneous tissue of the human finger and are considered to be **heat** receptors. They are oval in shape and have a tough connective tissue sheath. Inside the nerve fibers branch and end in small free knobs.

Tactile Corpuscles of Meissner (Fig. 18C). These are receptors for **light touch** and occur in the papillae of the corium and in many areas such as the hands, feet, lips, mucous membranes of tongue, skin of mammary papillae, and the front of the forearm. They are small and oval with a connective tissue sheath and what appear to be tiny plates placed one above the other. The nerve fiber penetrates the capsule and forms a spiral arrangement ending in globular structures among the plates.

Pacinian (lamellated) Corpuscles (Fig. 18D). These are large receptors, visible to the naked eye, from 2 to 4 millimeters in diameter. Each corpuscle is at the end of a nerve fiber and its structure consists of a number of concentric lamellae reminding one of a section of an onion. The nerve fiber passes into the central part of the corpuscle, losing first its myelin sheath and then the neuri-

lemma as it passes among the specialized layers. The Pacinian corpuscles are **deep pressure** receptors found in the subcutaneous, submucous and subserous connective tissue, being especially common in the palm of the hand, sole of the foot, genital organs, around the joints and in the mesentery around the pancreas.

Neuromuscular Spindles (Fig. 18F). These are present in most skeletal muscles and are the receptors for **muscle sense** (*proprioceptive*) which enable us to be aware of the position of the body and its parts without following movements with the eyes. It would be very awkward to have to watch the feet in order to walk or to know that you had your hand in your pocket only by seeing it there. These receptors consist of small bundles of delicate muscle fibers enclosed in a capsule within which two or three myelinated nerve fibers lose their sheath, branch out and end by flattened expansions or discs on the muscle fibers. When the muscle contracts, the nerve endings are stimulated.

Neurotendinous spindles are much like muscle spindles except that they are smaller and are formed in association with the tendons near their junctions with the muscle fibers. These too are **proprioceptive** in function. (Fig. 18E.)

General Visceral Receptors. Some of the receptors just described, the Pacinian corpuscles, for example, are found also in the viscera. Many visceral receptors consist of free nerve endings. The impulses from visceral receptors do not ordinarily reach the level of consciousness; some, of course, do. Impulses from visceral receptors in the thorax and abdomen travel over sympathetic nerves and rami communicantes to the spinal ganglia. Others by way of the vagus and glossopharyngeal nerves reach their sensory ganglia, while still others travel in the pelvic nerves to spinal ganglia in the sacral area. These relationships will become clearer if the autonomic nervous system is reviewed.

Special Visceral Receptors. These are the receptors for reflex control of respiration and circulation. The nerve fibers are carried in vagus and glossopharyngeal nerves. The carotid and aortic bodies and the carotid sinus mentioned under the circulatory system are among the special visceral receptors. The **carotid body** and **aortic body** are influenced by the concentration of carbon dioxide in the blood; the **carotid sinus,** an enlargement at the beginning of the internal carotid artery, is sensitive to changes in blood pressure and its receptors initiate reflex controls over circulation.

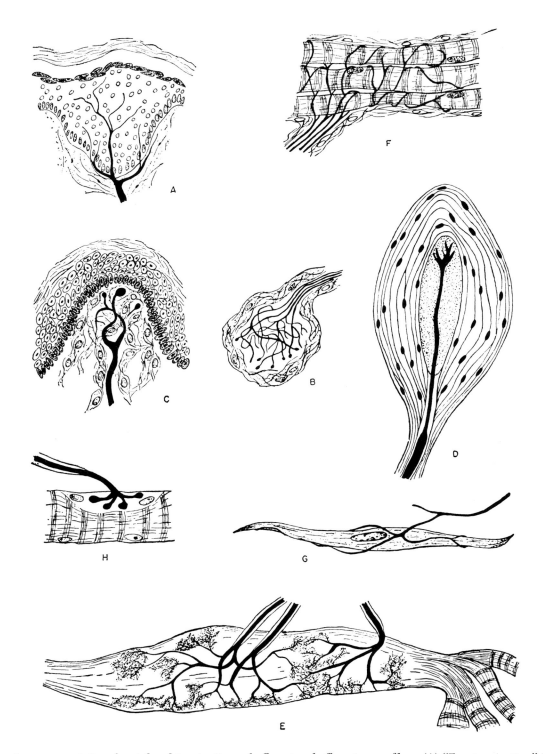

Fig. 18. *Showing some varieties of peripheral terminations of afferent and efferent nerve fibers.* (A) *"Free termination" in epithelium (after Retzius).* (B) *Krause's corpuscle from conjunctiva (after Dogiel).* (C) *Meissner's corpuscle from skin (after Dogiel).* (D) *Pacinian corpuscle (after Dogiel).* (E) *termination upon tendon-sheath (Huber and DeWitt).* (F) *neuromuscular spindle (after Ruffini).* (G) *motor termination upon smooth muscle fiber.* (H) *motor "end plate" on skeletal muscle fiber (after Böhm and von Davidoff).* (*From Edwards, L. F.:* Concise Anatomy. *2nd ed. New York, McGraw-Hill Book Co., Inc., 1956.*)

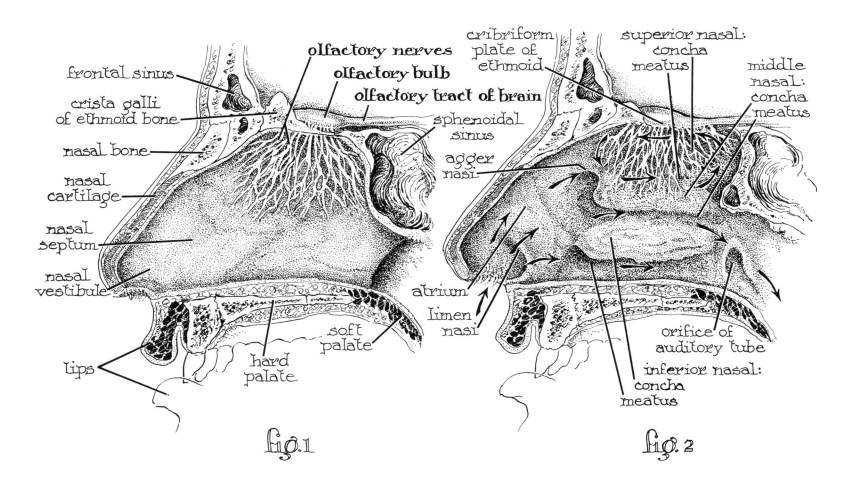

fig. 1

frontal sinus

crista galli of ethmoid bone

nasal bone

nasal cartilage

nasal septum

nasal vestibule

lips

olfactory nerves

olfactory bulb

olfactory tract of brain

sphenoidal sinus

hard palate

soft palate

fig. 2

cribriform plate of ethmoid

superior nasal: concha meatus

middle nasal: concha meatus

agger nasi

atrium

limen nasi

orifice of auditory tube

inferior nasal: concha meatus

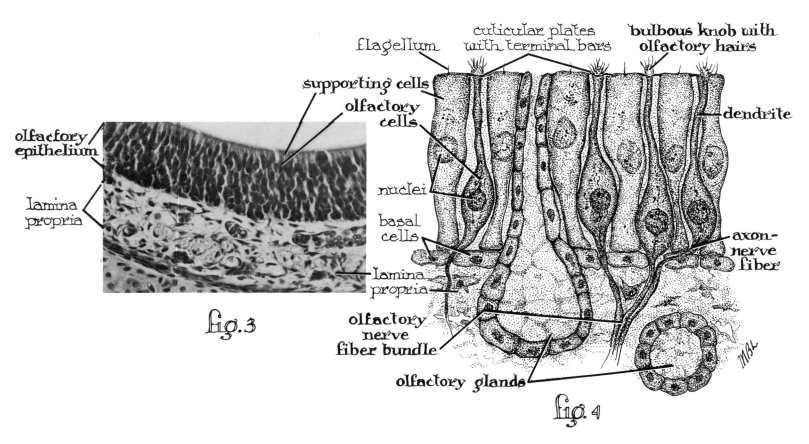

fig. 3

olfactory epithelium

lamina propria

supporting cells

olfactory cells

nuclei

lamina propria

fig. 4

flagellum

cuticular plates with terminal bars

bulbous knob with olfactory hairs

dendrite

basal cells

axon-nerve fiber

olfactory nerve fiber bundle

olfactory glands

MBL

The Olfactory Organ and Its Relationships
PLATE 110

(Crouch: Functional Human Anatomy, Lea & Febiger)

Figures 1 and 2 show the nose and nasal passageways of man but, except in general form, they differ in no important way from those in the cat. Refer to Plate 49 where these structures are shown in the cat. Figures 3 and 4 present the microscopic features of the olfactory epithelium.

Figure 1 shows the skull sectioned just to the left of the nasal septum. This gives one an opportunity to review skeletal and other features of this region. The hard and soft palate separate the mouth below from the nasal passageway above. The nasal vestibule, nasal septum, nasal cartilage, nasal bone, frontal bone and sinus, the crista galli of the ethmoid, the spenoid and sinus and a part of the cranial cavity are all clearly designated. Our primary concern is with the olfactory nerves which can be seen spreading out over the upper parts of the nasal septum; the olfactory bulb into which these nerves pass after passing through the foramina of the cribriform plate of the ethmoid and the olfactory tract leading to the brain proper.

Figure 2 is a section through the right nasal passageway which provides a good view of the lateral wall. The superior, middle and inferior nasal conchae and meatuses are shown. Arrows indicate the direction of flow of air through the nose and into the nasopharynx. Note the orifice of the auditory tube and the division of the nasal passageway into vestibule and atrium separated by a curved ridge, the limen nasi. Again, and most important, notice the olfactory nerves spread over the superior nasal concha. Like those on the nasal septum they pass through the foramina of the cribriform plate to reach the olfactory bulb.

Figure 3 is a photomicrograph of the olfactory membrane or mucosa to show the olfactory epithelium resting on the lamina propria. Figure 4 is an idealized drawing of the olfactory mucosa. Observe the olfactory glands extending into the subepithelial tissue and opening on the epithelial surface where their secretions keep the mucous membrane moist. This is important because the olfactory cells are chemoreceptors and can detect chemicals only when they are in solution. Musk in dilutions of one part in eight million will activate the olfactory cells or receptors. Notice too the olfactory cells. They are the least specialized of the sensory receptors. They are modified epithelial cells scattered among the other epithelial cells which here serve as supporting cells. The olfactory cells are bipolar with a small amount of cytoplasm and a large nucleus. A slender peripheral process extends from each olfactory cell body to the free surface of the mucous membrane and send out beyond the surface a bulbous knob with olfactory hairs. The axon fiber (*central process*) passes through the basement membrane and joins with other axon nerve fibers to form bundles of unmyelinated fibers the olfactory nerves. These bundles form a plexus in the submucosa from which about twenty olfactory nerves emerge, pass through the foramina of the cribriform plate and end by synapsing with the mitral cells in the glomeruli of the olfactory bulb of the brain.

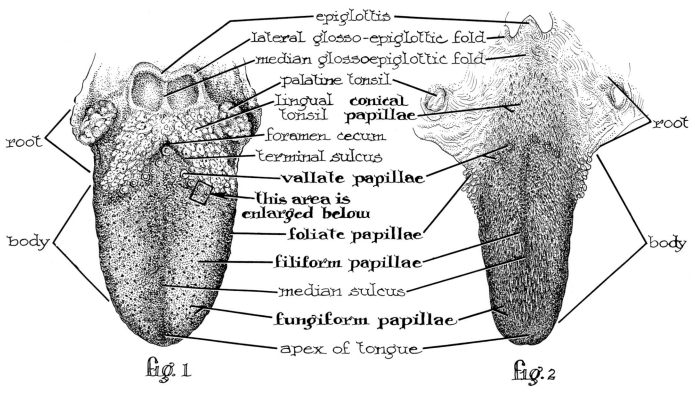

epiglottis
lateral glosso-epiglottic fold
median glossoepiglottic fold
palatine tonsil
lingual tonsil   conical papillae
foramen cecum
terminal sulcus
vallate papillae
this area is enlarged below
foliate papillae
filiform papillae
median sulcus
fungiform papillae
apex of tongue

root

body

root

body

fig. 1

fig. 2

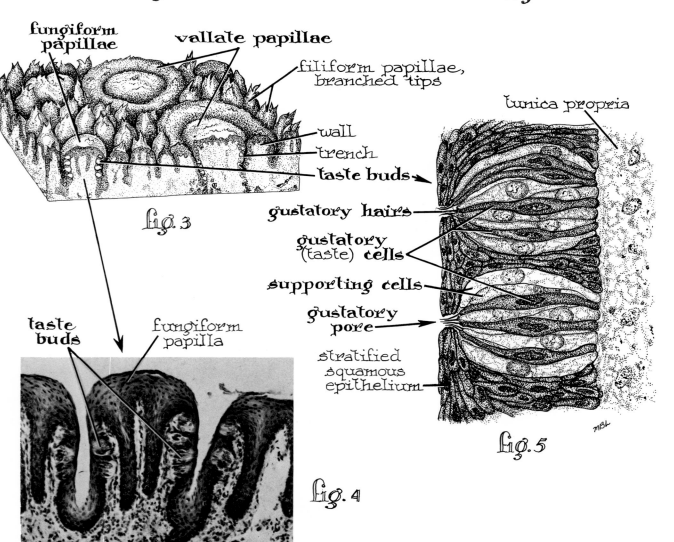

fungiform papillae
vallate papillae
filiform papillae, branched tips
wall
trench
taste buds

fig. 3

taste buds

fungiform papilla

fig. 4

tunica propria

taste buds
gustatory hairs
gustatory (taste) cells
supporting cells
gustatory pore
stratified squamous epithelium

fig. 5

The Tongues of Man and the Cat and the Gustatory Organs
PLATE 111

(In part from Crouch: Functional Human Anatomy, Lea & Febiger)

In this illustration figures 1 and 2 compare the superficial structures of the upper surface of the tongues of man and the cat respectively. Their relationship to palatine and lingual tonsils are also shown as well as the epiglottis. Figure 3 is an enlarged drawing of a small area cut from the human tongue as shown in figure 1. Figures 4 and 5 are studies of taste buds or gustatory "organs".

One should recall, before continuing with this study of the tongue as an organ of special sense, the basic structural make up of the tongue and its functions other than described here. Reference to Plates 24, 49, and 50 and the accompanying descriptions and page 133 in the text will provide some of this information.

Figure 1, the dorsal aspect of the human tongue, shows the lingual tonsil on the root of tongue and lateral to it the palatine tonsils resting in a fossa formed by the pillars of the fauces, page 133. The epiglottis is seen in back of the tongue with its median and lateral glossoepiglottic folds. The dorsum of the tongue shows a slight furrow down its center, the median sulcus. This surface and the margins of the tongue are provided with papillae of various types. Numerous **filiform papillae** are found scattered over the anterior two-thirds of the tongue. They are small, slender and have a scaly surface. **Fungiform papillae** are globular and highly vascular and hence have a pink color. They are most numerous on the sides and tip of the tongue and more irregularly spaced over its surface. On the margins of the body of the tongue posteriorly are a number of leaf-like papillae, often considered with the fungiform type, but called here the **foliate papillae.** Forming a V-shaped pattern at the posterior limit of the anterior two-thirds of the tongue are about twelve **vallate** (*circumvallate*) **papillae.** They lie just in front of the terminal sulcus, a V-shaped groove with the foramen cecum at its apex. The vallate papillae are flattened structures with a moat-like trough around them, outside of which is a raised ring. Taste buds are found on the fungiform, foliate and vallate papillae—mostly on their sides, though some are found on the free surface of fungiform papillae—especially in the cat.

Figure 2 is a similar view of the tongue of the cat. Note the palatine tonsils and the epiglottis and the distribution of papillae over the tongue surface. The papillae are similar to those of the human tongue except that the **filiform papillae** are more sharply pointed and hooked and point as much or more toward the root of the tongue. In addition, the cat has numerous **conical papillae** in back of the vallate papillae. The terminal sulcus and foramen cecum are not seen in the cat.

Figures 3, 4, and 5 are drawn from studies of the microscopic structure of the tongue. Figure 3 is a small block of tissue removed from the point indicated in figure 1 and very much enlarged. It shows magnified filiform papillae with their branched sharp tips; one **fungiform papilla** with its flat surface and **taste buds** shown along its sides; and two **vallate papillae,** showing the moat or trench with its outer wall and the taste buds on the sides of the cut section of the papilla.

Figure 4 is a photomicrograph of fungiform papillae showing the **taste buds** along their sides (130×).

Figure 5 is a drawing of two **taste buds** showing the **gustatory** (*taste*) **cells,** with **gustatory hairs** projecting from the **gustatory pores.** The gustatory cells are held in place by the supporting cells. Note that the taste buds, themselves epithelial in origin, are found within the stratified squamous epithelium which covers the surface of the tongue and its structures. The epithelium rests upon the tunica propria (*connective tissue*). Though not shown in this illustration it should be remembered that each gustatory cell at its central end comes into intimate contact with many fine terminations of nerves which enter the taste buds through their basement membranes. These are myelinated nerves which lose their sheaths as they enter the taste bud. These nerves are afferent neurons from the glossopharyngeal (**IX**) and the facial (**VII**) cranial nerves. The gustatory cells are stimulated by substances in solution in the mouth. Impulses are carried to the brain stem, are transferred to internuncial or association neurons which carry them to the hippocampal gyrus in the temporal lobe where they are interpreted as tastes. We taste with our brain—as does the cat. The taste buds are receptors only.

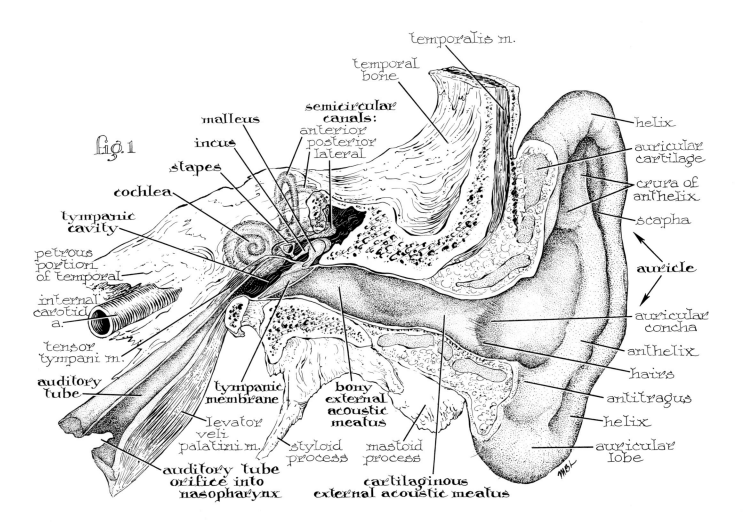

fig.1

temporalis m.

temporal bone

semicircular canals:
anterior
posterior
lateral

malleus

incus

stapes

cochlea

tympanic cavity

petrous portion of temporal

internal carotid a.

tensor tympani m.

auditory tube

tympanic membrane

levator veli palatini m.

styloid process

bony external acoustic meatus

mastoid process

auditory tube orifice into nasopharynx

cartilaginous external acoustic meatus

helix

auricular cartilage

crura of anthelix

scapha

auricle

auricular concha

anthelix

hairs

antitragus

helix

auricular lobe

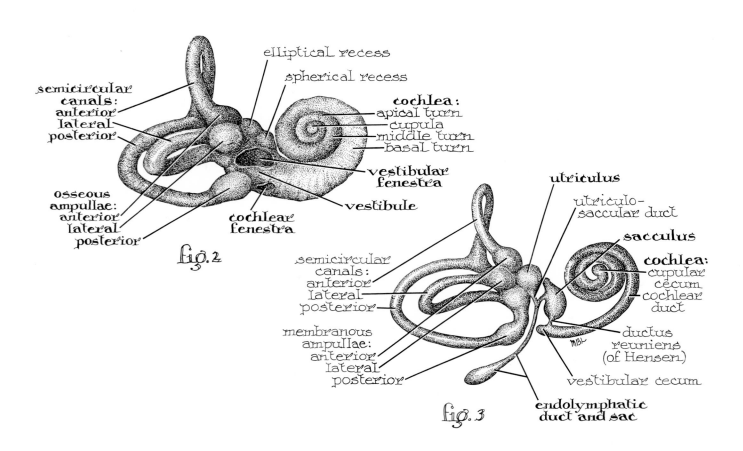

semicircular canals:
anterior
lateral
posterior

osseous ampullae:
anterior
lateral
posterior

fig.2

elliptical recess

spherical recess

cochlea:
apical turn
cupula
middle turn
basal turn

vestibular fenestra

cochlear fenestra

vestibule

semicircular canals:
anterior
lateral
posterior

membranous ampullae:
anterior
lateral
posterior

fig.3

utriculus

utriculo-saccular duct

sacculus

cochlea:
cupular cecum
cochlear duct

ductus reuniens (of Hensen)

vestibular cecum

endolymphatic duct and sac

(Crouch: Functional Human Anatomy, Lea & Febiger)

This plate illustrates in figure 1 the entire human ear showing pertinent relationships of external, middle and internal ears. Except for the shape of the auricle the human ear does not differ appreciably in gross structure from that of other mammals. Figure 2 shows the bony labyrinth and figure 3 the membranous labyrinth of the ear. Details of structure of these parts and of the cochlea can be studied in connection with Plate 113. A view of the external ear of the cat can be seen in Plate 75.

The **external ear** consists of the auricle (*pinna*) and the cartilaginous and bony external acoustic meatuses. The **auricle**, extending lateral to the temporal bone of the skull consists of a curved rolled-in edge called the helix. At its lower end is the auricular lobe. Inward from the helix and separated from it by a groove, the scapha, is the anthelix which divides above forming the crura of the anthelix. Inward from the anthelix is a depression the auricular concha and below it a prominence the antitragus. The tragus, a conspicuous process in front of the concha is cut away in this section. You can feel it on your own ear. The auricular cartilage is of the elastic or yellow type and is covered over by a thin skin. The cartilage gives flexibility to the auricle. It is lacking in the auricular lobe.

The outer part of the external acoustic meatus is supported by cartilage, the inner by bone, hence the terms **cartilaginous** and **bony external acoustic meatuses.** The whole meatus is about 2.5 centimeters long and leads from the "funnel" of the auricle inward to the tympanic membrane. It is lined with skin which is provided with many fine hairs and sebaceous glands near its orifice. In its upper wall are modified sweat glands, the ceruminous glands which secrete cerumen or ear wax.

The **tympanic membrane** lies between the external acoustic meatus and the middle ear cavity. It is tipped to form an angle of about 55 degrees with the floor of the meatus. It is composed of three layers—an external thin skin, an intermediate fibrous layer and an inner mucous membrane. The sound waves, collected by the auricle and external acoustic meatus, impinge upon the tympanic membrane and cause it to vibrate.

The middle ear or **tympanic cavity** is a laterally compressed space in the petrous portion of the temporal bone. It is separated laterally from the external acoustic meatus by the tympanic membrane and from the internal ear by a bony wall in which are found two small openings covered by membrane, the oval window or **vestibular fenestra** and the round window or **cochlear fenestra** (fig. 2). The posterior wall of the tympanic cavity opens into a large air space, the mastoid antrum, which in turn communicates with air cells in the mastoid process. Anteriorly the **auditory tube** (*Eustachean*) connects the tympanic cavity with the nasopharynx which enables one to equalize air pressure in the tympanic cavity with atmospheric pressure.

Three small bones the ear ossicles bridge the gap across the tympanic cavity from the tympanic membrane to the fenestra vestibuli (*oval window*). They are the malleus, incus and stapes and phylogenetically are related to the jaw structures of the lower vertebrates. The lateral bone, the **malleus,** (*hammer*) attaches to the upper part of the tympanic membrane; the **stapes** (*stirrup*) base fits into the fenestra vestibuli; and the **incus** (anvil) lies between the other two ossicles and is articulated to them by synovial joints. By them the small pressure on the large area of the tympanic membrane is concentrated upon the small area of the vestibular fenestra increasing 22-fold the pressure on the fluid in the internal ear.

The internal ear is the essential part of the ear containing the actual sensory mechanism by which impulses are set up in the two parts of the vestibulocochlear cranial nerve (Plate 113). It consists of an osseous and membranous labyrinth (figs. 2 and 3). The **osseous labyrinth** is hollowed out of the substance of the petrous portion of the temporal bone. It is lined with a thin fibroserous membrane which secretes a fluid called **perilymph.** The parts of the osseous labyrinth are the vestibule, semicircular canals and cochlea. It is shown in figure 2.

The **vestibule** is a chamber just medial to the tympanic cavity, separated from it by a thin partition of bone in which is found the **vestibular fenestra** which, in the living state, is closed by the base of the stapes and its annular ligament. It communicates posteriorly through five orifices with the semicircular canals and anteriorly through one orifice with the cochlea.

The **semicircular canals** are three in number one lying in each of three planes—the **anterior** (*superior*), **lateral** and **posterior.** Each one is provided with a swelling the **ampulla** at the end of one of its arms.

The bony **cochlea** is shaped like a snail's shell consisting of two and three quarter spiral coils. The basal coil is broad and it tapers as it spirals to a narrow apex. Inside, the coil is partially divided by an osseous spiral lamina into two parts. The division is completed by the basilar membrane and the two parts then communicate only at the apex where they are continuous one with the other. The upper part of the divided canal is called the scala vestibuli; the lower part the scala tympani. The scala vestibuli enters the vestibule at its basal end; the scala tympani ends at the **cochlear fenestra** which in the living subject is closed by the secondary tympanic membrane.

Figure 3 is the **membranous labyrinth** which lies within the osseous labyrinth and therefore has much the same form except that the vestibular portion is divided into two chambers the utriculus and sacculus. The space between the walls of the two labyrinths is filled with **perilymph,** while inside, the membranous labyrinth is filled with **endolymph.**

The **utriculus** is larger than the sacculus and lies in the upper back part of the vestibule. It communicates with the five openings of the semicircular canals and has in thickened areas in its walls the utricular filaments of the vestibular nerve (Plate 113). It communicates also with the endolymphatic duct and sac.

The **sacculus,** also a chamber within the vestibule, has a thickened area in its walls provided with saccular filaments of the vestibular nerve (Plate 113). It communicates with the cochlear duct through a small canal from its lower side and with the endolymphatic duct from its posterior wall.

The **endolymphatic duct** ends in a blind pouch on the posterior surface of the petrous temporal bone where it lies against the dura mater.

The **cochlear duct** lies between the scala vestibuli above, the scala tympani below. It rests on the basilar membrane and is limited above by the thin vestibular membrane. It is filled with endolymph and contains the spiral organ of Corti, the essential receptor for hearing (Plate 113).

The Vestibulocochlear Nerve and the Organ of Hearing and Equilibrium

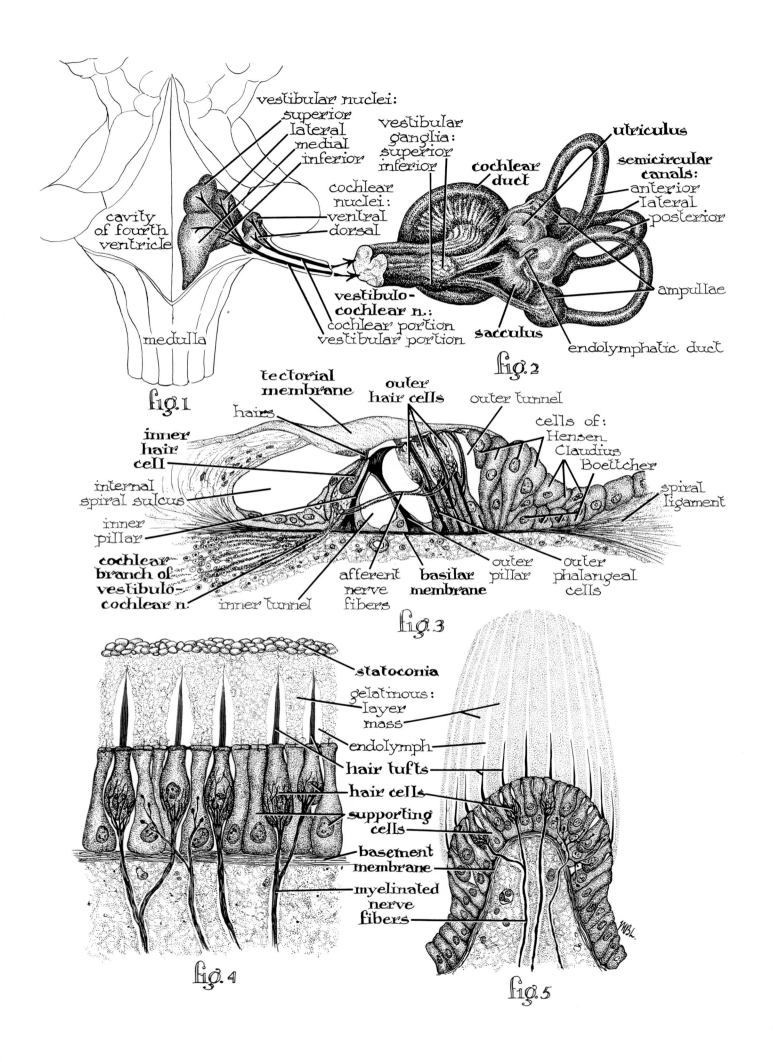

vestibular nuclei:
superior
lateral
medial
inferior

vestibular ganglia:
superior
inferior

cochlear nuclei:
ventral
dorsal

cavity of fourth ventricle

medulla

vestibulo-cochlear n.:
cochlear portion
vestibular portion

utriculus

cochlear duct

semicircular canals:
anterior
lateral
posterior

ampullae

sacculus

endolymphatic duct

fig. 1

fig. 2

tectorial membrane

hairs

outer hair cells

outer tunnel

cells of:
Hensen
Claudius
Boettcher

inner hair cell

internal spiral sulcus

inner pillar

cochlear branch of vestibulo-cochlear n.

inner tunnel

afferent nerve fibers

basilar membrane

outer pillar

outer phalangeal cells

spiral ligament

fig. 3

statoconia

gelatinous:
layer
mass

endolymph

hair tufts

hair cells

supporting cells

basement membrane

myelinated nerve fibers

fig. 4

fig. 5

MBL.

The Vestibulocochlear Nerve and the Organ of Hearing and Equilibrium

PLATE 113

(Crouch: Functional Human Anatomy, Lea & Febiger)

This illustration shows the eighth or vestibulocochlear (*statoacoustic; auditory*) cranial nerve, its receptors and central nuclei. Its receptors lie in the organ of hearing and equilibrium (*internal ear*); its central nuclei are in the brain stem.

Figure 1 is a diagram of the dorsal side of the brain stem showing the medulla and the cavity of the fourth ventricle. The cerebellar hemispheres have been cut away and the roof of the fourth ventricle removed. The vestibular and cochlear nuclei with their various lobes are drawn in to show their approximate position in the brain stem. The **vestibulocochlear nerve,** an afferent (*sensory*) nerve, consisting of vestibular and cochlear portions, is shown with each part entering its respective nucleus.

Figure 2 shows the distal part of the vestibulocochlear nerve and its relationship to the cochlea and its **cochlear duct** and to the **utriculus, sacculus,** and **semicircular canals.** The ampullae of the semicircular canals are shown and also the endolymphatic duct which arises from the posterior wall of the sacculus and ends as a blind pouch on the posterior surface of the petrous bone where it contacts the dura mater.

Figure 3 is a section of the organ of Corti, which rests on a **basilar membrane** and projects into the cochlear duct of the cochlea. On the left, the **cochlear branch of the vestibulocochlear nerve** is seen as it arises from afferent nerve fibers which come from the inner and outer hair cells. There is a single row of **inner hair cells,** about 3500 in all extending the length of the coiled cochlea and lateral to these about 12,000 **outer hair cells** arranged in three rows in the basal coil and four rows in the apical coil of the cochlea. The hair cells have long hair-like processes at their free ends and large basal nuclei. The inner hair cells are supported by columnar cells and the outer hair cells by the outer phalangeal cells (*cells of Deiters*). A reticular framework (*shown in black*) supported by specialized rods the inner and outer pillars "fits over" the free ends of the outer hair cells to give them additional support.

The spiral organ of Corti is covered by a very delicate and flexible membrane, the **tectorial membrane.** It is attached medially and extends roof-like over the hair cells, contacting their hair-like processes.

A brief statement of how the organ of Corti works is in order. Recall that the organ of Corti is surrounded by fluid, the endolymph, which fills the membranous labyrinth. This fluid is made to "vibrate" by sound waves transmitted to it by the structures of the external and middle ear. The basilar membrane is in turn set in motion by the vibrations in the endolymph. Since the basilar membrane varies in width with its narrowest portion and

its shortest fibers in the basal coil of the cochlea and its widest portion and longest fibers in the apical coil, the various tones produce maximal vibrations in different parts of the membrane. The basilar membrane also makes a frequency analysis for pitch. These variable vibrations of the basilar membrane are faithfully transmitted by the hair cells in contact with the tectorial membrane and which set up impulses in the peripheral fibers of the cochlear branch of the vestibulocochlear nerve. These impulses are carried to the spiral ganglion in which are found the cell bodies of these bipolar neurons. The impulses then travel through the central fibers of the neurons which pass through the internal acoustic meatus to terminate in the dorsal and ventral cochlear nuclei in the medulla (*see figs. 1 and 2*). In the cochlear nuclei the fibers of the cochlear nerve, in turn, synapse with neurons whose fibers reach the medial geniculate body in the midbrain. Here they also synapse with the third neurons in the chain which carry the impulses to the hearing center in the temporal cortex of the cerebrum where sound is perceived.

Figure 4 is a section of a sensory area like we find in both the utriculus and sacculus of the internal ear (fig. 2). It is called the **macula** and consists of supporting and hair cells. The columnar **supporting cells** have basal nuclei and a tiny flagellum (*not labeled*) projecting from the free surface. They rest on a **basement membrane.**

The **hair** (*sensory*) **cells** are flask-shaped with basal nuclei and do not extend to the basement membrane. Projecting from the free surface of each hair cell is a tuft of long non-motile cilia which are cemented together and taper to a point. A single flagellum extends beyond the point. These structures are called, on figure 4, the **hair tufts;** the flagella are illustrated but not labeled. The surface of the macula is covered with an otolithic membrane called in the illustration the gelatinous layer into which the hair tufts penetrate. Each tuft is surrounded by a space filled with endolymph. Beyond the hair tufts the gelatinous substance contains numerous crystalline bodies, the statoconia (*otoconia or otoliths*). They contain calcium carbonate and a protein. Any disturbance caused by an altered position of the head causes the statoconia to exert a pull on the hair cells resulting in their stimulation. The information received in this way from all four maculae in the two ears is one of the factors contributing to an animal's capacity to maintain the normal head position. The utriculus and sacculus appear therefore, to be involved in maintaining static equilibrium.

Figure 5 is an illustration of a structure, similar to a macula, which is found in the ampullae of the semicircular canals. It has a raised center where the **hair cells** are found

329

and from which it tapers off to the sides. The hair tufts project into the endolymph and since the semicircular canals are placed in all body planes any movement of the head will cause a pull on the hairs of one crista or another, due to the inertia of the fluid contained in the canals. This sets up impulses in the hair (*sensory*) cells of the crista; impulses travel to the brain and reflex adjustments are made.

Note in figure 2 that nerve fibers from the utriculus, sacculus and the ampullae of the semicircular canals all come together to form the vestibular portion of the vestibulocochlear nerve. These are the **myelinated nerve fibers** which are seen coming out of the hair cells of macula and crista in figures 4 and 5. They enter the vestibular ganglion in which are located the cell bodies of these sensory or afferent neurons. Their central processes follow the vestibular portion of the vestibulocochlear (*VIII*) nerve to enter the vestibular nuclei in the medulla. From here they may go directly to the cerebellum; be relayed to it through the vestibular nuclei; or they may be relayed upward to motor cranial nerves like those of the eye muscles; or downward by the vestibulospinal tract to the anterior gray columns of the spinal cord. The cerebellum thus receives "proprioceptive impulses" from the vestibular part of the ear as well as from the tendons and joints by way of the spinocerebellar tracts, which, with the vestibulocerebellar tracts enter through the inferior cerebellar peduncles. By this complicated mechanism the body is kept in equilibrium and the muscles coordinated for efficient action.

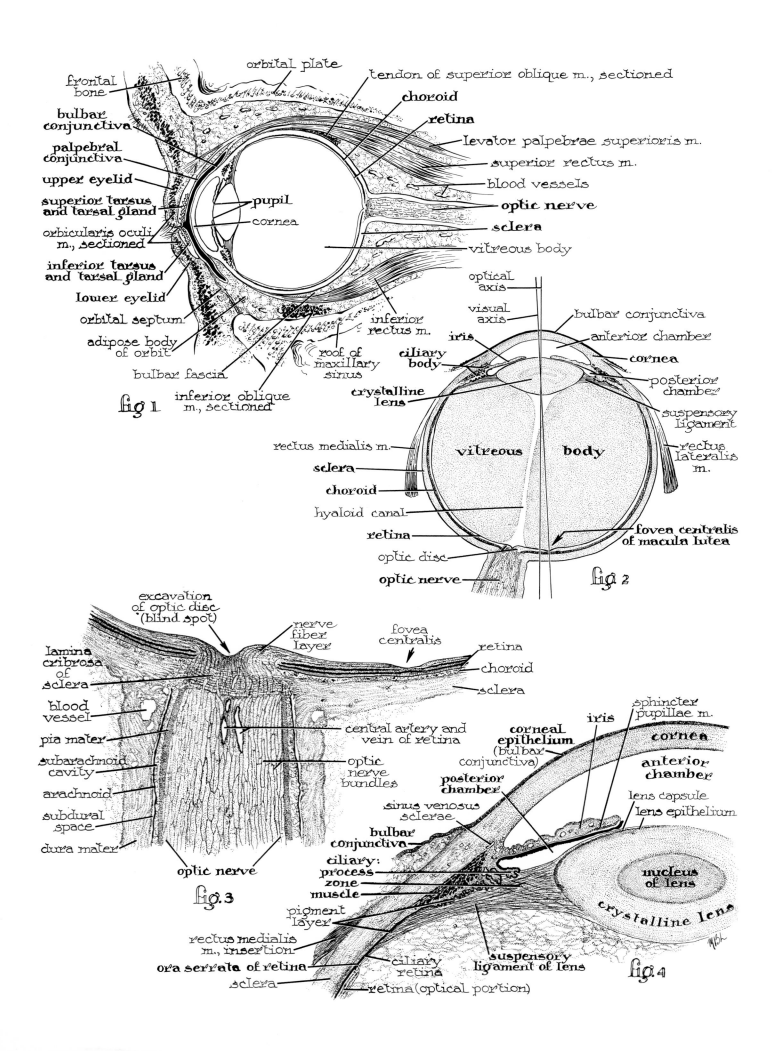

frontal bone
orbital plate
tendon of superior oblique m., sectioned
bulbar conjunctiva
choroid
retina
palpebral conjunctiva
levator palpebrae superioris m.
upper eyelid
superior rectus m.
superior tarsus and tarsal gland
pupil
blood vessels
cornea
optic nerve
orbicularis oculi m., sectioned
sclera
inferior tarsus and tarsal gland
vitreous body
lower eyelid
orbital septum
adipose body of orbit
inferior rectus m.
bulbar fascia
roof of maxillary sinus
fig 1
inferior oblique m., sectioned

optical axis
visual axis
bulbar conjunctiva
anterior chamber
iris
cornea
ciliary body
posterior chamber
crystalline lens
suspensory ligament
rectus medialis m.
rectus lateralis m.
vitreous body
sclera
choroid
hyaloid canal
retina
fovea centralis of macula lutea
optic disc
optic nerve
fig 2

excavation of optic disc (blind spot)
nerve fiber layer
fovea centralis
retina
lamina cribrosa of sclera
choroid
sclera
blood vessel
central artery and vein of retina
pia mater
subarachnoid cavity
optic nerve bundles
arachnoid
subdural space
dura mater
optic nerve
fig.3
rectus medialis m., insertion
ora serrata of retina
sclera

sphincter pupillae m.
iris
corneal epithelium (bulbar conjunctiva)
cornea
anterior chamber
posterior chamber
sinus venosus sclerae
lens capsule
lens epithelium
bulbar conjunctiva
ciliary: process zone muscle
nucleus of lens
pigment layer
crystalline lens
suspensory ligament of lens
ciliary retina
retina (optical portion)
fig.4

The eyes illustrated are those of man but differ in no important detail of gross anatomy from those of the cat or other mammals. Indeed, they do not vary a great deal from those of members of other vertebrate classes. Like the ear they are distance receptors informing us of objects and events far removed from the body, as well as those close at hand. They give to the living body a kind of continuity with the environment.

Figure 1 shows a vertical section of the orbital cavity containing the eye and related structures. The eyes are closed by the **upper and lower eyelids.** On the outer surfaces of the eyelids is a thin skin which extends over the edge of the lid to the inner surface where it becomes the **palpebral conjunctiva,** a thick, vascular membrane containing papillae and some lymphoid tissue in its deeper portions. It is continuous at the base of the eyelids with the **bulbar conjunctiva** which as a thin, transparent stratified squamous epithelium passes over the anterior part of the sclera and cornea. At the edges of the eyelids are the eyelashes which are modified hairs. The upper ones are longer than the lower ones and turn upward. The lower ones turn downward. Within the eyelids, just inside of the thin skin, is the orbicularis oculi muscle and in back of it the **tarsus and tarsal glands.** Inserted into the upper lid is the aponeurosis of the levator palpebrae superioris muscle which lifts the eyelid.

Note that other muscles of the eyeball (*bulb*) are shown; the superior and inferior rectus; the cut surface of the inferior oblique and the cut tendon of the superior oblique. The eyeball is surrounded in the orbital cavity by bulbar fascia and the adipose body.

The main features of the eyeball or bulb are clearly shown. The three layers of the eyeball are the **retina, choroid** and the outermost, the **sclera.** Anteriorly the sclera is continuous with the transparent cornea. The **pupil** is shown as an opening in the iris behind which is the lens. The optic nerve is seen at the back of the eyeball and the vitreous body fills the area behind the lens.

Figure 2 is a schematic section through the right eye in the horizontal plane. The remaining extrinsic muscles of the eyeball are shown—the lateral rectus and medial rectus. These muscles and those shown in figure 1 move the eyeballs in all directions giving animals the capacity to direct the eyes without moving the head. The optical axis of the eye running from the center of the cornea to the center of the larger sphere posteriorly is shown and also the visual axis passing through the fovea centralis of the macula lutea.

Since figures 3 and 4 show in greater detail the structures presented in figure 2 these illustrations can be considered together. Notice again the three tunics of the eyeball; (1) the outermost fibrous tunic consisting of the sclera and cornea; (2) the middle vascular tunic consisting, from the back to the front of the eye, of the choroid, ciliary body and iris; and (3) the nervous tunic, or retina.

The **sclera** constitutes about the posterior five sixths of the eyeball or bulb. It is composed of dense, hard, smooth, unyielding membrane which serves to maintain the form of the eyeball. Just mediad to the posterior pole of the eye it is pierced by fibers of the optic nerve and is continuous through the sheath of this nerve with the dura mater of the brain (fig. 3). Around the entrance of the optic nerve there are other small apertures for the entrance of ciliary vessels and nerves. The central artery and vein of the retina enter with the optic nerve. The sclera is continuous in front with the cornea at the sclerocorneal junction in which is found an encircling sinus, the sinus venosus sclerae (*canal of Schlemm*).

The transparent **cornea** makes up the anterior one-sixth of the fibrous tunic of the eyeball. Being a segment of a smaller sphere than the sclera it bulges forward from it. Its anterior surface is covered with a **corneal epithelium** or **bulbar conjunctiva** (fig. 4). It is nonvascular but richly supplied with sensory nerves. The cornea is an important part of the refracting system of the eye. Unequal curvature of the corneal surface causes a blurring of vision called astigmatism.

The choroid layer of the vascular tunic occupies the posterior five-sixths of the bulb. It is a thin, highly vascular membrane, dark brown in color due to the pigment cells in its outermost layer. It is loosely joined to the sclera except at the point where the optic nerve pierces it, where it is firmly attached. Internally, the choroid relates intimately to the pigment layer of the retina. It is analagous to the pia-arachnoid of the brain.

The **ciliary body** (figs. 2 and 4) is the thickest portion of the vascular coat of the eye, extending forward from the ora serrata of the retina to a point just behind the sclerocorneal junction. It forms a ring around this part of the eye to which the **suspensory ligament of the lens** attaches. It consists of the ciliary muscle and the ciliary process. The inner surface of the ciliary body is covered by an epithelium which is continuous with the retina, the ciliary retina. This area is thrown into 70 or 80 radiating folds, especially prominent at the free mesial edge of the ciliary body, which constitutes the **ciliary process.** The **ciliary muscle** consists of smooth muscle fibers arranged in meridonal, radial, and circular fashion. The meridonal fibers run from the sclera in front to the choroid behind and by contraction tense the choroid. The radial fibers run from the sclera into the ciliary body; the circular fibers form a sphincter-like muscle near the base of the iris. The ciliary muscle functions in the **accomodation** (*focusing*) of the eye.

The **iris,** the colored part of the eye, is a thin circular, muscular diaphragm which is attached at its periphery to the ciliary body and has at its center an opening, the **pupil** (figs. 2 and 4). It lies in front of the lens and divides the chamber between the front of the lens and the cornea into the **anterior** and **posterior chambers** of the eye. The muscle fibers of the iris are in two groups, circular and radiating. When the circular fibers contract, the pupil becomes smaller; when the radiating fibers contract, it is made larger.

The **retina** is a delicate nervous coat seen best in figures 1 and 2 with some of the details of its anterior parts in figure 4 and its posterior relations in figure 3. It is continuous behind with the optic nerve. It diminishes in thickness from back to front in the eye and near the ciliary body appears to end in a jagged margin, the **ora serrata.** The

nervous tissue of the retina does end at this point, but a thin pigmented portion of the membrane continues forward on to the back of the ciliary body and iris. In the center of the posterior part of the retina note the oval area (figs. 2 and 3) where the retina has thickened, the **macula lutea.** In the center of the macula lutea is a depression in which the retina is very thin and only cone cells remain, the receptors for bright light for color vision. This is the **fovea centralis,** the point of greatest visual acuity. Just to the nasal side of the fovea is the exit of the optic nerve, the **optic disc,** the margins of which are elevated. In its center is the central artery of the retina. This area has no light-sensitive cells. It is called the **blind spot** (*excavation of optic disc*). Note the pia mater, arachnoid and subarachnoid cavity in figure 3.

Study the microscopic structure of the retina as seen in Plate 115.

The Retina of the Human Eye
PLATE 115

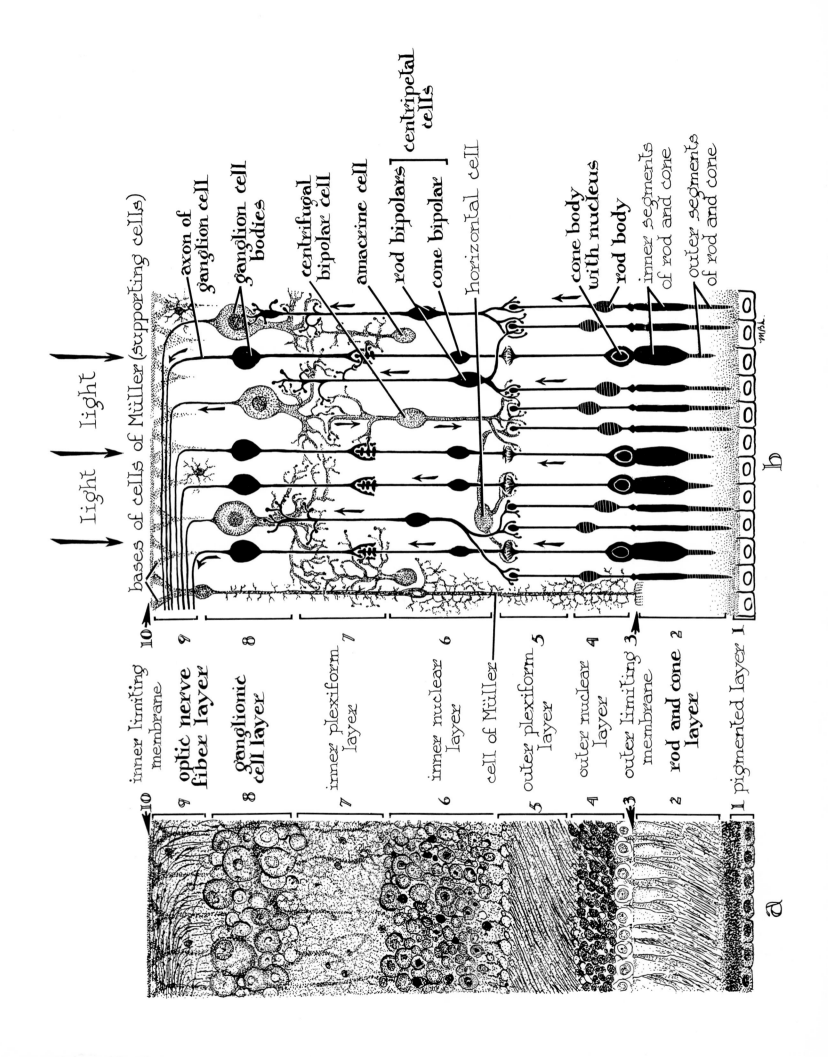

light

light

light

light

bases of cells of Müller (supporting cells)

axon of ganglion cell

ganglion cell bodies

centrifugal bipolar cell

amacrine cell

rod bipolars — centripetal cells

cone bipolar — centripetal cells

horizontal cell

cone body with nucleus

rod body

inner segments of rod and cone

outer segments of rod and cone

b

10 inner limiting membrane
9 **optic nerve fibex layer**
8 **ganglionic cell layer**
7 inner plexiform layer
6 inner nuclear layer
  cell of Müller
5 outer plexiform layer
4 outer nuclear layer
3 outer limiting membrane
2 **rod and cone layer**
1 pigmented layer

a

(Crouch: Functional Human Anatomy, Lea & Febiger)

In this illustration attention is given to the microscopic structure of the human **retina.** Though microscopic anatomy is not our major concern in this book, it has seemed important from time to time to turn to a study of it to satisfy our need for understanding. So it is with the retina of the eye on which we depend so much for information from our ever changing environment.

In **a** the retina is drawn as it is seen stained routinely and highly magnified. In **b** is seen a schematic presentation of the retina to show types of nerve cells and their relationships. The large arrows at the top of the drawing indicate light coming to the retina; the small arrows show the direction of passage of nerve impulses. Note that the light must penetrate most of the retina to reach the light sensitive rods and cones. The impulses then pass inward to the retina surface and to the optic nerve.

The retina consists of an outer pigmented layer, one cell thick seen at the bottom of the plate and a nervous portion of the retina proper. The retina proper, when sectioned perpendicular to its surface, shows several layers of supporting and of nervous structures. Three layers of cell bodies are of importance, the **rods and cones, bipolar neurons** and **ganglion** (*optic*) **cells.** Other nuclei are present but the above are most important. The nonnucleated layers in the retina are made up of the synapses between rods and cones and bipolar neurons (*outer plexiform layer*) and between the latter and the ganglion (*optic*) neurons (*inner plexiform layer*).

The **rods and cones** are the peripheral processes of the visual cells or rod and cone cells and extend outward through the external limiting membrane and constitute the receptor organs for vision. The cones are much more discriminating receptors in adequate light but in dim light the rods are more effective. The cones alone are found in the fovea centralis where vision and color discrimination are best (Plate 114).

The axons of the **ganglion cells** converge upon the region of the optic disc (Plate 114) where they emerge from the eye as the optic nerve.

Notice the amacrine cell body in the inner portion of the inner nuclear layer. This, of course, represents only one of many such cells in this area. They are called amacrine because no one as yet has shown that they have axons. Their dendrites undergo extensive ramifications in the inner plexiform layer.

Horizontal cells, of which one is indicated here, lie in the outer part of the inner nuclear layer. Their cell bodies are rather flattened; their dendrites form numerous branches in the outer plexiform layer; their axons run some distance horizontally and finally ramify in the same layer.

The above nervous constituents of the retina are held together by a supporting framework formed by the sustentacular fibers of Müller which pass through all of the nervous layers except the rods and cones. One of these is shown at the left side of **b.** Notice that the inner ends of other such cells are indicated along the inner limiting membrane which together they actually form. Their inner ends are bifurcate and united as shown. As the cells pass outward through the **ganglionic cell layer** they send off lateral branches; more numerous and longer branches being formed in the inner nuclear layer to support the bipolar cells. In the outer nuclear layer they form a network of branches around the rod and cone fibers and unite with other Müller cell fibers to form the outer limiting membrane. Each sustentacular Müller fiber contains a nucleus at the level of the inner nuclear layer.

The endocrine glands, sometimes called the glands of internal secretion or the ductless glands, are impossible to define as a system from the standpoint of their anatomy or their embryology. They are scattered about the body in relationship to other organs or systems and there is no anatomical continuity from one to another. They are diverse in their embryological development. They do have in common the negative characteristic of being without ducts to carry their secretions to a free surface. They are alike in introducing their secretions into the blood for distribution through the body, and in their basic epithelial nature clearly evident in the oldest endocrine gland, the thyroid. The secretions are known collectively as hormones and are in a sense "chemical messengers" which bring their influence to bear on certain cells, tissues or organs of the body. This indiscriminate broadcasting of hormones through the general circulation has been one of the factors which slowed the development of our knowledge of endocrine functions. With the development of more sophisticated research techniques and instrumentation the general, if not the specific, functions of these glands have become better known. With this knowledge of their functions there does seem to be justification for considering them a body system. Their hormones in some cases control other endocrine glands as well as cells, tissues and organs of other systems. In the body as a whole their control and coordination of functions is vital to the well being of the organism and often to the feeling or lack of feeling of well being of the organism. At least this appears to be the case in man and we can only speculate about the cat. Endocrine malfunction can be serious, sometimes fatal.

The assignment of organs or tissues to this system presents certain problems. Some organs are in part concerned with one function, but certain tissues within the organ may be endocrine. Some endocrine structures are transitory in the life cycle of the individual. With consideration of these difficulties the following are listed as endocrine organs or tissues or—if you wish—glands. Where possible a page or plate number will be given where each gland may be seen in its proper relationship.

(1) Hypophysis (*pituitary*)—essentially two glands, anterior and posterior, on the underside of the brain. Plates 101 and 102, figures 2.

(2) Thyroid—in neck region below or beside larynx —paired. Plate 55.

(3) Parathyroid—tiny bodies associated with the thyroid glands, often embedded in them. They are different in origin and function from the thyroid. They vary in number. Plate 55.

(4) Adrenal (*suprarenal*)—craniad of the kidney, essentially two glands as their cortex and medulla serve very diverse functions and have different embryological origins. Plate 63.

(5) Testes, interstitial tissue only. Plate 58 and Plate 61, figures 3 and 4.

(6) Ovary, ovarian follicles and corpus luteum only. Plate 63 and Plate 65, figure 1.

(7) Placenta—a transitory structure—Plates 67 and 68.

(8) Pancreas—islet cells (*islets of Langerhans*) only. Plate 53, figure 5.

(9) Mucosa of stomach and small intestine. Plate 53, figures 1 and 3.

(10) Kidney—special cells in cortex. Plate 60, figure 2.

Both the thymus gland and pineal body are often considered as endocrine glands. They are excluded as such here. They may be seen, however, in Plates 47 and 102.

REFERENCES

The references cited below include those mentioned in the text, and in addition, a sampling of textbooks on mammalian anatomy and related subjects and a few special papers. These should serve to introduce students to the vast literature of the field at various levels. More extensive bibliographies will be found in many of the references cited below.

NOMENCLATURE

Borror, Donald J. 1960. Dictionary of Word Roots and Combining Forms. N-P Publications, Palo Alto.

Jaeger, E. C. 1960. The Biologist's Handbook of Pronunciations. Charles C Thomas. Springfield, Illinois.

Nomina Anatomica. 2nd Ed. 1961. Revised by the International Anatomical Nomenclature Committee and approved by the Seventh International Congress of Anatomists, New York. 1960. x + 99 pages. Excerpta Medica Foundation, Amsterdam.

Skinner, H. A. 1963. The Origin of Medical Terms, 2nd Ed. x + 437 pages, illustrated. The Williams & Wilkins Co., Baltimore.

GROSS ANATOMY

Crouch, James E. 1961. Introduction to Human Anatomy, a Laboratory Manual. N-P Publications, Palo Alto.

Field, H. E. and M. E. Taylor. 1954. An Atlas of Cat Anatomy. Chicago University Press, Chicago.

Gray, H. 1966. Anatomy of the Human Body, edited by C. M. Goss. 28th Ed. Lea & Febiger, Philadelphia.

Miller, Malcolm E. assisted by George C. Christensen and Howard E. Evans. 1964. Anatomy of the Dog. W. B. Saunders Co., Philadelphia.

Reighard, Jacob and H. S. Jennings. 1935. Anatomy of the Cat. 3rd Ed. by R. Elliott. Holt, Rinehart & Winston, New York.

Stromsten, Frank A. 1947. Davison's Mammalian Anatomy with special reference to the cat. The Blakiston Co., Philadelphia.

Walker, Warren F. Jr. 1967. A Study of the Cat. W. B. Saunders Co., Philadelphia.

COMPARATIVE ANATOMY

Hyman, Libbie Henrietta. 1942. Comparative Vertebrate Anatomy. 2nd Ed. University of Chicago Press, Chicago.

Leach, W. James. 1961. Functional Anatomy Mammalian and Comparative. 3rd Ed. McGraw-Hill Book Co., New York.

Romer, A. S. 1962. The Vertebrate Body. 3rd Ed. W. B. Saunders Co., Philadelphia.

Smith, H. M. 1960. Evolution of Chordate Structure (An Introduction to Comparative Anatomy). Holt, Rinehart & Winston, Inc., New York.

Weichert, Charles K. 1958. Anatomy of the Chordates. 2nd Ed. McGraw-Hill Book Co., New York.

Wischnitzer, Saul. 1967. Atlas and Dissection Guide for Comparative Anatomy. W. H. Freeman & Co., San Francisco.

MICROSCOPIC ANATOMY

Bloom, W. and D. W. Fawcett. 1968. A Textbook of Histology. 9th Ed. W. B. Saunders Co., Philadelphia.

di Fiore, M. S. H. 1967. Atlas of Human Histology. 3rd Ed. Lea & Febiger, Philadelphia.

Porter, Keith R. and Mary A. Bonneville. 1968. An Introduction to the Fine Structure of Cells and Tissues. 3rd Ed. Lea & Febiger, Philadelphia.

Rhodin, Johannes A. G. 1963. An Atlas of Ultrastructure. W. B. Saunders Co., Philadelphia.

EMBRYOLOGY

Balinsky, B. I. 1961. Introduction to Embryology. W. B. Saunders Co., Philadelphia.

Nelson, Olin E. 1953. Comparative Embryology of the Vertebrates. The Blakiston Co., New York.

Patten, B. M. 1964. Foundations of Embryology. 2nd Ed. McGraw-Hill Book Co., New York.

Patten, B. M. 1953. Human Embryology. 2nd Ed. McGraw-Hill Book Co. (Blakiston Division), New York.

SPECIAL PAPERS

Adrian, R. W. 1964. Segmental anatomy of the cat's lung. Amer. Jour. Veterinary Res. 25(109):1724–1733.

Beattie, J., G. R. Brow, and C. N. H. Long. 1930. Physiological and anatomical evidence for the existence of nerve tracts connecting the hypothalamus with spinal sympathetic centres. Proc. Roy. Soc. B 106:253–275.

Benoit, Paul E. 1962. Sympathetic fibers in forelimb nerves of the cat. Anat. Rec. 142(4):531–535.

Biscoe, T. J. and Anne Bucknell. 1963. The arterial blood supply to the cat diaphragm with a note on the venous drainage. Quart. Jour. Exptle. Physiology 48(1):27–33.

Bradshaw, P. 1958. Arteries of the spinal cord in the cat. Jour. Neurol., Neurosurg., & Psychiat. 21(4):284–289.

Burnett, Macfarlane. 1962. The Thymus Gland. Sci. Am., 207, 5, 50–57.

Coulter, Calvin. 1909. Early Development of Aortic Arches in the Cat. Anat. Record. 3.

Davis, D. Dwight and H. Elizabeth Story. 1943–1944. Carotid Circulation in the Domestic Cat. Field Museum of Natural History. Zoological Series, 28:5–47. Chicago.

French, J. D. 1957. The Reticular Formation. Sci. Am. *196*, 30, 54–60.

Gilbert, P. W. 1947. The origin and development of the extrinsic ocular muscles in the domestic cat. J. Morph. *81:*151–194.

Huntington, G. S. and C. F. W. McClure. 1920. The development of the veins in the domestic cat. Anat. Record *20:*1–31.

Kerr, N. S. 1955. The homologies and nomenclature of the thigh muscles of the opossum, cat, rabbit, and Rhesus monkey. Anat. Rec. *121*(3):481–494.

McChesney, J. M. and H. M. Smith. 1944. The hepatic portal system in the cat. Ward's Nat. Sci. Bull. 18.

Nandy, K. and Geoffrey H. Bourne. A study of the morphology of the conducting tissue in mammalian hearts. Acta Anat. [Basel] *53*(3):217–226.

Straus, W. L. Jr. and J. M. Sprague. 1944. Interosseous muscles of the cat. Am. J. Physiol., *142*.

Strickland, James H. 1963. The integumentary system of the cat. Amer. Jour. Vet. Res. *24*(102):1018–1029.

Taber, Elizabeth. 1961. The cytoarchitecture of the brain stem of the cat. I. Brain stem nuclei of the cat. Jour. Comp. Neurol. *116*(1):27–69.

Thomas, Carolyn Eyster and C. Murphy Combs. 1962. Spinal cord segments. A. Gross structure in the adult cat. Am. Jour. Anat. *110*(1):37–47.

Towe, A. L., H. D. Patton, and Thelma T. Kennedy. 1963. Properties of the pyramidal system in the cat. Exp. Neurol. *8*(3):220–238.

Winkelmann, Richard K. 1958. The sensory endings in the skin of the cat. J. Comp. Neur. *109*(2):221–232.

MISCELLANEOUS

Everett, N. B. 1965. Functional Neuroanatomy. 5th Ed. Lea & Febiger, Philadelphia.

Gardner, E. 1958. Fundamentals of Neurology. 3rd Ed. W. B. Saunders Co., Philadelphia.

Hall, P. F. 1959. Functions of Endocrine Glands. W. B. Saunders Co., Philadelphia.

Kuntz, A. 1953. Autonomic Nervous System. 4th Ed. Lea & Febiger, Philadelphia.

Young, J. A. 1957. The Life of Mammals. Oxford University Press, New York.

MUSCLE TABLE

The muscles are arranged alphabetically for quick reference.

Name	Origin	Insertion	Action	Innervation
M. abductor auris brevis (Plate 23, fig. 2)	From the lambdoidal crest	Into the medial surface of the concha	Pulls the concha caudad	Facial n.
M. abductor auris longus (Plate 22, fig. 1)	On the sagittal crest	On the lateral surface of the eminentia conchae	Pulls the external ear caudad	Facial n.
M. abductor caudae externus (Plates 33 and 35, fig. 1)	From the medial side of the dorsal border of the ilium and from the dorsal surface of the sacrum	Into the transverse processes and lateral surfaces of the caudal vertebrae, as far back as the eighth or ninth	Bends the tail sideways	Dorsal and ventral rami of spinal nerves of caudal region
M. abductor caudae internus (Plates 33 and 35, fig. 2)	On the spine of the ischium	Into the transverse processes of the second to the fifth caudal vertebrae	Unilaterally, abducts the tail; bilaterally, flexes the tail and draws it between the legs	Dorsal and ventral rami of spinal nerves of caudal region
M. abductor cruris caudalis (tenuissimus) (Plate 40, fig. 2)	From the tip of the transverse process of the second caudal vertebra	With biceps femoris	Weak abductor and extensor of the thigh, flexor of the shank	Sciatic (ischiatic) n.
M. abductor digiti quinti (Plate 39, fig. 2)	From the distal surface of the pisiform bone and from the transverse ligament	Into the ulnar side of the base of the proximal phalanx of the fifth digit	Abduction of digit V	Ulnar n., deep br.
M. abductor digiti quinti medius (Plate 45)	Calcaneus and metatarsal V	Lateral side of base of proximal phalanx of digit V	Abduction of digit V	Tibial n.
M. abductor digiti secundi (Plate 39, fig. 2)	From the radial and palmar surfaces of the base of the second metacarpal and from the trapezium	Into the radial side of the base of the proximal phalanx of the second digit and into its sesamoid	Abduction of digit II	Ulnar n., deep br.
M. abductor digiti secundi medius (Plate 45)	From ventral processes of lateral cuneiform with four other parts of the interossei II	Medial side of the base of the proximal phalanx of digit II	Abducts digit II	Tibial n.
M. abductor brevis pollicis (Plate 38, fig. 1)	From the transverse ligament	Into the base of the proximal phalanx of the thumb	Flexion of first digit	Ulnar n., deep br.
M. abductor pollicis longus (Plate 37)	From the ventral half of the lateral surface of the shaft of the ulna; from the ulnar half of the dorsal surface of the radius; and from the interosseous membrane	Into the radial side of the base of the first metacarpal	Extends and abducts the pollex	Radial n.
M. anconeus (Plates 36 and 37)	At the distal end of the dorsal surface of the humerus	The lateral surface of the ulna from the distal margin of the semilunar notch to the proximal end of the olecranon	Keeps the capsule tense and probably rotates the ulna slightly so as to pronate the hand	Radial n.

Name	Origin	Insertion	Action	Innervation
M. adductor auris inferior (Plate 22, fig. 2; Plate 23, fig. 1)	On the orbital ligament	On the tip of the antitragus	Pulls the ear craniodorsad	Facial n.
M. adductor auris medius (Plates 22 and 23)	On the middle two thirds of the caudoventral edge of the scutiform cartilage	The fibers pass nearly ventrad and are inserted along the medial or caudal surface of the tragus	Pulls the concha dorsocraniad	Facial n.
M. adductor auris superior (Plates 22 and 23)	On the dorsal surface of the scutiform cartilage for its entire length	Into the spina helicis or craniomedial margin of the auricular cartilage	Draws the auricle craniad	Facial n.
M. adductor digiti quinti (Plate 39, fig. 2)	By a flat tendon from the ventral surface of the capitate on the ulnar side of the adductor pollicis	(1) Into nearly the whole of the radial surface of the fifth metacarpal, (2) into the base of its proximal phalanx	Adduction of digit V	Ulnar n., deep br.
M. adductor digiti quinti medius (Plate 45)	From the ligament covering the peroneal canal	Medial side of the base of the proximal phalanx of digit V	Adduction of digit V	Tibial n.
M. adductor digiti secundi (Plate 39, fig. 2)	From the ventral surface of the capitate	Into the ulnar side of the base of the proximal phalanx of the second digit	Adducts digit II	Ulnar n., deep br.
M. adductor digiti secundi medius (Plate 45)	Middle of the ligament covering the peroneal canal	Lateral side of the base of the proximal phalanx of digit II	Adducts digit II	Tibial n.
M. adductor femoris (magnus and brevis) (Plates 41–45)	By muscle fibers along the whole length of the pelvic symphysis and part of the ischial ramus	Into nearly the whole length of the lateral lip of the linea aspera on the caudal surface of the femur	Adduction and extension of the thigh	Obturator n.
M. adductor longus (Plate 40)	By muscle-fibers from the cranial border of the pubis, the line of origin forming a medial continuation of that of the pectineus	By a thin aponeurosis into the linea aspera of the femur	Abductor and flexor of the thigh	Obturator n.
M. adductor pollicis (Plate 39, fig. 2)	By fleshy fibers from the ventral border of the capitate	By fleshy fibers into the base of the proximal phalanx of the pollex on its ulnar side	Adducts digit I	Ulnar n.
M. antitragicus (Plate 23, fig. 1)	Attached at one end to the caudal border of the antitragus	The fibers pass mediodistad and are inserted on the tragus, in common with the fibers of the tragicus medialis	Constricts the external auditory opening	Facial n.
M. arytenoideus transversus	Muscular process of arytenoid cartilage	Lateral ends and dorsal surface of interarytenoid cartilage	Constricts glottis and adducts the vocal cords	Caudal laryngeal n.

344

Name	Origin	Insertion	Action	Innervation
M. auricularis externus (Plates 22 and 23)	On the eminentia conchae distad of the insertion of the abductor auris longus	On concha	Flexes the auricular cartilage	Facial n.
M. auricularis superior (Plate 22, fig 1)	The sagittal crest craniad of the interparietal bone	Into the auricular cartilage	Adducts the external ear	Facial n.
M. biceps brachii (Plates 38 and 39)	By a strong, round tendon from the supraglenoid tubercle of the scapula	By a rounded tendon on the radial tuberosity	Flexes the forearm and tends to supinate the hand	Musculocutaneous n.
M. biceps femoris (Plate 40)	From the ventral surface of the tuberosity of the ischium	Into rather more than the proximal ⅓ of the dorsal border of the tibia along its lateral margin and into the lateral margin of the patella	Abductor and extensor of the thigh and flexor of the shank	Cranial part: caudal gluteal n; middle and caudal parts: tibial n.
M. biventer cervicis (medial portion of M. semi-spinalis capitis) (Plate 33)	From the tips of the spinous processes of the last cervical and the first three thoracic vertebrae; the anterior processes of the second, third, fourth and fifth thoracic vertebrae; from the cervical ligament between the last cervical and the first thoracic vertebrae	By a strong tendon into the medial part of the lambdoidal crest	Raises the snout	Dorsal rami of spinal nerves
M. brachialis (Plates 36 and 37)	From a long V-shaped area 2 to 4 millimeters wide on the lateral surface of the humerus	On the lateral surface of the ulna just distad of the semilunar notch	Flexor of the antibrachium (forearm)	Musculotaneous n.
M. brachioradialis (Plates 36 and 37)	By a thin tendon from about the middle fifth of the dorsal border of the humerus	By a short tendon into the outer surface of the styloid process of the radius	Supinator of the hand	Radial n.
M. buccinator (Plate 22, fig. 2; Plate 23, fig. 2)	On the outer surface of the maxilla	Mingles with those fibers of the deep part of the orbicularis oris	Raises the upper lip	Facial n.
M. bulbocavernosus (accelerator urinae) (Plate 64)	From a median raphe, which passes from the bulbous portion of the urethra toward the distal end of the penis	Into the distal half of the lateral surface of the corpus cavernosum penis	Slow flow of blood through penal veins in erection of penis	Pudendal n., rectal brs.
M. calcaneometatarsalis (mostly tendon) (Plate 44, fig. 1)	The lateral and ventral surface of the calcaneus near the proximal end	The lateral side of the base of the fifth metatarsal, and the adjacent surfaces of the cuboid and calcaneus	Probably causes slight motion of the cuboid on the head of the calcaneus	Tibial n.

Name	Origin	Insertion	Action	Innervation
M. caninus or levator anguli oris (Plate 22, fig. 2; Plate 23, fig. 1)	On the lateral surface of the maxillary bone	Into the whisker pad	Retracts the whiskers (vibrissae) and raises the upper lip	Facial n.
M. capsularis (Plate 42)	By fleshy fibers from the surface of the iliopectineal eminence of the ilium	By fleshy fibers for about 1 centimeter in the middle line on the cranial surface of the femur distad of the great trochanter	Flexion of the thigh	Femoral n.
M. caudoanalis (Plate 64)	On the middle line of the ventral surface of the second and third caudal vertebrae	In the sphincter ani internus	Draws the anus cranio-dorsad	Pudendal n.
M. caudocavernosus (Plate 64)	On the median ventral line of the first two caudal vertebrae	It divides into two bands, one of which is inserted into the base of the corpus cavernosum, while the other extends farther caudad and is inserted at the distal extremity of the corpus cavernosum	Flexes the penis (bends it backward)	Pudendal n.
M. caudofemoralis (Plate 40)	By a flat tendon from the transverse processes of the second and third caudal vertebrae	Into the middle of the lateral border of the patella	Abducts the thigh and helps to extend the shank	Inferior gluteal n.
M. caudorectalis (Plate 64).	From the ventral surface of the sixth and seventh caudal vertebrae	Into the walls of the rectum	Pulls the rectum caudodorsad	Pudendal n.
M. caudovaginalis (Plate 64).	From the median line of the ventral surface of the first two caudal vertebrae	Into the ventral side of the urogenital sinus at the base of the clitoris	Elevates the urogenital sinus	Pudendal n.
M. ceratohyoideus (Plate 24, fig. 3)	Into the ceratohyal and epihyal	Into the caudal cornu of the hyoid bone	Draws the basihyal craniad	Hypoglossal n.
M. cleidomastoideus (Plate 29)	From the mastoid process	Into the clavicle and the raphe which continues laterad of the clavicle	Pulls the clavicle craniad when the head is fixed. Turns head and depresses the snout when the clavicle is fixed	Spinal accessory and ventral rami of cervicals 2–3
M. complexus (lateral portion of M. semi-spinalis capitis) (Plate 33)	From the anterior articular processes of the last five or six cervical and the first one, two or three thoracic vertebrae	By a flat tendon into the lambdoidal crest	Raises the head	Dorsal rami of spinal nerves
M. compressor urethrae membranaceae	A thick layer of striated muscle fibers which surrounds the urethra between Cowper's gland and the prostate. The fibers have a circular course, and the cranial ones are attached to the crura of the penis.			

346

Name	Origin	Insertion	Action	Innervation
M. conchaeus externus (Plate 23, fig. 2)	The cranial end of this muscle is attached to the concha a short distance distad of the antitragus. Again attached to the concha caudodorsad of origin.		Constricts the concha	Facial n.
M. constrictor pharyngis caudalis (Plate 24, fig. 1)	From the lateral surfaces of the thyroid and the cricoid cartilages	The median longitudinal raphe on the dorsum of the pharynx	Constrictor of the pharynx	Glossopharyngeal and vagus nn.
M. constrictor pharyngis cranialis or pterygopharyngeus (Plate 24, fig. 3)	The tip of the hamular process of the pterygoid bone. The muscle passes caudad, the fibers diverging, and dips beneath the cranial border of the constrictor medius	Into the median dorsal raphe of the pharynx	Constrictor of the pharynx and draws it forward	Glossopharyngeal and vagus nn.
M. constrictor pharyngis medius (Plate 24, fig. 2)	The ventral two pieces of the cranial horn and the whole of the caudal horn of the hyoid	Into the median dorsal raphe of the pharynx	Constrictor of the pharynx	Glossopharyngeal and vagus nn.
M. constrictor vestibule (Plate 64).	From the sides of the sphincter ani internus	Into the ventral surface of the urogenital sinus, caudad of the insertion of the ischiocavernosus	Constricts the vestibule (during copulation)	Pudendal n.
M. coracobrachialis (Plates 38 and 39)	By a round tendon from the tip of the coracoid process	On the medial side of the proximal end of the humerus	Adducts the humerus	Musculocutaneous n.
M. corrugator supercilii lateralis (Plate 22, fig. 1; Plate 23, fig. 1)	Arise from the fronto-scutularis and from the tendon lying just craniad of the external opening of the ear	At the caudolateral angle of the eye where it unites with the orbicularis oculi	Pulls the angle of the eye caudad; pulls the external ear craniad	Facial n.
M. corrugator supercilii medialis (Plate 22, fig. 1)	Near the midline of skull, passes laterad, then craniad	Into upper eyelid, especially near the caudal angle	Raises the upper eyelid	Facial n.
M. cricoarytenoideus lateralis	The lateral part of the cranial border of the cricoid cartilage	The caudal lip of the laterocaudal angle of the arytenoid cartilage	Turns the arytenoids inward to close the glottis	Caudal laryngeal n.
M. cricoarytenoideus posterior (dorsalis)	From the dorsal part of the caudal border of the cricoid cartilage and from its median dorsal crest	The dorsal border of the caudal end of the arytenoid cartilage	Separates the vocal cords so as to widen the glottis	Caudal laryngeal n.
M. cricothyreoideus (Plate 24, fig. 2)	The lateral half of the ventral surface of the cricoid cartilage	The ventral part of the caudal border of the thyroid cartilage	Retracts the thyroid cartilage—tenses the vocal cords	Cranial laryngeal n., external br.

Name	Origin	Insertion	Action	Innervation
M. cutaneus maximus (Plate 27)	Outer surface of latissimus dorsi near its ventral end; the bicipital arch; the linea alba; and the thorax between the axilla and xiphoid process	Into the skin, at or near the midline from shoulder and tail and thigh	Moves the skin	Cutaneous branches of ventral rami of spinal nerves
M. deltoideus 1. M. spinodeltoideus (Plates 29, 36, 37, fig. 1)	The middle third of the spine of the scapula	By a flat tendon upon the deltoid ridge by the humerus, nearly parallel to that of the pectoralis major	Flexes the humerus and rotates it outward	Axillary n.
2. M. acromiodeltoideus (Plates 29, 36, 37, fig. 1)	From the caudal border of the acromion and sometimes the adjacent metacromion	The outer surface of the spinodeltoideus, along a line ventrad of the line of insertion of the spinodeltoideus and extending further distally	Flexor and outer rotator of the humerus	Axillary n.
3. M. cleidobrachialis (Plates 28, 39, fig. 2) (cleidodeltoideus)	Its superficial fibers are continuations of the cleidotrapezius; other fibers have origin from the clavicle and from a raphe laterad of the clavicle	Joins the brachialis muscle to be inserted with it by a flat tendon on the medial surface on the ulna just distad of the semilunar notch	Flexor of the antibrachium	Axillary n.
M. depressor conchae (Plate 23, fig. 2)	Arises on ventral surface of the neck	Into the summit of the antitragus	Draws the external ear ventrad	Facial n.
M. digastricus (Plate 24, fig. 1)	From the outer surface of the jugular process of the occipital bone and from the tip of the mastoid process and from the intermediate ridge	The ventral border of the mandible near the molar tooth	Depresses the lower jaw	Facial n. and inferior alveolar branch of mandibular division of trigeminal
M. epicranius (Plate 22, fig. 2)	See frontalis and occipitalis			
M. epitrochlearis (Plate 38)	From the lateral or outer surface of the ventral border of the latissimus dorsi. Fibers are often attached to the teres major and the pectoralis minor	By a flat tendon to the olecranon process of ulna	Extends the antibrachium	Radial n.
M. extensor carpi radialis brevis (Plates 36 and 37)	From the distal part of the lateral supracondyloid ridge distad of the extensor longus	Into the radial side of the dorsal surface of the base of the third metacarpal	Extensor of the hand	Radial n.

Name	Origin	Insertion	Action	Innervation
M. extensor carpi radialis longus (Plates 36 and 37)	From the lateral supracondyloid ridge between the origin of the brachialis and the anconeus	Tendon is inserted onto the dorsal surface of the base of the second metacarpal	Extension of the hand	Radial n.
M. extensor carpi ulnaris (Plates 36 and 37)	By a short, broad tendon from the distal portion of the lateral epicondyle of the humerus	Into the tubercle on the ulnar side of the base of the fifth metacarpal	Extension of carpal joint-weak lateral rotation	Radial n.
M. extensor caudae lateralis (Plate 33; 35, fig. 1)	In many fleshy bundles from the articular processes of the sacral vertebrae, and the transverse processes of the caudal vertebrae	By many long slender tendons on the dorsal surfaces of the caudal vertebrae	Raises the tail	Ventral coccygeal n.
M. extensor caudae medialis (Plates 33; 34, fig. 1; 35, fig. 1)	From the spinous processes of the sacral and first caudal vertebrae	By tendons into the articular processes and the dorsal surface of the caudal vertebrae	Extends (raises) the tail	Dorsal coccygeal n.
M. extensor digitorum brevis (Plates 40–42)	From the distal border of the calcaneal ligament and from the proximal ends of the dorsal surface of the three lateral metatarsals	Lateral branch into the cartilaginous plate which lies in the metatarsophalangeal articulation of the digit. The medial branch joins the lateral side of the extensor longus tendon on the dorsum of the proximal phalanx	Extensor or the digits	Peroneal n.
M. extensor digitorum communis (Plates 36 and 37)	The distal surface of the lateral supracondyloid ridge dorsad of the origin of the extensor carpi radialis brevis	Mostly into the base of the second phalanx; a portion continues distad to be inserted into the third (distal) phalanx	Extensor of the four principal digits	Radial n.
M. extensor digitorum lateralis (Plates 36 and 37)	From the lateral supracondyloid ridge of the humerus distad of the origin of the extensor communis	The three tendons on the ulnar side join tendons of the extensor communis at their insertions. The radial tendon (not always present) joins the tendon of the extensor indicis and may also give a branch to the radial side of the base of the first phalanx to the third digit	Extension of the four digits	Radial n.

Name	Origin	Insertion	Action	Innervation
M. extensor digitorum longus (Plates 40–42)	By a thin, flat tendon from the lateral surface of the lateral epicondyle of the femur	Into the basal phalanx of digits 2–5	Extension of the digits of flexion of the foot	Peroneus n.
M. extensor indicis (proprius) (Plates 36 and 37)	By short, fleshy fibers from the lateral surface of the ulna	To the base of the second phalanx of the second digits; one may pass to the pollex also	Extends second digit and pollex; adducts pollex	Radial n.
M. extensor pollicis longus (Plates 36 and 37)	By short fleshy fibers from the lateral surface of the ulna	To the pollex	Extends and adducts the first digit or pollex	Radial n.
M. flexor carpi radialis (Plates 38 and 39)	From the tip of the medial epicondyle of the humerus	Into the bases of the second and third metacarpals	Flexor of the wrist	Median n.
M. flexor carpi ulnaris (Plates 38 and 39)	(1) humeral head from the median surface of the distal end of the humerus just distad of the medial epicondyle (2) ulnar head from the lateral surface of the olecranon	By fleshy and tendinous fibers into the proximal surface of the pisiform bone	Flexor of wrist	Ulnar n.
M. flexor caudae brevis (Plate 35, fig. 2)	On the ventral surface of the caudal vertebrae, from the first to the seventh or eighth	Each bundle inserted into the ventral surface of a vertebra some distance caudad of the origin	Flexes the tail	Ventral rami of caudal (coccygeal) spinal nerves
M. flexor caudae longus (Plates 33, 35, fig. 2)	On the ventral surface of the last lumbar vertebra, the sacrum, and of the transverse processes of the caudal vertebrae	On the ventral surface of the tail	Flexes the tail	Ventral rami of spinal nerves of sacral and caudal regions
M. flexor digitorum brevis (Plates 43 and 44)	Attached by an oblique tendinous band to the medial surface of the navicular and the medial cuneiform	Into the bases of the phalanges	Flexion of the digits	Medial plantar n.
M. flexor digitorum longus (Plate 44)	By muscle—and tendon—fibers from the caudal surface of the tibia over its proximal half and by tendon from the medial surface of the head of the fibula	To the medial border of the common tendon with flexor hallucis longus to distal phalanges of digits 2–5	Flexor of the digits and extensor of foot	Tibial n.

Name	Origin	Insertion	Action	Innervation
M. flexor digitorum profundus (Plates 38 and 39)	From the dorsal half of the medial surface of the ulna; from the distal end of the medial epicondyle of the humerus; from the adjacent parts of the interosseous membrane; and from the medial surface of the shaft of the ulna	Into the bases of the distal phalanges	Flexor of all the digits	Median and ulnar nn.
M. flexor digitorum superficialis (Plates 38 and 39)	From the tendon of the palmaris longus and the outer surface of a tendon of the flexor profundus	Bases of the middle phalanges of the digits	Flexion of the second phalanges of the digits 2–5	Median n.
M. flexor hallucis longus (Plates 42, 44 and 45)	By fleshy fibers from the caudal surface of the tibia; from the medial surface of the shaft and head of the fibula and from the whole interosseous ligament	By four tendons into the distal phalanges of digits 2–5	Flexor of the digits, extension of the foot	Tibial n.
M. flexor pollicis brevis (Plate 39, fig. 2)	By fleshy fibers from the adjacent borders of the capitate and scapholunar bones	By a short tendon into the base of the proximal phalanx of the pollex	Flexor of the thumb	Ulnar n., deep br.
M. frontalis (Plate 22, fig. 2)	From nasal and frontal region	Galea aponeurotica	Moves integument of dorsal surface of head	Facial n., ramus auriculopalpebralis
M. frontauricularis (Plate 22, fig. 1; Plate 23, fig. 1)	Along the upper eyelid	With fibers of the abductor auris superior at the craniomedial angle of the auricular cartilage	Pulls ear craniad—weak adductor	Facial n.
M. frontoscutularis (Plate 22, fig. 2; Plate 23, fig. 2)	On the frontal bone, along the supraorbital margin	Scutiform cartilage mostly along its ventrolateral border; the cartilage of the ear; the adductor auris superior	Pulls the ear craniad	Facial n.
M. gastrocnemius (Plates 40, 43 and 44)	By two heads proximal to the lateral and medial condyles on the caudal surface of the femur. Each head is provided with a sesamoid bone, the fabellae	Into the proximal end of the calcaneus	Extensor of foot, flexion of shank	Tibial n.

Name	Origin	Insertion	Action	Innervation
M. gemellus inferior (Plate 42, fig. 1)	From the dorsal one-half of the lateral surface of the ischium between the ischial spine and the ischial tuberosity	Into the inner surface of the tendon of the obturator internus by tendon and muscle fibers. Some of the muscle fibers are inserted into the capsule of the joint	Abductor and outer rotator of the thigh	Sciatic (ischiatic) n.
M. gemellus superior (Plate 42, fig. 1)	By fleshy fibers from an elongated area on the dorsal border of the ilium and ischium	Into a triangular area dorsad of the tip of the great trochanter	Outward rotation and abduction of the femur	Sciatic (ischiatic) n.
M. genioglossus (Plates 24 and 26	From the medial surface of the mandible near the symphysis	Cranial fibers in frenulum to tip of tongue; caudal fibers sweep in fan-like fashion upward and caudad of midventral surface of tongue and to basihyoid and ceratohoid bones	Depresses tongue, anterior fibers curl tip and draw it downward, posterior fibers draw it forward	Hypoglossal n.
M. geniohyoideus (Plate 24, Fig. 2)	From the ventral half of the inner surface of the mandible for about 1 millimeter from the symphysis	Into the lateral half of the ventral surface of the basihyal	Draws the hyoid forward	Hypoglossal n.
M. glossopharyngeus (Plate 24, fig. 3)	From the ventral and lateral part of the genioglossus and the midventral part of the styloglossus	Into the medial dorsal raphe of the pharynx	Constrictor of the pharynx	Glossopharyngeal and vagus nn.
M. glossoepiglotticus	The median fibrous septum of the tongue	The cranial surface of the epiglottic cartilage in the median line near its attached border	Draws the epiglottis craniad	Caudal laryngeal n.
M. gluteus maximus (superficialis) (Plate 40)	By fleshy fibers from the tips of the transverse processes of the last sacral and the first caudal vertebrae, and from the gluteal fascia.	By tendon and muscle fibers into a tubercle on the caudal side of the greater trochanter	Abducts and extends the thigh	Caudal gluteal n.
M. gluteus medius (Plate 40, fig. 2; Plate 41, fig. 1)	By fleshy fibers from the superficial sacral and caudal fascia; by tendon fibers from the dorsal half of the lateral surface of the ilium craniad of the auricular impression; from the tips of the transverse processes of the last sacral and the first caudal vertebrae	Into the proximal end of the great trochanter	Abducts and extends the thigh	Cranial gluteal n.

Name	Origin	Insertion	Action	Innervation
M. gluteus minimus (profundus) (Plate 42, fig. 1)	From the ventral half of the ilium, from near its cranial end to a point midway between the caudal iliac spine and the spine of the ischium	Into an oval facet at the base of the dorsal surface of the great trochanter on its lateral side	Outward rotation of femur and some abduction	Cranial gluteal n.
M. gracilis (Plate 43)	By a strong tendon from the caudal three-fourths of the pelvic symphysis	Into the medial surface of the tibia near its proximal end	Abduction and extension of the thigh	Obturator n.
M. helicis (Plate 23)	On the medial surface of the concha, just caudad of the proximal end of the tragus	On the auricular cartilage along with the caudal fibers of the adductor auris superior	Draws the cranial margin of the auricle proximad	Facial n.
M. hyoepiglotticus	The lateral end of the cranial surface of the basihyoid	With the glossoepiglotticus in the cranial surface of the epiglottis	Draws the epiglottis craniad	Caudal laryngeal n.
M. hyoglossus (Plates 24, 26, and 49)	From the ventral surface of the basihyoid laterad of the geniohyoid, and by a second head from the ceratohyal	Both heads penetrate into the root and posterior two-thirds of the tongue	Retracts the tongue and depresses it	Hypoglossal n.
M. iliocaudalis (Plate 35, fig. 2)	Along the ventral half of the medial surface of the ilium, caudad of the sacrum	By a flat tendon into the ventral surface of the caudal vertebrae, from the second or third to about the seventh	Together they flex the tail	
M. iliocostalis (Plate 32)	In bundles from the lateral surface of the ribs	On the lateral surface of the third or fourth rib craniad of the rib of origin	Draws the ribs together	Dorsal rami of spinal nerves
M. iliopsoas (Plates 44, 45)	From the 5 cranial tendons of origin of the psoas minor; the transverse process of the fifth lumbar vertebra; the 6, 7, 8, 9, and 10 heads arise from the centra of the last 4 lumbar vertebrae; iliacus arises by fleshy fibers from the ventral border of the ilium	By tendon and fleshy fibers into the apex of the lesser trochanter of the femur	Flexes the thigh; fixation and flexion of the spine	Lumbar nn. (ventral rami)
M. infraspinatus (Plate 36)	By fleshy fibers from the whole infraspinatus fossa, and by a raphe between it and the teres major	By a flat tendon which passes over the capsule of the joint, into the ventral half of the infraspinatus fossa on the greater tuberosity of the humerus	Abducts and rotates the humerus outward	Suprascapular n.

Name	Origin	Insertion	Action	Innervation
Mm. intercostales externi (Plate 33)	Caudal border of each rib, run caudo-ventrally	Cranial border of each rib	Inspiration, increasing the transverse diameter of the thorax. Protractors of the ribs	Muscular branches of the nn. intercostales 1–12
Mm. intercostales interni (Plate 33)	Cranial border of each rib, run cranioventrad	Caudal border of each rib	Expiration, retractors of the ribs	Intercostales nn.
M. intermedius scutulorum (Plate 22, fig. 1)	Connects the scutiform cartilages of the ears		Draws the two ears dorsad, toward the median line	Facial n.
Mm. interossei (forefoot) (Plate 39, fig. 2; Plate 37, fig. 1)	By fleshy fibers from the palmar surfaces of the bases of the metacarpals II–V	Partly onto the lateral surfaces of the bases of the proximal phalanges and their sesamoids, partly by a slender tendon which is continued dorsad to join the extensor tendon of the digit	Flexion of the metacarpophalangeal joints	Ulnar n., deep br.
Mm. interossei (hindfoot) (Plate 45)	A group of abductor and adductor muscles of the digits of the hindfoot. They are detailed on Plate 45. A few are listed individually in this table.			Tibial n.
Mm. interspinales (lumbar, thoracic and cervical) (Plate 34, fig. 2)	Run between contiguous edges of the spinous processes		Fixation of the vertebral column	Medial branch of the dorsal rami of the spinal nn.
Mm. intertransversarii dorsalis (Plate 34, fig. 2)	Connect accessory and mammillary processes of vertebrae extending over one or two, sometimes three, vertebrae.		Lateral flexion of vertebral column or bilaterally to stabilize the spine	Branches of the truncus coccygeus ventralis
M. intertransversarii ventralis (Plate 34, fig. 2)	Connect transverse processes of adjacent vertebrae		Lateral flexion of vertebral column or bilaterally to stabilize spine	Branches of the truncus coccygeus ventralis
M. ischiocavernosus (Plates 61 and 64)	From the caudal border of the ramus of the ischium, about 1 centimeter from the median line	In the male, into the whole outer surface of the crus penis, or bulb of the corpus cavernosum penis. In the female into the ventral surface of the urogenital sinus, at the base of the clitoris	Help to maintain erection of the penis by restricting flow of blood from veins	Pudendal n., rectal brs.
M. jugulohyoideus (Plate 24, fig. 3)	From the ventral border of the jugular process	Into the caudal side of the cartilaginous part of the stylohyoid	Draws the stylohyal backward giving a firmer surface of origin for the styloglossus	Hypoglossal n.
M. latissimus dorsi (Plates 29, 38 and 39)	From the tips of the neural spines of the vertebrae from the fourth or fifth thoracic to about the sixth lumbar	Through the bicipital arch into a rough elongated area on the medial surface of the shaft of the humerus	Pulls the arm caudo-dorsad	Anterior thoracic and caudal division of subscapular

354

Name	Origin	Insertion	Action	Innervation
M. levator ani (or pubiocaudalis) (Plates 61 and 64)	From the symphysis of the pelvis	Into the midventral line of the centra of the third, fourth and fifth caudal vertebrae, close to the muscle of the opposite side	Bends the tail and compresses the rectum	Third sacral and first coccygeal n., ventral brs.
M. levator auris longus (Plate 22, fig. 1)	From the midline of the neck dorsad of the atlas, and from the sagittal crest	To the scutiform cartilage, the surface of the auricle	Pulls the external ear dorsocaudad	Facial n.
M. levator costae (Plate 33, fig. 1)	On the transverse processes of the thoracic vertebrae	On the angle of the rib lying immediately caudad of the origin	Pulls the ribs dorso-craniad; inspiration	Intercostales 1–12
M. levator palpebrae superioris (Plate 114)	On the wall of the optic foramen close to the rectus dorsalis	Into the margin of the upper eyelid by wide, flat tendon	Retractor of upper eyelid	Oculomotor n.
M. levator scapulae (Plates 29–31, 38)	From the dorsal tubercles of the trans-verse processes of the last five cervical verte-brae and from the ligaments between them	Into a triangular area on the medial surface of the scapula near its vertebral border	Draws the scapula cranioventrad	Ventral rami of cervical 3–4
M. levator scapulae ventralis (Plates 29 and 36)	From the atlantal transverse process and from the ventral surface of the basioccipital	Into the outer surface of the metacromion and into the infraspinatus fossa	Pulls the scapula craniad	Ventral ramus of cervical 3
M. levator scroti (Plate 64)	From the sphincter ani externus	Onto the scrotum	Elevates the scrotum	Pudendal n.
M. levator veli palatini	From the surface of the body of the sphenoid mediad of the groove for the Eustachian tube, from the styliform process of the bulla tympani	Into the velum palatinum, some of the fibers meeting in the middle line	Raises the caudal part of soft palate	Trigeminal n.
M. levator vulvae (Plate 64)	From the external sphincter ani	Onto the vulva	Elevates vulva	Pudendal n.
M. longissimus capitis (Plate 32)	By four slips which are attached by strong tendons to the articular processes of the last four cervical vertebrae	Into the mastoid process of the temporal bone	Extends head unilater-ally it is a lateral flexor of the head	Dorsal rami of spinal nn.
M. longissimus cervicis (Plate 33)	From articular processes and laminae of the more cranial thoracic verte-brae	Cervical vertebrae	Extends neck-working separately, to raise the neck obliquely and turn it to one side	Dorsal rami of spinal nerves

Name	Origin	Insertion	Action	Innervation
M. longissimus dorsi (thoracic and lumbar portions) (Plates 33 and 35)	Medial division-sacral and caudal vertebrae; lateral division-crest and medial surface of ilium deep lumbodorsal fascia	Medial division-accessory and mammillary process of lumbar, sacral and caudal vertebrae; lateral division-transverse processes of lumbar and thoracic vertebrae	Powerful extensor of vertebral column	Dorsal rami of spinal nerves
M. longus capitis (Plate 33, fig. 2)	By 5 (or 6) heads from the ventral margins of the transverse processes of the second to sixth cervical vertebrae	Into the body of the occipital bone between the bulla and the middle line, and craniad onto the blasisphenoid	Lowers the snout	Ventral ramus of cervical nerves
M. longus colli (Plate 33, fig. 2) (1) Thoracic portion	By six heads from the ventral surface of the first five thoracic vertebrae	Separate heads join to form a band which passes craniad and is inserted for the most part into the processus costarius of the sixth cervical vertebra	Bends the neck downward	Ventral rami of cervical 3 through thoracic 5 or 6
(2) Cervical portion	Arises in small bundles from the transverse processes and centra of the cervical vertebrae	The muscles of opposite sides meet on the centra of the vertebrae in the midline, each pair of bundles forming a V-opening caudad. The most cranial insertion is on the ventral arch of the atlas	Bends the neck downward	Ventral rami of cervical 3 through thoracic 5 or 6
Mm. lumbricales (fore-foot) (Plate 39)	By fleshy fibers from the palmar surface of the tendon common to the ulnar four parts of the profundus	The four slips are inserted into the radial side of the bases of the proximal phalanges of digits 2–5	Bends the digits towards the radial side	Ulnar n., deep br.; median n.
Mm. lumbricales (hind-foot) (Plate 44) (1) flexores superficialis	From the outer surface of the expanded tendon of the flexor digitorum longus	By slender tendons which unite with tendons of the flexor digitorum brevis at their entrance to the proximal annular ligaments	Move digits III, IV and V mediad	Tibial n.
(2) flexores profundus	From three intervals between the four divisions of the flexor digitorum longus (profundus) muscle	Into the medial sides of the proximal phalanges of digits III, IV and V	Move digits III, IV and V mediad	Tibial n.

Name	Origin	Insertion	Action	Innervation
M. masseter (Plate 24, figs. 1 and 2)	The outer surface of the zygomatic arch	Into the mandible	A very powerful elevator of the lower jaw	Masseteric branch of mandibular division of trigeminal
M. "moustachier" (Plate 22, fig. 2; Plate 23, fig. 2)	Outer surface of the premaxilla near the suture along the ventral border of the narial opening	The skin of the upper lip	Carries the lip craniad	Facial n.
M. multifidus spinae (1) Cervical portion or semispinalis cervicis (Plate 34, fig. 3)	Articular processes of last four cervical vertebrae	Spinous processes of cervical vertebrae to axis	Extends the neck	Dorsal rami of spinal nn.
(2) thoracic portion (Plates 33 and 34) and 34)	Transverse processes of eleventh to third vertebrae	To the spinous processes of the ninth to the seventh cervical vertebrae	Extension or lateral bending of spine	Dorsal rami of spinal nn.
(3) lumbar portion (Plates 33 and 34	First coccygeal vertebra, sacrum, accessory and mamillary processes of lumbar and last three thoracic vertebrae	To the spinous processes of the second vertebra craniad of origin	Both sides working together extend spine, one side alone flexes it laterally	Dorsal rami of spinal nn.
M. mylohoideus (Plate 24, fig. 1; Plate 25)	From the middle of the medial surface of the body of the mandible	With the opposite muscle into a median raphe; some of the fibers into the basihyal	Raises the floor of the mouth and brings the hyoid forward	Branch from inferior alveolar of mandibular division of trigeminal
M. myrtiformis (Plate 22, fig. 2; Plate 23, fig. 2)	From the whisker pad	Upon the wing of the nose and into the skin of the upper lip	Dilator of the nares and elevator of the upper lip	Facial n.
M. obliquus abdominis externus (Plates 25 and 30)	From the last nine or ten ribs by tendons, which are interconnected to form arches that span the slips of the serratus anterior	Into the median raphe ventrad of the sternum; into the linea alba and into the tubercle and the cranial border of the pubis	Constrictor of the abdomen, flexion of vertebral column, lateral bending	Branches from intercostal nerves and from ventral rami of lumbars 1–3
M. obliquus abdominis internus (Plates 25, 26, and 31)	(1) Between the fourth and seventh lumbar vertebrae from the lumbar aponeurosis (2) By a similar aponeurosis from the ventral half of the iliac crest	The fibers pass cranioventrad and end along a longitudinal line in a thin aponeurosis of insertion which is united in the linea alba to those of the external oblique and transversus	Compressor of the abdomen, support of the abdominal viscera	Branches from intercostal nerves and from ventral rami of lumbars
M. obliquus capitis caudalis (Plate 34, figs. 3 and 4)	From the whole lateral surface of the spine of the axis	Into the dorsal surface of the transverse process of the atlas	Rotates the head	Suboccipital n.

Name	Origin	Insertion	Action	Innervation
M. obliquus capitis cranialis (Plate 34, fig. 3)	From the lateral border of the transverse process of the atlas	Into the caudal side of the mastoid process of the temporal bone and into a line parallel with the lambdoidal ridge and ventrad of it	Flexes the head laterally	Suboccipital n.
M. obliquus inferior (ventralis) (Plate 105)	Arises from the maxillary bone and curves over the ventral side of the eyeball along the lines of insertion of the recti	Near the ventral edge of the tendon of the lateral rectus m.	Rotates the eyeball moving ventral part medially	Oculomotor n.
M. obliquus superior (dorsalis) (Plate 105)	From the cranial border of the optic foramen. Passes through cartilaginous pulley on medial side of orbit	Into the eyeball along the caudal margin of the dorsal rectus tendon	Rotates eyeball moving the dorsal part medially and ventrally	Trochlearis n.
M. obturator externus (Plate 45, fig. 2)	By fleshy fibers from the median lip of the obturator foramen and from both dorsal and ventral surfaces of the rami of the pubis and ischium adjacent to the lip. Also from the outer surface of the ramus of the ischium as far as the area for the quadratus femoris	Into the proximal portion of the bottom of the trochanteric fossa	Flexion and outward rotation of the thigh	Obturator n.
M. obturator internus (Plate 41, fig. 2; Plate 42, fig. 1)	From the dorsal surface of the ramus of the ischium from the symphysis nearly to the tuberosity	Into the bottom of the trochanteric fossa of the femur	Abductor of the thigh and outward rotator of the hip joint	Sciatic (ischiatic) n.
M. occipitalis (Plate 22, fig. 2)	On the sagittal crest or suture (part of epicranius)	Galea aponeurotica	Moves integument of the dorsal surface of head	Facial n., post-auricular br.
M. occipitoscapularis (Plates 29 and 38, fig. 1)	Medial half of lambdoidal ridge not extending quite to midline	Into the scapula near the cranial angle	Draws forward and rotates the scapula	Ventral ramus of cervical 4
M. opponens digiti quinti (Plate 45)	Middle of ligament covering peroneal canal	Medial side of shaft of metatarsal V	Adductor	Tibial n.
M. orbicularis oculi (Plate 22, fig. 2; Plate 23, fig. 2)	By short tendon fibers from tubercle on frontal process of maxilla	Into the two eyelids	Closes the eye	Facial n.

358

Name	Origin	Insertion	Action	Innervation
M. orbicularis oris (Plate 23, fig. 1)	The part of the muscle in the lower lip has its fibers intermingled with those of the platysma. In the median line the fibers of the upper lip are interrupted by a raphe, and caudad of this are intermingled with those of the caninus		Closes mouth	Facial n.
M. palmaris longus (Plates 38 and 39)	From the distal part of the medial surface of the medial epicondyle of the humerus	In the trilobed pad in the palm; the base of the proximal phalanx	Flexor of the first phalanx of each of the digits	Median n.
M. pectineus (Plates 43–45)	By fleshy fibers from the cranial border of the pubis	By muscle fibers into an elongated area (5 millimeters in length) on the shaft of the femur just distad of the lesser trochanter	Adductor of the thigh	Obturator n.
M. pectoantibrachialis (Plates 25 and 74)	On the lateral surface of the manubrium	By a flat tendon into the superficial fascia of the dorsal border of the forearm, near the joint	Draws the arm mediad	Anterior thoracic n.
M. pectoralis major (Plates 25 and 39) (1) Superficial layer	From a raphe in the midventral line, along the cranial half of the manubrium and for 5 to 10 millimeters craniad of it	The pectoral ridge of the middle third of the shaft of the humerus, slightly dorsad of the line of insertion of the deep portion	Draws the arm mediad and turns the foot forward	Anterior thoracic n.
M. pectoralis major (Plates 25 and 39) Deep portion	Ventral surface of the manubrium and of the first three divisions of the sternum and the median raphe	Along a line which begins at the infraspinatus fossa of the great tuberosity, and runs parallel to the deltoid ridge until it reaches the pectoral ridge	Draws the arm mediad turns the foot forward	Anterior thoracic n.
M. pectoralis minor (Plates 25 and 39)	From the lateral half of the first six divisions of the body of the sternum and sometimes from the xiphoid process	Into the humerus along a line which forms the ventral border of the bicipital groove at the proximal end of the bone	Draws the arm toward the midline	Anterior thoracic n.
M. peroneus brevis (Plate 42)	By fleshy fibers from the distal half of the surface of the fibula	Into the tubercle on the lateral side of the base of the fifth metatarsal	Extensor of the foot	Peroneal n.
M. peroneus longus (Plates 42 and 45)	By tendon-fibers from the lateral surface of the head of the fibula and from the proximal half of the lateral surface of the shaft	Into the bases of the metatarsals	Flexion and rotation of the foot	Peroneal n.
M. peroneus tertius (Plates 41 and 42)	By fleshy fibers from the lateral surface of the fibula	Extensor tendon of fifth digit	Extensor and abductor of the fifth digit and flexor of the foot	Peroneal n.

Name	Origin	Insertion	Action	Innervation
M. plantaris (Plates 41 and 45) (M. flexor digitorum brevis (superficialis)) (Plate 43)	By a strong tendon from the middle of the lateral border of the patella and by fleshy fibers from the ventral border of the lateral sesamoid (fabella)	Into tendon of flexor digitorum brevis	Extensor of foot	Tibial n.
M. platysma (Plate 23, fig. 1)	From the fascia of mid-dorsal line of the neck	Skin and facial muscles	Moves the skin of the face and neck	Facial n.
M. popliteus (Plate 44)	By a strong tendon from the popliteal groove on the surface of the lateral epicondyle of the femur	Into the proximal end of the caudal surface of the tibial shaft on the medial side of the medial oblique ridge	Inward rotation of the leg, flexion of shank	Tibial n.
M. propria lingua	Longitudinal, transverse and vertical muscle bundles intrinsic to the tongue			Hypoglossal n.
M. pronator quadratus (Plate 39)	By fleshy fibers the distal half of the ventral surface of the ulna and from the interosseous membrane	By fleshy fibers into the ventral surface of the radius	Pronate forefoot	Median n.
M. pronator teres (Plates 38 and 39)	By a short, strong tendon from the extremity of the medial epicondyle of the humerus	Along the medial border of the radius	Pronates the hand	Median n.
M. psoas minor (Plates 35, fig. 2; 43)	Usually by five heads from the caudal border of the bodies of the last 2 (or one) thoracic and first 3 (or 4) lumbar vertebrae	By a slendon tendon on the iliopectineal line, just craniad of the acetabulum	Flexes the back in the lumbar region	Ventral rami of thoracic and lumbar spinal nerves
M. pterygoideus lateralis (Plate 24, fig. 3)	External pterygoid fossa and on the surface of the pterygoid process	Into the medial surface of the mandible near its ventral border	Elevator of the lower jaw	Pterygoid branch of mandibular division of trigeminal
M. pterygoideus medialis (Plate 24, fig. 3)	The internal pterygoid fossa	Into the ventral surface of the external pterygoid and its tendon, into the medial surface of the angular process of the mandible	Elevator of mandible	Pterygoid branch of mandibular division of trigeminal
M. pyriformis (Plate 41, fig. 2)	By fleshy fibers from the tips of the transverse processes of the last two sacral and the first caudal vertebrae	By a flat tendon into an elongated area on the proximal border of the great trochanter	Abductor and extensor of the thigh	Caudal gluteal n.
M. quadratus femoris (Plate 41, fig. 2)	By fleshy fibers on the lateral surface of the ischium near the tuberosity	Into the caudal border of the great trochanter	Extensor and outward rotator of the femur	Sciatic (ischiatic) n.

Name	Origin	Insertion	Action	Innervation
M. quadratus labii inferioris (Plate 23, fig. 2)	From the alveolar border of the mandible, between the molar tooth and the canine	Into the lower lip where they intermingle with those of the orbicularis oris	Depressor of the lower lip	Facial n.
M. quadratus labii superioris (Plate 22, fig. 1; Plate 23, fig. 1) (1) M. levator labii superioris alaeque nasi	From frontal portion of the epicranius; frontal process of the maxilla	Into the integument on the outer side of the wing of the nose; the whisker pad	Erects the whiskers and raises the upper lip	Facial n.
(2) Angular head (caput angulare) or levator labii superioris proprius	From a small tubercle at the cranial border of the orbit	Into whisker pad	Erects the whiskers and raises the upper lip	Facial n.
M. quadratus lumborum (Plate 98)	On the ventral surface of the last two thoracic vertebrae and by a few fibers from the last rib	By a strong flat tendon into the anterior inferior spine of the ilium	Bends the vertebral column sideways—or fixation of spine	Ventral rami of thoracic and lumbar spinal nerves
M. quadratus plantae (Plate 44)	From the dorsal part of the lateral surface of the calcaneus and the cuboid	Into the medial part of the outer surface of the tendon of the flexor digitorum longus	It holds the flexor longus tendon in place	Lateral plantar n.
M. quadriceps femoris (Plates 40–45) (1) M. rectus femoris (Plates 42 and 44)	By strong tendon from the base of the acetabulum to about 5 to 7 millimeters craniad of the acetabulum, along the ventral border of the ilium	Into the outer surface of the patella near its proximal border	An extensor of the shank	Femoral n.
(2) M. vastus lateralis (Plate 40)	On the dorsal and lateral surfaces of the shaft and the great trochanter of the femur	On the outer surface of the patella	An extensor of the shank	Femoral n.
(3) M. vastus medialis (Plate 43)	By fleshy fibers on the shaft of the femur between the medial lip of the linea aspera and the cranial aspect of the shaft below the head	Into the medial border of the patella and the ligamentum patellae	An extensor of the shank	Femoral n.
(4) M. vastus intermedius (Plate 45)	From nearly the whole of the cranial surface of the shaft of the femur between the areas for the vastus medialis and the vastus lateralis	By muscle fibers into the capsule of the joint	The quadriceps extensor is an extensor of the shank. The vastus intermedius acting separately is a tensor of the capsule of the knee joint	Femoral n.

Name	Origin	Insertion	Action	Innervation
Mm. recti (lateralis, medialis, dorsalis, ventralis) (Plate 105)	The four recti muscles arise from the bone about the optic foramen and pass toward the eyeball	By thin, flat tendons along a line which separates the darker caudal part of the sclerotic from the white zone of the sclerotic which borders the cornea—the line of insertion of the four tendons forming thus a circle about the eyeball	Move the eyeball as indicated by their names—lateral, medial, etc.	All but lateral by oculomotor n., lateral by abducens n.
M. rectocavernosus or retractor penis (Plate 64)	Arises in two parts from the ventral surface of the sphincter ani internus	Into the corpus cavernosum just proximad of the glans	Retracts penis	Pudendal n.
M. rectus abdominis (Plates 26, 31 and 33)	Into the first costal cartilage near its middle, into the second costal cartilage near its sternal end, and into the sternum between the first and fourth cartilages	By a strong tendon from the tubercle of the pubis	Compresses the abdomen, flexion of trunk	Ventral rami of lumbars 1–3
M. rectus capitis lateralis (Plate 33, fig. 2)	From the ventral surface of the transverse process of the atlas	Into the fossa laterad of the condyle of the occipital bone	Flexes the head laterally	Ventral rami of cervical 2 and 3
M. rectus capitis posterior major (Plate 34, fig. 3)	The spinous process of the axis; united by a raphe to the opposite muscle	Ventrad of the medial part of the lambdoidal crest	Raises the snout	Suboccipital n.
M. rectus capitis posterior medius (Plate 34, fig. 4)	From the cranial end of the axial spine	On the occipital bone, ventrad of the median half of the lambdoidal crest	Raises the snout	Suboccipital n.
M. rectus capitis posterior minor (Plate 34, fig. 5)	From the dorsal arch of the atlas laterad of the median line	Into the occipital bone, ventrad of the insertion of the rectus capitis posterior medius	Raises the snout	Suboccipital n.
M. rectus capitis ventralis (Plate 33, fig. 2)	From the ventral arch (body) of the atlas	Into a deep depression on the basioccipital, caudad of the insertion of the longus capitis	Depresses the snout by flexing atlanto-occipital joint	Ventral rami of cervical 2 and 3
M. retractor oculi (bulbi)	Arises about the apex of the orbit and divides into four heads which lie nearer the eyeball than the recti and are therefore partly covered by the latter	They alternate with the recti and are inserted into the eyeball at about its equator, except the inferior division, which is inserted on a line with the recti	Retraction of the eyeball, oblique movements of eyeball	Abducens n.

Name	Origin	Insertion	Action	Innervation
M. rhomboideus (Plates 29, 31, 36, and 38)	From the caudal part of the cervical supra-spinous ligament and from the first four thoracic vertebral spines and the interspinous ligaments	Into the vertebral border of the scapula and into the outer surface of the caudal angle of the scapula	Draws the scapula toward the vertebral column	Ventral rami of spinal nerves in region of origin
M. rotator auris or scutuloauricularis inferior (Plate 22, fig. 2)	On the scutiform cartilage, just caudad of the frontoscutularis of which this muscle seems to be a continuation	This muscle passes caudad, curving about the medial surface of the auricle and is inserted on the caudo-medial surface of the eminentia conchae	Rotates the external ear mediad and caudad	Facial n., temporal ramus
Mm. rotatores breves (Plate 34, fig. 2)	Extend between transverse processes and spinous processes of adjacent vertebrae from tenth to first thoracic		Unilateral—to rotate thoracic spine around longitudinal axis; bilaterally, to fix thoracic column	Thoracic nn., dorsal rami
Mm. rotatores longi (Plate 34, fig. 2)	Extend between the transverse process and spinous process of two alternate vertebrae—from tenth to first thoracic		Unilateral—to rotate thoracic spine around longitudinal axis; bilaterally, to fix thoracic column	Thoracic nn., dorsal rami
M. sartorius (Plates 40 and 43)	From the cranial half of the crest of the ilium and from medial half of the cranial border	Medial border of the proximal end of the tibia, the medial epicondyle, the patella and adjacent fascia	Flexes the thigh and shank	Femoral n.
M. scalenes (Plates 30 and 31)	(1) scalenus medius by thin tendons from the sixth-ninth ribs, (2) scalenus dorsalis by a very slender tendon from about the middle of the outer surface of the third or fourth rib, (3) scalenus ventralis by one or two minute tendons from the cartilages of the second and third ribs, (4) scalenus cervicalis from first rib and last six cervical vertebrae	Into the transverse processes of all the cervical vertebrae, including the axis and the atlas	Flexes the neck and draws the ribs craniad	Ventral rami of cervical 1–6
M. scaphocuneiformis (Plate 45, fig. 2)	The lateral tubercle of the navicular bone	The lateral surface of the medial cuneiform	Rotates the medial cuneiform on the navicular and would thus act as an opponens of the great toe if the great toe were present	Tibial n.

Name	Origin	Insertion	Action	Innervation
M. semimembranosus (Plates 40, 41 and 43)	By short tendon-fibers from the caudal border of the tuberosity and the ramus of the ischium	Into the medial surface of the femur and on the medial epicondyle	Extensor of the thigh, flexion of the shank	Tibial n.
M. semitendinosus (Plates 40, 41 and 43)	From the apex of the tuberosity of the ischium beneath the origin of the biceps femoris	Into the crest (dorsal border) of the tibia 1 or 2 centimeters from its proximal end	Flexor of the shank, extension of the thigh	Tibial n.
M. serratus dorsalis caudalis (Plate 31)	Into the lumbar spinous processes and the intervening interspinous ligaments	By 4 or 5 heads from the last 4 or 5 ribs	Draws the last 4 or 5 ribs dorsocaudad and aids in expiration	Branches from the intercostal nn. (from the trunk or the ramus medialis of the thoracic nn.
M. serratus dorsalis cranialis (Plate 31)	Into the median dorsal raphe between the axial spinous process and the tenth thoracic spinous process	By fleshy slips from the outer surfaces of the first nine ribs just ventrad of their angles. The origin may extend as far as the tenth or eleventh ribs	Draws the ribs craniad aiding in inspiration	Intercostal nn.
M. serratus ventralis (Plates 30, 31 and 38)	From the first nine or ten ribs by muscular slips	On the medial surface of the scapula near the vertebral border	Depressor of the scapula	Posterior thoracic n.
M. soleus (Plate 41)	The lateral surface of the head of the fibula, and from the proximal two-fifths of its ventral border	Joins the lateral border of the gastrocnemius tendon to form the tendon of Achilles which inserts into the proximal end of calcaneus	Extend the foot	Tibial n.
M. sphincter ani externus (Plate 64)	From the integument on the dorsum of the root of the tail dorsad of the fifth caudal vertebra	Dorsad of the anus the inner fibers of the muscle of the opposite sides are united. They then separate and surround the anus as a band 5 millimeters wide situated beneath the integument	Closes the anus	Pudendal n., rectal brs.
M. sphincter ani internus (Plate 64)	A caudal thickened portion of the circular muscle of the anal canal. Composed of smooth muscle		Closes anal canal	Pelvic n. (autonomic)
M. splenius (Plate 32)	First thoracic spine, ligamentum nuchae, fascia covering deep	By thin tendon into the lambdoidal ridge	Two together elevate head—one only, lateral flexion of head	Dorsal rami of spinal nerves

Name	Origin	Insertion	Action	Innervation
M. spinalis dorsi (Plates 32 and 33)	By strong tendons from the tips of the spinous processes of the tenth to the thirteenth thoracic vertebrae	By fleshy bundles into the sides of the spinous processes of the first nine or ten thoracic vertebrae and of the cervical vertebrae as far forward as the second	Extensor of the vertebral column	Dorsal rami of spinal nerves
M. stapedius	From a fossa in the lateral surface of the petrous bone caudad of that for the incus	Into the head of the stapes	Moves the anterior end of the base of the stapedius caudolaterally	Facial n., stapedial br.
M. sternohyoideus (Plate 24)	From the cranial border of the first costal cartilage and the manubrium of the sternum	Into the outer half of the ventral surface of the basihyal	Draws the hyoid caudad. Raises the ribs and sternum when the hyoid is fixed	Hypoglossal and ventral ramus of cervical 1
M. sternomastoideus (Plates 26 and 28)	From the manubrium and from the median raphe	Into the lateral half of the lambdoidal ridge and into the mastoid portion of the temporal bone as far as the mastoid process	One muscle turns the head and depresses the snout. Both together depress the snout	Spinal accessory and ventral rami of cervical 1–3
M. sternothyroideus (Plate 24, fig. 2; Plate 26)	From the first costal cartilage beneath the sternohyoid	Into the lateral part of the caudal border of the thyroid cartilage	Pulls the larynx caudad	Hypoglossal and ventral ramus of cervical 1
M. styloglossus (Plates 24 and 26)	From mastoid process, stylohyoid by three heads	From the base of the tongue along the ventro-lateral surfaces to the ventral half of the tongue near the median line	Retracts and raises tongue	Hypoglossal n.
M. stylohyoideus (Plate 24, fig. 1)	From the outer surface of the stylohyal bone	Into the middle of the ventral surface of the body of the hyboid bone	Raises the hyoid	Facial n.
M. stylopharyngeus (Plate 24, fig. 3)	From the tip of the mastoid process of the temporal bone and from the tymphanohyal and the stylohyal bones	Among the fibers of the middle and cranial constrictor	Constrictor, an elevator of the pharynx	Glossopharyngeal and vagus nn.
M. submentalis (Plate 23, fig. 2)	Near the midventral line	With the zygomaticus into the preauricular aponeurosis	Draws the external ear ventrad	Facial n.
M. subscapularis (Plates 30, 38 and 39)	From the subscapular fossa	By a strong, flat tendon into the dorsal border of the lesser tubercle of the humerus	Adducts and extends the shoulder joint	Subscapular (cranial division) n.

Name	Origin	Insertion	Action	Innervation
M. supinator (Plate 37)	By a short, strong tendon from the lateral side of the annular ligament of the radius and by tendinous fibers from the radial collateral ligament	The dorsal surface and part of the medial border of the proximal two-fifths of the radius	Supinator of the hand	Radial n.
M. supraspinatus (Plates 29, 36 and 37, fig. 1)	By fleshy fibers from the whole surface of the supraspinatus fossa	Passes over the capsule of the shoulder joint, to which it is closely attached, and is inserted into the free border of the greater tubercle	Extends the humerus after it has been flexed on the scapula	Suprascapular n.
M. temporalis (Plate 22, fig. 2; Plate 24, figs. 1 and 2)	From the temporal fossa	The inner and outer surface of the coronoid process of the mandible	Elevator of the lower jaw	Trigeminal n., mandibular br.
M. tensor fasciae latae (Plate 40)	By fleshy fibers from the outer margin of the ventral border of the ilium craniad of the auricular impression	Into the fascia lata	Tension of the fascia lata and thus flexion of the hip joint; extension of the leg	Cranial gluteal n.
M. tensor tympani (Plate 112)	In a small fossa in the petrous bone dorso-craniad of the fenestra vestibuli	Into a projection on the neck of the malleus	Tenses the tympanic membrane	Trigeminal n., mandibular br.
M. tensor veli palatini	From the ventral surface of the body of the sphenoid	By spreading out in the soft palate into an aponeurosis	Stretches the palate	Trigeminal n., mandibular br.
M. teres major (Plates 28, 38 and 39)	From the dorsal one-third of the caudal border of the scapula	By a tendon common to it and the latissimus dorsi on medial surface of shaft of humerus	Rotates the humerus inward and flexes it in opposition to the infraspinatus, teres minor and the deltoides	Subscapular (middle division) n.
M. teres minor (Plates 36 and 37)	By a sheet of tendinous fibers from the caudal border of the scapula	By a short tendon into the tubercle just distad of the infraspinatus fossa on the greater tubercle of the humerus	Assists the infraspinatus to rotate the humerus outward, flexes shoulder joint	Axillary n.
M. thyreoarytenoideus	The internal midline of the thyroid cartilage	Into the cranial lip of the laterocaudal angle of the arytenoid cartilage	Turns the arytenoid on its oblique articulation so as to close the glottis and relax the vocal cords	Caudal laryngeal n.
M. thyreohyoideus (Plate 24, fig. 2)	On the lateral part of the caudal border of the thyroid cartilage	On the medial two-thirds of the caudal border of the thyro-hyoid bone (thyrohyal)	Draws hyoid caudad and dorsad	Cervical and accessory nn.

Name	Origin	Insertion	Action	Innervation
M. tibialis anterior (cranialis) (Plates 40–42)	By fleshy fibers from the proximal lateral surface of the shaft of the tibia, from the proximal third of the medial border of the shaft and head of the fibula and from the intervening interosseous membrane	Into the outer surface of the first metatarsal	Flexor and lateral rotator of foot	Peroneal n.
M. tibialis posterior (caudalis) (Plate 45)	By fleshy fibers from nearly the whole medial surface of the head of the fibula and by a few fleshy fibers from the caudal surface of the tibia	Onto the outer tuberosity on the surface of the navicular and onto the proximal end of the ventral surface of the medial cuneiform	Extensor of the foot	Tibial n.
M. tragicus lateralis (mandibuloauricularis) (Plate 22, fig. 2; Plate 24, Fig. 2)	On the caudal end of the mandibula, in the cavity found between the condyloid process and the angular process	On the caudal margin of the tragus and in the depression on the concha just caudad of the tragus	Pulls the ear ventrad and probably rotates it outward	Facial n., post-auricular br.
M. tragicus medialis (Plate 22, fig. 2)	On the ventral end of the tragus	The cranial fibers insert on the cranial surface of the concha; the caudal fibers on the medial surface of the concha	Flexes the concha	Facial n.
M. transversus abdominis (Plates 26 and 38)	(1) By fleshy fibers or by a thin aponeurosis from the cartilages of all the false and floating ribs, by interdigitation with the fibers of the diaphragm. (2) From the tips of all the lumbar transverse processes. (3) From the ventral border of the ilium.	Into the linea alba	Constrictor of the abdomen	Ventral rami of lumbars 1–3
M. transversus auriculae (Plate 22)	On the medial surface of the concha, just proximad of the furrow which corresponds to the antihelix	On the auricular cartilage on a line which forms a caudal continuation of the line of insertion of the levator auris longus	Flexes the scapha mediad on the concha, thus enlarging the external opening of the concha	Facial n.
M. transversus costarum (Plate 30)	By tendon from the side of the sternum between third and sixth ribs	On the first rib and the lateral portion of its costal cartilage	Draws the sternum forward	Lateral branch of the intercostal n.
M. transversus perinei	Small bundle of fibers which arises from the medial surface of the ischium	On central tendon of peroneum or on sphincter ani internus	Supports perineum	Pudendal n.

Name	Origin	Insertion	Action	Innervation
M. transversus thoracis (Plate 79, fig. 1)	On the lateral borders of the dorsal face of the sternum, opposite the attachments of the cartilages of the third to the eighth ribs	Into the cartilages of the ribs near their junction with the ribs and into the fascia which covers the inner surface of the internal intercostals in this region	Draws the ventral ends of the ribs caudad	Intercostal n.
M. trapezius—m. acromiotrapezius (Plates 28 and 29)	Along the mid-dorsal line from the spinous process of the axis to a point between the spinous processes of the first and fourth thoracic vertebrae	Into the metacromion, the scapular spine to the tuberosity, and onto the surface of the spinotrapezius	Two muscles hold the scapulae together and pull them closer	Spinal accessory and ventral rami of cervicals 1–4
M. trapezius—m. cleidotrapezius (Plate 28)	The medial half of the lambdoidal crest and the mid-dorsal line between the crest and the caudal end of the spine of the axis	Into the clavicle and into a raphe between the cleidotrapezius and the cleidobrachialis muscles	Draws the scapula craniodorsad	Spinal accessory and ventral rami of cervical 1–4
M. trapezius—m. spinotrapezius (Plates 28 and 29)	From the tips of the spinous processes of the thoracic vertebrae and from the intervening supraspinous ligament	The tuberosity of the scapular spine and from the fascia of the supraspinatus and infraspinatus muscles	Draws the scapula dorsocaudad	Spinal accessory and ventral rami of cervical 1–4
M. triceps brachii (Plates 36–39)	By three heads, two from the humerus, one —the long head, from the caudal part of the scapula near the glenoid	Olecranon process of ulna	Powerful extensor of the elbow joint	Radial n.
M. urethralis	Partly on the caudal part of the symphysis of the ischium, partly from the ventral surface of the urogenital sinus, where the fibers are attached to the corpora cavernosa clitoridis	Inserted into the sides of the vagina and the dorsal surface of the urogenital sinus	Constricts urogenital sinus	Pudendal n.
M. xiphihumeralis (Plates 25, 38 and 39)	A median raphe along the xiphoid process or at an angle to the median line on the rectus abdominis muscle	In a flat tendon which is connected by a strong fascia with the tendon of the latisimus to be inserted with its cranial fibers near the ventral border of the bicipital groove	Assists the pectoralis minor	Cranial pectoral nn. and branches from cervicals 7 and 8
M. zygomaticus (Plate 23, fig. 2)	Connects angle of the mouth with scutiform cartilage		To fix or draw angle of mouth dorsocaudad; to fix or draw external ear ventrocraniad	Facial n.
M. zygomaticus minor (Plate 23, fig. 1)	Among the fibers of the orbicularis oculi in the lower eyelid	Among the fibers of the orbicularis oris, at the angle of the mouth	Pulls the angle of the mouth dorsad	Facial n.

OUTLINE OF THE CRANIAL NERVES (*Gray's Anatomy*, Lea & Febiger)

Nerves	Components	Function	Central Connection	Cell Bodies	Peripheral Distribution
I. Olfactory	Afferent Special visceral	Smell	Olfactory bulb and tract	Olfactory epithelial cells	Olfactory nerves
II. Optic	Afferent Special somatic	Vision	Optic nerve and tract	Ganglion cells of retina	Rods and cones of retina
III. Oculomotor	Efferent Somatic	Ocular movement	Nucleus III	Nucleus III	Branches to Levator palpebrae, Rectus superior, medius, inferior, Obliquus inferior
	Efferent General Visceral	Contraction of pupil and accommodation	Nucleus of Edinger-Westphal	Nucleus of Edinger-Westphal	Ciliary ganglion; Ciliaris and Sphincter pupillae
	Afferent Proprioceptive	Muscular sensibility	Nucleus mesencephalicus V	Nucleus mesencephalicus V	Sensory endings in ocular muscles
IV. Trochlear	Efferent Somatic	Ocular movement	Nucleus IV	Nucleus IV	Branches to Obliquus superior
	Afferent Proprioceptive	Muscular sensibility	Nucleus mesencephalicus V	Nucleus mesencephalicus V	Sensory endings in Obliquus superior
V. Trigeminal	Afferent General somatic	General sensibility	Trigeminal sensory nucleus	Trigeminal ganglion (Gasserian)	Sensory branches of ophthalmic maxillary and mandibular nerves to skin and mucous membranes of face and head
	Efferent Special visceral	Mastication	Motor V nucleus	Motor V nucleus	Branches to Temporalis, Masseter, Pterygoidei, Mylohyoideus, Digastricus, Tensores tympani and palatini
	Afferent Proprioceptive	Muscular sensibility	Nucleus mesencephalicus V	Nucleus mesencephalicus V	Sensory endings in muscles of mastication
VI. Abducent	Efferent Somatic	Ocular movement	Nucleus VI	Nucleus VI	Branches to Rectus lateralis
	Afferent Proprioceptive	Muscular sensibility	Nucleus mesencephalicus V	Nucleus mesencephalicus V	Sensory endings in Rectus lateralis
VII. Facial	Efferent Special visceral	Facial expression	Motor VII nucleus	Motor VII nucleus	Branches to facial muscles, Stapedius, Stylohyoideus, Digastricus
	Efferent General visceral	Glandular secretion	Nucleus salivatorius	Nucleus salivatorius	Greater petrosal nerve, pterygopalatine ganglion, with branches of maxillary V to glands of nasal mucosa. Chorda tympani, lingual nerve, submandibular ganglion, submandibular ganglion, and sublingual glands
	Afferent Special visceral	Taste	Nucleus tractus solitarius	Geniculate ganglion	Chorda tympani, lingual nerve, taste buds, anterior tongue

OUTLINE OF THE CRANIAL NERVES (*Gray's Anatomy*, Lea & Febiger) (continued)

Nerves	Components		Function	Central Connection	Cell Bodies	Peripheral Distribution
VIII. Acoustic	Afferent	General visceral	Visceral sensibility	Nucleus tractus solitarius	Geniculate ganglion	Great petrosal, chorda tympani and branches
	Afferent	General somatic	Cutaneous sensibility	Nucleus spinal tract of V	Geniculate ganglion	With auricular branch of vagus to external ear and mastoid region
	Afferent	Special somatic	Hearing	Cochlear nuclei	Spiral ganglion	Organ of Corti in cochlea
	Afferent	Proprioceptive	Sense of equilibrium	Vestibular nuclei	Vestibular ganglion	Semicircular canals, saccule, and utricle
IX. Glossopharyngeal	Afferent	Special visceral	Taste	Nucleus tractus solitarius	Inferior ganglion IX	Lingual branches, taste buds, posterior tongue
	Afferent	General visceral	Visceral sensibility	Nucleus tractus solitarius	Inferior ganglion IX	Tympanic nerve to middle ear, branches to pharynx and tongue, carotid sinus nerve
	Efferent	General visceral	Glandular secretion	Nucleus salivatorius	Nucleus salivatorius	Tympanic, lesser petrosal nerves, otic ganglion, with auriculotemporal V to parotid gland
	Efferent	Special visceral	Swallowing	Nucleus ambiguus	Nucleus ambiguus	Branch to Stylopharyngeus
X. Vagus	Efferent	General visceral	Involuntary muscle and gland control	Dorsal motor nucleus X	Dorsal motor nucleus X	Cardiac nerves and plexus; ganglia on heart. Pulmonary plexus; ganglia, respiratory tract. Esophageal, gastric, celiac plexuses; myenteric and submucous plexuses, muscle and glands of digestive tract down to transverse colon
	Efferent	Special visceral	Swallowing and phonation	Nucleus ambiguus	Nucleus ambiguus	Pharyngeal branches, superior and inferior laryngeal nerves
	Afferent	General visceral	Visceral sensibility	Nucleus tractus solitarius	Inferior ganglion X	Fibers in all cervical, thoracic, and abdominal branches; carotid and aortic bodies
	Afferent	Special visceral	Taste	Nucleus tractus solitarius	Inferior ganglion X	Branches to region of epiglottis and taste buds
	Afferent	General somatic	Cutaneous sensibility	Nucleus spinal tract V	Superior ganglion X	Auricular branch to external ear and meatus
XI. Accessory	Efferent	Special visceral	Swallowing and phonation	Nucleus ambiguus	Nucleus ambiguus	Bulbar portion, communication with vagus, in vagus branches to muscles of pharynx and larynx
	Efferent	Special somatic	Movements of shoulder and head	Lateral column of upper cervical spinal cord	Lateral column of upper cervical spinal cord	Spinal portion, branches to Sternocleidomastoideus and Trapezius
XII. Hypoglossal	Efferent	General somatic	Movements of tongue	Nucleus XII	Nucleus XII	Branches to extrinsic and intrinsic muscles of tongue

GLOSSARY

If you cannot find the word you want in the glossary, refer to the index for a page reference.

abducens (ăbdū′sĕnz) [L. abduceu, to lead away.] The sixth cranial nerve.

abduction (ăbdŭkt′shŭn) [L. abductus, led away.] Movement away from the central axis of the body or part.

absorption (ăbsôrp′shŭn) [L. absorbere, to suck in.] Passage of materials into or through living cells.

acetabulum (ăsĕtăb′ūlŭm) [L. acetabulum, vinegar-cup.] The socket in the pelvic girdle for the head of the femur.

adduction (ădŭk′shŭn) [L. ad, to; ducere, to lead.] Movement toward the central axis of the body or part.

adipose capsule (ăd′ipōs) [L. adips, fat.] A mass of fat around the kidney giving it protection and support.

adrenal (ădrē′năl) [L. ad, to; renes, kidneys.] An endocrine gland located on the superior surface of the kidney.

afferent (ăf′ĕrĕnt) [L. affere, to bring.] Conveying to.

after discharge. A phenomenon in which the response lasts after the termination of the stimulus.

allantois (ălăn′tŏĭs) [Gk. allas, sausage.] A fetal membranous sac arising from posterior part of alimentary canal.

alveolus (ălvē′ŏlŭs) [L. alveolus, small pit.] A tooth socket or small depression.

amnion (ăm′nĭŏn) [Gk. amnion, fetal membrane.] A fetal membrane enclosing amniotic fluid and embryo.

amniotic cavity (ămnĭŏt′ĭk) [Gk. amnion, fetal membrane.] Pertaining to the amnion. A cavity enclosed in the amnion within the ectoderm cells of the blastocyst.

ampulla (ămpŭl′a) L. ampulla, flask.] A membranous vesicle.

anal glands (ā′năl) [L. anus, anus.] Large modified sweat glands in the stratified squamous epithelium located above the anal opening.

anastomosis (ănăs′tŏmō′sĭs) [Gk. ana, up; stoma, mouth.] Interconnecting of blood vessels or nerves to form network.

anatomy (ănăt′ŏmĭ) [Gk. ana, up; tome, cutting.] The science which deals with the structure of the body.

anthropoids (an′thrō pŏĭd) [Gk. anthopos, man; eidos, form.] The tailess apes, including chimpanzee, gibbon, gorilla and orang-utan.

aortic sinuses (āôr′tĭk sī′nŏsĭs) [Gk. aorte, the great artery; L. sinus, curve.] Dilated pockets between the cusps and the aortic wall.

apex (ā′pĕks) [L. apex, summit.] Tip or summit.

apical foramen (ăp′īkăl fŏră′mĕn) [L. apex, summit; foramen, opening.] The opening of the root of the tooth to the root canal.

apocrine (ap′o-krīn) [Gk. apo, from, away from; krino, to separate.] A type of gland in which the secretions gather at the outer end of the gland cells, which are then pinched off to form the secretion. Example—mammary glands.

aponeurosis (ap′onūro′sĭs) [Gk. apo, from; neuron, tendon.] A white, flattened, sheet-like tendon.

aqueous humor (ā′kwĕŭs hū′mŏr) [L. aqua, water; humor, moisture.] A dilute alkaline solution filling the anterior and posterior chambers of eyes.

arachnoid (ărăk′noid) [Gk. arachne, spider; eidos, form.] The intermediate meninx which is thin, web-like, and transparent.

arbor vitae (âr′bōr vī′tē) [L. arbor, tree; vita, life.] Arborescent appearance of cerebellum in midsagittal section.

arteriole (ârtē′rĭōl) [L. arteriola, small artery.] An artery of under 0.5 millimeter in diameter.

arteriosclerosis (ârtē′rĭō′sklĕ′ro′sĭs) [L. arteria, artery; sclerosis, hardness.] Abnormal thickening and hardening of the arteries.

artery (âr′tĕrĭ) [L. arteria, artery.] A vessel which carries blood away from the heart.

arthritis (ârthrĭt′ĭs) [Gk. arthron, joint.] Inflammation of a joint.

arthrology (âr′thrŏl′ŏjĭ) [Gk. arthron, joint; logos, discourse.] Study of joints.

articulation (ârtĭkūlā′shŭn) [L. articulus, joint.] A joint by which bones are held together.

aryepiglottic fold (ar′yep′iglot′tic) Upper border of the vestibular membranes found between the epiglottis and the apex of arytenoids.

arytenoid (ăr′īte′noid) [Gk. arytoina, pitcher; eidos, form.] A pair of small cartilages of the larynx articulating with cricoid cartilage.

atrium (ā′trĭŭm) [L. atrium, chamber.] A superior cavity of the heart which acts as the receiving chamber; also a part of the tympanic cavity of the ear.

atrophy (ăt′rŏfĭ) [Gk. a, without; trophe, nourishment.] Disappearance or diminution in size and function.

auricle (ôr′ĭkl) [L. auricula, small ear.] Any ear-like lobed appendage.

autonomic (ôt′ŏnŏm′ĭk) [Gk. autos, self; nomos, province.] Self-governing, spontaneous; as the involuntary nervous system.

axon (ak′son) [G. axon, axle.] Nerve cell process limited to one per cell which are involved in conducting away from cell.

axon hillock. Elevated area of perikaryon at which axon originates which is nearly devoid of Nissl bodies.

bipennate (bīpĭn′ăt) [L. bi, twice; penna. contour feather.] Muscle fiber arrangement in which the fiber are attached to both sides of a tendon.

blood (blŭd) [A. S. blód, blood.] The fluid tissue of the vascular system of animals

bone (bōn) [A. S. bon, bone.] Connective tissue whose ground substance contains salts of lime.

boutons terminaux (boo-taw′ tăr-mino-o) Bulb-like enlargements of terminal branches of axons which are in relation to the dendrites and cell bodies of other neurons.

Bowman's capsule [Sir William Bowman, English physician 1816–1892.] The vesicle of a renal tubule; capsula glomeruli.

branchial arch (brăng′kĭăl) [Gk. brangchia, gills.] A cartilaginous or bony arch on the side of the pharynx supporting the gill bars.

bronchiole (brŏng′kĭōl) [Gk. brongchos, windpipe.] One of the finer sub-divisions of the bronchiolar tree.

bronchus (brŏng′kŭs) [Gk. brongchos, windpipe.] Short connecting tube between trachea and lungs.

Brunner's glands [Johann Konrad Brunner, Swiss anatomist, 1653–1727.] Compound tubuloalveolar glands found in the submucosa of the duodenum.

brushes of Ruffini [A. Angelo Ruffini, Italian anatomist, 1864–1929.] Cylindrical end-bulbs in subcutaneous tissue of finger.

buccal frenula (bŭk′ăl frĕn′ū lah) [L. bucca, cheek.] Small folds of membrane between the cheeks and the gums.

buccal glands (bŭk′ăl) [L. bucca, cheek.] Small glands of the submucosa of the cheeks which secrete mucus into the vestibule.

bulbo-urethral glands (bŭl′bōūrē′thrăl) [L. bulbus, bulb; Gk. ourethra, urethra.] Glands lying in pelvic floor to either side of the membranous urethra of the male; open into cavernous urethra.

bundle of His [W. His, German anatomist.] Band of specialized muscle fibers in the interventricular septum of the heart—a part of the conducting mechanism.

bursa (bŭr′să) [L. bursa, purse.] A fluid filled sac-like cavity situated in the tissues at points of friction or pressure—mostly around joints.

calyx (kāl′ĭks) [Gk. kalyx, calyx.] Cup-like extensions of pelvis of kidney.

cancellous bone (kăn′sĕlŭs) [L. cancellous, chambered.] Inner, more spongy, portion of bony tissue.

canine (kănīn) [L. canis, dog.] One of the teeth primarily for tearing, found on either side of the incisors.

capitulum (kăpĭt′ūlŭm) [L. caput, head.] A knob-like swelling at end of a bone.

cardiac glands (kâr′dĭăk) [Gk. kardia, heart.] One of the three gastric glands found in the immediate area of the cardiac orifice.

carpus (kâr′pŭs) [L. carpus, wrist.] Collective term for the eight bones which support the wrist.

cartilage (kâr′tĭlĕj) [L. cartilago, cartilage.] A form of connective tissue usually bluish-white, firm and elastic; cells placed in groups in spaces called lacunae.

caruncula (kăr-ung′ku lah) [L. small piece of flesh.] Small pinkish elevation at the inner angle of the eye.

cauda equinia (kô′dă e-kwin′a) [L. cauda, tail.] A tail-like collection of spinal nerves at the end of the spinal cord.

caudal (kô′dăl) [L. cauda, tail.] Of or pertaining to the tail end of the animal.

cavernosus body (kăv′ĕrnō′sus) [L. cavernosus, chambered.] A structure of the penis and clitoris containing blood spaces; involved in erection of these organs.

cecum (sē′kŭm) [L. caecus, blind.] A large blind pouch found at the beginning of the large intestines.

celom (sē′lŏm) (also coelom) [Gr. koilōma, a hollow, fr. koilos, hollow] A cavity formed within the mesoderm and generally lined by mesothelium.

cementum (sĕmĕnt′ŭm) [L. caementum, mortar.] The covering of the root of the tooth which is connected to the alveolar bone.

cerebellum (sĕr′ĕbĕl′ŭm) [L. cerebrum, brain.] Expanded, folded dorsal outgrowth of hindbrain.

cerebral aqueduct (sĕr′ĕbrăl ăk′wĕdŭkt) [L. cerebrum, brain; aqua, water; ducere, to lead.] A narrow canal passing through the mesencephalon connecting third and fourth ventricles.

cerebral peduncle (sĕr′ĕbrăl pĕdŭng′kĕl) [L. cerebrum, brain; pedunculus, small foot.] Large bundles of nerve fibers which form the inferior portion of the midbrain.

cerebrospinal fluid (sĕr′ĕbrö spī′năl) [L. cerebrum, brain; spina, spine.] A fluid produced in the choroid plexuses of the ventricles of the brain.

choana (kō′ănă) [Gk. choane, funnel.] A funnel-shaped opening; the internal naris.

chordae tendineae (chor′dae tendi′neae) [chorde, string; tendene, to stretch.] Fine tendinous strings connecting the ventricular walls of the heart to the valve cusps or flaps.

chordata (kor-da′-tah) [Gk. chorde, string.] A phylum which includes the vertebrates and the other animals which have a notochord.

chorion (kō′rion) [Gk. chorion, skin.] An embryonic membrane external to and enclosing the amnion.

choroid (kōr′oid) [Gk. chorion, skin; eidos, form.] The middle layer of the eyeball between the retina and sclera.

choroid plexus (kō′roid plĕk′sŭs) [Gk. chorion, skin; eidos, form; L. plexus, interwoven.] Vascular structures in the roofs of the four brain ventricles which produce cerebrospinal fluid.

circle of Willis [Thomas Willis, English anatomist and physician, 1621–1675.] A circular system of arteries inferior to the brain.

circumduction (sĕr′kŭmdŭk shŭn) [L. circum, around; ductus, led.] An action which involves flexion, extension, abduction, adduction and rotation.

cisterna (sĭs′tĕrnăh) [L. cistern.] Any closed space serving as a resevoir, especially one of the enlarged subarachnoid spaces.

class (klâs) [L. classes, division.] A division of a phylum and divided into orders.

clitoris (klĭ'tōrĭs) [Gk. kleiein, to enclose.] An erectile organ which is homologous to the penis.

cochlea (kŏk'lĕä) [Gk. kochlias, snail.] Anterior part of labyrinth of the ear, spirally coiled as a snail shell.

collagenous fibers (kŏl'ăjĕn'ŏus) [Gk. kolla, gule; genos, offspring.] Strong, inelastic fibers composed of many parallel fibrils.

collateral (kŏlăt'ĕrăl) [L. cum, with; latera, sides.] Fine lateral branches of the axon.

colon (kō'lŏn) [Gk. kolon, colon.] Portion of the large intestine between the cecum and rectum.

conductivity (kŏn'dŭktĭv'ĭtĭ) [L. conducere, to lead together.] Power of transmitting stimuli from receptor to other parts of body.

condyle (kŏn'dĭl) [Gk. Kondylos, knuckle.] A rounded process on a bone for articulation.

cone One of the photopic and color receptors of the retina.

contractility (kŏn'trăktĭl'ĭtĭ) [L. cum, together; trahere, to draw.] The capacity to change form.

conus arteriosus (kō'nŭs, arterio'sus) [L. conus, cone; arteria, to keep air.] A structure between ventricle and aorta in fishes and amphibians; the upper and anterior part of the right ventricle of the human heart from which the pulmonary artery arises.

conus medullares (kō'nŭs mĕd'ŭ lā'rĭs) [Gk. konos, cone; L. medullaris, marrow.] Terminal, tapering portion of the spinal cord.

coordination A special function of the central organs of the nervous system by which impulses are sorted and channeled for favorable response.

cornea (kôr'nĕă) [L. corneus, horny.] The anterior, transparent and bulging portion of the outer fibrous coat of the eye.

corniculate cartilage (kôrnĭk'ūlăt) [L. cornu, horn.] One of two small, conical, elastic cartilages articulating with apex of arytenoid.

coronal (kŏr'ŏnăl) [L. corona, crown.] (same as frontal) A plane vertical to the median plane which divides the body into anterior and posterior parts.

corpora quadrigemina (kor'po-rah kwod-re-jem'-i-nah) [L. corpus, body; quad, four.] Four small lobes on dorsal region of mesencephalon associated with visual and auditory functions.

corpus callosum (kôr'pŭs kălō-sum) [L. corpus, body; callosus, hard.] Broad sheet of white matter uniting the two cerebral hemispheres below the longitudinal fissure.

cortex (kôr'tĕks) [L. cortex, bark.] Outer or more superficial part of an organ, as the cortex of the adrenal gland or of the cerebrum.

cranial (krā'nĭăl) [Gk. kranion, skull.] Referring to the head end of the body.

crest (krĕst)[L. crista, crest.] A prominent ridge or border of bone.

cricoid cartilage (krĭk'oid) [Gk. krikos, ring; lidos, form.] Thick ring-like cartilage in larynx, articulating with the thyroid and arytenoid cartilages.

crista (krĭs'tă) [L. crista, crest.] Hair cells found on ampulla of semicircular ducts.

crown (krown) [L. corona, crown.] The grinding surface of a tooth.

crypts of Lieberkuhn [Johann Nathaniel Lieberkuhn, German anatomist, 1711–1756.] Tubular glands of the small intestine.

cuneate (ku'ne'āt) [L. cuneus, wedge.] A wedge-shaped body.

cuneiform cartilage (kūnē'ĭfôrm) [L. cuneus, wedge; forma, shape.] One of two small, elongated pieces of elastic cartilage found in the aryepiglottic folds.

cutaneous plexus (kūtā'nĕus plĕk'sŭs) [L. cutis, skin; plexus, interwoven.] Network of arteries at the interface of the corium and subcutaneous layers.

cuticle (kū'tĭkl) [L. cutis, skin.] A layer of more or less solid substance secreted by and covering the surface of an epithelium—sharply delimited from the cell surface.

cystic duct (sĭs'tĭk dŭkt) [Gk. kystis, bladder; L. ducere, to lead.] Duct of the gallbladder which empties into the common bile duct.

dartos (dâr'tŏs) [Gk. dartos, flayed.] Involuntary muscle fibers which lie within the superficial fascia of the scrotum.

deciduous (dēsĭd'ūŭs) [L. de, away; cadere, to fall.] Falling at end of growth period or at maturity.

deglutition (dēglootĭsh'ŭn) [L. deglutire, to swallow down.] The process of swallowing.

dendrite (dĕn'drīt) [Gk. dendron, tree.] Nerve cell process which normally conducts impulses toward cell body.

dentin (dĕn'tĭn) [L. dens, tooth.] A hard, highly elastic substance constituting the greater part of the tooth.

dermis (dĕrm'ĭs) [Gk. derma, skin.] A layer of dense connective tissue derived from the mesoderm germ layer, the "inner" skin.

desquamation (dĕs'kwămā' shŭn) [L. de, away; squama, scale.] Shedding of cuticle or epidermis in flakes.

diabetes (dī'a-bē'-tēz) [Gk. diatetes, fr. diabainein, to pass through.] A deficiency condition marked by habitual discharge of an excessive quantity of urine.

diaphragm (dī'ă frăm) [Gk. diaphragma, midriff.] A partition partly muscular, partly tendinous, separating cavities of chest from abdominal cavity; a most important organ of breathing.

diaphysis (dĭăf'ĭsĭs) [Gk. dia, through; phyein, to bring forth.] Shaft of bone.

diastole (dīăs'tōlē) [Gk. diastole, difference.] Relaxation phase of the heart beat.

diencephalon (dī'ĕnsĕf'ălŏn) [Gk. dia, between; enakephalos, brain.] Hind part of forebrain.

digestion (dījĕs'chŭn) [L. digestio, digestion.] Mechanical and chemical breakdown of food whereby it may be absorbed.

distal (dĭs'tăl) [L. distare, to stand apart.] End of any structure farthest from midline or from point of attachment.

dorsal (dôr'săl) [L. dorsum, back.] Pertaining to or lying near the back.

ductus deferens (dŭk'tŭs def'-er-ens) [L. ducere, to lead, deferens, to carry away.] The excretory duct of the testis leading from the testis to the ejaculatory duct.

duodenum (dū'ödē'nŭm) [L. duodeni, twelve each.] The short upper portion of the small intestines.

dura mater (dū'rǎ mā'tër) [L. dura, hard; mater, mother.] The outermost and toughest meninx.

ectoderm (ĕk'tö dĕrm) [Gk. ektos, outside; derma, skin.] The outer germ layer of a multicellular animal.

effectors (ĕffĕk tŏrz) [L. efficere, to carry out.] Muscles and glands which respond to impulses carried to them by nerves.

efferent (ĕf'ĕrĕnt) [L. ex, out; ferre, to carry.] Conveying from.

efferent ductules (ĕf'ĕrĕnt dŭk'tūlz) [L. ex, out; ferre, to carry; ducere, to lead.] Tubes from testes to the head of epididymis carrying spermatozoa.

egestion (ējĕst'shŭn) [L. ex, out; gerere, to carry.] Elimination at the inferior end of the digestive tube.

ejaculatory duct (ejak ūlătörī dŭkt) [L. ex, out; jacere, to throw.] A continuation of the ductus deferens from the point of entrance of the seminal vesicles and the prostatic urethra.

eleidin (ĕlē ĭdĭn) [Gk. elaia, olive.] Substance related to keratin found in the stratum lucidum of the skin.

embryo (ĕm brĭö) [Gk. embryon, embryo.] A young organism in early stages of development, before it becomes self-supporting (in human embryology the first eight weeks of development).

embryology (ĕmbrĭöl'öjĭ) [Gk. embryon, embryo; logos, discourse.] Science of development from egg to birth or hatching.

enamel (ĕnăm'ĕl) [O. F. esmaillier, to coat with enamel.] The hard material which forms a cap over dentine.

endocardium (ĕn'dokar'dium) [Gk. endon, within; kardia, heart.] The inner layer of the heart wall.

endochondral (en'dökôn drăl) [Gk. endon, within; chondros, cartilage.] Bones formed by the replacement of hyaline cartilage.

endocrine (ĕn'dökrĭn) [Gk. endon, within; krinein, to separate.] A ductless gland which conveys its secretions into the blood for distribution.

endolymph (ĕn'dölĭmf) [Gk. endon, whithin; L. lympha, water.] The fluid found inside the membranous labyrinth.

endomysium (ĕn'dömĭz'ĭŭm) [Gk. endon, within; mys, muscle.] Sheath-like covering of connective tissue around each muscle fiber.

endoneurium (ĕn'dönū'rĭŭm) [Gk. endon, within; neuron, nerve.] The delicate connective tissue holding together and supporting nerve fibers within fasciculi.

endothelium (ĕn'döthē'lĭŭm) [Gk. endon, with; thele, nipple.] A simple squamous epithelium which lines cavities of the heart, blood and lymphatic vessels.

enzyme (ĕn'zīm) [Gk. en, in; zyme, leaven.] Organic catalysts which act only upon specific substances and under specific conditions.

epaxial (ĕpăk'sĭăl) [Gk. epi, upon; L. axis, axle.] Above the axis; dorsal.

epicardium (ep'ĭkâr'dĭŭm) [Gk. epi, upon; kardia, heart.] The thin transparent outer layer of the heart wall.

epicondyle (ĕp'ikŏn dīl) [Gk. epi, upon; knodylos, knob.] A projection above or upon a condyle.

epididymis (ĕp'ĭdĭd'ĭmĭs) [Gk. epi, upon; didymos, testicle.] A convoluted duct found on the posterior surface of the testis.

epiglottis (ĕp'ĭglŏt'īs) [Gk. epi, upon; glotta, tongue.] A leaf-shaped elastic cartilage, between root of tongue and entrance to larynx.

epinephrine (ĕp inĕf'rēn) [Gk. epi, upon, nephros, kidney.] The hormone of the adrenal medulla.

epineurium (ĕp'ĭneū'rĭŭm) [Gk. epi, upon; neuron, nerve.] The external connective tissue sheath of a nerve.

epiphyses (ĕpĭf'īsĭs) [Gk. epi, upon; phyein, to grow.] Enlarged ends of bones, formed from separate centers of ossification.

epiploic foramen (ĕp'ĭplo'ĭk fŏrā'mĕn) [Gk. epiploon, caul of entrails; L. foramen, opening.] The opening above the duodenum from the greater into the lesser peritoneal cavity (*foramen of Winslow*).

epithalamus (ĕp'ĭ thăl'ămŭs) [Gk. epi, upon; thalamos, chamber.] Thin roof of third ventricle.

epithelial (ĕp'ĭthē'lĭăl) [Gk. epi, upon; thele, nipple.] Characterized by cells closely joined together and found on free surfaces of the body.

eponychium (ĕp'önīk ĭŭm) [Gk. epi, upon; onyx, nail.] The fold of stratum corneum which overlaps the lunula of the nail.

Eustachian tube (ūstā'kĭăn) [B. Eustachio, Italian physician.] Tube connecting the middle ear with the nasopharynx.

evolution (ĕv'ölū'shŭn) [L. evolvere, to unrole.] The process of development of organisms from pre-existing forms.

excretion (ĕkskrē'shŭn) [L. ex, out; cernere, to sift.] The passage of waste products from the internal environment through living membranes to the external environment.

expiration (ĕk'spĭrā'shŭn) [L. ex, out; spirare, to breathe.] The act of emitting air from lungs.

extension (ĕkstĕn'shŭn) [L. ex, out; tendere, to stretch.] A motion which increases the angle between two bones.

exteroceptor (ĕk'stĕrösĕp'tŏr) [L. exter, outside; capere, to take.] A receptor which receives stimuli from the external environment.

extraembryonic celom (ĕk'strâĕm brĭŏn'ik) [L. extra, outside; Gk. embryon, fetus.] The cavity between the layers of extraembryonic mesoderm.

extraembryonic mesoderm (ĕk'strâĕm'brĭŏn'ĭk) [L. extra, outside; Gk. embryon, fetus.] A layer of cells lying between the yolk sac and the trophoblast; outside of the embryo proper.

facet (făs'ĕt) [L. facies, face.] A smooth, flat, or rounded surface for articulation.

family (făm'ĭlĭ) [L. familia, household.] A group of related genera. Families being grouped into orders.

fasciculi (făsĭk'ūlŭs) [L. fasciculus, little bundle.] Bundles of fibers.

fenestra cochleae (fĕnĕs'tră) [L. fenestra, window.] A round window below the fenestra vestibuli in the bony wall between the middle ear and the cochlea of the inner ear.

fenestra vestibuli (fĕnĕs'tră) [L. fenestra, window.] An oval window above the fenestra cochlea in the bony wall between the middle ear and the vestibule of the inner ear.

fertilization (fĕr'tĭlĭză shŭn) [L. fertilis, fertile.] The union of male and female pronuclei.

fetus (fē'tŭs) [L. fetus, offspring.] Product of conception after the second month of gestation.

fibril (fī'brĭl) [L. fibrilla, small fibre.] Fine thread-like structures which give cell stability.

filum terminale (fī'lŭm tĕr'mĭnā'lē) [L. a thread; terminal.] The non-nervous terminal thread extending from the conus medularis of the spinal cord to the coccyx.

fimbria (fĭm'brĭă) [L. fimbria, fringe.] Any fringe-like structure; as on the infundibulum of the uterine tube.

fissure (fĭsh-ūr) [L. fissus, cleft.] A cleft, deep groove, or furrow dividing an organ into lobes.

fixators (fĭks-a'-tors) [L. fixa-tis, to hold.] Muscles which maintain the position of the body; which fix one part to support the movement of another.

flexion (flĕk'shun) [L. flexus, bent.] A movement which decreases the angle between two bones.

folia cerebelli [L. folium, leaf.] More or less parallel ridges of the gray cortex of the surface of cerebellum.

fontanels (fŏn'tănĕl) [F. fontanelle, little fountain.] Membranous areas of the head of the fetus and infant.

foramen (fŏră'mĕn) [L. foramen, opening.] A hole to allow passage of blood vessels or nerves.

foramen of Monro [John Cummings Munro, Boston surgeon, 1858–1910.] The opening between the third ventricle and a lateral ventricle of the cerebral hemisphere.

fossa (fŏs'ă) [L. fossa, ditch.] A depressed area; usually broad and shallow.

fovea centralis (fō'vĕă cen tra'lis) [L. fovea, pit; centrum, center.] A depression in the center of the macula lutea which marks the point of keenest vision, contains only cone cells.

frontal (frŭn'tăl) [L. frons, fore-head.] A plane, verticle to the median plane which divides the body into anterior and posterior parts.

frontonasal process (frŭn'tŏnā zăl) [L. frons, forehead; nasus, nose.] Process of frontal region and nose lying anterior to the pharyngeal pouches.

fundic glands (fŭn'dĭk) [L. fundus, bottom.] Principal glands of the stomach found throughout body and fundus.

gallbladder (gōlblăd'ĕr) [A. S. gealla, gall; blaedre, bag.] The structure for storing bile located on the inferior surface of the liver.

ganglion (găng'glĭŏn) [Gk. gangglion, little tumour.] Collections of nerve cells occurring outside of central nervous system.

genus (jē'nŭs) [L. genus, race.] A group of closely related species.

gill (gĭl) [M. E. gille, gill.] Filamentous or plate-like outgrowths serving as respiratory organs in many aquatic animals.

gill slits (gĭl) [M. E.] A series of openings in the walls of the pharynx—persistent, in lower chordates; transitory in higher groups.

gingival (jĭnjĭ'val) [L. gingivae, gums.] The gums which attach to the maxillae and mandibles.

glands of Littré [A. Littré, French surgeon.] Branching tubules or glands in the walls of the cavernous urethra.

glia (glī'ă) [Gk. glia, glue.] A cell of the neuroglia.

glomerulus (glōb'ūlŭs) [L. globulus, small globe.] A network of blood capillary within Bowman's capsule.

glossopharyngeal (glŏs'ŏfărĭn'jĕăl) [Gk. glossa, tongue, pharynx, gullet.] The ninth cranial nerve.

glottis (glŏt'ĭs) [Gk. glotta, tongue.] The slit between the vocal folds which marks the opening of the larynx.

Golgi material [Golgi, Italian histologist.] Network of fibrils or a series of membranous structures commonly found close to cell center.

gonad (gŏn'ăd) [Gk. gone, birth.] A sexual gland, either ovary or testis.

gray matter. The "substance" of the spinal cord and brain composed of cell bodies, dendrites, unmyelinated nerve fibers and neuroglia.

gubernaculum (gū'bĕrnăk'ūlŭm) [L. gubernaculum, rudder.] A ligament which connects the testis and epididymis to the inside of the scrotal swelling.

gustatory (gŭs'tātŏrĭ) [L. gustare, to taste.] Pertaining to sense of taste.

gut (gŭt) [A. S. gut, channel.] Intestine or part thereof, according to structure of animal.

gyrus (jī'rŭs) [L. gyrus, circle.] A cerebral convolution; a ridge between two grooves.

hair (hār) [A. S. haer.] A thread-like or filamentous outgrowth of epidermis of animals.

hamulus (hăm'ūlŭs) [L. hamulus, little hook.] A bony process shaped like a hook.

haustrum (haws'trum) [L. haustor, drawer.] The recess made by one of the sacculations of the intestinal wall.

Henle's loop [Friedrich Gustav Jakob G. Henle, German anatomist, 1809–1885.] Loop in a kidney tubule within apical portion of pyramid.

hepatic ducts (hĕpăt'ĭk dŭkt) [L. hepar, liver; ducere, to lead.] Paired ducts of the major liver lobes.

hernia (her'ne-ah) [L.] A rupture or separation of some part of the abdominal wall; the protrusion of a viscus from its normal position.

hilus (hī'lŭs) [L. hilum, trifle.] Small notch, opening or depression, usually where blood vessels enter.

holocrine (hol'o-krin) [Gk. holos, whole; krino, to separate.] A type of gland in which the secreting cells and their secretions constitute the glandular product. Example—sebaceous glands.

homo (hō'mö) [Gk. homos, alike, same.] Meaning the same or alike; the genus of man.

horizontal (hŏr'ĭzŏn'tăl) [Gk. horizon, bounding.] A plane at right angles to both the sagittal and the frontal planes, dividing the body into superior and inferior portions.

hormones (hôrmōn'z) [Gk. hormaein, to excite.] Secretions of ductless glands.

hydrocephalus (hī-dro-sĕfah lus) [hydro, water, kephalus, head.] Accumulation in excess of cerebrospinal fluid as a result of closure of the median and lateral apertures of the fourth ventricle.

hypodermis (hī pŏdĕr mĭs) [Gk. hypo, under; L. dermis, skin.] Subcutaneous tissue separating integument from body.

hyponychium (hī'pŏnĭk'ĭum) [Gk. hypo, under, onyx, nail.] A thickened layer of stratum corneum at the distal end of digit under the free edge of nail.

hypophysis (hīpōf'ĭsĭs) [Gk. hypo, under, physis, growth.] An endocrine gland which rests in the hypophyseal fossa of the sella turcica.

hypothalamus (hī'pŏthăl'ămŭs) [Gk. hypo, under, thalamos, chamber.] Region below thalamus; structures forming greater part of floor of third ventricle.

ileum (ĭl'ĕŭm) [L. ileum, groin.] The terminal two-thirds of the small intestine.

ilium (ĭl'ĭ-ŭm) [L. ile, ileum, ilium, pl. ilia, groin.] The dorsal and upper of the bones of the pelvis.

incisor (ĭnsī'zŏr) [L. incisus, cut into.] One of the front, chisel-shaped teeth primarily for cutting.

infundibulum (ĭnfŭndĭb'ŭlĭm) [L. infundibulum, funnel.] A stalk of the hypophysis by which it is attached to the hypothalamus of the diencephalon.

ingestion (ĭnjĕs'chŏn) [L. ingestus, taken in.] The taking in of food at the mouth.

inhibition (ĭn'hĭbĭsh'ŏn) [L. inhibere, to prohibit.] Prohibition or checking of an action already commenced.

inspiration (ĭnspīrā'shŭn) [L. inspirare, to inhale.] The act of drawing air into the lungs.

insulin (ĭn'sūlĭn) [L. insula, island.] Product of island cells of pancreas which regulates carbohydrate metabolism and blood-sugar levels.

integument (ĭntĕg'ūmĕnt) [L. integumentum, covering.] The protective covering over entire body, the skin and its derivatives.

intercalated discs (in-ter'-kah-lāt-ed) [L. intercalatus, interposed, inserted.] Short lines or stripes extending across the fibers of heart muscles.

intermediate cell mass (ĭn'tĕrmē'dĭāt) [L. inter, between; medium, middle.] A constricted mass of mesoderm between the somites and the lateral mesoderm.

internode (ĭn'tĕrnōd) [L. inter, between; nodus, knot.] Segments between nodes as of nerve fibers.

interoceptor (ĭn'tĕrösĕp'tŏr) [L. internus, inside; capere, to take.] A receptor which receives stimuli from the internal environment.

interstitial cells (ĭn'tĕrstĭsh'ĭăl) [L. inter, between; sistere, to set.] Hormone producing cells found in mature testis.

intervertebral discs (ĭn'tĕrvĕr'tĕbrăl dĭsk) [L. inter, between; vertebra, vertebra; Gk. diskos, disc.] Pads of fibrocartilage between the bodies of the vertebrae.

intramembranous (ĭn'tra'mĕm'brănŭs) [L. intra, within; membrane, membrane.] Bones formed directly in a fibrous membrane.

iris (ī'ris) [L. iris, rainbow.] A thin, circular, muscular diaphragm of the eye; contains the eye color.

irritability (ĭr'ĭtăbĭl'ĭtĭ) [L. irritare, to provoke.] The capacity to respond to stimulation.

Islands of Langerhans (lahng'er hanz) [Paul Langerhans, German pathologist and anatomist 1839–1915.] Endocrine glands of the pancreas which secrete the hormone insulin.

isthmus (ĭsth'mŭs) [Gk. isthmos, neck.] A narrow structure connecting two large parts.

jejunum (jējoon'ŭm) [L. jejunus, empty.] The central portion of the small intestine.

labial frenula (lā'bĭăl fren'ulum) [L. labium, lip; frenum, bridle.] Medial folds of mucous membrane between the gums and the inner surface of the lips.

labial glands (lā'bĭăl) [L. labium, lip.] Glands of the submucosa of the lips which secrete mucus into the vestibule.

labia majora (lā'bĭă majōrā) [L. labium, lip; magnus, great.] Two large folds of skin which constitute the outer lips of the vulva; homologous to scrotum of male.

labia minora (lā'bĭă minōrā) [L. labium, lip; minor, less, smaller.] Two small folds lying between the labia major.

labyrinth (lăb'ĭrĭnth) [L. labyrinthus, labyrinth.] The complex internal ear; bony and membranous. Comprise vestibule, cochlea, and semicircular canals.

lacrimal gland (lăk'rĭmăl) [L. lacrima, tear.] One of the glands of the eye found on upper lateral side of the orbit.

lacteals (lak'tealz) [L. lac, milk.] The lymphatic capillaries in the villi which transport chyle.

lanugo (lănū′gō) [L. lanugo, wool.] The first hair to appear in the fetus.

lateral (lăt′ĕrăl) [L. latus, side.] Those structures farther to the sides, away from the midline.

ligament (lĭg′amĕnt) [L. ligamentum, bandage.] A fibrous band of connective tissue holding two or more bones in articulation.

lingual frenulum (ling′gwal frĕn′ūlŭm) [L. lingua, tongue; frenulum, bridle.] A fold of mucous membrane which connects the tongue to the floor of the mouth.

lumen (lūmĕn) [L. lumen, light.] The cavity of a tubular structure or hollow organ.

lymph (lĭmf) [L. lympha, water.] An alkaline colorless fluid contained in lymphatic vessels.

lymphocyte (lĭm′fōsīt) [L. lympha, water; Gk. kytos, hollow.] A small mononuclear colorless corpuscle of blood and lymph.

lymphoid (**tissue**) (lĭm′foĭd) [L. lympha, water; Gk. eidos, form.] A reticular connective tissue infiltrated with lymphocytes.

lymph nodes (lĭmf nōd) [L. lympha, water; nodus, knob.] Small oval collections of lymphatic tissue interposed in the course of lymphatic vessels.

macroglia (măk′rōglĭ′ă) [Gk. makros, large; glia, glue.] Supporting cells of neuroglia, which are ectodermal in origin.

macula (măk′ūlă) [L. macula, spot.] One of the sensitive areas in the walls of the saccule and utricle.

macula lutea (măk′ūlă lū′teă) [L. macula, spot; luteus, orange-yellow.] An oval, yellowish thickened area found in the center of the posterior part of retina; contains the fovea centralis, the area of keenest vision.

mammary (măm′ărĭ) [L. mamma, breast.] Specialized integumentary glands characteristic of class Mammalia.

mastication (măs′tĭkā′shŭn) [L. masticare, to chew.] Process of chewing food with teeth till reduced to small pieces or pulp.

meatus (meā′tŭs) [L. meatus, passage.] A short canal.

medial (mē′dĭăl) [L. medius, middle.] Structures of the body nearer the midline.

mediastinum (mē′dĭăstĭ′nŭm) [L. mediastinus, servant.] Space between right and left pleura in and near median sagittal thoracic plane.

mediastinum testis (mē′dĭăstĭ′nŭm) [L. mediastinus, servant.] Incomplete vertical septum of testis.

medulla (mĕdŭl′ă) [L. medulla, marrow, pith.] Central part of an organ or tissue as the medulla of the adrenal gland.

medulla oblongata (mĕdŭl′ă) [L. medulla, marrow, pith.] Posterior portion of brain, continuous with spinal cord, which houses fourth ventricle.

medullary sheath (mĕdŭl′ărĭ shĕth) [L. medulla, pith; A. S. sceth, shell or pod.] The thick covering of myelin which surrounds myelinated nerve fibers.

meninges (mĕnĭn′jēz) [Gk. meningx, membrane.] Sin.(meninx.) Membranes which enclose the spinal cord and continue through foramen magnum to cover the brain.

meniscus (mēnĭs′kus) [Gk. meniskos, little moon.] Interarticular fibrocartilage found in joints exposed to violent concussion, as the knee joint.

merocrine (mer′o-krīn) [Gk. meros, part; krino, to separate.] A type of gland in which there is no cell destruction in the production of the secretion.

mesencephalon (mĕs′ĕnsĕf′ălŏn) [Gk. mesos, middle; en, in; kephale, head.] The midbrain.

mesenchyme (mĕsĕng′kĭm) [Gk. mesos, middle; engchein, to pour in.] A primitive diffuse, embryonic tissue derived largely from the mesoderm.

mesentery (mĕs′ĕnterĭ) [L. mesentirium, mesentery.] A peritoneal fold serving to hold viscera in position; specifically it supports the hollow organs of the digestive tube.

mesonephric duct (mĕs′ŏnĕf′rĭk; mēz) [Gk. mesos, middle; nephros, kidney.] The duct of the early vertebrate kidney which in the cat contributes to the ductus deferens.

mesosalpinx (mĕs′ŏsăl′pĭngks) [Gk. mesos, middle; salpinx, trumpet.] The portion of the broad ligament enclosing and supporting the uterine tube.

mesothelium (mĕs′ŏthē′lĭŭm) [Gk. mesos, middle; thele, nipple.] A simple squamous epithelium of mesodermal origin which lines the celomic cavities; a part of a somite giving rise to muscular tissue.

mesovarium (mĕs′ōvā′rĭŭm) [Gk. mesos, middle; L. ovarium, ovary.] The mesentery which supports the ovary.

metabolism (mĕtăb′ōlĭzm) [Gk. metabola, change.] All chemical changes constructive and destructive by which protoplasm uses and transforms materials.

metacarpus (mĕt′ăkâr′pŭs) [Gk. meta, after, karpos, wrist.] The collective name for the five bones which support the palm of the hand.

metatarsus (mĕt′ătâr′sŭs) [Gk. meta, after; L. tarsus, ankle.] Collective name for the five bones supporting the region of the foot between the ankle and the digits.

metencephalon (mĕt′ĕnsĕf′ălŏn) [Gk. meta, after; en, in; kephale, head.] The forward portion of hindbrain.

microglia (mīkrŏg′lĭă) [Gk. mikro, small; glia, glue.] Small phagocytic cells of mesodermal origin in gray and white matter of central nervous system.

microvilli (mī′krö vīlĭ) [Gk. mikros, small; L. villus, shaggy hair.] Very tiny protoplasmic projections of epithelial cells; visible individually only with the use of the electron microscope. Make up striated and brush borders.

middle ear cavity. Small air chamber within the temporal bone, houses malleus, incus, and stepes.

midsagittal. Divides the body into equal right and left halves.

modiolus (mŏdī′ŏlŭs) [L. modiolux, small measure.] The conical-shaped central axis of the cochlea of the ear.

mucin (mū′sĭn) [L. mucus, mucus.] Protein material produced by mucous cells.

mucous membrane (mū′kŭs) [L. mucus, mucus.] Lines all hollow organs and cavities which open upon the skin surface of the body.

muscle (mŭs′ĕl) [L. musculus, muscle.] Generally characterized by high degree of contractility.

myelencephalon (mī ĕlĕnsĕf′ălŏn) [Gk. myelos, marrow; en, in; kephale, head.] The lower part of the hindbrain.

myelin (mī′ĕlĭn) [Gk. myelos, marrow.] A white fatty material forming medullary sheath of nerve fibers.

myocardium (mī′ōkar′dĭum) [Gk. mys, muscle; kardia, heart.] The thick muscular layer of the heart wall.

myofibril (mī′ōfĭbrĭl) [Gk. mys, muscle; L. fibrilla, small fiber.] Contractile fibril of muscular tissue.

myology (mīŏl′ŏjĭ) [Gk. mys, muscle; logos, discourse.] The study of the muscular system.

myoneural junction (mī′ŏnū′răl) [Gk. mys, muscle; neuron, nerve.] That point at which terminal nerve branches make connection with muscle fibers.

nares (nā′rēz) [L. nares, nostrils.] The openings into the nasal cavities.

nephron (nĕf′rŏn) [Gk. nephros, kidney.] Structural and functional unit of the kidney, including the renal corpuscle, convoluted tubules and Henle's loop.

nerve (nerv) [L. nervus, sinew.] Bundles of nerve fibers coursing together outside central nervous system.

nervous (tissue) (nĕr′vŭs) [L. nervus, sinew.] Composed of cells specialized in the properties of irritability and conductivity.

neural tube (nū′răl) [Gk. neuron, nerve.] The enclosure made by the union of the neural folds.

neurofibrils (nū′rōfĭ′brĭlz) [Gk. neuron, nerve; L. fibrilla, fine fiber.] Delicate structures which extend through the nerve cell and into the processes—also involved in conduction.

neuroglia (nūrŏg′lĭă) [Gk. neuron, nerve; glia, glue.] Supporting and protecting tissue of the central nervous system consisting, in part, of macroglial and microglial cells and their processes.

neurolemma (nū′rŏ′lĕm′a) [Gk. neuron, nerve; lemm, husk, sheath.] Single layer of flattened cells found only on fibers of peripheral and autonomic system (also sheath of Schwann).

neuromere (nū′rōmēr) [Gk. neuron, nerve; meros, part.] Spinal segment.

neuromuscular spindles (nū′rōmŭs′kūlăr) [Gk. neuron, nerve; L. musculus, muscle.] Proprioceptors present in most skeletal muscles.

neuron (nū′rŏn) [Gk. neuron, nerve.] A complete nerve cell with outgrowths constituting the basic structural unit of nervous system.

neurotendinous spindles (nū′rōtĕn′dinus) [Gk. neuron, nerve; L. tendere, to stretch.] Proprioceptors associated with tendons near their junctions with muscle fibers.

Nissl bodies [Franz Nissl, neurologist in Heidelberg, 1860–1919.] Angular protein particles found in cytoplasm of nerve cell; related to cell metabolism.

nodes of Ranvier [Louis Antoine Ranvier, French pathologist, 1835–1922.] Constrictions at intervals of the medullary sheath of a nerve fiber.

notochord (nō′tōkôrd) [Gk. noton, back; chorde, cord.] Semistiff axial rod in the middorsal line between the chordate nerve cord and the dorsal aorta.

nucleolus (nūklē′ōlŭs) [L. nucleolus, little kernel.] A rounded mass occurring in nucleus.

nucleus (nū′klĕŭs) [L. nucleus, kernel.] The controlling center of the cell.

nucleus pulposus (nū′-klĕus) [L. nucleus, kernel.] The soft core of an intervertebral disc; remnant of notochord.

omentum (ōmĕn′tŭm) [L. omentum, fold.] A fold of mesentery either free or acting as a connection between organs.

omocervical (ō mō sûr′vĭ kăl) [G. omos, shoulder; L. cervix, neck.] Arteries and veins supplying and draining blood in shoulder and neck regions.

ontogeny (ŏntŏj′ĕnĭ) [Gk. on, being; genesis, descent.] The study of the entire life history of the individual.

optic chiasma (ŏp′tĭk kĭăz′mă) [Gk. opsis, sight; chiasma, cross.] The point of decussation of optic nerves anterior to the infundibulum.

ora serrata (o′rah ser-a′tah) [L. oralis, mouth; serratus, saw-like.] The jagged anterior margin of the retina near the ciliary body where its nervous portions cease.

Organ of Corti [A. Alfonso Corti, Italian histologist 1822–1888.] The spirale organ on the inner portion of the basilar membrane; contains the vital acoustic cells and their supporting cells; the true receptor for hearing.

orgasm (ôr′găzm) [Gk. organ, to swell.] The crisis of sexual excitement.

orifice (ŏr′ĭfĭs) [L. os, mouth; facere, to make.] An opening or aperture of a tube or duct.

ossicle (ŏs′ĭkl) [L. os, bone.] Any small bone; specifically the small bones of the middle ear; malleus, incus and stapes.

osteology (ŏs′tĕŏl′ojĭ) [Gk. osteon, bone; logos, discourse.] The study of the structure, nature, and development of bones.

otoconium (ō′tōkō′nĭŭm) [Gk. ous, ear; knois, sand.] One of crystals of calcium carbonate attached to hairs of maculae.

ovary (ō′vărĭ) [L. ovarium, ovary.] The female reproductive and endocrine organ producing ova and hormones.

ovulation (ōvūlā′shŭn) [L. ovum, egg; latum, borne away.] The emission of the egg from the ovary.

ovum (ō′vŭm) [L. ovum, egg.] A female germ cell; mature egg-cell.

Pacinian corpuscles [F. Pacini, Italian anatomist.] Laminated connective tissue capsule serving as deep pressure receptors, found in subcutaneous, submucous and subserous tissue.

palate (păl′ĕt) [L. palatum, palate.] The roof of the mouth.

palmar (păl′măr) [L. palma, palm of hand.] Pertaining to the palm of the hand; the palmar surface.

pampiniform plexus (pămpĭn′ĭfôrm) [L. pampinus, tendril; forma, shape.] A large plexus formed by the somatic veins of the testes.

pancreas (păn′krēăs) [Gk. pan, all; kreas, flesh.] A compound tubuloalveolar gland, with exocrine and endocrine functions.

papilla (ae) (păpĭl′ă) [L.] A small nipple-shaped elevation.

papillary layer (păp′ĭlărĭ) [L. papilla, nipple.] Outer layer of the dermis, characterized by numerous projections into the epidermis.

papillary plexus (păp′ĭlărĭ plĕk′sŭs) [L. papilla, nipple; plexus, interwoven.] Network of arteries at the level of the papillary layer.

paranasal sinus (par-ah-nāzăl) [Gk. para, beside; L. nasus, nose.] Spaces in the maxillary, frontal, sphenoid and ethmoid bones; open into nasal passageways.

parasagittal (pără′săjĭt′ăl) [L. para, beside; sagitta, arrow.] Any plane parallel to the median plane.

parasympathetic (păr′ăsĭmpăthĕ-ĭk) [Gk. para, beside; syn, with; pathos, feeling.] One of the two divisions of the visceral efferent system of the autonomic nervous system; craniosacral portion.

parathyroid (părăthĭ′roid) [Gk. para, beside; thyreos, shield; eidos, form.] One of four small endocrine glands embedded in posterior side of thyroid.

parotid gland (par-ot′id) [L. para, beside; A. S. eare, ear.] Paired salivary glands lying below and in front of ear and opening into mouth.

pedicle (pĕd′ĭkĕl) [L. pediculus, small foot.] A vertebral process which forms the base of the vertebral arch.

pelvis (pĕl′vĭs) [L. pelvis, basin.] The bony cavity formed by pelvic girdle along with coccyx and sacrum; also the cavity in the kidney at the superior end of the ureter.

penis (pē′nĭs) [L. penis, penis.] The male copulatory organ.

perforations (pĕr′fŏr′ăshŭn) [L. perforare, to bore through.] Pores or openings.

perilymph (pĕr′ĭlĭmf) [Gk. peri, round; L. lympha, water.] A fluid separating membranous from osseous labyrinth of ear.

perimysium (pĕr′ĭmĭzĭŭm) [Gk. peri, around; mys, muscle.] Layer of loose connective tissue covering fasciculi.

perineum (pĕr′ĭnē′ŭm) [Gk. perinaion, part between anus and scrotum.] The region of the outlet of the pelvis.

perineurium (pĕr′ĭnū′rĭŭm) [Gk. peri, round; neuron, nerve.] Areolar connective tissue forming outer wrapping of fasciculi.

periosteum (pĕr′ĭŏs′tĕŭm) [Gk. peri, around; osteum, bone.] A fibrous membrane around bone.

peristalsis (pĕr′ĭstăl′sĭs) [Gk. peri, round; stellein, to place.] Progressing waves of contraction along a muscular tube by action of circular muscles; moves material through tube.

Peyer's patches [Johann Konrad Peyer, Swiss anatomist, 1653–1712.] Oval patches of aggregated lymph follicles on walls of ileum.

phagocyte (făg′ösīt) [Gk. phagein, to eat; kytos, hollow.] A colorless blood corpuscle, or other cell, which ingests foreign particles.

phalanx, phalanges, plu. (făl′ăngks; fălăn′jēz) [Gk. phalangx, line of battle.] The bones of the digits.

phallus (făl′ŭs) [Gk. phallos, penis.] The embryonic structure which becomes penis or clitoris.

pharyngeal pouches (fărĭn′jēăl) [Gk. pharyngx, gullet.] Evaginations of the lateral pharyngeal walls.

pharynx (făr′ĭngks) [Gk. pharyngx, gullet.] The anterior part of alimentary canal following the buccal canal.

phylogeny (fīlŏj′ĕnĭ) [Gk. phyllon, race; genesis, origin, descent.] History of development of species or race.

pia mater (pī′ă mā′tĕr) [L. pia mater, kind mother.] The innermost meninx which is a delicate membrane closely investing brain, and spinal cord.

pineal body (pin′e-al) [L. pinea, pine cone.] A median outgrowth from roof of diencephalon.

pinna (pĭn′ă) [L. pinne, feather.] auricle or outer ear.

pituitary gland (pĭtū′ĭtarĭ) [L. pituita, phlegm.] The hypophysis.

placenta (plăsĕn′tă) [L. placenta, flat cake.] A double structure derived in part from maternal tissue and in part from the embryo.

plantar (plăn′tăr) [L. planta, sole of foot.] Refers to the sole of the foot.

pleura (ploor′ă) [Gk. pleura, side.] A serous membrane lining thoracic cavity and investing lung.

plexus (plĕk′sŭs) [L. plexus, interwoven.] A network of interlacing blood vessels or nerves.

plica circularis (plī′ka) [L. plicare, to fold.] Valve-like fold of the mucosa and submucosa which project into the intestinal lumen.

pons (pŏnz) [L. pons, bridge.] A structure connecting two parts, as the pons of the hindbrain.

premolar (prēmō′lăr) [L. prae, before; mola, mill.] Also bicuspid, which are found between the canines and molar.

prepuce (pre′pūs) Part of the integument of the penis which leaves the surface at the neck and is folded upon itself.

process (pros′es) [L. proces′sus, process.] A broad designation for any bony prominence or prolongation.

proprioception (prö′prĭösĕp′shŭn) [L. proprius, one's own; capere, to take.] Sensations which convey position and movements of joints and muscles and hence of the parts of the body.

proprioceptor (prö′prĭösĕp′tŏr) [L. proprius, one's own; capere, to take.] A receptor which receives stimuli from the muscle tissue and tendons and enables us to orient the body and its parts.

prosencephalon (prŏ'ĕnsĕf'ălŏn) [Gk. pro, before; engkephalos, brain.] The most anterior enlargement of the brain, the forebrain.

prostate (prŏs'tāt) [L. pro, before; stare, to stand.] A muscular and glandular organ found ventral to the rectum.

prostatic utricle (prŏstăt'ĭk ū'trĭkl) [L. pro, before; stare, to stand; utriculus, small bag.] Small recess in urethral crest homologous to uterus of female.

protoplasm (prŏ'tŏplăzm) [Gk. protos, first; plasma, form.] The living substance in all plant and animal bodies.

proximal (prŏk'sĭmăl) [L. proximus, next.] Nearest body, center or base of attachment.

pudendum (pūdĕn'dŭm) [L. pudere, to be ashamed.] External female genitalia.

Purkinje fibers (pur-kin'jex) [Johannes Evangelista Purkinje, Bohemian physiologist, 1787–1869.] Specialized muscular fibers in the subendocardial tissue forming an important part of the intrinsic conduction mechanism of the heart.

pyloric glands (pīlŏr'ik) [Gk. pyloros, gate-keeper.] Short, branched, tubular gastric glands found in the pyloric portion of the stomach.

pyramid (pĭr'ămĭd) [L. pyramis, pyramid.] A conical structure, protuberance or eminence.

ramus (rā'mŭs) [L. ramus, branch.] Any branch-like structure.

raphe (rā'fē) [Gk. rhaphe, seam.] A ridge or seam-like suture.

Rathke's pouch [Martin H. Rathke, German anatomist 1793–1860.] A diverticulum of ectoderm from the roof of the stomadaeum.

receptors (rĕsĕp'tŏrz) [L. recipere, to receive.] Sense organs which receive stimuli from the environment.

rectal columns (rĕk'tăl) [L. rectus, straight.] Folds of mucosa and muscle tissue of the upper portion of the anal canal.

rectum (rĕk'tŭm) [L. rectus, straight.] The continuation of the digestive tract from the pelvic colon to the anal orifice.

reflex (rē'flĕks) [L. reflectere, to turn back.] An involuntary response to stimulus.

renal corpuscle (rē'năl) [L. ren, kidney.] The glomerulus and Bowman's capsule of a nephron.

renal fascia (rē'năl făsh'ĭā) [L. ren, kidney; fascia, band.] A part of the subserous fascia supporting the kidney.

respiration (rĕs'pĭrā'shŭn) [L. re, again; spirare, to breathe.] Interchange of oxygen and carbon dioxide between an organism and its surrounding medium.

rete testis (rē'tē) [L. rete, net.] Network of tubes formed by the tubuli recti.

reticular formation (rĕtĭk'ūlōs) [L. reticulum, small net.] Minute nerve network extending through central part of brain stem.

retina (rĕt'ĭnă) [L. rete, net.] The nervous coat which forms the inner layer of the eyeball; contains rod and cone cells.

retroperitoneal (rĕt'rōpĕr'ĭtōnē'ăl) [L. retro, backwards; Gk. peri, round; teinein, to stretch.] Behind peritoneum.

rhombencephalon (rômb'ĕnsĕf'ălŏn) [Gk. rhombos, wheel; engkephalos, brain.] Hindbrain.

root (root) [A. S. wyrt, root.] That part of the tooth which embeds in the bony alveolus.

rotation (rōtā'shŭn) [L. rota, wheel.] Movement of a bone around an axis, either its own or that of another.

ruga (roog'ă) [L. ruga, wrinkle.] Prominent fold of the mucosa and submucosa of the stomach lining.

saccule (săk'ūl) [L. sacculus, small bag.] The lowest and smallest of the two chambers of the vestibular portion of the membranous labyrinth.

sagittal (săjĭt'ăl) [L. sagitta, arrow.] The median plane or any plane parallel to it which divides the body into right and left parts.

salivary glands (săl'ĭvărĭ) [L. saliva, spittle.] The three glands of the mouth region involved in the production and secretion of saliva.

scala tympani (skā'lă) [L. scala, ladder.] The lower portion of the divided canal of the cochlea.

scala vestibuli (skā'lă) [L. scala, ladder.] The upper portion of the divided canal of the cochlea.

sclera (sklē'ră) [Gk. skleros, hard.] The outer fibrous tunic of the eyeball.

scrotum (skrō'tŭm) [L. scrotum.] A medial pouch of loose skin which contains testes in mammals.

sebaceous (sēbā'shus) [L. sebum, tallow.] Epithelial gland which secretes sebum.

segmentation (sĕg'mĕntā'shun) [L. segmentum, piece.] The division or splitting into segments or parts.

semicircular canals (sĕm'īsĕr'kūlăr) [L. semi, half; circulus, circle.] Three bony canals in mammals lying posterior to the vestibule which serve in maintaining equilibrium.

seminal vesicle (sĕm'ĭnăl vĕs'ĭkl) [L. semen, seed.] A convoluted and saccular outgrowth of the ductus deferens, behind bladder, produces fluid for sperms.

seminiferous tubules (sēmĭnĭf'ĕrŭs) [L. semen, see; ferre, to carry.] The structure in which the spermatozoa and seminal fluids are produced.

septum (sĕp'tŭm) [L. septum, partition.] A partition of connective tissue separating two cavities or masses.

serous membrane (sē'rŭs) [L. serum, serum.] Lines the celomic cavities, contributes to mesenteries and omenta, and covers the outer surfaces of related organs.

sesamoid (sĕs'ămoid) [Gk. sesamon, sesame; eidos, form.] A bone developed within a tendon and near a joint.

sinus venosus (sī'nŭs veno'sus) [L. sinus, curve; vena, vein.] A receiving chamber for the veins entering the heart in the lower vertebrates and in the embryos of higher vertebrates.

species (spē'shēz) [L. species, particular kind.] A systematic unit including geographic races and varieties, included in a genus.

spermatozoon (spĕr′mătōzō′ŏn) [Gk. sperma, seed; zoon, animal.] A male reproductive cell.

sphincter (sfĭng′ktĕr) [Gk. sphinggein, to bind tight.] A muscle which contracts and closes an orifice.

spinal ganglion (spī′năl găng′glĭŏn) [L. spina, spine; Gk. gangglion, little tumour.] An aggregate of nerve cell bodies on the dorsal root of the spinal nerve.

spine (spīn) [L. spine, spine.] A more or less sharp projection.

splanchnocranium (splăngk′nōkrā′nĭŭm) [Gk. splangchnon, entrail; kranion, skull.] Jaws and visceral arches of the skull.

stimulus (stĭm′ūlŭs) [L. stimulare, to incite.] An environment change or an act which produces reaction in a receptor or in an irritable tissue.

stomodeum (stŏm′ōdē′ŭm) [Gk. stoma, mouth; odaios, (pert) way.] Anterior invaginated portion of embryonic gut.

stratum corneum (strā′tŭm kor′ne-um) [L. stratum, layer; cornu, horn or horny.] The outermost layer of the epidermis of the skin.

stratum granulosum (strā′tum gran-u-lo′sum) [L. stratum, layer; granulum, small grain.] A layer of the epidermis below the stratum lucidum; granular in appearance.

stratum lucidum (strā′tŭm lu′sid um) [L. stratum, layer; lucidus, clear.] The clear layer between the stratum corneum and stratum granulosum of the skin.

stratum spinosum (strā′tum spī′nōs sum) [L. stratum, layer; spinose, spinous.] The inner, growing layer of the epidermis which rests on the papillary surface of the dermis (germinativum).

subarachnoid space (sŭbărăk′noid) [L. sub, under; Gk. arachne, spider's web; eidos, form.] A wide space surrounding the spinal cord and its pia mater; under the arachnoid.

subdural space (sŭbdū′răl) [L. sub, under; durus, hard.] A space containing a small amount of fluid below the dura mater.

sublingual gland (sŭblĭng′gwăl) [L. sub, under; lingua, tongue.] A salivary gland found in a fold of mucous membrane in the floor of mouth.

submandibular gland (sub′măndĭb′ūler) [L. sub, under; mandibula, jaw.] A salivary gland which lies below the body of the mandible and myohoid muscle; opens into the mouth.

subserous fascia (sŭbsē′rŭs) [L. sub, under; serum, whey.] Fascia present beneath a serous membrane.

sudoriferous (sō′dorĭf′erus) [L. sudar, sweat; ferre, to carry.] Simple coiled tubules commonly called sweat glands.

sulcus (sŭl′kŭs) [L. sulcus, furrow.] A groove.

supination (sūpīnă′shŭn) [L. supinus, bent backward.] Lateral rotation of the forearm which brings the palm of the hand upward.

suture (sū′tūr) [L. sutura, seam.] Line of junction of two bones immovably connected; as in the skull.

sympathetic (sĭmpăthĕt′ĭk) [Gk. syn, with; pathos, feeling.] One of the divisions of the visceral efferent system of the autonomic nervous system.

symphysis (sĭm′fĭsĭs) [Gk. symphysis, a growing together.] Permanent cartilaginous joints.

synapse (sĭn ăps) [Gk. snyapsis, union.] The area of functional continuity between neurons.

synchondrosis (sĭn′kŏndrō′sĭs) [Gk. syn, with; chondros, cartilage.] Temporary cartilaginous joint.

syndesmosis (sĭn′dĕsmō′sĭs) [Gk. syndesmos, ligament.] Articulations with fibrous tissue between the bones.

synostosis (sin-os-to′sis) [Gk. syn; with, together; osteon, bone.] The union of adjacent bones by means of osseous matter.

synovial joints (sĭnō′vĭal) [Gk. syn, with; L. ovum, egg.] Joints characterized by one or more synovial cavities.

systole (sĭs′tōlē) [Gk. systole, drawing together.] Contraction phase of the heart beat.

tactile corpuscles of Meissner (Meissner's corpuscles) [G. Georg Meissner, German histologist, 1829–1905.] Receptors for light touch occurring in the papillae of the corium.

tarsal (târ′săl) [Gk. tarsos, sole of foot.] Pertains to tarsus bones, or to certain glands of the tarsal region of the eyelids.

tarsus (târ′sŭs) [Gk. tarsos, sole of foot.] Collective name for the seven bones of the ankle.

telencephalon (tĕl′ĕnsĕf′ălŏn) [Gk. tele, far; engkephalos, brain.] Anterior terminal segment of brain.

tendon (tĕn′dŏn) [L. tendere, to stretch.] A white fibrous cord connecting a muscle with another structure, usually bone.

testis (tĕs′tĭs) [L. testis, testicle.] Male reproductive and endocrine organ producing spermatozoa and male sex hormones.

thalamus (thăl′ămŭs) [Gk. thalamos, receptacle.] One of two large nuclear masses which form lateral walls of diencephalon forming an important sensory center of brain.

thyroid (thī′roid) [Gk. thyra, door; eidos, form.] An endocrine gland which lies in the neck region.

thyroid cartilage (thī′roid kâr′tĭlĕj) [Gk. thyra, door; lidos, form; L. cartilago, cartilage.] The largest single cartilage of the larynx.

tonsil (tŏn′sĭl) [L. tonsilla, tonsil.] Aggregates of lymphatic follicles in the pharynx.

tonus (tōn′ŭs) [Gk. tonos, tension.] A constant state of partial contraction or tension.

trabeculae (trăbĕk′ūlē) [L. trabecula, little beam.] Septa of connective tissue of muscle extending from a capsule or wall into the enclosed substance or cavity of an organ as in lymph nodes, trabeculae carneae of heart, etc.

trachea (trăkē′ă) [L. trachia, wind pipe.] A fibroelastic tube found at the level of the sixth cervical vertebrae to the fifth thoracic vertebrae; carries air to and from lungs.

tracheal rings (trăkē′ăl) [L. trachia, windpipe.] Cartilaginous rings in the mesenchyme of the trachea which prevent collapsing of the tube.

transverse (trănsvĕrs) [L. transversus, across.] (same as horizontal) A plane at right angles to both the sagittal and frontal planes, dividing the body into superior and inferior portions.

trigone (trīgōn) [Gk. trigonon, triangle.] A small triangular area in the urinary bladder between the orifices of the ureters and urethra.

trochanter (trōkăn′tĕr) [Gk. trochanter, runner.] A very large, usually blunt, process.

trochlea (trŏk′lëä) [Gk. trochilia, pulley.] A pulley-like structure through which a tendon passes.

trochlear (trŏk′lëär) [Gk. trochilia, pulley.] The fourth cranial nerve; also a pulley.

tuber cinereum (tū′bër sin-e′re-um) [L. tuber, knob; cinereus, ashen-hued.] A rounded eminence of gray matter forming part of the inferior surface of the hypothalamus between the mammillary bodies and the optic chiasma; the infundibulum arises from its under-surface.

tubercle (tū′bërkël) [L. tuberculum, small hump.] Usually a small rounded eminence.

tuberosity (tū′bërŏs′ĭtĭ) [L. tuber, hump.] Usually a large rounded eminence.

tubuli recti (tu′buli) [L. tubulus, small tube; rectus, straight.] The less convoluted, nearly straight ducts of the seminiferous tubules.

tunica adventitia (tū′nĭkă, ad-ven-tish′-e-ah) [L. tunica, coating; ad, to; venine, to come.] The outer tunic of various tubular structures as arteries, veins esophagus, uterine tubes, and ductus deferens.

tunica externa (tu′-nik-ah ex′terna) Outer layer of wall of artery or vein.

tunica intima (tu′-nik-ah in′tima) The innermost coat of wall of artery or vein.

tunica media (tu′-nik-ah me′-dia) Intermediate coat of the wall of an artery or vein.

tunica vaginalis (tū′nĭkă văj′ĭnăl-is) [L. tunica, coating; vagina, sheath.] Tough, fibrous outer layer of the testis.

tympanum (tĭm′pănŭm) [Gk. tympanon, drum.] Strong thin membrane between the external auditory meatus and the middle ear.

umbilical cord (ŭm′bĭlĭ′kăl) [L. umbilicus, naval.] The cord formed from the yolk sac and the body stalk, which connects the embryo with the placenta and carries the umbilical arteries and veins.

ureter (ūrē′tĕr) [Gk. oureter, ureter.] Duct conveying urine from kidney to bladder or cloaca.

urethra (ūrē′thră) [Gk. ourethra, urine.] Duct from the urinary bladder to body surface.

uterine tube (ū′tĕrĭn) [L. uterus, womb.] The upper portion of oviduct (*Fallopian tube.*).

uterus (ū′tĕrŭs) [L. uterus, womb.] Single, hollow, muscular organ which lies between urinary bladder and the sigmoid colon.

utricle (ū′trikl) [L. utriculus, small bag.] The larger of the two chambers of the vestibular portions of the membranous labyrinth.

vagina (văjĭnă) [L. vagina, sheath.] A sheath-like tube which extends from the cervix of the uterus to the vestibule.

vasa vasorum (va′sah vaso′rum) [L. vas, vessel; vasorum, genit. of vas.] Nutrient vessels for larger arteries and veins.

veins (vānz) [L. vena, vein.] Vessels which convey blood to or toward the heart.

ventricle (vĕn′trĭkl) [L. ventriculus, belly.] A cavity or chamber, as in heart or brain; the dispensing chamber of the heart.

ventricular folds (vĕntrĭk′ūlăr) [L. ventriculus, belly.] Lower free border of the vestibular membranes attaching to inside angle of thyroid cartilage and to arytenoids; "false" vocal cords.

venule (vĕn′ūl) [L. venula, vein.] Small vessel conducting venous blood from capillaries to vein.

vermis (vĕr′mĭs) [L. vermis, worm.] Narrow median portion of the cerebellum separating the two cerebellar hemispheres.

vestibule (vĕs′tĭbūl) [L. vestibulum, passage.] A cavity leading into another cavity or passage; the vestibule of the internal ear or of the mouth.

villus (vil′us) [L. villus, shaggy hair.] Minute finger-like projections of the mucosa into the lumen of the small intestine.

vitreous body (vĭt′rëŭs) [L. vitreus, glossy.] A transparent, semigelatinous substance filling large cavity of the eye behind the lens.

vocal folds (vō′kăl) [L. vox, voice.] Mucous membrane folds involved in sound production located on the inferior margin of the vestibule of the larynx, "true" vocal cords.

volar (vō′lăr) [L. vola, palm of hand.] Anterior surfaces of the hands or forearms.

vulva (vŭl′vă) [L. volva, vulva.] The external female genitalia.

382

INDEX

Page numbers of an entry defined or treated in some detail are in **boldface;** illustrations are indicated by *italics.* Many structures which are illustrated only are indexed in the belief that much can be learned from a good picture.

M. caudoanalis, *178*, 179
M. caudofemoralis, 81, *106*, **107**
M. caudorectalis, *178*, 179
M. ceratohyoideus, *68*, 69
M. cleidobrachialis, 72, 73, *80*, **81**, *208*
M. cleidomastoideus, 73, 82, **83**, *208*, 209
Muscle, cleidotrapezius, 73, *80*, **81**, *208*
Muscle, coccygeus, **97**, 179
Muscle, common dorsal extensor, 89
M. complexus, 90, 91
M. conchaeus externus, *66*, 67, *210*
M. constrictor pharyngis cranialis, *68*, 69
Muscle, constrictor pharyngis caudalis, *68*, 70
M. constrictor pharyngis medius, *68*, 69
M. coracobrachialis, *102*, 103
M. corrugator supercilii lateralis, *64*, *66*, 67
M. corrugator supercilii medialis, *64*, *66*, 67
M. cricothyroideus, *68*
M. cutaneous maximus, 78, **79**, *182*, 183
Muscles, deep trunk, **91-93**
Muscles, of deglutition, **69-70**
Muscle, depressor, 63
Muscle, depressor conchae, *66*, 67
Muscle, digastricus, *68*, **69**, 72, *208*, *310*
Muscle, dorsal oblique, *304*, 305
Muscle, elevator, 63
Muscle, endomysium, 63
Muscle, epicranius, *64*, 65, *66*
Muscle, epimysium, 63
M. epitrochlearis, 72, **73**, *102*, 103
Muscle, extensor, 63
M. extensor brevis digitorum, *110*
Muscle, extensor brevis pollicis, 101
M. extensor carpi radialis, 98, 99
M. extensor carpi ulnaris, 98, 99
M. extensor caudae lateralis, 90, 91, *96*, 97
Muscle, extensor caudi medialis, 90, **91**, 93, *96*, 97
Muscle, extensor digitorum, 98, **99**
Muscle, extensor digitorum brevis, *110*, **111**
M. extensor digitorum communis, 98, 99, *100*
M. extensor digitorum lateralis, 98, 99, *100*
M. extensor digitorum longus, *106*, 107
Muscle, extensor ossis metacarpi pollicis, 101
M. extensor pollicis longus, *100*, 101
M. external intercostals, *90*
Muscle, external oblique, 72, **73**, 76, 79, 80, **85**
Muscles, extrinsic appendicular, 73
Muscles, of eye, *304*, 305
Muscle, fasciculus, 62, 63
Muscles, female pelvis, 179
Muscle, fibularis, 113
Muscle, flexor, 63
M. flexor carpi radialis, *104*, 105
M. flexor carpi ulnaris, *102*, 103
M. flexor caudae brevis, *96*, 97
M. flexor caudae longus, *96*, 97
Muscle, flexor digitorum brevis, *114*, 115
Muscle, flexor digitorum longus, *114*, *116*, 117
M. flexor digitorum profundus, *104*, 105
Muscle, flexor digitorum sublimis, 103
M. flexor digitorum superficialis, *102*, **103**, **115**
M. flexor hallucis longus, 112, 113, *114*, *116*, **117**, *118*, 119
M. flexor pollicis brevis, *104*, 105
M. flexores profundus, *116*, 117
M. flexores superficialis, *114*, *116*, 117
Muscles, forefoot, *240*, 241, 242
Muscles, forelimb, *234*, 235, 236, *238*, 239, *240*, 241, 242

Muscle, frontoauricularis, *64*, 65, *66*
Muscle, frontoscutularis, *64*, 65, *66*
Muscle, galea aponeurotica, *64*, 65
Muscle, gastrocnemius, *106*, 107, *114*
M. gemellus inferior, *112*, 113
M. gemellus superior, *112*, 113
M. genioglossus, *68*, 69, 76, 141, *310*
Muscle, geniohyoid, 141, *242*
M. geniohyoideus, *68*, 69, 76, *310*
M. glossopharyngeus, *68*, 70
Muscle, gluteus maximus, 81, *106*, **107**
M. gluteus medius, *106*, 107, *110*
Muscle, gluteus minimus (profundus), *112*, 113
Muscle, gracilis, *114*, 115
Muscles, of groin and thigh, *244*, 245
M. helicus, *66*, 67
Muscles, of hindlimb, *246*, **247**, *248*, **249**, **250**, *252*, **253**, **254**
M. hyoglossus, *68*, 69, 76, *310*
Muscles, of hyoid, *310*
Muscle, hyoideus group, 69
Muscle, hyopharyngeus, *310*
M. iliocaudalis, *96*, 97
M. iliocostalis, *88*, 89
M. iliopsoas, *96*, 97, *118*, **119**, *222*, **223**, *284*, 285
M. indicis proprius, *100*, 101
Muscle, inferior oblique, *332*, 333
Muscle, inferior rectus, *332*, 333
M. infraspinatus, 98, 99
Muscle, insertion, 63
Muscles, integumentary, 61, **79**
M. intermedius scutulorum, *64*, 65
M. internal intercostals, *90*, *220*, 221
M. internal oblique, 72, **74**, 76, 84, 86, 87, *88*, 89
Muscle, interosseus, 101, *104*, **105**, *118*, **119**
Muscle, interspinales, 92, **93**, 94, *96*
Muscle, intertransversarii, 92, 93
Muscle, intertransversarii cervicis ventralis, *90*, 94
Muscle, ischiocavernosus, *170*, **171**, *178*, 179
M. jugulohyoideus, *68*, 69, 76
Muscle, lateral rectus, *332*, 333
M. latissimus dorsi, 72, 76, **77**, *80*, **81**, *82*, 83
M. levator ani, *96*, *178*, 179
Muscle, levator auris, *64*, 65
Muscle, levator auris longus, *64*, 65
M. levator costae, *90*
Muscle, levator labii superioris proprius, 67
Muscle, levator palpebrae dorsalis, *304*, 305
Muscle, levator palpebrae superioris, *332*, 333
Muscle, levator scapulae, 83, *84*
Muscle, levator scapulae ventralis, 73, *82*, **83**
M. levator scroti, *178*
Muscle, levator veli palatini, *310*, 324
Muscle, levator vulvae, 179
M. longissimus capitis, *88*, 89
Muscle, longissimus cervicis, *88*, 89, *90*, **91**
Muscle, longissimus dorsi, *88*, **89**, *90*, *96*, 97
M. longus atlantis, 92, 94
Muscles, longus capitis, *88*, **89**, *90*, **91**
M. longus colli, *90*, **91**, *220*, 221
Muscles, lumbricales, *104*, **105**, *114*, **115**, *116*
Muscles, male pelvis, *178*, 179
Muscle, masseter, *68*, 69, 72, *208*
Muscles, of mastication, *68*, **69-70**

Muscle, medial rectus, *332*, 333
M. moustachier, *64*, *66*, 67
M. multifidus, 92, *96*, 97
Muscle, mylohyoideus, *68*, 69, 72, *208*, *310*
M. myrtiformis, *64*, *66*, 67
Muscles, naming of, 63
Muscle, obliquus capitis, 92, 94
M. obliquus capitis caudalis, 92, 94
M. obliquus capitis ventralis, 92
M. obturator externus, *118*, 119
M. obturator internus, *96*, *110*, 111, *112*
Muscle, occipitofrontalis, 65
Muscle, occipitoscapularis, 82, 83, *84*
Muscle, opponens digiti quinti, *118*, 119
Muscle, orbicularis oculi, 67, *332*, 333
M. orbicularis oris, *64*, *66*, 67
Muscle, organ, 63
Muscle, origin, 63
M. palmaris longus, *102*, 103
Muscle, papillary, *198*, 199, 200
Muscles, pectinate, *198*, 199
Muscle, pectineus, *114*, 115
M. pectoantibrachialis, 72, 73, *208*
M. pectoralis major, 72, 73, *208*
Muscle, pectoralis minor, 72, 73
Muscle, perimysium, 62, 63
M. peroneus brevis, *112*, 113
M. peroneus longus, *112*, 113, *118*
M. peroneus tertius, *112*, 113
Muscles, pharyngeal, **69, 70**
Muscles, of pharynx, *310*
M. plantaris, *110*, 111, *118*
Muscle, platysma, *64*, **65**, *66*, 78, **79**, *208*
M. popliteus, *114*, *116*, 117
M. pronator quadratus, *104*, 105
M. pronator teres, *104*, 105
M. psoas major, *158*, 164
Muscle, psoas minor, *158*, 159, *164*, *220*, 221, *222*, **223**, *284*, 285
M. pterygoideus lateralis, *68*, 69, *310*
Muscle, pterygoideus med., 69, *310*
Muscle, pterygopharyngeus, 69, *310*
M. pyriformis, *110*, 111
Muscle, quadratus femoris, *110*, **111**, *118*, 119
Muscle, quadratus labii inferioris, *66*, 67
M. quadratus labii superioris alaeque nasi, *64*, *66*, 67
Muscles, quadratus labii superioris, angular, 67
M. quadratus labii superioris proprius, *64*, *66*, 67
M. quadratus lumborum, *118*, 119, *158*, 159, *220*, 221, *222*, **223**, *284*, 285
M. quadratus plantae, *114*, *116*, 117
Muscle, quadriceps femoris, 111
Muscle, rectovaginalis, 179
Muscle, rectus abdominis, **74**, 76, *182*
Muscle, rectus capitis, 92, 93
M. rectus capitis dorsalis, 92, 93
M. rectus capitis lateralis, *90*, 91
M. rectus capitis ventralis, *90*, 91
Muscle, rectus dorsalis, *304*, 305
Muscle, rectus femoris, *112*, 113, *114*
Muscle, rectus lateralis, *304*, 305
Muscle, rectus medialis, *304*
Muscle, rectus ventralis, *304*
Muscle, retractor bulbi, *304*, 305
M. retractor penis, *178*, 179
Muscles, rhomboideus, *86*, 87
M. rhomboideus capitis, *82*, 83
M. rhomboideus major, *82*, 83
M. rhomboideus minor, *82*, 83
Muscle, rotator, 63, 92
M. rotator auris, *64*, 65
Muscle, rotatores brevi, 92, 93
Muscle, rotatores longus, 92, 93